Handbook of Experimental Pharmacology

Volume 230

More information about this series at
http://www.springer.com/series/164

Philip K. Moore • Matt Whiteman
Editors

Chemistry, Biochemistry and Pharmacology of Hydrogen Sulfide

Springer

Editors
Philip K. Moore
National University of Singapore
Singapore

Matt Whiteman
St. Luke's Campus
University of Exeter Medical School
Exeter
United Kingdom

ISSN 0171-2004 ISSN 1865-0325 (electronic)
Handbook of Experimental Pharmacology
ISBN 978-3-319-18143-1 ISBN 978-3-319-18144-8 (eBook)
DOI 10.1007/978-3-319-18144-8

Library of Congress Control Number: 2015945344

Springer Cham Heidelberg New York Dordrecht London

Printed on acid-free paper

Springer International Publishing AG Switzerland is part of Springer Science+Business Media (www.springer.com)

Contents

Part I

Biochemistry, Molecular Biology and Chemistry

H_2S Synthesizing Enzymes: Biochemistry and Molecular Aspects

Caleb Weihao Huang and Philip Keith Moore

Contents

Abstract

Hydrogen sulfide (H_2S) is a biologically active gas that is synthesized naturally by three enzymes, cystathionine γ-lyase (CSE), cystathionine β-synthetase (CBS) and 3-mercaptopyruvate sulfurtransferase (3-MST). These enzymes are constitutively present in a wide array of biological cells and tissues and their expression can be induced by a number of disease states. It is becoming

C.W. Huang
Neurobiology Group, Life Sciences Institute and Department of Pharmacology, Yong Loo Lin School of Medicine, National University of Singapore, Singapore, Singapore

NUS Graduate School for Integrative Sciences and Engineering, Singapore, Singapore

P.K. Moore (✉)
Neurobiology Group, Life Sciences Institute and Department of Pharmacology, Yong Loo Lin School of Medicine, National University of Singapore, Singapore, Singapore
e-mail: dprmpk@nus.edu.sg

P.K. Moore, M. Whiteman (eds.), *Chemistry, Biochemistry and Pharmacology of Hydrogen Sulfide*, Handbook of Experimental Pharmacology 230,
DOI 10.1007/978-3-319-18144-8_1

increasingly clear that H_2S is an important mediator of a wide range of cell functions in health and in disease. This review therefore provides an overview of the biochemical and molecular regulation of H_2S synthesizing enzymes both in physiological conditions and their modulation in disease states with particular focus on their regulation in asthma, atherosclerosis and diabetes. The importance of small molecule inhibitors in the study of molecular pathways, the current use of common H_2S synthesizing enzyme inhibitors and the relevant characteristics of mice in which these enzymes have been genetically deleted will also be summarized. With a greater understanding of the molecular regulation of these enzymes in disease states, as well as the availability of novel small molecules with high specificity targeted towards H_2S producing enzymes, the potential to regulate the biological functions of this intriguing gas H_2S for therapeutic effect can perhaps be brought one step closer.

Keywords
H_2S synthesizing enzymes • CSE • CBS • 3-MST • Hydrogen sulfide

1 Introduction

Research into the biology of hydrogen sulfide (H_2S) over the last decade has exponentially increased our understanding of the way in which this gasotransmitter influences physiological and pathophysiological (i.e. disease) processes in a wide range of biological systems. With the characteristic and undoubtedly obnoxious odour of rotten eggs and an early history which appeared to mark H_2S as nothing more than a poisonous gas (Beauchamp et al. 1984), it is perhaps surprising that H_2S is synthesized endogenously in mammalian cells by both enzymatic and non-enzymatic mechanisms and that, once formed, H_2S produces the wide array of biological effects which are now being reported (Li et al. 2011). H_2S has therefore followed an interesting 'development curve' in the minds of researchers starting as an agent which was of toxicological (sometime devastatingly so) impact, via a gas of interest to pharmacologists to a molecule which now occupies a very much more central position in gas physiology and with growing therapeutic potential.

The last decade has seen multiple advances in our knowledge of the mechanisms by which H_2S is synthesized, how these are regulated, the tissue and cellular distribution of H_2S synthesizing enzymes in tissues and how this distribution changes in disease. These various areas form the central themes of this present review.

H_2S can be produced naturally in the body by at least three enzymes. Cystathionine γ-lyase (CSE), cystathionine β-synthetase (CBS) and 3-mercaptopyruvate sulfurtransferase (3-MST) are known to be significant producers of endogenous H_2S. Collectively they have been termed 'H_2S synthesizing enzymes'. Both CSE and CBS are pyridoxal-5′-phosphate (PLP) dependent and utilize L-cysteine as substrate whilst 3-MST is non-PLP dependent and uses 3-mercaptopyruvate as substrate (Shibuya et al. 2009a; Steegborn et al. 1999). Endogenous H_2S can also be produced non-enzymatically. For example, erythrocytes, supplemented with glucose and sulfur and containing electron carriers such as NADH, NADPH and reduced glutathione (GSH), spontaneously react with sulfur to produce H_2S (Searcy and Lee 1998). Iron-sulfur cluster containing proteins carrying Fe_2S_2, Fe_3S_4 or Fe_4S_4 clusters are another source of non-enzymatically generated H_2S. These include proteins such as ferredoxins and 'Rieske' proteins among others (Beinert et al. 1997). In the presence of reducing agents such as glutathione, H_2S can also be released from bound sulfur in, for example, persulfides in neurons and astrocytes (Ishigami et al. 2009). Although non-enzymatic routes for H_2S generation have been identified, it seems likely that the bulk, if not all, of the H_2S which is generated for the purpose of carrying out biological functions in the body is derived enzymatically. This no doubt explains the huge emphasis in recent years in understanding as much as possible about the biochemistry, molecular biology and physiological relevance of the H_2S synthesizing enzymes.

The list of reported biological effects of both endogenous and exogenously administered H_2S is far reaching and expanding on an almost weekly basis. Numerous molecular targets for this gas have also been identified. The biological effects of H_2S are outside the scope of this present review and are dealt with elsewhere in this volume. Rather, we will provide here an overview of the H_2S synthesizing enzymes. This will encompass a discussion of the way in which H_2S synthesizing enzymes are regulated in health and in disease, as well as a consideration of the limitations of the currently available H_2S synthesizing enzyme inhibitors which have been widely used to probe the biological roles of this gas.

2 H_2S Synthesizing Enzymes: The Beginnings

The detection of sulfide in brain tissue in 1989 prompted the suggestion that H_2S, produced endogenously, may have a role to play in physiology (Goodwin et al. 1989). The subsequent demonstration that H_2S produced by the enzyme CBS in the mammalian brain selectively enhanced NMDA receptor-mediated responses and facilitated the induction of hippocampal long-term potentiation led to the first proposed biological role of this gas as a neuromodulator (Abe and Kimura 1996). Shortly thereafter, the second H_2S producing enzyme, CSE, was found to be expressed in guinea pig ileum, rat portal vein and thoracic aorta, and the smooth muscle relaxant effect of H_2S was noted for the first time (Hosoki et al. 1997). This seminal work led to a number of detailed investigations which rapidly established the vascular (Zhao et al. 2001a, b) and gastrointestinal (Teague et al. 2002) muscle relaxant properties of this gas. Several years later, the third H_2S synthesizing enzyme (3-MST) was discovered when brain homogenates of CBS knockout mice were shown to produce H_2S at levels similar to that of wild-type mice in the absence of PLP, suggesting redundancy of CBS and the existence of yet another H_2S producing enzyme. This new enzyme, 3-MST, produced H_2S from 3-mercaptopyruvate, which is derived from cysteine in the presence of α-ketoglutarate. This latter reaction is catalysed by the enzyme cysteine

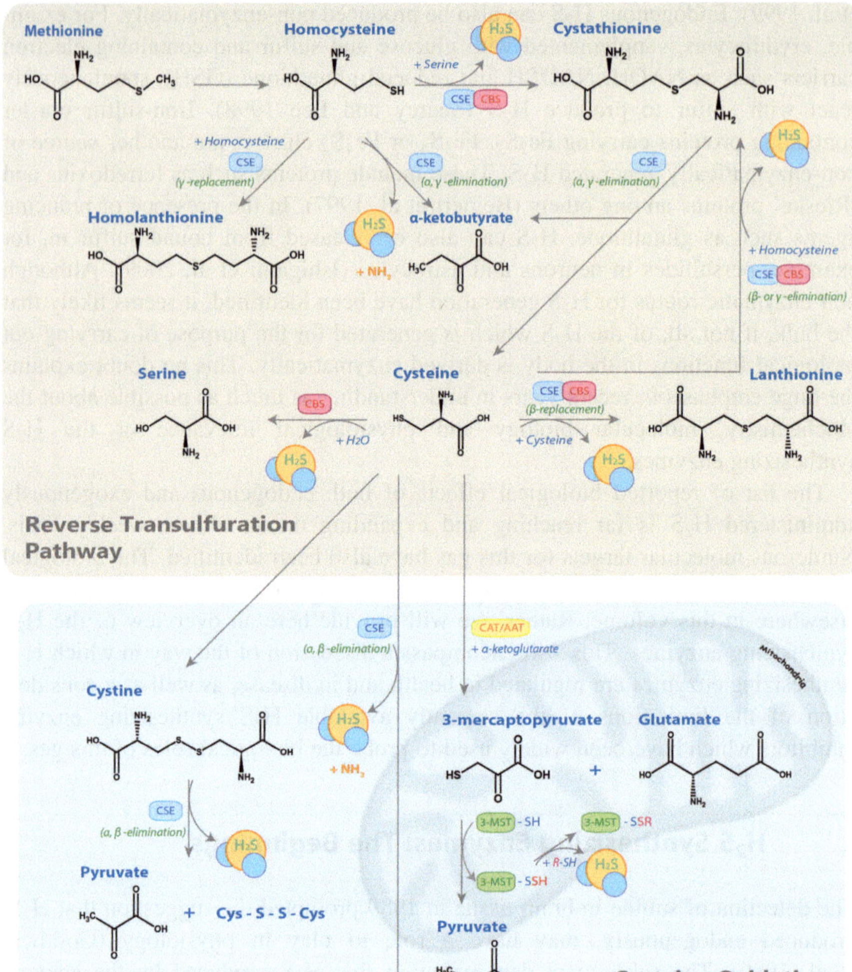

Fig. 1 Pathways for H_2S biosynthesis in the cell. The endogenous production of H_2S occurs via two main pathways—reverse transsulfuration and cysteine oxidation which take place partly inside mitochondria. 3-MST, 3-mercaptopyruvate sulfurtransferase; CBS, cystathionine β-synthetase; CSE, cystathionine γ-lyase; H_2S, hydrogen sulfide; NH_3, ammonia

aminotransferase (CAT), which is identical with aspartate aminotransferase (AAT) (Fig. 1) (Shibuya et al. 2009b).

Over recent years, considerable attention has been focused on the biochemistry and molecular biology of the H_2S synthesizing enzymes. Much is now known about the mechanics of H_2S biosynthesis in terms of substrate specificity, cofactor requirement and tissue and cell distribution, and the molecular topology of these enzymes has also been described. Interested readers are referred to several excellent reviews on this topic (e.g. Kabil and Banerjee 2014; Kabil et al. 2014; Nagahara et al. 2007; Singh and Banerjee 2011). In contrast, there have been few concerted attempts to review changes in the expression or activity of the H_2S synthesizing enzymes with disease states and this review will thence, in part, concentrate on this aspect.

3 Endogenous Synthesis of H_2S and Regulation of Its Synthesizing Enzymes Under Physiological Conditions

CBS catalyses the formation of cystathionine via two separate pathways. The first pathway involves linking homocysteine with serine (Stipanuk 1986) and the second pathway links homocysteine with cysteine (Chen et al. 2004). H_2S is produced only from the second reaction. Besides generating cystathionine, CBS also converts cysteine to serine and lanthionine, and in the process yields H_2S. These reactions are part of the reverse transsulfuration pathway of H_2S biosynthesis (See Fig. 1).

Although found in many organs including brain, kidney, liver, ileum, uterus, placenta (Patel et al. 2009) and pancreatic islets (Kaneko et al. 2006), mouse tissue-specific quantification of CBS revealed that this enzyme is predominantly expressed in the brain and kidney (Kabil et al. 2011a).

The regulation of CBS activity is dependent on the presence of several cofactors and activators. Besides requiring PLP to function, the NH_2 terminal end of CBS contains the binding site of a heme cofactor group that further regulates the activity of CBS by functioning as a redox-sensitive gas sensor (Banerjee and Zou 2005). Because CBS contains iron, the oxidation state of CBS can be toggled, viz. ferric (Fe^{3+}) to ferrous (Fe^{2+}) CBS by reducing agents such as NADPH in the presence of diflavin methionine synthase reductase (Kabil et al. 2011b). In the ferrous (Fe^{2+}) state, CBS can be bound by other gasotransmitters like CO (Puranik et al. 2006) and NO (Taoka and Banerjee 2001), resulting in downregulation of the activity of CBS. Subsequently, the oxidation of the ligand-bound CBS (i.e. CO-CBS or NO-CBS) by air to the ferric (Fe^{3+}) state restores the catalytic activity of CBS. Such a regulatory mechanism suggests a complex interplay between the different gasotransmitters in controlling endogenous H_2S biosynthesis at least via the CBS route. This would further imply that in conditions such as inflammation or hypoxia in which cellular NO and CO levels may be either increased or decreased, the precise effect on CBS activity and hence H_2S generation by this enzyme will be difficult to predict.

CBS activity is also regulated by allosteric binding of S-adenosylmethionine (SAM) (Finkelstein et al. 1975), which is a co-substrate that is involved in a multitude of other metabolic pathways including transmethylation, transsulfuration and aminopropylation (Cantoni 1952). Using mice in which the *MAT1A* gene (which encodes for SAM) has been disrupted and human hepatocellular carcinoma cells which have the *MAT1A* gene silenced, it was demonstrated that the binding of SAM to CBS stabilizes the protein against proteolysis (Prudova et al. 2006). In vitro, SAM increased CBS enzyme activity and thence H_2S production (Finkelstein et al. 1975).

CBS activity is not only regulated by cofactors and activators but also by its location within the cell. Using yeast cells as a model system, human CBS has been shown to be post-translationally modified by the small ubiquitin-like modifier-1 (SUMO-1) protein at its C-terminal regulatory domain. SUMOylation enables CBS to be shuttled into the nucleus where it associates with the nuclear scaffold (Kabil et al. 2006). SUMOylation of CBS was enhanced by human polycomb group protein 2 (hPc2). It has been proposed that nuclear localization of CBS occurs when local glutathione demand is high, since CBS generates cysteine, which is the rate-limiting amino acid in glutathione synthesis (Agrawal and Banerjee 2008). Coupled with the observation that regulation of CBS was correlated with the rate of proliferation in human and yeast cells through a redox-sensitive mechanism (Maclean et al. 2002), it is possible that the presence of CBS in the nucleus provides a source of an antioxidant (i.e. H_2S) in the cell when, and if, required. This possibility warrants further investigation to determine whether such a mechanism does indeed exist in cells either in vitro or in vivo.

Like CBS, cystathionine γ-lyase (CSE) uses homocysteine as substrate to produce H_2S, α-ketobutyrate, ammonia and homolanthionine. CSE can also generate H_2S from cysteine, forming cystathionine and pyruvate (See Fig. 1). Although both homocysteine and cysteine are substrates of CSE, at physiological concentrations of cysteine and homocysteine, kinetic simulations employing physiologically relevant concentrations of homocysteine and cysteine using the dithiobisnitrobenzene (DTNB) assay showed that α- and β-elimination of cysteine catalysed by CSE accounts for approximately 70 % of the H_2S produced while the α,γ-elimination of homocysteine contributes about 29 %. However, once homocysteine levels are elevated to those apparent in hyperhomocysteinemia, then about 90 % of the H_2S generated by CSE occurs via α,γ-elimination and γ-replacement reactions of homocysteine (See Fig. 1) (Chiku et al. 2009). This indicates that the identity of the main substrate for CSE-dependent generation of H_2S in cells is likely to depend on the relative availability of cysteine and homocysteine.

CSE is located in a wider range of tissues than CBS. CSE activity is several-fold higher in human brain than in mouse brain (Diwakar and Ravindranath 2007). The mouse small intestine, stomach (Ishii et al. 2004), rat portal vein and thoracic aorta (Hosoki et al. 1997), mouse pancreas (Kaneko et al. 2006), rat uterus, human placenta, myometrium, amnion and chorion (Patel et al. 2009) all express CSE. As compared to CBS, CSE is more abundant in mouse liver where it is the principal enzyme responsible for H_2S generation (Kabil et al. 2011a).

Like CBS, the activity of CSE is also regulated by a number of cofactors. CSE was initially reported to be activated physiologically by calcium-calmodulin after stimulation of muscarinic cholinoceptors in vascular endothelial cells in culture, as determined by an increase in CSE activity and H_2S production. Through co-immunoprecipitation experiments, recombinant CSE was found to bind directly to calmodulin (Yang et al. 2008). Although Ca^{2+} has been suggested to activate CSE, the precise concentration of Ca^{2+} is critical as too much Ca^{2+} inhibits purified rat CSE-mediated H_2S production in the presence of PLP. At physiologically relevant Ca^{2+} concentrations (i.e. ~100 nM) H_2S was efficiently produced in the presence of PLP. However, H_2S production was decreased at 300 nM Ca^{2+} and reduced even further at concentrations up to 3 μM. The involvement of calmodulin in regulating CSE activity however is not yet clear since a recent study showed that neither calmodulin nor a calmodulin antagonist, W-7, affected the production of H_2S by purified rat liver CSE (Mikami et al. 2013).

SUMO, besides decreasing CBS activity, also targets CSE in vitro, in that recombinant human CSE can also be SUMOylated (Agrawal and Banerjee 2008). However, whether the SUMOylation of CSE occurs within the cell and has a physiological effect in consequence is not clear. If this is indeed the case then the re-localization of both CBS and CSE into the nucleus may be a strategy to increase cysteine levels inside the nucleus when the demand for glutathione is increased such as during early phases of cell proliferation (Markovic et al. 2007) or when telomerase activity is high as observed in 3T3 fibroblast cells (Borrás et al. 2004).

As an important intracellular antioxidant protecting the cell against oxidative stress, physiological glutathione levels are well controlled via glutathione synthesis in the transsulfuration pathway. Without glutathione, numerous cell types including lymphocytes (Franco and Cidlowski 2009), embryos (Dalton et al. 2004) and hepatocytes (Ghibelli et al. 1999) undergo apoptosis due to excess reactive oxygen species (ROS) generation. Since activation transcription factor 4 (ATF4) controls expression of genes responsible for amino acid metabolism and redox status, ATF4 might, by increasing glutathione synthesis through the transsulfuration pathway, be a key cellular survival factor. Indeed, the absence of ATF4 in mouse embryonic fibroblasts resulted in a substantial fall in glutathione levels due to downregulation of CSE expression (Dickhout et al. 2012). Thus, if the amount of H_2S synthesizing enzymes in the nucleus are reduced, the levels of glutathione would subsequently be reduced, thereby affecting the fate of the cell.

The most recently discovered H_2S synthesizing enzyme 3-MST, unlike CBS and CSE, is PLP independent. 3-MST catalyses the conversion of 3-mercaptopyruvate to pyruvate by degrading cysteine. In this pathway, cysteine is first converted by cysteine aminotransferase (CAT), also known as aspartate aminotransferase (AAT), to 3-mercaptopyruvate via the incorporation of α-ketoglutarate into the reaction. 3-MST then forms a persulfide by transferring a sulfur from 3-mercaptopyruvate onto itself which in the presence of a reductant like thioredoxin produces pyruvate and H_2S (Nagahara et al. 2007; Yadav et al. 2013). Dihydrolipoic acid has also been identified to associate with 3-MST to release H_2S (Mikami et al. 2011a).

Like CBS and CSE, 3-MST is found in many tissues with high activity in the proximal tubular epithelium of the kidney, pericentral hepatocytes in the liver, cardiac cells and neuroglial cells (Shibuya et al. 2009b). Within the cell, 3-MST is located both in the cytoplasm as well as the mitochondria. However, because the concentration of cysteine is much higher in the mitochondria, as compared to the cytoplasm, it is plausible that most of the H_2S generated by 3-MST occurs in mitochondria (Nagahara et al. 1998).

H_2S production by 3-MST can be significantly inhibited by aspartate which is a potent inhibitor of CAT (Akagi 1982). H_2S produced by 3-MST and CAT in retinal neurons is also regulated by Ca^{2+}. A negative feedback mechanism has been proposed whereby H_2S produced after activation by Ca^{2+} in turn suppresses Ca^{2+} influx into photoreceptor cells by activating the vacuolar-type $H(_+)$-ATPase (V-ATPase). The activated V-ATPase then releases protons into the synaptic cleft thereby suppressing Ca^{2+} channels found in the photoreceptor cells (Mikami et al. 2011b).

Unlike CBS and CSE, the activity of which is regulated for the most part by binding of other factors, 3-MST activity is probably regulated intrinsically by its redox state. Structural studies have shown that 3-MST contains catalytic site cysteines (Cys^{247}, Cys^{154} and Cys^{263}) each of which is redox active (Nagahara 2013). Cys^{154} is unique to rat 3-MST whilst Cys^{263} is conserved in mammalian 3-MST (Nagahara et al. 2007). The activity of 3-MST is regulated by both intermolecular and intramolecular redox-sensing switches as reviewed elsewhere (Nagahara 2013). Nevertheless, it is important to note that the redox-dependent modulation of 3-MST may not be relevant in human MST since the cysteine residues (i.e. Cys^{154}) which are involved in the cross-linking in rat MST are not conserved in the human enzyme (Nagahara et al. 2007).

4 Molecular Regulation of H_2S Synthesizing Enzymes

Hydrogen sulfide contributes to cell signalling by regulating the function of a number of molecular targets including ion channels (Sun et al. 2008; Zhao et al. 2001b), kinases (Hu et al. 2007) and transcription factors such as signal transducer and activator of transcription (STAT) (Calvert et al. 2009; Li et al. 2009a) as well as nuclear factor-kappaB (NFκB) (Li et al. 2011; Oh et al. 2006). However, the expression of the enzymes that are responsible for producing H_2S (i.e. CSE, CBS and 3-MST) are also frequently regulated at a transcriptional and subsequently at a protein level by multiple molecular factors. In many disease states, the expression of these enzymes is both inducible and dynamic. With a better understanding of how the H_2S synthesizing enzymes are regulated in disease states, a more targeted effort could be developed to identify drugs to affect endogenous production of H_2S in disease states with a view to restoring physiological H_2S levels.

To date, changes in expression (mRNA and/or protein) of H_2S synthesizing enzymes have been identified in a number of cells and tissues in a variety of different disease states and involving the intermediacy of a range of different signalling pathways (summarized in Table 1). Most evidence collected to date suggests a pivotal role for NFκB in the control of cellular CSE and CBS expression, and this interaction will therefore be described in some detail below.

In an early study seeking to identify the molecular targets of H_2S, it was shown that H_2S inhibited the activation of NFκB in lipopolysaccharide (LPS)-stimulated mouse macrophages. The mechanism is believed to be via increased heme-oxygenase-1 (HO-1) expression under the control of an ERK1/2-dependent pathway. Increased HO-1 expression then led to reduced IκB phosphorylation and degradation and impaired translocation of NFκB into the nucleus

Table 1 Examples of modulated H_2S producing enzymes in selected disease states

Disease/stress	Model	Factors/signalling involved	Location	Regulation of H_2S producing enzyme	Reference
Atherosclerosis	RAW264.7 macrophages	Akt signalling	–	↑ CSE (mRNA & protein)	Xu et al. (2014b)
	High-fat diet in mice	–	Liver, lung Liver, kidney	↓ CSE (protein) ↑ CBS (protein) ↓ 3-MST (protein)	Peh et al. (2014)
	High-fat diet in mice	–	Liver	↑ CSE & CBS (mRNA)	Hwang et al. (2013)
Asthma	Ovalbumin-treated mice	–	Lung	↓ CSE (protein)	Zhang et al. (2013)
	Ovalbumin-treated rats	–	Lung	↓ CSE (protein)	Chen et al. (2009)
Bowel syndromes	Visceral hyperalgesia in rats	Voltage-gated sodium channel	Colon	↑ CBS (protein)	Wang et al. (2012b)
	Chronic visceral hyperalgesia in rats	NFκB	Colon	↑ CBS (protein)	(Li et al. 2012)
Cancer	Colon cancer	–	Colon	↓ 3-MST (protein)	Ramasamy et al. (2006)
	Human hepatocellular carcinoma (HCC) cell lines	Sp1, PI3K/Akt (PKB)	–	↑ CSE (protein)	Yin et al. (2012)
	Down's syndrome megakaryocytic leukaemia cell line (CMK)	Sp1, Sp3	–	↑ CBS (mRNA)	Ge et al. (2003)
Diabetes	Streptozotocin-treated rats	–	Liver	↓ CSE (protein)	Manna et al. (2014)
	INS-1E cells, rat pancreatic islets	Glucose	–	↓ CSE (mRNA & protein)	Zhang et al. (2011)
	Streptozotocin-treated rats	Insulin	Liver	↑ CBS (mRNA)	Ratnam et al. (2002)

(continued)

Table 1 (continued)

Disease/stress	Model	Factors/signalling involved	Location	Regulation of H_2S producing enzyme	Reference
Down's syndrome	Patients with Down's syndrome	–	Astrocytes	↑ CBS (protein)	Ichinohe et al. (2005)
Hypoxia	U87-MG human glioblastoma cells, PC12 rat cells, rats exposed to hypoxic conditions	Hypoxia-inducible factors (HIFs)	Cerebral cortex	↑ CBS (mRNA & protein)	Takano et al. (2014)
Inflammation	LPS-treated mammalian cell lines (HEK-293, COS-7, J774.1A, RAW264.7)	NFκB	–	↑ CSE (mRNA & protein)	Wang et al. (2014)
	Temporomandibular joint pain in rats	Inhibition of voltage-gated potassium currents	Trigeminal ganglion neuron	↑ CBS (mRNA & protein)	Miao et al. (2014)
	LPS-induced inflammation in rats	Dexamethasone	Neutrophils	↓ CSE (protein)	Li et al. (2009b)
	LPS-induced inflammation in mice	–	Liver, kidney	↑ CSE (mRNA)	Li et al. (2005)
Oxidative stress	Mammalian cells (A549, SMMC-7721)	H_2O_2	–	↑ CSE (mRNA & protein)	Wang et al. (2012a)
	Rat mesangial cells	Platelet-derived growth factor (PDGF)-BB, STAT, Nrf2	–	↑ CSE (mRNA & protein)	Hassan et al. (2012)
Renal ischaemia	Renal ischaemia/reperfusion in rats	Sp1	Kidney	↓ CBS (mRNA & protein)	Wu et al. (2010)
Ulcerations	Gastric ulcers in rats	–	Stomach	↑ CSE & CBS (protein)	Wallace et al. (2007)

3-MST 3-mercaptopyruvate sulfurtransferase, *Akt/PKB* protein kinase B, *CBS* cystathionine β-synthetase, *CSE* cystathionine γ-lyase, H_2S hydrogen sulfide, H_2O_2 hydrogen peroxide, *Nrf2* nuclear factor like 2, *PI3K* phosphoinositide 3-kinase, *Sp1* specificity protein 1, *STAT* signal transducer and activator of transcription

(Oh et al. 2006). Later studies confirmed that H_2S affects NFκB but by a different mechanism. In this case, H_2S was shown to sulfhydrate NFκB on the p65 subunit at the cysteine-38 residue. This molecular event then increased NFκB binding to a co-activator ribosomal protein, S3 (RPS3), which triggers the transcription of antiapoptotic genes (Sen et al. 2012). Interestingly, recent work has shown that NFκB also regulates CSE expression by binding to its promoter region in macrophages challenged with LPS (Wang et al. 2014). In a separate study, rats with chronic visceral hyperalgesia displayed upregulated colonic CBS expression. An inhibitor of the p65 subunit of NFκB, pyrrolidine dithiocarbamate (PDTC), reduced the expression of CBS, suggesting that activation of the p65 subunit of NFκB upregulates CBS expression (Li et al. 2012). Thus, it appears that H_2S may act to maintain the levels of CSE and CBS inside the cell in check during an inflammatory response by regulating the levels of NFκB.

Drugs have also been shown to regulate CSE expression. In the case of atherosclerosis, statins, a class of drugs used to lower cholesterol levels by inhibiting the enzyme 3-hydroxy-3-methyl-glutaryl-CoA reductase (HMG-CoA reductase), were recently reported to upregulate CSE transcription and subsequent H_2S generation. This effect was mediated by activation of Akt signalling (Xu et al. 2014b). In inflammation, dexamethasone, a glucocorticoid steroid, reduced CSE expression in neutrophils isolated from LPS-challenged rats. Dexamethasone, in this model, was proposed to inhibit the formation of pro-inflammatory H_2S, most probably through reduced NFκB-mediated CSE expression (Li et al. 2009b).

In certain disease conditions, the effect of the disease itself could also have a role to play in directly regulating the expression of H_2S synthesizing enzymes. For example in diabetes, high glucose significantly reduced CSE mRNA and CSE activity levels in freshly isolated rat pancreatic islets and in the rat pancreatic cell line, INS-1E. This was mediated by stimulating phosphorylation of the transcription factor SP1 via p38 MAPK activation which subsequently led to decreased CSE promoter activity (Zhang et al. 2011).

5 H_2S Synthesizing Enzymes in Diseases States

5.1 Asthma

Asthma is a chronic inflammatory disease characterized by shortness of breath, wheezing and recurring cough. It is caused by a combination of excess airway mucus production and reversible airflow obstruction. Immune cells such as eosinophils and neutrophils play a major role in this inflammatory state although different patients with severe asthma may exhibit different phenotypes that need to be properly defined if the most effective treatment is to be provided (Gibeon and Chung 2012). Current methods for distinguishing different phenotypes of asthma involve detecting sputum eosinophils (eosinophilic asthma), serum periostin levels which is a biomarker for eosinophilic asthma (Jia et al. 2012) and breath nitric oxide (NO) levels (Dweik et al. 2011). H_2S is one such potential biomarker of asthma given that it was found to be elevated in the sputum of asthmatics (Chung 2014; Saito et al. 2013).

Endogenous serum H_2S levels were found to be altered in chronic obstructive pulmonary disease (COPD) in man almost a decade ago (Chen et al. 2005). This group reported that endogenous serum H_2S levels were higher in patients with COPD as compared to patients with acute exacerbation of COPD (AECOPD). Serum levels decreased in patients with stable COPD as the condition became clinically more severe. Subsequently, serum H_2S levels in asthmatic children were reportedly decreased compared with a matched healthy control group (Tian et al. 2012).

A more recent clinical study examined levels of H_2S in serum and sputum from 40 patients with varying degrees of severity of asthma and compared the data obtained with that of 15 healthy subjects (Saito et al. 2013). H_2S levels in sputum of asthma patients were significantly higher than those in sputum from healthy subjects. Serum H_2S concentration in asthmatics was also 10 times higher than that in sputum. Sputum measurements are more likely to be indicative of asthma as measurement of serum H_2S will be confounded by the presence of H_2S generated by non-respiratory tissues. Moreover, a positive correlation was identified between sputum H_2S levels and sputum neutrophil number. These observations raise the possibility that sputum H_2S concentration may perhaps be a biomarker of neutrophilic asthma.

Since endogenous H_2S levels are modulated during asthma, it is conceivable that the activity or expression of H_2S synthesizing enzymes may also be altered at one or other time in the course of the disease (See Table 1). To date, there have been few studies to investigate this possibility, although there are clues in the literature that this may be the case. For example, it has recently been shown, using mouse lung slices, that exogenous H_2S impeded airway contraction of smooth muscle cells by inhibiting intracellular Ca^{2+} release evoked by inositol-1,4,5-triphosphate (InsP3). This effect was also apparent when murine lung tissues was treated with L-cysteine (an endogenous H_2S precursor) which effect was reversed by the irreversible CSE inhibitor, DL-propargylglycine (PAG) (Castro-Piedras and Perez-Zoghbi 2013). More direct evidence comes from experiments using an ovalbumin (OVA)-induced mouse model of acute asthma. In this case, CSE expression was downregulated in the lungs of wild-type mice challenged with OVA, whilst mice in which CSE was knocked out ($CSE^{-/-}$) displayed augmented airway inflammation following ovalbumin treatment as indicated by higher levels of Th2 cytokines such as IL-5, IL-13 and eotaxin-1. Exogenous H_2S administration, in the form of NaHS, effectively 'rescued' $CSE^{-/-}$ mice from the exacerbated symptoms of asthma strongly implying that an upregulated CSE/H_2S system has a protective role during the development, or perhaps the maintenance, of asthma in this model (Zhang et al. 2013). Whether such a mechanism also occurs in man is not yet known. Considerable evidence now exists, reviewed elsewhere in this volume, that H_2S exhibits both pro- and anti-inflammatory effects in a range of animal models. Thus, whether an H_2S donor administered may prove beneficial or deleterious in asthma remains an open question.

5.2 Atherosclerosis

H_2S exhibits pronounced vasodilator activity both in vitro and in vivo, and it is hence no surprise that deranged H_2S biosynthesis has been implicated in a number of cardiovascular diseases including atherosclerosis. H_2S dilates blood vessels by various mechanisms, including the opening of potassium-activated ATP channels (K_{ATP} channels) in vascular smooth muscle (Zhao et al. 2001a) as well as potentially via intracellular acidification by activating of the Cl^-/HCO_3^- exchanger (Lee et al. 2007). Atherosclerosis is a chronic and complex inflammatory condition. Its pathogenesis involves an intricate tapestry of prolonged immune cell recruitment and cytokine/chemokine secretion coupled with the presence of reactive oxygen species (ROS), as well as vascular smooth muscle cell proliferation and migration (Ross 1999). Intriguingly, each of these separate events has been reported to be regulated by H_2S (Du et al. 2004; Yan et al. 2006). However based on the evidence available thus far, the link between H_2S and atherosclerosis and the involvement of the H_2S synthesizing enzymes in the process is far from clear.

In a recent study, Peh and colleagues reported reduced expression of CSE and CBE protein in liver, kidneys and lungs of mice fed a high-fat diet. 3-MST expression was also reduced in liver. In addition, CSE/CBS and 3-MST enzyme activity was also diminished as determined by ex vivo H_2S synthesis using either cysteine or 3-mercaptopyruvate as substrates. It should be noted that these animals showed no evidence of frank atherosclerosis in that plasma levels of serum amyloid A (SAA) and C-reactive protein (CRP) were normal and the histological appearance of blood vessels revealed no ongoing disease. Changes in the expression of these enzymes were therefore likely to be a consequence of fat feeding and not a reaction to vascular disease. If this is the case, then changes in H_2S synthesis may be presumed to occur before the onset of atherosclerosis (Peh et al. 2014). Whether measurement of these enzymes in susceptible individuals might be a surrogate biomarker of impeding vascular damage due to atherosclerosis remains to be determined. Intriguingly, in a separate study, mice fed a high-fat diet leading to the development of fatty liver exhibited higher levels of CSE and CBS in the liver (Hwang et al. 2013). The mechanisms which control CSE, CBS and 3-MST expression in animals consuming fat in their diet warrant further study.

Other researchers have studied H_2S synthesizing enzyme expression and activity in apolipoprotein E (ApoE) knockout mice. These animals have increased plasma levels of cholesterol, triglyceride and low-density lipoprotein cholesterol, develop a frank atherosclerotic state and also show evidence of defective H_2S biosynthesis including lower production of H_2S in aorta and reduced plasma H_2S concentration. Interestingly, CSE mRNA expression in the aorta was actually increased. Administration of exogenous H_2S (NaHS) elevated plasma H_2S and reduced aortic CSE mRNA thereby suggesting an inverse correlation between CSE expression and H_2S levels in the aorta which may be a consequence of a positive feedback mechanism in which decreased H_2S production (due to vascular disease) is compensated for by increased CSE gene expression (Wang et al. 2009). Interestingly, elevated plasma H_2S has also been reported in patients with atherosclerosis (Peter et al. 2013), whilst diminished production of endogenous H_2S which is apparent in $CSE^{-/-}$ mice is associated with accelerated atherosclerosis (Mani et al. 2013). In this latter work, the rate of development of atherosclerosis in these animals could be reduced by replacement therapy with NaHS. The current consensus is therefore that lack of H_2S contributes to accelerated progression of atherosclerosis (Szabó et al. 2011; Xu et al. 2014a; Zhang et al. 2012).

5.3 Diabetes

Altered expression of H_2S synthesizing enzymes as well as endogenous H_2S levels has been observed in diabetic animals. In an early study, plasma H_2S and the expression of CSE and CBS were elevated in the pancreas of streptozotocin (STZ)-induced diabetic rats (Yusuf et al. 2005) which suggests a possible correlation between Type 1 diabetes (Schnedl et al. 1994) and increased pancreatic and liver H_2S synthesis. To evaluate more precisely the role of H_2S in diabetes, a subsequent study made use of $CSE^{-/-}$ mice injected with STZ to provoke a diabetic state. Since the pathogenesis of diabetes is associated with reduced number, but increased activity, of K_{ATP} channels in β-pancreatic cells, these authors sought to compare the mass and K_{ATP} channel activity of β-pancreatic cells in $CSE^{-/-}$ mice with that of wild-type mice (Yang et al. 2011). CSE deficiency protected mice from STZ-induced diabetes as well as the damage and dysfunction of pancreatic islets—as determined by measurement of the changes in insulin secretion and electrophysiological recording of K_{ATP} channel currents after induction of a diabetic state.

Curiously, other studies have reported diametrically opposite results. For example, it was previously noted that circulating H_2S levels were lower (c.f. control, non-diabetic animals) in both STZ-treated rats and non-obese diabetic (NOD) models of Type 1 diabetic mice (Brancaleone et al. 2008; Jain et al. 2010). A similar conclusion was reached from the examination of blood H_2S levels in humans with diabetes (Jain et al. 2010). CSE enzyme activity was also reduced in the liver of Type 1 diabetic rats as well as in peripheral blood mononuclear cells from Type 1 diabetic patients (Manna et al. 2014).

In contrast to Type 1 diabetes which is caused by the death of β-pancreatic cells, Type 2 diabetes arises, usually later in life, with the development of insulin resistance. To study the possible role of H_2S in this process, 6-month-old $CSE^{-/-}$ mice were fed for 8 weeks with a high-fat diet in order to induce hyperglycaemia (Okamoto et al. 2013). The mice developed impaired glucose tolerance and decreased insulin content in their pancreatic cells as compared to their wild-type counterparts. It was noted that the lack of CSE in these animals promoted thioredoxin-binding protein-2 (TBP-2) gene expression which increases insulin insensitivity and causes glucose intolerance (Yoshihara et al. 2010). These results indicate that changes in expression or activity of CSE (and perhaps other H_2S synthesizing enzymes) are implicated in the outcome and severity of both Type 1 and 2 diabetes.

6 Frequently Used Inhibitors and Mouse Knockout Models of H_2S Synthesizing Enzymes

To study the biological effects of endogenously synthesized H_2S, as well as to study the characteristics and behaviour of the H_2S synthesizing enzymes, the use of selective inhibitors of H_2S synthesizing enzymes is crucial. Currently, several inhibitors of CSE and CBS have been used (See Fig. 2) although to date, pharmacological inhibitors of the enzyme 3-MST have yet to be identified. In this context, L-aspartate inhibits the production of H_2S by 3-MST indirectly by inhibiting CAT/AAT (Akagi 1982). Mice with genetic deletion of one or other of the H_2S synthesizing enzymes are also beginning to show their value to the field as well. Of the three known H_2S synthesizing enzymes only $CSE^{-/-}$ mice have been widely used as 90 % of $CBS^{-/-}$ mice died during the first 2 weeks of neonatal life because of growth retardation and severe hepatopathy (Watanabe et al. 1995) and only lately has it been possible to produce 3-MST knockout mice (Nagahara 2013). However, recently a non-neonatal fatal mouse model with the *cbs* gene inactivated but instead expresses low levels of the human CBS transgene and exhibits classical homocystinuria could provide a useful, alternative to a $CBS^{-/-}$ mouse model (Maclean et al. 2010).

Some of the commonly used pharmacological inhibitors of CSE are DL-propargylglycine (PAG) (Marcotte and Walsh 1975) which irreversibly inhibits CSE by physically obstructing the access of the substrate to the active site of the enzyme (Sun et al. 2009) and the reversible inhibitor β-cyanoalanine (BCA) which has also been proposed as a competitive inhibitor of this enzyme (Pfeffer and Ressler 1967). However, these compounds have frequently been claimed to exhibit poor selectivity, require high concentrations and have limited ability to permeate the cell membrane (Szabó 2007). For example, PAG was reported many years ago to cause an irreversible inactivation of both aspartate aminotransferase (Tanase and Morino 1976) and alanine transaminase (Burnett et al. 1980). Aminooxyacetic acid (AOAA) is also very commonly used and is often stated to be a selective inhibitor of CBS (d'Emmanuele di Villa Bianca et al. 2009; Oh et al. 2006; Roy et al. 2012). However, experiments using recombinant human CSE and CBS enzymes have revealed that AOAA inhibits both CSE and CBS. Indeed, AOAA appeared to be

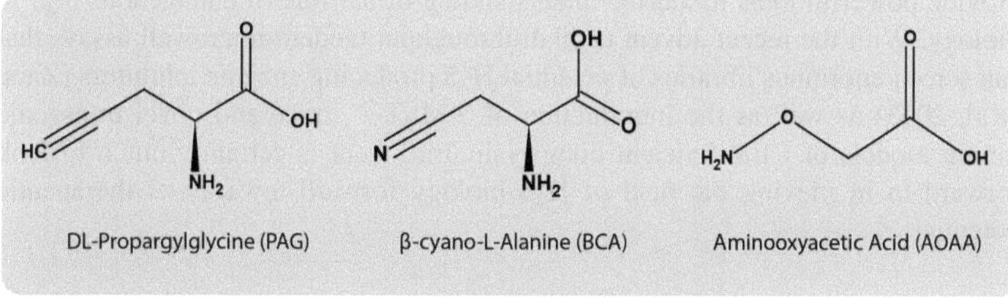

Fig. 2 Chemical structures of compounds that inhibit CSE or CBS

an even more potent inhibitor of CSE than CBS at least under these experimental conditions (Asimakopoulou et al. 2013). Moreover, AOAA, like PAG, is a general inhibitor of pyridoxal phosphate-dependent enzymes and has been reported to inhibit enzymes such as 4-aminobutyrate aminotransferase (GABA-T) (WALLACH 1961) and aspartate aminotransferase (Kauppinen et al. 1987).

7 Conclusion

Much has been learnt about the physiological and pathophysiological implications of H_2S since it was realized that this evanescent gas is formed naturally in cells and tissues in the 1990s. Indeed, research in the last two decades has thrown H_2S very much into the spotlight and it is now widely referred to as the so-called third gasotransmitter after CO and NO. Many mechanisms governing the regulation of H_2S synthesizing enzymes both in homeostatic and disease conditions have been elucidated. However, many questions still remain to be answered. For example, the quantitative contribution that each of the three H_2S synthesizing enzymes makes to the levels of H_2S found naturally in cells in health and disease remains unclear. This is important since, as detailed in this review, the presence and/or activity of such enzymes may be useful biomarkers of disease progression in some cases and it is critical to understand which H_2S synthesizing enzyme is important in which cell at which time point and how this is affected by disease. Moreover, whilst the biological significance of H_2S has been under intense scrutiny for more than two decades, there are still no reliably potent and selective inhibitors of any of the three H_2S synthesizing enzymes. For comparison, potent and selective inhibitors of both cyclooxygenase and nitric oxide synthase are available and have played crucial roles in understanding the complex biological roles of prostanoids and nitric oxide, respectively. Potent and targeted inhibitors of CSE, CBS and 3-MST could be expected to fill a similar niche. Finally, the precise intracellular signalling pathways which determine the degree of expression of H_2S synthesizing enzymes in healthy and in diseased cells have yet to be fully elucidated.

While many advances have been made, more selective and potent H_2S synthesizing enzyme inhibitors as well as cells in which the H_2S synthesizing enzymes (both alone and in combination) have been knocked out would certainly provide powerful tools to aid the understanding of the role of endogenous H_2S in biology. With the recent advent of high-throughput tandem-microwell assays that can screen enormous libraries of potential H_2S producing enzyme inhibitors (Zhou et al. 2013) as well as the introduction of 3-MST$^{-/-}$ mice and novel transgenic mouse models of CBS-deficient homocystinuria, there is certainly much to look forward to in moving the field of H_2S biology forward towards its therapeutic potential.

References

Abe K, Kimura H (1996) The possible role of hydrogen sulfide as an endogenous neuromodulator. J Neurosci 16:1066–1071

Agrawal N, Banerjee R (2008) Human polycomb 2 protein is a SUMO E3 ligase and alleviates substrate-induced inhibition of cystathionine β-synthase sumoylation. PLoS ONE 3:e4032

Akagi R (1982) Purification and characterization of cysteine aminotransferase from rat liver cytosol. Acta Med Okayama 36:187–197

Asimakopoulou A, Panopoulos P, Chasapis CT, Coletta C, Zhou Z, Cirino G, Giannis A, Szabo C, Spyroulias GA, Papapetropoulos A (2013) Selectivity of commonly used pharmacological inhibitors for cystathionine beta synthase (CBS) and cystathionine gamma lyase (CSE). Br J Pharmacol 169:922–932. doi:10.1111/bph.12171

Banerjee R, Zou C-G (2005) Redox regulation and reaction mechanism of human cystathionine-β-synthase: a PLP-dependent hemesensor protein. Arch Biochem Biophys 433:144–156

Beauchamp R, Bus JS, Popp JA, Boreiko CJ, Andjelkovich DA, Leber P (1984) A critical review of the literature on hydrogen sulfide toxicity. CRC Crit Rev Toxicol 13:25–97

Beinert H, Holm RH, Münck E (1997) Iron-sulfur clusters: nature's modular, multipurpose structures. Science 277:653–659

Borrás C, Esteve JM, Viña JR, Sastre J, Viña J, Pallardó FV (2004) Glutathione regulates telomerase activity in 3T3 fibroblasts. J Biol Chem 279:34332–34335. doi:10.1074/jbc.M402425200

Brancaleone V, Roviezzo F, Vellecco V, De Gruttola L, Bucci M, Cirino G (2008) Biosynthesis of H2S is impaired in non-obese diabetic (NOD) mice. Br J Pharmacol 155:673–680. doi:10.1038/bjp.2008.296

Burnett G, Marcotte P, Walsh C (1980) Mechanism-based inactivation of pig heart L-alanine transaminase by L-propargylglycine. Half-site reactivity. J Biol Chem 255:3487–3491

Calvert JW, Jha S, Gundewar S, Elrod JW, Ramachandran A, Pattillo CB, Kevil CG, Lefer DJ (2009) Hydrogen sulfide mediates cardioprotection through Nrf2 signaling. Circ Res 105:365–374. doi:10.1161/CIRCRESAHA.109.199919

Cantoni G (1952) The nature of the active methyl donor formed enzymatically from l-methionine and adenosinetriphosphate1, 2. J Am Chem Soc 74:2942–2943

Castro-Piedras I, Perez-Zoghbi JF (2013) Hydrogen sulphide inhibits Ca2+ release through InsP3 receptors and relaxes airway smooth muscle. J Physiol 591:5999–6015. doi:10.1113/jphysiol.2013.257790

Chen X, Jhee K-H, Kruger WD (2004) Production of the neuromodulator H2S by cystathionine β-synthase via the condensation of cysteine and homocysteine. J Biol Chem 279:52082–52086

Chen YH, Yao WZ, Geng B, Ding YL, Lu M, Zhao MW, Tang CS (2005) Endogenous hydrogen sulfide in patients with COPD. Chest 128:3205–3211. doi:10.1378/chest.128.5.3205

Chen YH, Wu R, Geng B, Qi YF, Wang PP, Yao WZ, Tang CS (2009) Endogenous hydrogen sulfide reduces airway inflammation and remodeling in a rat model of asthma. Cytokine 45:117–123. doi:10.1016/j.cyto.2008.11.009

Chiku T, Padovani D, Zhu W, Singh S, Vitvitsky V, Banerjee R (2009) H2S biogenesis by human cystathionine γ-lyase leads to the novel sulfur metabolites lanthionine and homolanthionine and is responsive to the grade of hyperhomocysteinemia. J Biol Chem 284:11601–11612

Chung KF (2014) Hydrogen sulfide as a potential biomarker of asthma. Expert Rev Respir Med 8:5–13. doi:10.1586/17476348.2014.856267

Dalton TP, Chen Y, Schneider SN, Nebert DW, Shertzer HG (2004) Genetically altered mice to evaluate glutathione homeostasis in health and disease. Free Radic Biol Med 37:1511–1526. doi:10.1016/j.freeradbiomed.2004.06.040

D'Emmanuele di Villa Bianca R, Sorrentino R, Maffia P, Mirone V, Imbimbo C, Fusco F, De Palma R, Ignarro LJ, Cirino G (2009) Hydrogen sulfide as a mediator of human corpus cavernosum smooth-muscle relaxation. Proc Natl Acad Sci USA 106:4513–4518. doi:10.1073/pnas.0807974105

Dickhout JG, Carlisle RE, Jerome DE, Mohammed-Ali Z, Jiang H, Yang G, Mani S, Garg SK, Banerjee R, Kaufman RJ, Maclean KN, Wang R, Austin RC (2012) Integrated stress response modulates cellular redox state via induction of cystathionine γ-lyase: cross-talk between integrated stress response and thiol metabolism. J Biol Chem 287:7603–7614. doi:10.1074/jbc.M111.304576

Diwakar L, Ravindranath V (2007) Inhibition of cystathionine-gamma-lyase leads to loss of glutathione and aggravation of mitochondrial dysfunction mediated by excitatory amino acid in the CNS. Neurochem Int 50:418–426. doi:10.1016/j.neuint.2006.09.014

Du J, Hui Y, Cheung Y, Bin G, Jiang H, Chen X, Tang C (2004) The possible role of hydrogen sulfide as a smooth muscle cell proliferation inhibitor in rat cultured cells. Heart Vessels 19:75–80. doi:10.1007/s00380-003-0743-7

Dweik RA, Boggs PB, Erzurum SC, Irvin CG, Leigh MW, Lundberg JO, Olin AC, Plummer AL, Taylor DR, Applications ATSCoIoENOLFfC (2011) An official ATS clinical practice guideline: interpretation of exhaled nitric oxide levels (FENO) for clinical applications. Am J Respir Crit Care Med 184:602–615. doi:10.1164/rccm.9120-11ST

Finkelstein JD, Kyle WE, Martin JJ, Pick A-M (1975) Activation of cystathionine synthase by adenosylmethionine and adenosylethionine. Biochem Biophys Res Commun 66:81–87

Franco R, Cidlowski JA (2009) Apoptosis and glutathione: beyond an antioxidant. Cell Death Differ 16:1303–1314. doi:10.1038/cdd.2009.107

Ge Y, Jensen TL, Matherly LH, Taub JW (2003) Transcriptional regulation of the cystathionine-beta -synthase gene in Down syndrome and non-Down syndrome megakaryocytic leukemia cell lines. Blood 101:1551–1557. doi:10.1182/blood-2002-07-2337

Ghibelli L, Coppola S, Fanelli C, Rotilio G, Civitareale P, Scovassi AI, Ciriolo MR (1999) Glutathione depletion causes cytochrome c release even in the absence of cell commitment to apoptosis. FASEB J 13:2031–2036

Gibeon D, Chung KF (2012) The investigation of severe asthma to define phenotypes. Clin Exp Allergy 42:678–692. doi:10.1111/j.1365-2222.2012.03959.x

Goodwin LR, Francom D, Dieken FP, Taylor JD, Warenycia MW, Reiffenstein R, Dowling G (1989) Determination of sulfide in brain tissue by gas dialysis/ion chromatography: postmortem studies and two case reports. J Anal Toxicol 13:105–109

Hassan MI, Boosen M, Schaefer L, Kozlowska J, Eisel F, von Knethen A, Beck M, Hemeida RA, El-Moselhy MA, Hamada FM, Beck KF, Pfeilschifter J (2012) Platelet-derived growth factor-BB induces cystathionine γ-lyase expression in rat mesangial cells via a redox-dependent mechanism. Br J Pharmacol 166:2231–2242. doi:10.1111/j.1476-5381.2012.01949.x

Hosoki R, Matsuki N, Kimura H (1997) The possible role of hydrogen sulfide as an endogenous smooth muscle relaxant in synergy with nitric oxide. Biochem Biophys Res Commun 237:527–531

Hu LF, Wong PT, Moore PK, Bian JS (2007) Hydrogen sulfide attenuates lipopolysaccharide-induced inflammation by inhibition of p38 mitogen-activated protein kinase in microglia. J Neurochem 100:1121–1128. doi:10.1111/j.1471-4159.2006.04283.x

Hwang SY, Sarna LK, Siow YL, O K (2013) High-fat diet stimulates hepatic cystathionine β-synthase and cystathionine γ-lyase expression. Can J Physiol Pharmacol 91:913–919. doi:10.1139/cjpp-2013-0106

Ichinohe A, Kanaumi T, Takashima S, Enokido Y, Nagai Y, Kimura H (2005) Cystathionine beta-synthase is enriched in the brains of Down's patients. Biochem Biophys Res Commun 338:1547–1550. doi:10.1016/j.bbrc.2005.10.118

Ishigami M, Hiraki K, Umemura K, Ogasawara Y, Ishii K, Kimura H (2009) A source of hydrogen sulfide and a mechanism of its release in the brain. Antioxid Redox Signal 11:205–214

Ishii I, Akahoshi N, Yu XN, Kobayashi Y, Namekata K, Komaki G, Kimura H (2004) Murine cystathionine gamma-lyase: complete cDNA and genomic sequences, promoter activity, tissue distribution and developmental expression. Biochem J 381:113–123. doi:10.1042/BJ20040243

Jain SK, Bull R, Rains JL, Bass PF, Levine SN, Reddy S, McVie R, Bocchini JA (2010) Low levels of hydrogen sulfide in the blood of diabetes patients and streptozotocin-treated rats causes vascular inflammation? Antioxid Redox Signal 12:1333–1337. doi:10.1089/ars.2009.2956

Jia G, Erickson RW, Choy DF, Mosesova S, Wu LC, Solberg OD, Shikotra A, Carter R, Audusseau S, Hamid Q, Bradding P, Fahy JV, Woodruff PG, Harris JM, Arron JR, Group BERSoBiC-rABS (2012) Periostin is a systemic biomarker of eosinophilic airway inflammation in asthmatic patients. J Allergy Clin Immunol 130:647–654.e10. doi:10.1016/j.jaci.2012.06.025

Kabil O, Banerjee R (2014) Enzymology of H2S biogenesis, decay and signaling. Antioxid Redox Signal 20:770–782. doi:10.1089/ars.2013.5339

Kabil O, Zhou Y, Banerjee R (2006) Human cystathionine β-synthase is a target for sumoylation. Biochemistry 45:13528–13536

Kabil O, Vitvitsky V, Xie P, Banerjee R (2011a) The quantitative significance of the transsulfuration enzymes for H2S production in murine tissues. Antioxid Redox Signal 15:363–372

Kabil O, Weeks CL, Carballal S, Gherasim C, Alvarez B, Spiro TG, Banerjee R (2011b) Reversible heme-dependent regulation of human cystathionine β-synthase by a flavoprotein oxidoreductase. Biochemistry 50:8261–8263

Kabil O, Motl N, Banerjee R (2014) HS and its role in redox signaling. Biochim Biophys Acta 1844:1355–1366. doi:10.1016/j.bbapap.2014.01.002

Kaneko Y, Kimura Y, Kimura H, Niki I (2006) L-cysteine inhibits insulin release from the pancreatic β-cell possible involvement of metabolic production of hydrogen sulfide, a novel gasotransmitter. Diabetes 55:1391–1397

Kauppinen RA, Sihra TS, Nicholls DG (1987) Aminooxyacetic acid inhibits the malate-aspartate shuttle in isolated nerve terminals and prevents the mitochondria from utilizing glycolytic substrates. Biochim Biophys Acta 930:173–178

Lee SW, Cheng Y, Moore PK, Bian JS (2007) Hydrogen sulphide regulates intracellular pH in vascular smooth muscle cells. Biochem Biophys Res Commun 358:1142–1147. doi:10.1016/j.bbrc.2007.05.063

Li L, Bhatia M, Zhu YZ, Zhu YC, Ramnath RD, Wang ZJ, Anuar FB, Whiteman M, Salto-Tellez M, Moore PK (2005) Hydrogen sulfide is a novel mediator of lipopolysaccharide-induced inflammation in the mouse. FASEB J 19:1196–1198. doi:10.1096/fj.04-3583fje

Li L, Salto-Tellez M, Tan CH, Whiteman M, Moore PK (2009a) GYY4137, a novel hydrogen sulfide-releasing molecule, protects against endotoxic shock in the rat. Free Radic Biol Med 47:103–113. doi:10.1016/j.freeradbiomed.2009.04.014

Li L, Whiteman M, Moore PK (2009b) Dexamethasone inhibits lipopolysaccharide-induced hydrogen sulphide biosynthesis in intact cells and in an animal model of endotoxic shock. J Cell Mol Med 13:2684–2692. doi:10.1111/j.1582-4934.2008.00610.x

Li L, Rose P, Moore PK (2011) Hydrogen sulfide and cell signaling. Annu Rev Pharmacol Toxicol 51:169–187. doi:10.1146/annurev-pharmtox-010510-100505

Li L, Xie R, Hu S, Wang Y, Yu T, Xiao Y, Jiang X, Gu J, Hu CY, Xu GY (2012) Upregulation of cystathionine beta-synthetase expression by nuclear factor-kappa B activation contributes to visceral hypersensitivity in adult rats with neonatal maternal deprivation. Mol Pain 8:89. doi:10.1186/1744-8069-8-89

Maclean KN, Janošík M, Kraus E, Kožich V, Allen RH, Raab BK, Kraus JP (2002) Cystathionine β-synthase is coordinately regulated with proliferation through a redox-sensitive mechanism in cultured human cells and Saccharomyces cerevisiae. J Cell Physiol 192:81–92

Maclean KN, Sikora J, Kožich V, Jiang H, Greiner LS, Kraus E, Krijt J, Overdier KH, Collard R, Brodsky GL, Meltesen L, Crnic LS, Allen RH, Stabler SP, Elleder M, Rozen R, Patterson D, Kraus JP (2010) A novel transgenic mouse model of CBS-deficient homocystinuria does not incur hepatic steatosis or fibrosis and exhibits a hypercoagulative phenotype that is ameliorated by betaine treatment. Mol Genet Metab 101:153–162. doi:10.1016/j.ymgme.2010.06.010

Mani S, Li H, Untereiner A, Wu L, Yang G, Austin RC, Dickhout JG, Lhoták Š, Meng QH, Wang R (2013) Decreased endogenous production of hydrogen sulfide accelerates atherosclerosis. Circulation 127:2523–2534. doi:10.1161/CIRCULATIONAHA.113.002208

Manna P, Gungor N, McVie R, Jain SK (2014) Decreased cystathionine-γ-lyase (CSE) activity in livers of type 1 diabetic rats and peripheral blood mononuclear cells (PBMC) of type 1 diabetic patients. J Biol Chem 289:11767–11778. doi:10.1074/jbc.M113.524645

Marcotte P, Walsh C (1975) Active site-directed inactivation of cystathionine gamma-synthetase and glutamic pyruvic transaminase by propargylglycine. Biochem Biophys Res Commun 62:677–682

Markovic J, Borrás C, Ortega A, Sastre J, Viña J, Pallardó FV (2007) Glutathione is recruited into the nucleus in early phases of cell proliferation. J Biol Chem 282:20416–20424. doi:10.1074/jbc.M609582200

Miao X, Meng X, Wu G, Ju Z, Zhang HH, Hu S, Xu GY (2014) Upregulation of cystathionine-β-synthetase expression contributes to inflammatory pain in rat temporomandibular joint. Mol Pain 10:9. doi:10.1186/1744-8069-10-9

Mikami Y, Shibuya N, Kimura Y, Nagahara N, Ogasawara Y, Kimura H (2011a) Thioredoxin and dihydrolipoic acid are required for 3-mercaptopyruvate sulfurtransferase to produce hydrogen sulfide. Biochem J 439:479–485. doi:10.1042/BJ20110841

Mikami Y, Shibuya N, Kimura Y, Nagahara N, Yamada M, Kimura H (2011b) Hydrogen sulfide protects the retina from light-induced degeneration by the modulation of Ca2+ influx. J Biol Chem 286:39379–39386. doi:10.1074/jbc.M111.298208

Mikami Y, Shibuya N, Ogasawara Y, Kimura H (2013) Hydrogen sulfide is produced by cystathionine γ-lyase at the steady-state low intracellular Ca(2+) concentrations. Biochem Biophys Res Commun 431:131–135. doi:10.1016/j.bbrc.2013.01.010

Nagahara N (2013) Regulation of mercaptopyruvate sulfurtransferase activity via intrasubunit and intersubunit redox-sensing switches. Antioxid Redox Signal 19:1792–1802. doi:10.1089/ars.2012.5031

Nagahara N, Ito T, Kitamura H, Nishino T (1998) Tissue and subcellular distribution of mercaptopyruvate sulfurtransferase in the rat: confocal laser fluorescence and immunoelectron microscopic studies combined with biochemical analysis. Histochem Cell Biol 110:243–250

Nagahara N, Yoshii T, Abe Y, Matsumura T (2007) Thioredoxin-dependent enzymatic activation of mercaptopyruvate sulfurtransferase. An intersubunit disulfide bond serves as a redox switch for activation. J Biol Chem 282:1561–1569. doi:10.1074/jbc.M605931200

Oh GS, Pae HO, Lee BS, Kim BN, Kim JM, Kim HR, Jeon SB, Jeon WK, Chae HJ, Chung HT (2006) Hydrogen sulfide inhibits nitric oxide production and nuclear factor-kappaB via heme oxygenase-1 expression in RAW264.7 macrophages stimulated with lipopolysaccharide. Free Radic Biol Med 41:106–119. doi:10.1016/j.freeradbiomed.2006.03.021

Okamoto M, Yamaoka M, Takei M, Ando T, Taniguchi S, Ishii I, Tohya K, Ishizaki T, Niki I, Kimura T (2013) Endogenous hydrogen sulfide protects pancreatic beta-cells from a high-fat diet-induced glucotoxicity and prevents the development of type 2 diabetes. Biochem Biophys Res Commun 442:227–233. doi:10.1016/j.bbrc.2013.11.023

Patel P, Vatish M, Heptinstall J, Wang R, Carson RJ (2009) The endogenous production of hydrogen sulphide in intrauterine tissues. Reprod Biol Endocrinol 7:10. doi:10.1186/1477-7827-7-10

Peh MT, Anwar AB, Ng DS, Atan MS, Kumar SD, Moore PK (2014) Effect of feeding a high fat diet on hydrogen sulfide (H2S) metabolism in the mouse. Nitric Oxide 41:138–145. doi:10.1016/j.niox.2014.03.002

Peter EA, Shen X, Shah SH, Pardue S, Glawe JD, Zhang WW, Reddy P, Akkus NI, Varma J, Kevil CG (2013) Plasma free H2S levels are elevated in patients with cardiovascular disease. J Am Heart Assoc 2:e000387. doi:10.1161/JAHA.113.000387

Pfeffer M, Ressler C (1967) Beta-cyanoalanine, an inhibitor of rat liver cystathionase. Biochem Pharmacol 16:2299–2308

Prudova A, Bauman Z, Braun A, Vitvitsky V, Lu SC, Banerjee R (2006) S-adenosylmethionine stabilizes cystathionine β-synthase and modulates redox capacity. Proc Natl Acad Sci USA 103:6489–6494

Puranik M, Weeks CL, Lahaye D, Kabil Ö, Taoka S, Nielsen SB, Groves JT, Banerjee R, Spiro TG (2006) Dynamics of carbon monoxide binding to cystathionine β-synthase. J Biol Chem 281:13433–13438

Ramasamy S, Singh S, Taniere P, Langman MJ, Eggo MC (2006) Sulfide-detoxifying enzymes in the human colon are decreased in cancer and upregulated in differentiation. Am J Physiol Gastrointest Liver Physiol 291:G288–G296. doi:10.1152/ajpgi.00324.2005

Ratnam S, Maclean KN, Jacobs RL, Brosnan ME, Kraus JP, Brosnan JT (2002) Hormonal regulation of cystathionine beta-synthase expression in liver. J Biol Chem 277:42912–42918. doi:10.1074/jbc.M206588200

Ross R (1999) Atherosclerosis–an inflammatory disease. N Engl J Med 340:115–126. doi:10. 1056/NEJM199901143400207

Roy A, Khan AH, Islam MT, Prieto MC, Majid DS (2012) Interdependency of cystathione γ-lyase and cystathione β-synthase in hydrogen sulfide-induced blood pressure regulation in rats. Am J Hypertens 25:74–81. doi:10.1038/ajh.2011.149

Saito J, Zhang Q, Hui C, Macedo P, Gibeon D, Menzies-Gow A, Bhavsar PK, Chung KF (2013) Sputum hydrogen sulfide as a novel biomarker of obstructive neutrophilic asthma. J Allergy Clin Immunol 131:232–234.e1-3. doi:10.1016/j.jaci.2012.10.005

Schnedl WJ, Ferber S, Johnson JH, Newgard CB (1994) STZ transport and cytotoxicity. Specific enhancement in GLUT2-expressing cells. Diabetes 43:1326–1333

Searcy DG, Lee SH (1998) Sulfur reduction by human erythrocytes. J Exp Zool 282:310–322

Sen N, Paul BD, Gadalla MM, Mustafa AK, Sen T, Xu R, Kim S, Snyder SH (2012) Hydrogen sulfide-linked sulfhydration of NF-κB mediates its antiapoptotic actions. Mol Cell 45:13–24. doi:10.1016/j.molcel.2011.10.021

Shibuya N, Mikami Y, Kimura Y, Nagahara N, Kimura H (2009a) Vascular endothelium expresses 3-mercaptopyruvate sulfurtransferase and produces hydrogen sulfide. J Biochem 146:623–626. doi:10.1093/jb/mvp111

Shibuya N, Tanaka M, Yoshida M, Ogasawara Y, Togawa T, Ishii K, Kimura H (2009b) 3-Mercaptopyruvate sulfurtransferase produces hydrogen sulfide and bound sulfane sulfur in the brain. Antioxid Redox Signal 11:703–714

Singh S, Banerjee R (2011) PLP-dependent H(2)S biogenesis. Biochim Biophys Acta 1814:1518–1527. doi:10.1016/j.bbapap.2011.02.004

Steegborn C, Clausen T, Sondermann P, Jacob U, Worbs M, Marinkovic S, Huber R, Wahl MC (1999) Kinetics and inhibition of recombinant human cystathionine gamma-lyase. Toward the rational control of transsulfuration. J Biol Chem 274:12675–12684

Stipanuk MH (1986) Metabolism of sulfur-containing amino acids. Annu Rev Nutr 6:179–209

Sun YG, Cao YX, Wang WW, Ma SF, Yao T, Zhu YC (2008) Hydrogen sulphide is an inhibitor of L-type calcium channels and mechanical contraction in rat cardiomyocytes. Cardiovasc Res 79:632–641. doi:10.1093/cvr/cvn140

Sun Q, Collins R, Huang S, Holmberg-Schiavone L, Anand GS, Tan CH, van-den-Berg S, Deng LW, Moore PK, Karlberg T, Sivaraman J (2009) Structural basis for the inhibition mechanism of human cystathionine gamma-lyase, an enzyme responsible for the production of H(2)S. J Biol Chem 284:3076–3085. doi:10.1074/jbc.M805459200

Szabó C (2007) Hydrogen sulphide and its therapeutic potential. Nat Rev Drug Discov 6:917–935. doi:10.1038/nrd2425

Szabó G, Veres G, Radovits T, Gero D, Módis K, Miesel-Gröschel C, Horkay F, Karck M, Szabó C (2011) Cardioprotective effects of hydrogen sulfide. Nitric Oxide 25:201–210. doi:10.1016/j. niox.2010.11.001

Takano N, Peng YJ, Kumar GK, Luo W, Hu H, Shimoda LA, Suematsu M, Prabhakar NR, Semenza GL (2014) Hypoxia-inducible factors regulate human and rat cystathionine β-synthase gene expression. Biochem J 458:203–211. doi:10.1042/BJ20131350

Tanase S, Morino Y (1976) Irreversible inactivation of aspartate aminotransferases during transamination with L-propargylglycine. Biochem Biophys Res Commun 68:1301–1308

Taoka S, Banerjee R (2001) Characterization of NO binding to human cystathionine β-synthase: possible implications of the effects of CO and NO binding to the human enzyme. J Inorg Biochem 87:245–251

Teague B, Asiedu S, Moore PK (2002) The smooth muscle relaxant effect of hydrogen sulfide in vitro: evidence for a pysiological role to control intestinal contractility. Br J Pharmacol 137:139–145

Tian M, Wang Y, Lu YQ, Yan M, Jiang YH, Zhao DY (2012) Correlation between serum H2S and pulmonary function in children with bronchial asthma. Mol Med Rep 6:335–338. doi:10.3892/mmr.2012.904

Wallace JL, Dicay M, McKnight W, Martin GR (2007) Hydrogen sulfide enhances ulcer healing in rats. FASEB J 21:4070–4076. doi:10.1096/fj.07-8669com

Wallach DP (1961) Studies on the GABA pathway. I. The inhibition of gamma-aminobutyric acid-alpha-ketoglutaric acid transaminase in vitro and in vivo by U-7524 (amino-oxyacetic acid). Biochem Pharmacol 5:323–331

Wang Y, Zhao X, Jin H, Wei H, Li W, Bu D, Tang X, Ren Y, Tang C, Du J (2009) Role of hydrogen sulfide in the development of atherosclerotic lesions in apolipoprotein E knockout mice. Arterioscler Thromb Vasc Biol 29:173–179. doi:10.1161/ATVBAHA.108.179333

Wang M, Guo Z, Wang S (2012a) Cystathionine gamma-lyase expression is regulated by exogenous hydrogen peroxide in the mammalian cells. Gene Expr 15:235–241

Wang Y, Qu R, Hu S, Xiao Y, Jiang X, Xu GY (2012b) Upregulation of cystathionine β-synthetase expression contributes to visceral hyperalgesia induced by heterotypic intermittent stress in rats. PLoS ONE 7:e53165. doi:10.1371/journal.pone.0053165

Wang M, Guo Z, Wang S (2014) The binding site for the transcription factor, NF-κB, on the cystathionine γ-lyase promoter is critical for LPS-induced cystathionine γ-lyase expression. Int J Mol Med 34:639–645. doi:10.3892/ijmm.2014.1788

Watanabe M, Osada J, Aratani Y, Kluckman K, Reddick R, Malinow MR, Maeda N (1995) Mice deficient in cystathionine beta-synthase: animal models for mild and severe homocyst(e) inemia. Proc Natl Acad Sci 92:1585–1589

Wu N, Siow YL, O K (2010) Ischemia/reperfusion reduces transcription factor Sp1-mediated cystathionine beta-synthase expression in the kidney. J Biol Chem 285:18225–18233. doi:10.1074/jbc.M110.132142

Xu S, Liu Z, Liu P (2014a) Targeting hydrogen sulfide as a promising therapeutic strategy for atherosclerosis. Int J Cardiol 172:313–317. doi:10.1016/j.ijcard.2014.01.068

Xu Y, Du HP, Li J, Xu R, Wang YL, You SJ, Liu H, Wang F, Cao YJ, Liu CF, Hu LF (2014b) Statins upregulate cystathionine γ-lyase transcription and H2S generation via activating Akt signaling in macrophage. Pharmacol Res 87C:18–25. doi:10.1016/j.phrs.2014.06.006

Yadav PK, Yamada K, Chiku T, Koutmos M, Banerjee R (2013) Structure and kinetic analysis of H2S production by human mercaptopyruvate sulfurtransferase. J Biol Chem 288:20002–20013. doi:10.1074/jbc.M113.466177

Yan SK, Chang T, Wang H, Wu L, Wang R, Meng QH (2006) Effects of hydrogen sulfide on homocysteine-induced oxidative stress in vascular smooth muscle cells. Biochem Biophys Res Commun 351:485–491. doi:10.1016/j.bbrc.2006.10.058

Yang G, Wu L, Jiang B, Yang W, Qi J, Cao K, Meng Q, Mustafa AK, Mu W, Zhang S, Snyder SH, Wang R (2008) H2S as a physiologic vasorelaxant: hypertension in mice with deletion of cystathionine gamma-lyase. Science 322:587–590. doi:10.1126/science.1162667

Yang G, Tang G, Zhang L, Wu L, Wang R (2011) The pathogenic role of cystathionine γ-lyase/hydrogen sulfide in streptozotocin-induced diabetes in mice. Am J Pathol 179:869–879. doi:10.1016/j.ajpath.2011.04.028

Yin P, Zhao C, Li Z, Mei C, Yao W, Liu Y, Li N, Qi J, Wang L, Shi Y, Qiu S, Fan J, Zha X (2012) Sp1 is involved in regulation of cystathionine γ-lyase gene expression and biological function by PI3K/Akt pathway in human hepatocellular carcinoma cell lines. Cell Signal 24:1229–1240. doi:10.1016/j.cellsig.2012.02.003

Yoshihara E, Fujimoto S, Inagaki N, Okawa K, Masaki S, Yodoi J, Masutani H (2010) Disruption of TBP-2 ameliorates insulin sensitivity and secretion without affecting obesity. Nat Commun 1:127. doi:10.1038/ncomms1127

Yusuf M, Kwong Huat BT, Hsu A, Whiteman M, Bhatia M, Moore PK (2005) Streptozotocin-induced diabetes in the rat is associated with enhanced tissue hydrogen sulfide biosynthesis. Biochem Biophys Res Commun 333:1146–1152. doi:10.1016/j.bbrc.2005.06.021

Zhang L, Yang G, Tang G, Wu L, Wang R (2011) Rat pancreatic level of cystathionine γ-lyase is regulated by glucose level via specificity protein 1 (SP1) phosphorylation. Diabetologia 54:2615–2625. doi:10.1007/s00125-011-2187-4

Zhang H, Guo C, Wu D, Zhang A, Gu T, Wang L, Wang C (2012) Hydrogen sulfide inhibits the development of atherosclerosis with suppressing CX3CR1 and CX3CL1 expression. PLoS ONE 7:e41147. doi:10.1371/journal.pone.0041147

Zhang G, Wang P, Yang G, Cao Q, Wang R (2013) The inhibitory role of hydrogen sulfide in airway hyperresponsiveness and inflammation in a mouse model of asthma. Am J Pathol 182:1188–1195. doi:10.1016/j.ajpath.2012.12.008

Zhao W, Zhang J, Lu Y, Wang R (2001a) The vasorelaxant effect of H2S as a novel endogenous gaseous KATP channel opener. EMBO J 20:6008–6016

Zhao W, Zhang J, Lu Y, Wang R (2001b) The vasorelaxant effect of H(2)S as a novel endogenous gaseous K(ATP) channel opener. EMBO J 20:6008–6016. doi:10.1093/emboj/20.21.6008

Zhou Y, Yu J, Lei X, Wu J, Niu Q, Zhang Y, Liu H, Christen P, Gehring H, Wu F (2013) High-throughput tandem-microwell assay identifies inhibitors of the hydrogen sulfide signaling pathway. Chem Commun (Camb) 49:11782–11784. doi:10.1039/c3cc46719h

Yoshihara E, Fujimoto S, Inagaki N, Okawa K, Masaki S, Yodoi J, Masutani H (2010) Disruption of TBP-2 ameliorates insulin sensitivity and secretion without affecting obesity. Nat Commun 1:127. doi:10.1038/ncomms1127

[illegible]

Part II

Molecular Mechanisms of Action

Persulfidation (S-sulfhydration) and H_2S

Milos R. Filipovic

Contents

M.R. Filipovic (✉)
Department of Chemistry and Pharmacy, Friedrich-Alexander University of Erlangen-Nuremberg, Egerlandstrasse 1, 91058 Erlangen, Germany
e-mail: milos.filipovic@fau.de

P.K. Moore, M. Whiteman (eds.), *Chemistry, Biochemistry and Pharmacology of Hydrogen Sulfide*, Handbook of Experimental Pharmacology 230,
DOI 10.1007/978-3-319-18144-8_2

Abstract

The past decade has witnessed the discovery of hydrogen sulfide (H_2S) as a new signalling molecule. Its ability to act as a neurotransmitter, regulator of blood pressure, immunomodulator or anti-apoptotic agent, together with its great pharmacological potential, is now well established. Notwithstanding the growing body of evidence showing the biological roles of H_2S, the gap between the macroscopic descriptions and the actual mechanism(s) behind these processes is getting larger. The reactivity towards reactive oxygen and nitrogen species and/or metal centres cannot explain this plethora of biological effects. Therefore, a mechanism involving modification of protein cysteine residues to form protein persulfides is proposed. It is alternatively called S-sulfhydration. Persulfides are not particularly stable and show increased reactivity when compared to free thiols. Detection of protein persulfides is still facing methodological limitations, and mechanisms by which H_2S causes this modification are still largely scarce. Persulfidation of protein such as K_{ATP} could contribute to H_2S-induced vasodilation, while S-sulfhydration of GAPDH and NF-κB inhibits apoptosis. H_2S regulates endoplasmic reticulum stress by causing persulfidation of PTP-1B. Several other proteins have been found to be regulated by this posttranslational modification of cysteine. This review article provides a critical overview of the current state of the literature addressing protein S-sulfhydration, with particular emphasis on the challenges and future research directions in this particular field.

Keywords

Hydrogen sulfide • Polysulfides • Sulfenic acids • Persulfidation • S-sulfhydration • S-nitrosation

1 H_2S as a Signalling Molecule

In order to maintain life, nature actually uses a limited number of chemical reactions, one of which is sulfur-based chemistry, mainly exploited for the control of intracellular redox homeostasis and redox-based signalling. Hydrogen sulfide (H_2S) is the simplest of the thiols found in the cells, and ever since the first report of its potential physiological role (Abe and Kimura 1996), there has been a growing literature on the subject of H_2S signalling. Very fast, H_2S joined the other two gases, nitric oxide (NO) and carbon monoxide (CO), as the third gasotransmitter (Wang 2002; Li et al. 2009; Mustafa et al. 2009a).

Numerous are the physiological functions assigned to be exclusively or partly regulated by H_2S, some of which are vasodilation (Yang et al. 2008; Mustafa et al. 2011), neurotransmission (Abe and Kimura 1996; Kimura et al. 2005), angiogenesis (Papapetropoulos et al. 2009; Szabó and Papapetropoulos 2011), inflammation (Li et al. 2005; Whiteman and Winyard 2011), hypoxia sensing

(Olson et al. 2008; Peng et al. 2010), etc. In addition, H_2S showed a tremendous pharmacological potential in preventing ischemia–reperfusion injury (Calvert et al. 2009, 2010). Furthermore, H_2S is able to induce suspended animation-like state in mice (Blackstone et al. 2005). Several pharmacological donors of H_2S have been developed with hope of their eventual use in disease treatment (Sparatore et al. 2008; Zhou et al. 2012; Szczesny et al. 2014; Zhao et al. 2013).

H_2S is produced by the action of at least three enzymes, cystathionine beta synthase (CBS), cystathionine gamma lyase (CSE) and mercaptopyruvate sulfurtransferase (MST) (Kabil et al. 2014; Kabil and Banerjee 2014). Differently expressed in different tissues (and even cellular compartments), these enzymes control H_2S production with different efficiencies. How are these enzymes regulated to maintain spatio-temporal production/distribution of H_2S is still unclear.

The intracellular levels of H_2S are also a matter of debate, with values spanning from nondetectable to >100 μM, although it seems more probable that the steady-state levels are at low micromolar and/or submicromolar levels, depending on the tissue (Olson 2012; Olson et al. 2014; Kabil et al. 2014). Conversely, the flux of H_2S production is huge, almost as that of glutathione, suggesting that the removal of H_2S is an efficient and tightly regulated process (Vitvitsky et al. 2012; Kabil et al. 2014; Kabil and Banerjee 2014).

Hydrogen sulfide is a weak acid and immediately ionizes in aqueous solution reaching the equilibrium between $H_2S/HS^-/S^{2-}$ species:

$$H_2S \rightleftharpoons H^+ + HS^- \rightleftharpoons 2H^+ + S^{2-}$$

The pK_{a1} is around 6.9, while pK_{a2} is estimated to be >12, which means that under physiological conditions, approximately two thirds of H_2S are in the form of bisulfide (hydrogen(sulfide)(1-)), with negligible amounts of sulfide anion (sulfide(2-)). H_2S diffuses freely through the membranes (Mathai et al. 2009; Cuevasanta et al. 2012). Although the diffusion coefficient profile of H_2S is systematically lower than that of H_2O, the differences in the transmembrane Gibbs energy profiles are more dominant. Because of its hydrophobicity, H_2S experiences no barrier to permeation, so it can partition into the interior of the membrane readily (Riahi and Rowley 2014).

Sulfur atom in H_2S is in −2 oxidation state, but sulfur is very versatile in its ability to accept or donate electrons. It can cycle between −2 and +6 oxidation state, due to the six valence electrons and completely empty 3d orbital. Although the standard redox potential often cited in the literature for the two electron oxidation of H_2S to sulfur is 0.144 V (which makes H_2S a weaker reducing agent than cysteine or glutathione), another standard redox potential could be found for the alkaline conditions, which suggests that H_2S is a stronger reducing agent:

$$S_{(s)} + 2e^- + 2H^+ \rightarrow H_2S_{(aq)} \quad E^0 = 0.144\ \text{V}$$
$$S_{(s)} + 2e^- + H_2O \rightarrow HS^- + OH^- \quad E^0 = -0.476\ \text{V}$$

H_2S does not react readily with oxygen; however, the solutions of H_2S undergo oxidation, just like the solutions of other thiols (such as cysteine and glutathione).

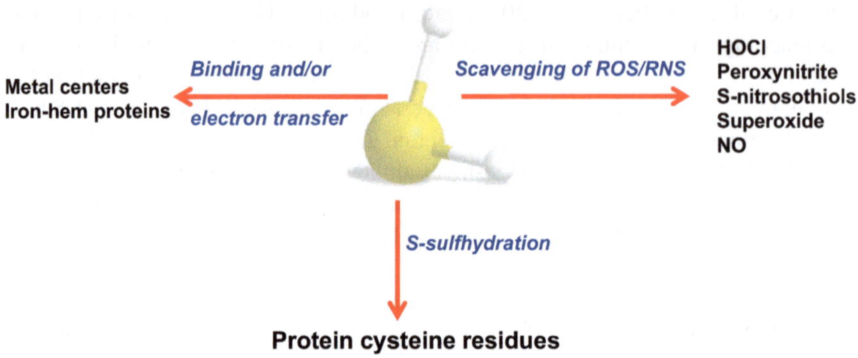

Fig. 1 Three main biochemical ways for direct H_2S signalling. Hydrogen sulfide can either react directly with reactive oxygen and nitrogen species (*arrow to the right side*) or react with metal centres by binding and/or electron transfer (*arrow to the left side*), or it can modulate cysteine residues of the proteins in a reaction called S-sulfhydration (*central arrow*)

This process is believed to be catalysed by the traces of metal ions present in the solution (Kotronarou and Hoffmann 1991); therefore, the thorough cleaning of solutions from heavy metals and removal of oxygen can keep H_2S solution stable for a while (Wedmann et al. 2014). Oxidation products of H_2S in solution are polysulfides, sulfites, thiosulfites and eventually elemental sulfur.

In biological systems, reactivity of H_2S could be divided in three groups of reactions: (1) reaction with/scavenging of reactive oxygen (ROS) and reactive nitrogen species (RNS), (2) binding to and/or subsequent redox reactions with metal centres and (3) reaction with proteins, herein called persulfidation (alternatively S-sulfhydration) (Fig. 1).

1.1 Reactions with ROS and RNS

H_2S reacts readily with hypochloric acid (HClO), produced by neutrophils, leading to the formation of polysulfides (Nagy and Winterbourn 2010). H_2S also reacts with peroxynitrite (Carballal et al. 2011; Filipovic et al. 2012a), in a reaction that generates thionitrate ($HSNO_2$) isomer, which can decompose and serve as an NO donor (Filipovic et al. 2012a). In addition, H_2S can also scavenge superoxide (Wedmann et al. 2014). Although the rates of these reactions are higher than those found for cysteine or glutathione, they are not so much higher that can overcome the difference in concentration (particularly when compared to glutathione which is present in millimolar steady-state levels) suggesting that H_2S cannot really serve as an antioxidant. Nonetheless, H_2S has been shown to have antioxidant and immediate protective effects in the cells exposed to ROS and RNS, something that cannot simply be explained by the modulation of proteins and gene expression. Since soluble macromolecules occupy a significant fraction of the total cell volume (Fulton 1982), within such crowded medium relative size and shape of a molecule and probability of its successful diffusion, placement and effective contact with a potential target become crucial factors that significantly alter its reaction rates (Minton 1998). This speaks in favour of hydrogen sulfide, when compared to GSH, despite the difference in steady-state concentrations (Filipovic et al. 2012a).

H_2S reacts with NO signalling pathways as well (Whiteman et al. 2006; Ali et al. 2006; Yong et al. 2010, 2011; Filipovic et al. 2012b, 2013; Eberhardt et al. 2014). With protein S-nitrosothiols, H_2S reacts to form the smallest S-nitrosothiols, thionitrous acid (HSNO), which can freely diffuse through the membrane and serve as *trans*-nitrosating agent (Filipovic et al. 2012b). More importantly, H_2S can react directly with NO giving nitroxyl (HNO), the one-electron-reduced sibling of NO which possesses the signalling properties of its own (Eberhardt et al. 2014). 2 μM combination of NO and H_2S gives the same rate of HNO generation as ~1 mM Angeli's salt, the commonly used pharmacological source of nitroxyl. Eberhardt et al. (2014) recently showed that co-localization of H_2S and NO production facilitate intracellular generation of HNO which then activates transient receptor potential A1 channel (TRPA1) leading to the Ca^{2+} influx and to the release of calcitonin gene-related peptide (CGRP), the strongest known vasodilator.

1.2 Reactions with Metal Centres

One of the main biological targets for H_2S would be metal centres. H_2S can coordinate and then additionally reduce the metal centre. In haemproteins, polar active site favours the reduction, while nonpolar centres favour coordination only (Pietri et al. 2009). Indeed, the presence of positive charges around the haem centres leads to a complete change in the haems' reactivity, allowing the design of efficient catalysts for H_2S removal (Ivanovic and Filipovic 2012).

H_2S binds to both haem a_3 and Cu_B centres of cytochrome c oxidase, with K_i being 0.2 μM for the purified enzyme (Hill et al. 1984; Nicholls et al. 2013). This suggests that cytochrome c oxidase should be permanently inhibited under physiological conditions (considering the micromolar steady-state concentrations of H_2S), which is obviously not the case. In fact H_2S shows biphasic effects on whole cell respiration, stimulating it at low doses and then completely inhibiting it at higher concentrations (Koenitzer et al. 2007). Binding of H_2S to haem has an important function in mollusc, *Lucina pectinata*, where H_2S binds to haemoglobin I and is then transported to the symbiotic chemoautotrophic bacteria living in their gills (Ríos-González et al. 2014).

Reaction of oxyhaemoglobin with H_2S leads to the green pigment, sulfhaemoglobin, which is in fact a chemically modified porphyrin centre. This subsequently affects oxygen-binding capacity of that haemoglobin. H_2S also binds and reduces cytochrome *c*, a reaction that in the presence of oxygen leads to superoxide production (Wedmann et al. 2014). In addition, Pálinkás et al. (2014) have recently investigated interactions of H_2S with human myeloperoxidase

(MPO), a major contributor to inflammatory oxidative stress, to show that H_2S inhibits the enzyme by reducing iron centre and by binding to the reduced Fe^{2+}.

It is still, however, unclear to which extent is the coordination and/or reduction of metal centres involved in signalling by H_2S. Miljkovic et al. (2013) demonstrated that metal centres in mitochondria are responsible for the H_2S-stimulated haem centre-catalysed reduction of nitrite, a reaction which can explain the use of nitrite as an antidote for acute H_2S poisoning.

1.3 Protein Persulfides

The third way of direct H_2S signalling would be modulation of proteins by modification which is named S-sulfhydration, although the more correct term should be persulfidation. Protein persulfides add up to the list of oxidative posttranslational modifications (oxPTMs) of cysteine, such as S-nitrosation, S-sulfenylation and S-glutathionylation (Fig. 2). Modification of proteins by H_2S could explain the plethora of effects that H_2S exhibit and several proteins have been identified to be indeed controlled by this modification (Mustafa et al. 2009a, b, 2011; Paul and Snyder 2012). However, this field of research is at its beginning, and it is still facing difficulties/challenges such as the proper choice of detection method, understanding of the mechanism(s) by which persulfidation takes place and the actual impact it has on the cellular functions.

Based on the calculation of the bond energies of GSSG and GSSH, the latter has ~18 kJ mol^{-1} lower bond energy (Filipovic et al. 2012b) so due to their inherent instability, there is limited information about persulfide reactivity to date. Francoleon et al. (2011) were among the first who reported preparation of the glutathione and papain persulfides, while Pan and Carroll (2013) successfully prepared persulfide on glutathione peroxidase 3. Zhang et al. (2014) also reported

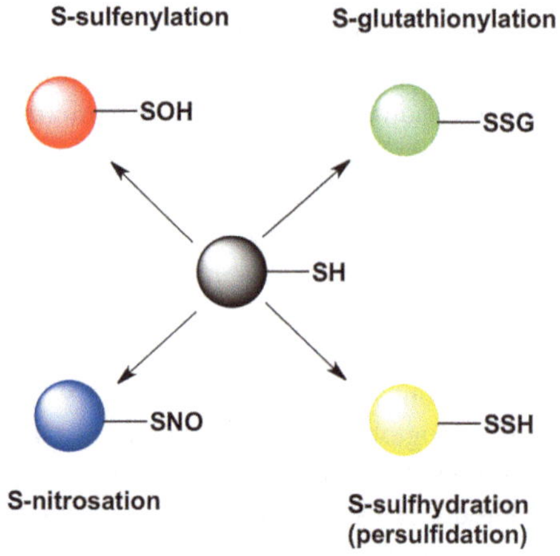

Fig. 2 Oxidative posttranslational modifications of protein cysteine residues known to regulate protein structure/function

facile preparation of persulfide of bovine serum albumin (BSA), which can be used as a model to study protein persulfide reactivity.

Electronegativity of sulfur is almost identical to that of carbon; therefore in R-S-SH, the sulfur atom covalently bound to carbon could be considered as sulfane sulfur. Sulfane sulfur is sulfur with six valence electrons and a formal charge 0, often represented by S^0. This sulfane sulfur could be a good target for nucleophilic attack, while the other sulfur atom, which is formally −1, could react with electrophiles. The pK_a of persulfides is lower than that of corresponding thiols, suggesting that at physiological conditions, majority of persulfide would be in deprotonated form (R-S-S^-), making the persulfide "super" nucleophilic. This dramatically increases persulfides' reactivity when compared to the corresponding thiols, as recently demonstrated for the reaction with H_2O_2 (Ida et al. 2014) and previously for the reduction of cytochrome c^{3+} (Francoleon et al. 2011).

Artaud and Galardon (2014) synthesized a persulfide analogue of the nitrosothiol SNAP, which opens up new possibilities for examining metastable low molecular weight (LMW) persulfides. The authors clearly demonstrate that spontaneous decay of LMW persulfides does not lead to H_2S release, but when mixed with other thiols, such as glutathione, an immediate H_2S release could be observed. Additionally, Bailey et al. (2014) characterized tritylhydrosulfide (TrtSSH), another LMW persulfide, to show that protonated form does not react with nucleophiles, while it readily reacts with electrophiles and reducing agents. Protonation leads to no change in substance stability, while the deprotonation stimulates decomposition with the elimination of elemental sulfur.

Based on the current literature data, the following characteristics of persulfides could be described:

1. Persulfides are metastable species which decompose in solution in a complex manner, leading to the formation of elemental sulfur, among other molecules.
2. Persulfides are much better nucleophiles than corresponding thiols which can explain their better reactivity.
3. Persulfides are better reducing agents than corresponding thiols.
4. Persulfides readily react with electrophiles.
5. Persulfides could transfer sulfane sulfur to other thiols, leading to *trans*-persulfidation (vide infra).

2 Persulfide Formation

Although there is a growing interest for persulfidation of proteins, only a few studies actually addressed the issue of the mechanism(s) behind protein S-sulfhydration. The original misconception (which tends to overtake this whole field) was that thiolate on the protein can react directly with H_2S to form protein persulfide (Fig. 3a). However, that reaction is impossible due to the thermodynamic constrains. Both sulfur atoms get oxidized in the reaction so the electrons will have to end up on protons, leading to elimination of hydrogen as a gas. Incubation of

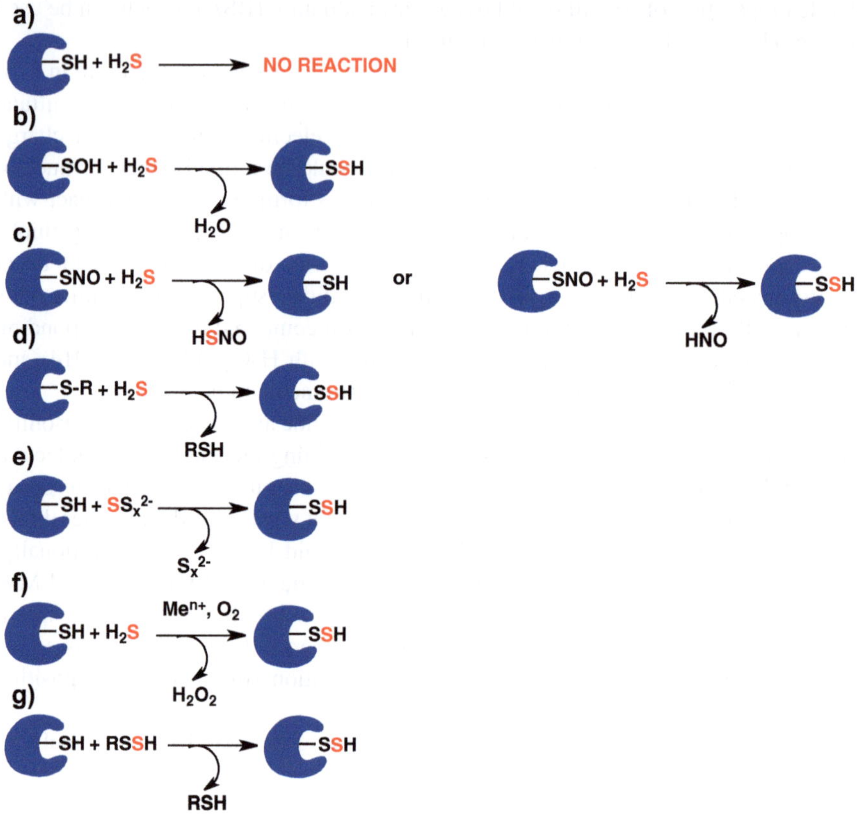

Fig. 3 Proposed reaction mechanisms for persulfide formation. (**a**) A direct reaction between protein thiols and H_2S is not possible, but H_2S can react with sulfenic acids (**b**). (**c**) Reaction of *S*-nitrosated cysteines with H_2S leads to the formation of HSNO, but depending on the protein environment surrounding the thiol, it is also possible to generate protein persulfides in this reaction. (**d**) H_2S could react with already existing inter- or intramolecular disulfides, while sulfane sulfur in polysulfides could react directly with protein thiols and give persulfide (**e**). (**f**) Metal centres could act as oxidants for the formation of protein persulfides from H_2S and protein thiols. (**g**) Persulfides could serve as carriers of sulfane sulfur and engage in "*trans*-S-sulfhydration" reaction

proteins, such as GAPDH, BSA or immunoglobulins with H_2S, led to no detectable protein S-sulfhydration (Zhang et al. 2014; Wedmann et al. 2014) confirming the theory.

So how are the proteins modified by H_2S? Getting the answer(s) to this question is of the utmost importance for our understanding of H_2S signalling and also for the interpretation of the vast amount of data accumulated to date.

2.1 Enzymatic Generation of Persulfides

Despite the recent interest for protein persulfidation, sparkled by the discovery of H_2S as a signalling molecule, protein persulfides formed as intermediates that facilitate sulfur delivery in several biosynthetic pathways have been known for a while (Mueller 2006). Persulfides are found to be formed in sulfurtransferases and cysteine desulfurases. Particularly interesting examples are the two enzymes involved in H_2S production and its oxidation, mercaptopyruvate sulfurtransferase (MST) and sulfide–quinone oxidoreductase (SQR), respectively.

MST is expressed in both mitochondria and cytoplasm and could be found in kidney cells, liver and cardiac cells, neuroglial cells, etc. (Kabil and Banerjee 2014; Kimura 2014). Recent studies suggest that MST is an important source of H_2S in some organisms and tissues (Mikami et al. 2011; Módis et al. 2013). Although it was known for a while that during the reaction protein persulfide is formed in the catalytic site (MST-Cys-S-SH), the mechanism of H_2S release has been only recently discovered. Yadav et al. (2013) reported the first crystal structure of MST with its cysteine residue in the form of persulfide. This allowed them to propose a detailed reaction mechanism, while kinetic analysis of the reaction led to conclusion that thioredoxin is likely to be the major physiological persulfide acceptor for MST (Fig. 4a).

The same group, led by Ruma Banerjee, elucidated the full mechanism of SQR (Libiad et al. 2014). The sulfide oxidation pathway begins with SQR, and it also includes a sulfur dioxygenase, rhodanese and sulfite oxidase (Fig. 4b). By consuming H_2S and its persulfide products, SQR and sulfur dioxygenase are important switch-off regulators of sulfide signalling (Bouillaud and Blachier 2011). In the first step, SQR catalyses the oxidation of H_2S to sulfane sulfur, which remains covalently attached to the enzyme. In the second step, this sulfane sulfur could be transferred to sulfite, to form thiosulfate (Jackson et al. 2012), but recently, Libiad et al. (2014) showed that in fact glutathione is a more probable acceptor of sulfane sulfur. This leads to the generation of glutathione persulfide, which can be consumed by rhodanese to actually form thiosulfate rather than to use it as a substrate (Fig. 4b).

Recent work on reactive cysteine persulfides and S-polythiolation has shaken our understanding of the action of CBS and CSE (Ida et al. 2014). Namely, the authors demonstrated that persulfide formation by CSE and CBS-mediated CysSSCys metabolism are facile and more likely a source of biological persulfides (Fig. 4c). Therefore, the direct enzymatic production of per- and polysulfide that is highly prevalent has been suggested. In fact, the authors question the role of H_2S as a major signalling molecule suggesting that Cys-based persulfides may be the actual signalling species (Ono et al. 2014).

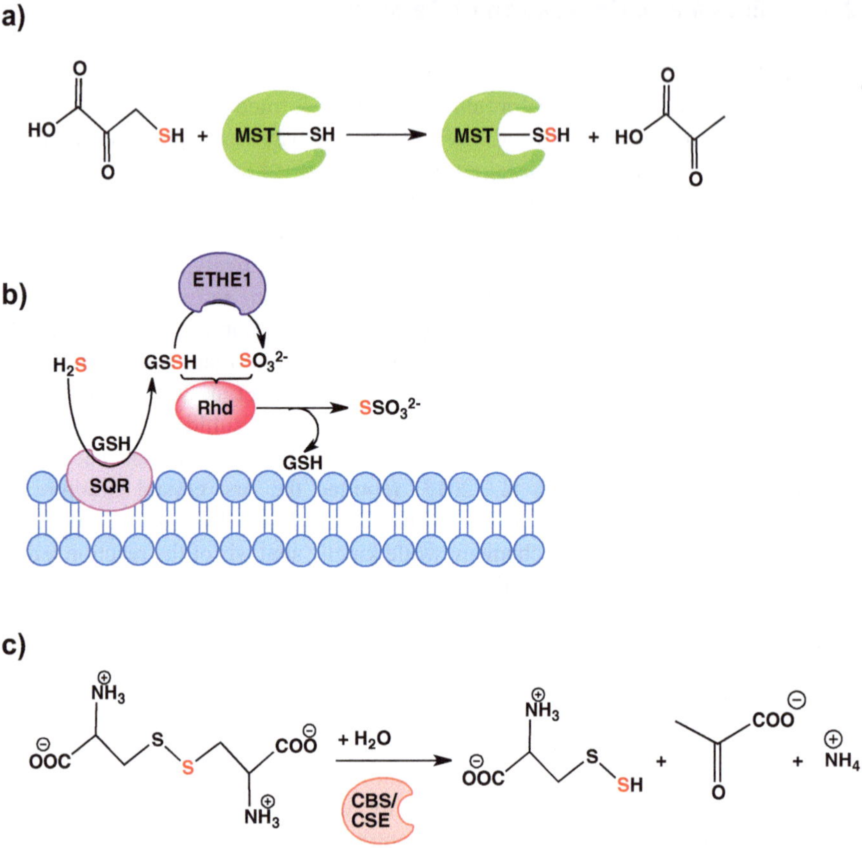

Fig. 4 Enzymatic production of protein persulfides. (**a**) Generation of protein persulfide on MST enzyme during its catalytic cycle. (**b**) Generation of glutathione persulfide during the enzymatic oxidation of H_2S. In the first step, SQR catalyses the oxidation of H_2S to sulfane sulfur with glutathione as probable acceptor of this sulfane sulfur. This leads to generation of glutathione persulfide, which can be consumed by rhodanese to actually form thiosulfate. (**c**) Cysteine persulfides are formed in the enzymatic cleavage of cystine by CBS or CSE

2.2 Direct Nonenzymatic Generation of Persulfides by H_2S

An obvious similarity between protein S-sulfhydration and S-glutathionylation could be drawn. Based on known mechanisms for the generation of glutathionylated proteins, it is possible to assume that the same reactions could lead to generation of protein persulfides.

2.2.1 Reaction with Sulfenic Acids

Protein S-sulfenylation (Fig. 2) is the reversible oxidation of protein thiols to sulfenic acids (R-SOH) that has been recognized as an important oxidative post-translational modification of cysteines (Paulsen and Carroll 2013; Gupta and Carroll 2014). Most recently, an extensive study was published identifying the whole cell sulfenylome, i.e. site-specific mapping and quantification of protein S-sulfenylation in the cells (Yang et al. 2014). More than 1000 proteins have been characterized as modified by sulfenylation. Peroxiredoxins are the most abundant antioxidant enzymes in the cytosol, and their catalytic cycle relies on sulfenic acid formation (Poynton and Hampton 2014). Sulfenic acid formation is shown to be involved in H_2O_2-mediated inactivation of protein tyrosine phosphatases (PTPs) (Paulsen et al. 2011). Also, reversible sulfenylation is shown to switch on or off the activity of transcription factors such as OxyR, OhrR or Orp1-Yap1 (for extensive review, see Paulsen and Carroll 2012). Epidermal growth factor receptor and the phosphatases SHP2, PTEN and PTP1B were all found to be sulfenylated upon regular signalling by epidermal growth factor (Paulsen et al. 2011).

The biochemistry of sulfenic acid formation, reactivity and functions has been extensively reviewed recently (Paulsen and Carroll 2013; Gupta and Carroll 2014). Sulfenic acids are known to react with thiols to form disulfides. The same could be said for the reaction of sulfenic acids with H_2S (Fig. 3b). Namely, in the recent study, Zhang et al. (2014) showed that formation of sulfenic acids on GAPDH and subsequent reaction with H_2S do indeed lead to the formation of protein persulfides. Furthermore, they used sulfenic acid of bovine serum albumin (BSA), known as an example of a relatively stable protein sulfenic acid (Carballal et al. 2003), to successfully generate S-sulfhydrated BSA. Although the steady-state concentration of H_2S is orders of magnitude lower than that of glutathione (Olson et al. 2014; Kabil et al. 2014), which makes it difficult for H_2S to compete for the reaction with protein sulfenic acids, the high flux of H_2S generation (Vitvitsky et al. 2012), its free diffusion (Mathai et al. 2009; Cuevasanta et al. 2012) and therefore its ability to reach deeper parts of the proteins suggest that the reaction of H_2S with sulfenic acids could still be a major source of protein persulfide formation (Fig. 3b). Indeed, Zhang et al. (2014) find that intracellular persulfidation co-localizes with endoplasmic reticulum, an organelle rich with sulfenic acids.

It is worth noting that sulfenic acids can be formed from the thiols just by the presence of traces of metal ions in the buffers and oxygen (Paulsen and Carroll 2013). Considering that R-SOH could react with H_2S to form protein persulfides, these reaction steps could be an explanation for the misconception that H_2S reacts directly with thiols.

2.2.2 Reaction with S-Nitrosothiols

Protein S-nitrosation is considered by some to be the second most important posttranslational modification of proteins. The number of proteins found to be controlled by this modification is constantly increasing. To date, S-nitrosation has been implicated in the regulation of proteins involved in muscle contractility, neuronal transmission, host defence, cell trafficking, apoptosis, etc. (Hess and Stamler 2012; Seth and Stamler 2011; Foster et al. 2009; Lima et al. 2010). S-nitrosation of haemoglobin has been proposed to regulate its ability to release oxygen (Reynolds et al. 2013). S-nitrosation of ryanodine receptors regulates intracellular Ca^{2+} levels (Xu et al. 1998) as does S-nitrosation of transient receptor potential cation channels (TRPs) (Yoshida et al. 2006). The role of S-nitrosation in controlling the protein function has been extensively reviewed elsewhere (Hess and Stamler 2012; Seth and Stamler 2011).

The mechanism of protein S-nitrosation is still the matter of debate (Broniowska and Hogg 2012). Like in the case of S-sulfhydration, direct reaction of NO with thiols to lead to the S-nitrosothiol formation is thermodynamically unfavourable. The mechanism usually involves either the reaction of thiols with higher nitrogen oxides, such as N_2O_3, or reactions with metal centre-catalysed one-electron oxidation product of NO (Broniowska and Hogg 2012). *Trans*-nitrosation, transfer of NO^+ moiety, from one protein to another is also a subject of a lively debate. Cysteine and glutathione have been proposed as common carriers. However, H_2S also reacts with S-nitrosothiols leading to the formation of the smallest S-nitrosothiol, HSNO (Fig. 3c, Filipovic et al. 2012b). HSNO can freely diffuse through the membranes serving as a carrier of NO^+ and conveying further trans-nitrosation. In addition, HSNO is very unstable and highly reactive so it could react further with H_2S-forming polysulfides and/or sulfur, which could be a source of protein S-sulfhydration (Fig. 3c). The reaction of S-nitrosothiols with H_2S to directly give persulfides and nitroxyl (HNO) is thermodynamically unfavourable ($\Delta_{rxn1}G° \approx +40$ kJ mol^{-1}) (Filipovic et al. 2012b). Although it seems that *trans*-nitrosation is a generally favoured pathway, a recent computational study pointed out that the surrounding of S–NO bond within a protein could significantly affect the thermodynamic feasibility of the thiolation reaction, making it possible for certain proteins (depending on the surrounding of the cysteine residue) to get directly S-sulfhydrated (Fig. 3c, Talipov and Timerghazin 2013).

2.2.3 Reaction with Disulfides

Reaction of thiols with disulfides normally proceeds until the equilibrium is established (Moriarty-Craige and Jones 2004). Following the same logic, H_2S should be able to react with disulfides leading to persulfide formation (Fig. 3d). Protein disulfides are always formed between cysteine residues (of the same or different polypeptide chain), or they appear as mixed disulfides between the cysteine residues of the protein and glutathione (in S-glutathionylated proteins). Based on the calculation of the bond energies of GSSG and GSSH, the latter has ~18 kJ mol^{-1} lower bond energy (Filipovic et al. 2012b) which makes the reaction of H_2S with oxidized thiols very slow. Francoleon and colleagues demonstrated that glutathione persulfide could indeed be formed in a reaction mixture containing oxidized glutathione and H_2S, but the product was unstable (Francoleon et al. 2011). Formation of protein persulfides by the H_2S-induced reduction of intra- or intermolecular protein disulfides seems highly unlikely in the cells, due to the low levels of H_2S when compared to other thiols (such as glutathione or cysteine) and the very slow reaction rate. Zhang et al. (2014) did not observe any S-sulfhydration as a consequence of the reaction of H_2S with disulfides of BSA, and a study by Wedmann et al. (2014) confirmed this observation on immunoglobulins purified from human blood. The use of very high, non-physiological concentrations of H_2S or its donors in experimental setups could, nonetheless, overcome this limitation and lead to protein modification.

2.3 Nonenzymatic S-sulfhydration by Species Originating from H_2S

2.3.1 Reaction with Polysulfides

Polysulfides (HS_x^-) are the products of incomplete H_2S oxidation on its way to total oxidation to elemental sulfur. Contrary to a widely spread opinion in a biological community that H_2S easily oxidizes, H_2S is known as a very persistent and difficult-to-get-rid-off pollutant in industrial processes. The oxidation that indeed does occur on air is most probably caused by the impurities present in the solutions such as traces of metal ions, which are inevitable contaminants of all standard buffer solutions (Kotronarou and Hoffmann 1991).

The length of polysulfide chain can be from 2 to 7. It is worth noting that older chemical literature clearly points out that H_2S_2 is a molecule that can be prepared under extreme conditions (Parsons and Walton 1921). It is a substance of pungent smell that cannot even stand the humidity from air, and it immediately decomposes to give H_2S and elemental sulfur (Parsons and Walton 1921). Therefore, assuming that H_2S_2 is an important player in the biological system would be wrong. If formed, H_2S_2 would immediately decompose.

Longer polysulfides are more stable, with $x = 4$ and/or 5 being the most abundant species. They are formed even in biological milieu and could be responsible for the part of the effects assigned to H_2S (Fig. 3e). Indeed, Kimura et al. (2013) demonstrated that polysulfides have a 300 times higher potency in activating TRP channels. In addition, Greiner et al. (2013) linked polysulfides to protein thiol oxidation suggesting that all sources of H_2S (salts and H_2S donors) inevitably contain polysulfides, which in turn oxidize cysteine residues leading to persulfide formation. All inorganic polysulfides with more than two sulfur atoms contain sulfane sulfur atoms, which could undergo nucleophilic attack by free protein thiols leading to the formation of protein persulfides. However, polysulfides could also serve as reducing agents. Francoleon et al. (2011) suggested that persulfides are much more potent reducing agents than H_2S, while Wedmann et al. (2014) recently showed that polysulfides are capable of completely cleaving intramolecular disulfides present in immunoglobulins. H_2S solutions prepared with care, however, could not do the same. Only when mixed with traces of metal ions, they achieved the same effect (Wedmann et al. 2014).

Although there is very little doubt that polysulfides are formed in the cells, it is still unclear how they could serve as signalling molecules. Signalling molecules should be able to achieve the effect at low doses, but at the same time, their production should be tightly controlled, and their reactivity should be specific, just like in the case of NO or H_2O_2. Formation of polysulfides cannot be controlled; it is guided by stochasticity as the reactions that polysulfide undergo are highly unpredictable and will largely depend on the availability of oxygen and metal centres on the one hand and the protein thiols/disulfides on the other hand. In addition, polysulfides are charged, and it is almost impossible to achieve their fully protonated form under physiological condition, making the diffusion through the membrane impossible without a facilitator.

2.3.2 Reaction with Metal Centres and Generation of HS•

Although not studied into too much detail, metal centres could play an important role in catalysing formation of protein persulfides (Fig. 3f). Namely, some iron haem centres are able to oxidize hydrogen sulfide forming HS• (Miljkovic et al. 2013), which could in turn react with free thiols to finally generate protein persulfides (Zhang et al. 2014). Cytochrome c, for example, readily reacts with H_2S (Wedmann et al. 2014). Iron porphyrins in general could be a good source of HS•. HS• reacts further with H_2S, in a diffusion-controlled reaction, to give H_2S_2•$^-$ (Das et al. 1999) which should be a powerful persulfidation agent when formed intracellularly, something that is yet to be confirmed. As demonstrated by Zhang et al. (2014), incubation of GAPDH with iron porphyrin and H_2S leads to the strongest generation of protein persulfides. This complemented with their observation that a large portion of intracellular S-sulfhydration co-localizes with mitochondria, the organelle richest in metallo-proteins.

2.3.3 "*Trans*-persulfidation" by Polythiolated Cysteine or Glutathione

A recent study by Ida et al. (2014) suggested very high levels of circulatory LMW persulfides. Francoleon et al. (2011) noticed that treatment of papain with GSSH leads to the enzyme inhibition in the same manner as it does the persulfidation of this enzyme. The authors recently went a step further suggesting that cysteine persulfide and/or glutathione persulfide could be the main persulfidating agents in the cells (Ono et al. 2014). Cysteine persulfides could even be transported through the membrane. S-nitrosocysteine does the same and is considered as one of the main carriers of "NO^+" moiety in *trans*-nitrosation reactions (Broniowska and Hogg 2012). Making a parallel with S-nitrosothiol chemistry, the term "*trans*-persulfidation" could be coined to describe this process (Fig. 3g).

The mechanism of this process is still unclear. As mentioned above, the persulfides would react with electrophiles rather than with nucleophiles (Artaud and Galardon 2014; Bailey et al. 2014), and when they do react with nucleophiles (such as other free thiols), this reaction leads to the release of H_2S (which represents the basis of one of the methods for persulfide detection: reduction with DTT). The reaction with nucleophiles could only go as an attack to sulfane sulfur, but that leads to the formation of a mixed disulfide and elimination of H_2S. Some literature data exist suggesting that persulfides could exist in the tautomeric thiosulfoxide form (Scheme 1), which would then act as a perfect donor of sulfane sulfur (Kutney and Turnbull 1982; Steudel et al. 1997). However, it is worth noting that although this may be true for the extreme case of F_2S_2, neither experimental nor computational data support the existence of this tautomeric form (Steudel et al. 1997). Therefore, the elucidation of this mechanism is to be done in some future studies.

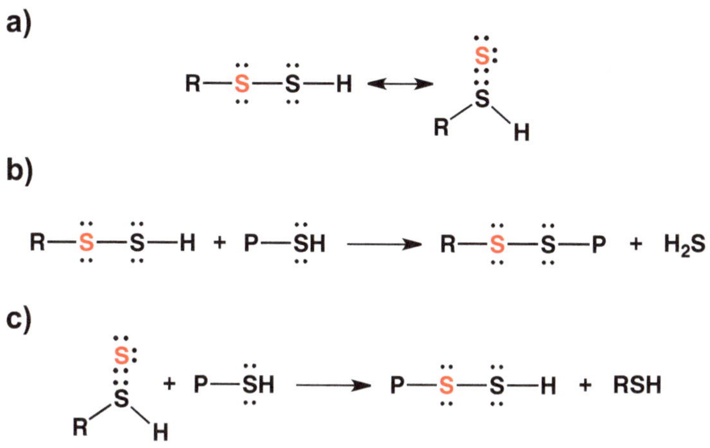

Scheme 1 Possible mechanism of *trans*-persulfidation. (**a**) The tautomeric forms of protein or LMW persulfides. Sulfane sulfur is marked *red*. (**b**) Reaction of LMW persulfides with protein thiols leads to the H_2S elimination and formation of mixed disulfide. (**c**) Reaction of thiosulfoxide tautomer, however, could lead to *trans*-persulfidation

3 Detection of Protein Persulfides

Detection of protein *S*-sulfhydration represents a certain challenge as the persulfide group exhibits the reactivity similar to other, free thiols (Flavin 1962; Heimer 1981; Mueller 2006; Pan and Carroll 2013; Zhang et al. 2014). Four distinctive approaches have been proposed (Fig. 5) for the persulfide detection, all of which are summarized here with a particular emphasis on the potential experimental problems that might occur.

The original method for protein persulfide detection was suggested by Mustafa et al. (2009b), and it was based on a premise that protein persulfides would not react with electrophilic thiol-blocking reagent *S*-methyl methanethiosulfonate (MMTS). In the subsequent step, persulfides were labelled with N-[6-(biotinamido)hexyl]-3′-(2′-pyridyldithio)propionamide (biotin-HPDP) (Fig. 5a). This method allowed the simultaneous labelling of *S*-sulfhydration and *S*-nitrosation. Using this method, Mustafa and colleagues suggested that a large number of proteins were a target for H_2S signalling and that basal protein persulfidation is as high as 25 % (Mustafa et al. 2009b).

To date, this is the most used methodological approach in reporting protein *S*-sulfhydration of different proteins. However, certain methodological limitations arise. Although MMTS has been widely used to study *S*-nitrosation (Forrester et al. 2009) and has been an efficient tool in trapping mixed disulfides in vivo (Peaper et al. 2005), Karala and Ruddock (2007) were able to show that in vitro MMTS treatment of both peptides and proteins resulted in the artificial formation of intramolecular and intermolecular protein disulfide bonds which could lead to

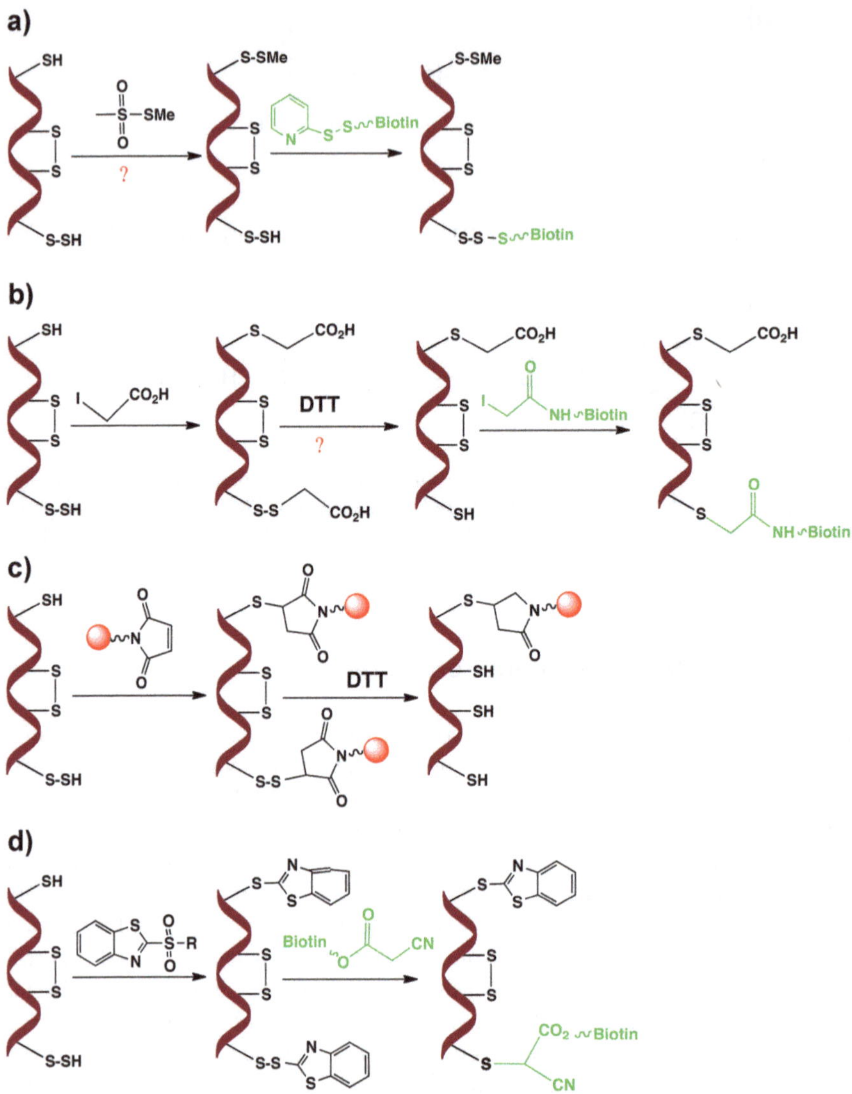

Fig. 5 An overview of the methods currently employed for the detection of protein persulfides. (**a**) The first method was based on a chemically wrong premise that protein persulfides would not react with electrophilic thiol-blocking reagent S-methyl methanethiosulfonate (MMTS). In the subsequent step, persulfides were labelled with N-[6-(biotinamido)hexyl]-3′-(2′-pyridyldithio) propionamide (biotin-HPDP). (**b**) In the second method, iodoacetic acid (IAA) is used to initially block both free thiols and protein persulfides. In the subsequent steps, alkylated persulfide is cleaved with DTT and then labelled with iodoacetamide-linked biotin (IAP). Although DTT would indeed cleave this adduct, it is unclear how this method distinguishes the persulfides from intra- and intermolecular disulfides and S-nitrosothiols, which would also be reduced by DTT. (**c**) In this method, both persulfide and free thiol would be blocked by the thiol fluorescently labelled N-ethyl maleimide (Cy5-conjugated maleimide). The adduct of persulfide and Cy5-maleimide is a disulfide that will be then cleaved by the DTT leading to a decrease of the in-gel fluorescence signal in the samples containing persulfides. (**d**) Finally, the use of the methylsulfonyl benzothiazole (MSBT) to block thiols and persulfides in the first step, followed by the tag switch with cyanoacetate derivatives in the second step, leads to the efficient labelling of persulfides, by the method called tag-switch technique

general data misinterpretation. In addition, Pan and Carroll (2013) studied how persulfides react with both electrophilic and nucleophilic species, reaffirming the nucleophilic properties of the persulfide sulfane sulfur. They unambiguously showed that persulfides react with MMTS (and its analogue S-4-bromobenzyl methanethiosulfonate) as readily as free thiols, questioning the interpretation of the data obtained by modified biotin-switch technique (MBST) approach. Two possible models were proposed to explain the data generated by MBST: (1) free thiols may be incompletely blocked in the first MMTS alkylation step and subsequently react with the pyridyldisulfide biotin reagent; (2) alternatively or in addition, labelling may be achieved via stepwise thiol-disulfide exchange in a reaction catalysed by trace free thiols (RSH).

In their attempt to identify persulfidation of protein tyrosine phosphatase 1B (PTP1B), Krishnan et al. (2011) proposed an approach for persulfide detection based on a completely opposite chemical premise. Namely, they proposed that thiol-blocking reagent, iodoacetic acid (IAA), will react with both free thiols and protein persulfides (Fig. 5b), which is in agreement with the persulfides' reactivity similar to that of free thiols. In the subsequent steps, however, they proposed the cleavage of the alkylated persulfide with DTT and then labelling of that particular cysteine with iodoacetamide-linked biotin (IAP). Although DTT would indeed cleave this adduct, it is unclear how this method distinguishes the persulfides from intra- and intermolecular disulfides and S-nitrosothiols, which would also be reduced by DTT.

Sen et al. (2012) proposed an alternative method for the persulfide detection. This method is based on the fact that both persulfide and free thiol would be blocked by the thiol-blocking reagent N-ethyl maleimide. The authors used Cy5-conjugated maleimide in the first step followed by the used of DTT in the second (Fig. 5c). The adduct of persulfide and Cy5-maleimide is a disulfide that will be cleaved by the DTT leading to a decrease of the in-gel fluorescence signal in the samples containing persulfides. This method has its advantage of being relatively simple and available to every researcher (because the reagents are commercially available). The only limitation of the method is that it does not offer actual persulfide labelling which would allow wide proteomic analysis. Particular care should be taken to ensure that all thiols and persulfides are indeed blocked in the initial step, as that is the crucial step for the subsequent data interpretation.

Most recently, Zhang et al. (2014) proposed a different approach for protein persulfide detection, named tag-switch assay (Fig. 5d). This method was based on an idea that thiol-blocking reagent should be introduced in the first step, which would tag both free thiols and persulfides. If an appropriate tag is employed, the disulfide bond in persulfide adducts might show much enhanced reactivity to certain nucleophiles than common disulfides in proteins. Therefore, a tag-switching reagent (containing both the nucleophile and a reporting molecule such as biotin) could be introduced to label only the persulfide adducts. It should be noted that thiol adducts from the first step are thioethers, which are not expected to react with the nucleophile. A major challenge in this technology was for the newly generated disulfide linkage from persulfide moieties to display a unique reactivity for a suitable nucleophile to an extent that distinguishes them from common disulfides. Using the methylsulfonyl benzothiazole (MSBT) to block thiols in the first step in combination with the tagged-cyanoacetate derivatives in the second step, persulfides could be efficiently labelled. Neither free thiols, intramolecular disulfides, S-glutathionylated or sulfenylated proteins were tagged by this approach (using BSA as a model protein). This method was further adapted for visualization of intracellular S-sulfhydration by fluorescence microscopy. Despite the lack of reactivity of BSA sulfenic acid derivative with the method, it is expected that sulfenic acids should react with cyanoacetate derivatives. Protein sulfenic acids are, in general, very unstable and prone to further oxidation, so it is improbable that any of them will remain in the solution after 30–45 min of initial incubation with the MSBT. Nonetheless, it is better to treat the cell lysates with dimedone, prior to the incubation with the first component of tag-switch assay, in order to eliminate any doubts of potential cross-reactivity (Park et al. 2015).

3.1 Detection of Sulfane Sulfurs

Although the methods for sulfane sulfur detection detect more than just protein persulfides, they could serve as a useful tool for the fast and easy proof-of-concept type of experiments. With recent advances in making these methods quantitative, it is also possible to assume that the results suggesting an increase or decrease of the sulfane sulfur levels do suggest the same trends for protein persulfides.

The traditional method for sulfane sulfur detection is cyanolysis (Wood 1987). The method is based on the reaction between sulfane sulfurs and cyanide under alkaline conditions ($pH > 8.5$). In this reaction, thiocyanate (SCN^-) is formed, which in the reaction with Fe^{3+} generates the $Fe(SCN)^{2+}$ complex with characteristic absorbance maximum at 460 nm.

Recently, two new methods have been proposed. The first one is designed to be used for the detection of sulfane sulfur levels in the cells. Chen et al. (2013) based their discovery on an assumption that sulfane sulfurs are likely to react with the nucleophile components of a fluorescent probe, which could then undergo spontaneous and fast cyclization to release fluorophore. They successfully produced two probes, named SSP1 and SSP2, which could be easily used for intracellular detection of sulfane sulfur. The same group improved the probes designing a new DSP-3 fluorescence sensor for hydrogen polysulfides (Liu et al. 2014a).

Most recently, the group led by Ming Xian established isotope dilution mass spectrometric method for the quantification of sulfane sulfurs (Liu et al. 2014b). Accurate and reliable measurements of sulfane sulfurs in biological samples are required in order to understand the impact of H_2S signalling. The method that Liu et al. (2014b) proposed employs a triphenylphosphine derivative to capture sulfane sulfurs as a stable phosphine sulfide product. The concentration of this product can be determined by isotope dilution mass spectrometry using a $^{13}C_3$-labelled phosphine sulfide as an internal standard. Using this method, the authors found that average concentrations of sulfane sulfur were 57.0 (liver), 150.9 (kidney), 46.0 (brain), 61.8 (heart), 56.1 (spleen) and 20.8 nmol/g (lungs).

4 Persulfidation in Action

As already mentioned, H_2S regulates the plethora of biological functions from neurotransmission and blood pressure to cardioprotection. To date, several key protein targets have been identified which do undergo oxPTM of the cysteine residue, suggesting that this modification can be responsible for some of the H_2S effects such as vasodilation, prevention of cell death and senescence, cell differentiation, etc.

4.1 S-sulfhydration of K_{ATP} Regulates Vasodilation

Seminal work by Yang et al. (2008) showed that CSE knockout mice develop hypertension, confirming previous assumptions about H_2S being the endogenous regulator of blood pressure. Several other studies have been published recently showing that majority of H_2S-induced vasodilation goes via its interaction with NO. This effect can be dual, via direct reaction with NO that leads to the nitroxyl (HNO) formation and subsequent activation of HNO–TRPA1–CGRP signalling cascade (Eberhardt et al. 2014) or by inhibiting the phosphodiesterase 5 and increasing the cGMP levels (Colleta et al. 2012). Nonetheless, the part of H_2S-induced vasodilation could be assigned to the activation of K_{ATP} channels. Persulfidation of C34 on the Kir6.1 subunit of K_{ATP} channel on smooth muscle cells prevents its association with ATP and promotes its binding to phopathidylinositol-4,5-bisphosphate (PIP2). This leads to the channel opening and K^+ influx and subsequently to the smooth muscle cells' relaxation. Several other Ca^{2+} channels have been also implicated in the H_2S-induced vasodilation.

4.2 Persulfidation of Electrophilic Messengers and Cardioprotection

Recently, direct persulfidation of several electrophilic messengers (such as 8-nitro-cGMP) by sulfide has been reported offering additional mechanism for H_2S-mediated signalling (Nishida et al. 2012). Redox signalling by electrophilic by-products, such as nitrated cyclic nucleotides and nitro- or keto-derivatives of unsaturated fatty acids, all generated by the inflammation-related enzymes, ROS and/or NO, has attracted much interest lately (Nishida et al. 2014; Fujii and Akaike 2013). S-alkylation of cysteine residues by 8-nitroguanosine 3′,5′-cyclic monophosphate mediates several redox signalling pathways. Nishida et al. (2012)

showed that HS^- could directly attack the above-mentioned electrophiles, forming derivatives that have biological effect of themselves. Namely, formation of 8-SH-cGMP blocks the S-guanylation of H-Ras, a modification which normally activates H-Ras to signal cell senescence as a response to stress (Fujii and Akaike 2013). The authors link the protective cardiovascular effects of H_2S and its donors to increased formation of 8-SH-cGMP and inhibition of H-Ras signalling. It is worth mentioning, however, that Terzić et al. (2014) showed that nucleophilic attack of the hydrosulfide anion to 8-nitro-cGMP cannot take place, as previously proposed. Instead, the formation of reactive species containing sulfane sulfur, like persulfides, is required (Terzić et al. 2014).

4.3 S-sulfhydration of GAPDH and NF-κB Protects Against Apoptosis

GAPDH was the first protein characterized as S-sulfhydrated in the study that sparkled the whole research field (Mustafa et al. 2009a). GAPDH has been known for a while as regulator of a cell death cascade (Hara et al. 2005). S-nitrosation of catalytic C152 abolishes its catalytic activity but makes it able to bind to Siah1, an E3 ubiquitin ligase. Siah1, which possesses a nuclear localization tag, helps the translocation of GAPDH to a nucleus where it enables Siah1 to degrade nuclear proteins, leading to cell death (Hara et al. 2005). Mustafa et al. showed that GAPDH is endogenously S-sulfhydrated at C150 which increases its enzymatic activity severalfold (Mustafa et al. 2009a). The authors also demonstrated that DTT treatment of GAPDH decreases its activity, suggesting that endogenous persulfidation regulates its function. CSE knockout mice showed ~35 % reduced activity of GAPDH when compared to control mice. Zhang and associates confirmed recently, using selective tag-switch assay, that GAPDH is indeed endogenously S-sulfhydrated (Zhang et al. 2014). The fact that S-sulfhydration of GADPH increases its enzymatic activity means that it also prevents its interaction with Siah1 protecting the cells from apoptosis.

Nuclear factor-κB (NF-κB) is an anti-apoptotic transcription factor, which is under basal conditions kept in cytosol via interaction with the inhibitor-κBα (Napetschnig and Wu 2013). During the inflammation, cells produce tumour necrosis factor-α (TNF-α), which could lead to cell death (Aggarwal et al. 2012). H_2S is known to have protective effects in inflammation but without distinctive mechanism that can explain it (Li et al. 2005; Whiteman and Winyard 2011). Sen and colleagues offered such mechanism in a study describing the persulfidation of NF-κB (Sen et al. 2012). Namely, they showed that C38 of p65 subunit of NF-κB is persulfidated which promotes its binding to the co-activator ribosomal protein S3, augmenting its binding to the promoters of anti-apoptotic genes (Fig. 6a). In addition, TNF-α stimulated transcription of CSE, increasing the total amount of S-sulfhydrated NF-κB. Conversely, Du et al. 2014 showed recently that hydrogen sulfide suppresses oxidized low-density lipoprotein-induced macrophage inflammation by inhibiting NF-κB. This study suggests that persulfidation of C38 on p65 in fact prevents the NF-κB from leaving the cytosol, therefore completely inhibiting its DNA binding activity. Further studies are obviously needed to give more conclusive answer about the role of H_2S in inflammation.

4.4 S-sulfhydration of Parkin Protects Against Parkinson's Disease

Parkinson's disease (PD) is a neurodegenerative disease caused by the death of dopamine-generating cells in the substantia nigra (Shulman et al. 2011). One of the proteins considered responsible for this cell death is parkin. Parkin is an E3 ubiquitin ligase that catalyses ubiquitination of diverse substrates. Mutations in parkin, which lead to the loss of its activity, are one of the causes of PD (Shulman et al. 2011; Moore et al. 2005). Parkin has reactive cysteine residues, which can be subjected to oxidative posttranslational modifications. S-nitrosation of parkin, for example, inhibits its E3 ubiquitin ligase activity contributing to the Parkinson's disease (Chung et al. 2004).

Recently, Solomon Snyder and colleagues demonstrated that cysteines C59, C95 and C182 could undergo S-sulfhydration (Vandiver et al. 2013). This leads to the increase of parkin's activity (Fig. 6b). Persulfidation of parkin is markedly decreased in PD brains, whereas S-nitrosation is increased. Development of H_2S donors opens up a possibility of their use in the early treatment of PD. Increase of parkin's activity could salvage the neurons from the cell death by removing damaged proteins. Parkin is also an important regulator of mitophagy, leading to the removal of damaged mitochondria, particularly in ischemia–reperfusion injury. As H_2S is known to have great pharmacological potential in preventing ischemia–reperfusion injury, it is possible that part of this effect goes via persulfidation of parkin and increased removal of damaged mitochondria.

4.5 S-sulfhydration of PTP-1B Regulates ER Stress

Protein tyrosine phosphatases (PTPs) in conjunction with protein tyrosine kinases are important controllers of various biological functions. PTPs are particularly sensitive to oxidative posttranslational modifications of cysteine, as cysteine is present, is in the active site and is important for their function (van Montfort et al. 2003). H_2O_2 is known to modulate function of these enzymes by forming an inactive adduct which contains sulfenic acid in the active site (van Montfort et al. 2003). PTP-1B is one of the members of this class of enzymes, located in the cytoplasmic face of the endoplasmic reticulum, where it plays an important role in ER stress signalling (Paulsen et al. 2012).

Persulfidation of C215 leads to the loss of enzymatic activity, which increases the phosphorylation of Y619 and therefore the activation of PERK in response to ER stress (Krishnan et al. 2011). PERK activation leads to global inhibition of protein translation. The mechanism by which H_2S causes persulfidation of PTP-1B,

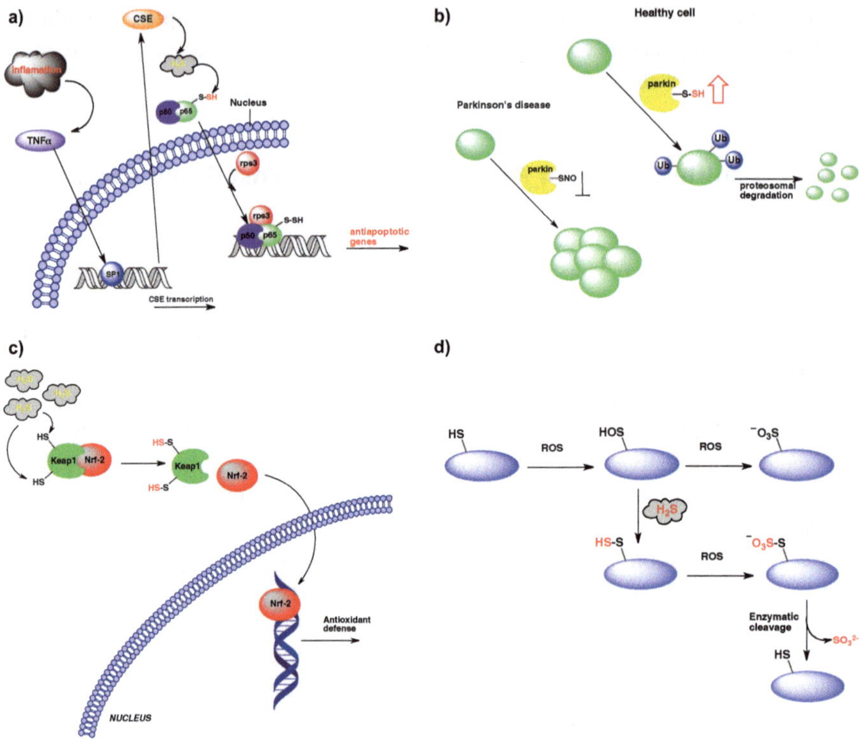

Fig. 6 Biological role of protein persulfidation. (**a**) As a response to inflammation, tumour necrosis factor alpha (TNF-α) migrates to nucleus where it binds to the transcription factor specificity protein 1 (SP1) causing the transcription and overexpression of CSE. H_2S produced by CSE induces persulfidation of p65 subunit of NF-κB which translocates to nucleus where its interaction with co-activator ribosomal protein S3 (rps3) is enhanced, resulting in the transcription of anti-apoptotic genes. (**b**) S-nitrosation of parkin inhibits its E3 ubiquitin ligase activity which leads to accumulation of toxic proteins. This is found in Parkinson's disease where H_2S production is reduced. Persulfidation of parkin, on the other hand, increases parkin's enzymatic activity, which could be used as a therapeutic approach for Parkinson's disease treatment. (**c**) Keap-1 keeps Nrf-2 in the cytoplasm where it can be ubiquitinated and proteolytically degraded. Persulfidation of critical cysteines on Keap1 disturbs its interaction with Nrf-2 leading to the release of Nrf-2 which can now move into the nucleus where it binds to antioxidant response element (ARE) causing the transcription of various antioxidant defence genes. (**d**) The hypothetical role of persulfidation in protection of protein function during oxidative stress. When exposed to hydrogen peroxide, proteins undergo oxidation to form sulfenic acids (P–SOH), sulfinic acids (P–SO_2H) and sulphonic acids (P–SO_3H), which cause the irreversible inactivation of the protein. Sulfenic acids could react with H_2S to form persulfides. In addition, protein persulfides could be formed by other mechanisms, and when exposed to ROS, they will readily react with it, forming P–S–SO_3^-, which can be enzymatically cleaved to restore free thiol on the protein

as in many other cases, is not clear. In the case of PTP-1B though, it is possible that ER stress in fact leads to PTP-1B inactivation by initial formation of sulfenic acid in the active site, which then reacts with H_2S to give persulfide. Although both modifications are inhibitory, persulfide could be then removed easier and the enzyme's activity restored. Therefore, H_2S could serve to protect the enzyme from a longer-lasting and potentially irreversible inhibition.

4.6 S-sulfhydration of Keap1 and p66Shc Increases Cellular Antioxidative Defence and Prevents Senescence

A major mechanism in which antioxidant enzymes are induced involves the activation of the antioxidant response element (ARE) by the oxidative-stress sensor protein Kelch-like ECH-associated protein 1 (Keap1) and the transcription factor nuclear factor (erythroid-derived 2)-like 2 (Nrf-2) (Hybertson et al. 2011; Kaspar et al. 2009). Under basal conditions, Keap1 sequesters Nrf-2 in the cytoplasm by binding to its Neh2 domain (Hybertson et al. 2011; Kaspar et al. 2009; Wakabayashi et al. 2004). Chemical inducers, such as sulforaphane, are known to react with Keap1 cysteine residues, thereby promoting Nrf-2 nuclear accumulation and hence ARE activation (Wakabayashi et al. 2004). A widely accepted model for Nrf-2 nuclear accumulation describes that a modification of the Keap-1 cysteines leads directly to the dissociation of the Keap1–Nrf-2 complex and to the translocation of Nrf-2 into the nucleus.

H_2S cardioprotective effects in ischemia–reperfusion injury are partly caused by Nrf-2 nuclear translocation and activation of antioxidant defence enzymes (Calvert et al. 2009, 2010). Recently, two independent studies suggested that Keap-1 gets S-sulfhydrated when cells were exposed to H_2S. Modification of C151 on Keap1 stimulates dissociation of Nrf-2 enabling its translocation to nucleus where it regulates expression of cytoprotective genes and delays cellular senescence (Fig. 6c; Yang et al. 2013). In second study, Hourihan et al. (2013) found that H_2S stabilizes Nrf2 via covalent modification of amino acids C226 and C613 in the Keap1. The authors show that H_2S leads to the production of H_2O_2, which inhibits Keap1 by stimulating the formation of an intramolecular disulfide bond between C226 and C613. The Keap1 C226 and C613 residues are also persulfidated. This may be explained either by the ability of H_2S to reduce C226–C613 disulfides originally formed by H_2O_2 or by direct reaction of H_2S with sulfenylated residues formed by H_2O_2. More importantly, the authors observed that Nrf-2 controls CBS, CSE and sulfide–quinone reductase-like enzyme, suggesting that a feedback loop exists between Nrf-2 and H_2S.

P66Shc belongs to the ShcA family of proteins whose members share three common functionally identical domains: the C-terminal Src homology 2 domain (SH2), the central collagen homology domain (CH1) and the N-terminal phosphotyrosine-binding domain (PTB) (Giorgio et al. 2005). In response to oxidative stress such as UV exposure or H_2O_2, p66Shc gets activated by phosphorylation at Ser-36. The activated p66Shc is then dephosphorylated and translocates to mitochondria, where it binds to cytochrome c helping in electron transport process (Giorgio et al. 2005). P66Shc$^{-/-}$ mice show 30 % increase of the lifespan. It has been shown recently that sulfhydration of p66Shc impaired the association of $PKC_{\beta II}$

and p66Shc and attenuated H_2O_2-induced p66Shc phosphorylation, a critical step in p66Shc-mediated mitochondrial ROS generation (Xie et al. 2014). H_2S is known to have dramatic effects on inhibiting oxidative stress, something that cannot be simply explained by its direct redox chemistry. A study by Xie and associates suggests that H_2S may inhibit mitochondrial reactive oxygen species production via a p66Shc-dependent mechanism.

4.7 MEK1/PARP-1 Activation and DNA Damage Repair

The salvage of DNA damage is essential for normal cell function. DNA damage stimulates a complex and highly concerted DNA damage repair response, which includes poly(ADP-ribose)ation catalysed by poly(ADP-ribose)ation polymerases (PARPs). Upon DNA damage, PARPs bind to DNA strand breaks and catalyse poly (ADP-ribose)ation which attracts other DNA damage repair proteins (D'Amours et al. 1999). The activation of PARPs is regulated by several kinases, of which MEK/ERK signalling cascade plays an important role (Cohen-Armon et al. 2007). Zhao et al. (2014) reported recently that H_2S attenuates DNA damage in human endothelial cells by causing S-sulfhydration of cysteine 341 on MEK1. This facilitates the translocation of phosphorylated ERK1/ERK2 into nucleus where it activates PARP-1 and increases the DNA damage repair yield, protecting cells from senescence.

4.8 TRP Channel S-sulfhydration Regulates Osteogenic Differentiation

Bone marrow mesenchymal stem cells (BMMSCs) are nonhaematopoietic multipotent stem cells and play an important role in the maintenance of the bone marrow homeostasis (Prockop 1997; Pittenge et al. 1999). BMMSCs and BMMSC-derived osteoblasts are responsible for bone formation and balancing osteoclast-mediated bone resorption in order to maintain bone mineral density (BMD) (Pittenge et al. 1999). CBS-deficient patients exhibit a variety of phenotypes, including osteoporosis. Osteoporosis is characterized by low bone mass and deterioration of osseous microarchitecture, resulting in decreased bone strength and increased risk of fragility fractures. This phenotype is often observed in patients with hyperhomocysteinemia (Herrmann et al. 2005; Melton 2003).

It has been demonstrated recently that H_2S deficiency causes aberrant intracellular Ca^{2+} influx because of reduced persulfidation of cysteine residues on multiple TRP channels (Liu et al. 2014c). Decreased Ca^{2+} influx downregulates PKC-/Erk-mediated Wnt/beta-catenin signalling which controls osteogenic differentiation of BMMSCs. Therefore, the authors suggest that bone marrow mesenchymal stem cells produce H_2S in order to regulate their self-renewal and osteogenic differentiation and that H_2S deficiency results in defects in BMMSC differentiation. This study is in agreement with the previous observation that hydrogen sulfide protects MC3T3-E1 osteoblastic cells against H_2O_2-induced oxidative damage (Xu et al. 2011) suggesting that development of new H_2S-releasing drugs could be a potential therapeutic route for the treatment of osteoporosis.

5 Placing Persulfidation in a Broader Biological Context and Future Directions

The biochemical properties of protein persulfides discussed here suggest a specific and enhanced reactivity, which can be used for the regulation of protein's function. As mentioned above, persulfidation increases the nucleophilicity of protein thiols, and, as shown in several examples of proteins such as GAPDH and parkin, it increases their enzymatic activity. Further, the increased reducing power of persulfides suggests that they could act as efficient antioxidants. For example, GSSH reacts rapidly with H_2O_2, neutralizing its toxic effects. In fact, one of the possible roles of protein persulfidation could be the protection of a particular protein from irreversible damage induced by ROS and/or RNS (Fig. 6d). Thiol oxidation, which initially starts with the formation of sulfenic acids (still reversible modification), could proceed further with the formation of irreversible sulfonic acids. H_2S could react with sulfenic acid preventing this oxidation. In addition, persulfidated protein, even when exposed to ROS/RNS, will form an adduct that could be cleaved by the action of certain enzymes restoring free thiol. Therefore, spatio-temporal distribution of sulfenylation vs. persulfidation could help us to understand the relative ratio and importance of these two oxPTMs of cysteine.

S-nitrosation of proteins is a very important posttranslational modification of proteins. S-nitrosation of parkin leads to its inactivation (Chung et al. 2004). Conversely, persulfidation of parkin stimulates the enzyme increasing its ligase activity, which prevents accumulation of toxic proteins (Vandiver et al. 2013). On the other hand, intracellular S-nitrosothiol levels were found to be lower in cells with lower H_2S production, suggesting its role in *trans*-nitrosation reactions as well (Filipovic et al. 2012a). Therefore, the understanding of the NO and H_2S crosstalk, with particular emphasis on S-nitrosation vs. persulfidation, should be one of the future tasks.

All these could be done with proper tools for persulfide detection. Further development of assays, which would allow easy labelling of protein persulfides and their subsequent proteomic analysis, would accelerate the progression of the field. It is worth mentioning that most of the protein persulfidation reports to date used the methodology, which has been shown to generate artefacts.

Very little is known about the mechanisms of protein persulfidation. *Trans*-persulfidation of proteins by low molecular weight persulfides is an exciting concept that warrants further exploring. The role of metal centres in facilitating persulfidation is also of potential interest. Elucidation of all these mechanisms (the role of H_2S vs. polysulfides vs. LMW persulfide) could help us in getting a few steps closer to the understanding of actual contribution of different H_2S-producing enzymes in the regulation of the intracellular S-sulfhydration levels.

Finally, the question that warrants an equally important attention is the extent to which this modification is regulatory and to which it is just a consequence of stochastic events. Answering this could lead to the identification of specific targets that could later on be pharmacologically tempered with. Therefore, this field of research will remain one of the hot topics for many years to come.

Acknowledgement The author is grateful to professors Ivana Ivanovic-Burmazovic, Jon Fukuto and Ruma Banerjee for the helpful discussion. This work is supported by the FAU Erlangen-Nuremberg intramural grant within Emerging Fields Initiative (MRIC).

References

Abe K, Kimura H (1996) The possible role of hydrogen sulfide as an endogenous neuromodulator. J Neurosci 16:1066–1071

Aggarwal BB, Gupta SC, Kim JH (2012) Historical perspectives on tumor necrosis factor and its superfamily: 25 years later, a golden journey. Blood 119:651–665

Ali MY, Ping CY, Mok YY et al (2006) Regulation of vascular nitric oxide in vitro and in vivo; a new role for endogenous hydrogen sulphide? Br J Pharmacol 149:625–634

Artaud I, Galardon E (2014) A persulfide analogue of the nitrosothiol SNAP: formation, characterization and reactivity. Chembiochem 15:2361–2364

Bailey TS, Zakharov LN, Pluth MD (2014) Understanding hydrogen sulfide storage: probing conditions for sulfide release from hydrodisulfides. J Am Chem Soc 136:10573–10576

Blackstone E, Morrison M, Roth MB (2005) H_2S induces a suspended animation-like state in mice. Science 308:518

Bouillaud F, Blachier F (2011) Mitochondria and sulfide: a very old story of poisoning, feeding, and signaling? Antioxid Redox Signal 15:379–391

Broniowska KA, Hogg N (2012) The chemical biology of S-nitrosothiols. Antioxid Redox Signal 17:969–980

Calvert JW, Jha S, Gundewar S et al (2009) Hydrogen sulfide mediates cardioprotection through Nrf2 signaling. Circ Res 105:365–374

Calvert JW, Elston M, Nicholson CK et al (2010) Genetic and pharmacologic hydrogen sulfide therapy attenuates ischemia-induced heart failure in mice. Circulation 122:11–19

Carballal S, Radi R, Kirk MC, Barnes S, Freeman BA, Alvarez B (2003) Sulfenic acid formation in human serum albumin by hydrogen peroxide and peroxynitrite. Biochemistry 42:9906–9914

Carballal S, Trujillo M, Cuevasanta E et al (2011) Reactivity of hydrogen sulfide with peroxynitrite and other oxidants of biological interest. Free Radic Biol Med 50:196–205

Chen W, Liu C, Peng B, Zhao Y, Pacheco A, Xian M (2013) New fluorescent probes for sulfane sulfurs and the application in bioimaging. Chem Sci 4:2892–2896

Chung KK, Thomas B, Li X et al (2004) S-nitrosylation of parkin regulates ubiquitination and compromises parkin's protective function. Science 304:1328–1331

Cohen-Armon M, Visochek L, Rozensal D, Kalal A, Geistrikh I, Klein R, Bendetz-Nezer S, Yao Z, Seger R (2007) DNA-independent PARP-1 activation by phosphorylated ERK2 increases Elk1 activity: a link to histone acetylation. Mol Cell 25:297–308

Coletta C, Papapetropoulos A, Erdelyi K et al (2012) Hydrogen sulfide and nitric oxide are mutually dependent in the regulation of angiogenesis and endothelium-dependent vasorelaxation. Proc Natl Acad Sci USA 109(23):9161–9166

Cuevasanta E, Denicola A, Alvarez B, Moller MN (2012) Solubility and permeation of hydrogen sulfide in lipid membranes. PLoS ONE 7:e34562
D'Amours D, Desnoyers S, D'Silva I, Poirier GG (1999) Poly(ADP-ribosyl)ation reactions in the regulation of nuclear functions. Biochem J 342:249–268
Das TN, Huie RE, Neta P, Padmaja S (1999) Reduction potential of the sulfhydryl radical: pulse radiolysis and laser flash photolysis studies of the formation and reactions of center dot SH and HSSH center dot(-) in aqueous solutions. J Phys Chem A 103:5221–5226
Du J, Huang Y, Yan H et al (2014) Hydrogen sulfide suppresses oxidized low-density lipoprotein (ox-LDL)-stimulated monocyte chemoattractant protein 1 generation from macrophages via the nuclear factor κB (NF-κB) pathway. J Biol Chem 289:9741–9753
Eberhardt M, Dux M, Namer B et al (2014) H2S and NO cooperatively regulate vascular tone by activating a neuroendocrine HNO-TRPA1-CGRP signalling pathway. Nat Commun 5:4381
Filipovic MR, Miljkovic J, Allgäuer A et al (2012a) Biochemical insight into physiological effects of H_2S: reaction with peroxynitrite and formation of a new nitric oxide donor, sulfinyl nitrite. Biochem J 441:609–621
Filipovic MR, Miljkovic J, Nauser T et al (2012b) Chemical characterization of the smallest S-nitrosothiol, HSNO; cellular cross-talk of H2S and S-nitrosothiols. J Am Chem Soc 134: 12016–12027
Filipovic MR, Eberhardt M, Prokopovic V et al (2013) Beyond H_2S and NO interplay: hydrogen sulfide and nitroprusside react directly to give nitroxyl (HNO). A new pharmacological source of HNO. J Med Chem 56:1499–1508
Flavin M (1962) Microbial transsulfuration: the mechanism of an enzymatic disulfide elimination reaction. J Biol Chem 237:768–777
Forrester MT, Foster MW, Benhar M, Stamler JS (2009) Detection of protein S-nitrosylation with the biotin-switch technique. Free Radic Biol Med 46:119–126
Foster MW, Hess DT, Stamler JS (2009) Protein S-nitrosylation in health and disease: a current perspective. Trends Mol Med 15:391–404
Francoleon NE, Carrington SJ, Fukuto JM (2011) The reaction of H_2S with oxidized thiols: generation of persulfides and implications to H_2S biology. Arch Biochem Biophys 516: 146–153
Fujii S, Akaike T (2013) Redox signaling by 8-nitro-cyclic guanosine monophosphate: nitric oxide- and reactive oxygen species-derived electrophilic messenger. Antioxid Redox Signal 19:1236–1246
Fulton AB (1982) How crowded is the cytoplasm? Cell 30:345–347
Giorgio M, Migliaccio E, Orsini F et al (2005) Electron transfer between cytochrome *c* and p66Shc generates reactive oxygen species that trigger mitochondrial apoptosis. Cell 122:221–233
Greiner R, Palinkas Z, Basell K et al (2013) Polysulfides link H2S to protein thiol oxidation. Antioxid Redox Signal 19:1749–1765
Gupta V, Carroll KS (2014) Sulfenic acid chemistry, detection and cellular lifetime. Biochim Biophys Acta 1840:847–875
Hara MR, Agrawal N, Kim SF et al (2005) S-nitrosylated GAPDH initiates apoptotic cell death by nuclear translocation following Siah1 binding. Nat Cell Biol 7:665–674
Heimer NE (1981) Biologically oriented organic sulfur chemistry. 21. Hydrodisulfide of a penicillamine derivative and related compounds. J Org Chem 46:1374–1377
Herrmann M, Widmann T, Colaianni G, Colucci S, Zallone A, Herrmann W (2005) Increased osteoclast activity in the presence of increased homocysteine concentrations. Clin Chem 51: 2348–2353
Hess DT, Stamler JS (2012) Regulation by S-nitrosylation of protein post-translational modification. J Biol Chem 287:4411–4418
Hill BC, Woon TC, Nicholls P, Peterson J, Greenwood C, Thomson AJ (1984) Interactions of sulphide and other ligands with cytochrome c oxidase. An electron-paramagnetic-resonance study. Biochem J 224:591–600

Hourihan JM, Kenna JG, Hayes JD (2013) The gasotransmitter hydrogen sulfide induces nrf2-target genes by inactivating the keap1 ubiquitin ligase substrate adaptor through formation of a disulfide bond between cys-226 and cys-613. Antioxid Redox Signal 19:465–481

Hybertson BM, Gao B, Bose SK, McCord JM (2011) Oxidative stress in health and disease: the therapeutic potential of Nrf2 activation. Mol Aspects Med 32:234–246

Ida T, Sawa T, Ihara H et al (2014) Reactive cysteine persulfides and S-polythiolation regulate oxidative stress and redox signaling. Proc Natl Acad Sci USA 111:7606–7611

Ivanovic-Burmazovic I, Filipovic MR (2012) WO2012/175630

Jackson MR, Melideo SL, Jorns MS (2012) Human sulfide:quinone oxidoreductase catalyzes the first step in hydrogen sulfide metabolism and produces a sulfane sulfur metabolite. Biochemistry 51:6804–6815

Kabil O, Banerjee R (2014) Enzymology of H_2S biogenesis, decay and signaling. Antioxid Redox Signal 20:770–782

Kabil O, Motl N, Banerjee R (2014) H_2S and its role in redox signaling. Biochim Biophys Acta. doi:10.1016/j.bbapap.2014.01.002

Karala AR, Ruddock LW (2007) Does s-methyl methanethiosulfonate trap the thiol-disulfide state of proteins? Antioxid Redox Signal 9:527–531

Kaspar JW, Niture SK, Jaiswal AK (2009) Nrf2:INrf2 (Keap1) signaling in oxidative stress. Free Radic Biol Med 47:1304–1309

Kimura H (2014) Hydrogen sulfide and polysulfides as biological mediators. Molecules 19: 16146–16157

Kimura H, Nagai Y, Umemura K, Kimura Y (2005) Physiological roles of hydrogen sulfide: synaptic modulation, neuroprotection, and smooth muscle relaxation. Antioxid Redox Signal 7:795–803

Kimura Y, Mikami Y, Osumi K, Tsugane M, Oka J, Kimura H (2013) Polysulfides are possible H_2S-derived signaling molecules in rat brain. FASEB J 27:2451–2457

Koenitzer JR, Isbell TS, Patel HD et al (2007) Hydrogen sulfide mediates vasoactivity in an O-2-dependent manner. Am J Physiol Heart Circ Physiol 292:H1953–H1960

Kotronarou A, Hoffmann MR (1991) Catalytic autooxidation of hydrogen sulfide in wastewater. Environ Sci Technol 25:1153–1160

Krishnan N, Fu C, Pappin DJ, Tonks NK (2011) H_2S-induced sulfhydration of the phosphatase PTP1B and its role in the endoplasmic reticulum stress response. Sci Signal 4(203):ra86

Kutney GW, Turnbull K (1982) Compounds containing the sulfur-sulfur double bond. Chem Rev 82:333–357

Li L, Bhatia M, Zhu YZ et al (2005) Hydrogen sulfide is a novel mediator of lipopolysaccharide-induced inflammation in the mouse. FASEB J 19:1196–1198

Li L, Hsu A, Moore PK (2009) Actions and interactions of nitric oxide, carbon monoxide and hydrogen sulphide in the cardiovascular system and in inflammation–a tale of three gases! Pharmacol Ther 123:386–400

Libiad M, Yadav PK, Vitvitsky V, Martinov M, Banerjee R (2014) Organization of the human mitochondrial H_2S oxidation pathway. J Biol Chem pii: jbc.M114.602664

Lima B, Forrester MT, Hess DT, Stamler JS (2010) S-nitrosylation in cardiovascular signaling. Circ Res 106:633–646

Liu C, Chen W, Shi W et al (2014a) Rational design and bioimaging applications of highly selective fluorescence probes for hydrogen polysulfides. J Am Chem Soc 136:7257–7260

Liu C, Zhang F, Munske G, Zhang H, Xian M (2014b) Isotope dilution mass spectrometry for the quantification of sulfane sulfurs. Free Radic Biol Med 76C:200–207

Liu Y, Yang R, Liu X et al (2014c) Hydrogen sulfide maintains mesenchymal stem cell function and bone homeostasis via regulation of Ca(2+) channel sulfhydration. Cell Stem Cell 15:66–78

Mathai JC, Missner A, Kugler P et al (2009) No facilitator required for membrane transport of hydrogen sulfide. Proc Natl Acad Sci USA 106:16633–16638

Melton LJ 3rd (2003) Adverse outcomes of osteoporotic fractures in the general population. J Bone Miner Res 18:1139–1141

Mikami Y, Shibuya N, Kimura Y, Nagahara N, Ogasawara Y, Kimura H (2011) Thioredoxin and dihydrolipoic acid are required for 3-mercaptopyruvate sulfurtransferase to produce hydrogen sulfide. Biochem J 439:479–485
Miljkovic JL, Kenkel I, Ivanovic-Burmazovic I, Filipovic MR (2013) Generation of HNO and HSNO from nitrite by heme-iron-catalyzed metabolism with H_2S. Angew Chem Int Ed Engl 52:12061–12064
Minton AP (1998) Molecular crowding: analysis of effects of high concentrations of inert cosolutes on biochemical equilibria and rates in terms of volume exclusion. Methods Enzymol 295:27–149
Módis K, Coletta C, Erdélyi K, Papapetropoulos A, Szabo C (2013) Intramitochondrial hydrogen sulfide production by 3-mercaptopyruvate sulfurtransferase maintains mitochondrial electron flow and supports cellular bioenergetics. FASEB J 27:601–611
Moore DJ, West AB, Dawson VL, Dawson TM (2005) Molecular pathophysiology of Parkinson's disease. Annu Rev Neurosci 28:57–87
Moriarty-Craige SE, Jones DP (2004) Extracellular thiols and thiol/disulfide redox in metabolism. Annu Rev Nutr 24:481–509
Mueller EG (2006) Trafficking in persulfides: delivering sulfur in biosynthetic pathways. Nat Chem Biol 2:185–194
Mustafa AK, Gadalla MM, Sen N et al (2009a) H_2S signals through protein S-sulfhydration. Sci Signal 2:ra72
Mustafa AK, Gadalla MM, Snyder SH (2009b) Signaling by gasotransmitters. Sci Signal 2:re2
Mustafa AK, Sikka G, Gazi SK et al (2011) Hydrogen sulfide as endothelium-derived hyperpolarizing factor sulfhydrates potassium channels. Circ Res 109:1259–1268
Nagy P, Winterbourn CC (2010) Rapid reaction of hydrogen sulfide with the neutrophil oxidant hypochlorous acid to generate polysulfides. Chem Res Toxicol 23:1541–1543
Napetschnig J, Wu H (2013) Molecular basis of NF-κB signaling. Annu Rev Biophys 42:443–468
Nicholls P, Marshall DC, Cooper CE, Wilson MT (2013) Sulfide inhibition of and metabolism by cytochrome c oxidase. Biochem Soc Trans 41:1312–1316
Nishida M, Sawa T, Kitajima N et al (2012) Hydrogen sulfide anion regulates redox signaling via electrophile sulfhydration. Nat Chem Biol 8:714–724
Nishida M, Toyama T, Akaike T (2014) Role of 8-nitro-cGMP and its redox regulation in cardiovascular electrophilic signaling. J Mol Cell Cardiol 73:10–17
Olson KR (2012) A practical look at the chemistry and biology of hydrogen sulfide. Antioxid Redox Signal 17:32–44
Olson KR, Healy MJ, Qin Z et al (2008) Hydrogen sulfide as an oxygen sensor in trout gill chemoreceptors. Am J Physiol Regul Integr Comp Physiol 295:R669–R680
Olson KR, DeLeon ER, Liu F (2014) Controversies and conundrums in hydrogen sulfide biology. Nitric Oxide 41:11–26
Ono K, Akaike T, Sawa T et al (2014) Redox chemistry and chemical biology of H2S, hydropersulfides, and derived species: implications of their possible biological activity and utility. Free Radic Biol Med. doi:10.1016/j.freeradbiomed.2014.09.007
Pálinkás Z, Furtmüller PG, Nagy A et al (2014) Interactions of hydrogen sulfide with myeloperoxidase. Br J Pharmacol. doi:10.1111/bph.12769
Pan J, Carroll KS (2013) Persulfide reactivity in the detection of protein S-sulfhydration. ACS Chem Biol 8:1110–1116
Papapetropoulos A, Pyriochou A, Altaany Z et al (2009) Hydrogen sulfide is an endogenous stimulator of angiogenesis. Proc Natl Acad Sci USA 106:21972–21977
Park CM, Macinkovic I, Filipovic MR, Xian M (2015) Use of the "Tag-Switch" method for the detection of protein S-Sulfhydration. Methods Enzymol. doi:10.1016/bs.mie.2014.11.033
Parsons LB, Walton JH (1921) Preparation and properties of the persulfides of hydrogen. J Am Chem Soc 43:2539–2548
Paul BD, Snyder SH (2012) H_2S signalling through protein sulfhydration and beyond. Nat Rev Mol Cell Biol 13:499–507

Paulsen CE, Carroll KS (2013) Cysteine-mediated redox signaling: chemistry, biology, and tools for discovery. Chem Rev 113:4633–4679

Paulsen CE, Truong TH, Garcia FJ et al (2011) Peroxide-dependent sulfenylation of the EGFR catalytic site enhances kinase activity. Nat Chem Biol 8:57–64

Peaper DR, Wearsch PA, Cresswell P (2005) Tapasin and ERp57 form a stable disulfide-linked dimer within the MHC class I peptide-loading complex. EMBO J 24:3613–3623

Peng YJ, Nanduri J, Raghuraman G et al (2010) H2S mediates O_2 sensing in the carotid body. Proc Natl Acad Sci USA 107:10719–10724

Pietri R, Lewis A, León RG et al (2009) Factors controlling the reactivity of hydrogen sulfide with hemeproteins. Biochemistry 48:4881–4894

Pittenge MF, Mackay AM, Beck SC et al (1999) Multilineage potential of adult human mesenchymal stem cells. Science 284:143–147

Poynton RA, Hampton MB (2014) Peroxiredoxins as biomarkers of oxidative stress. Biochim Biophys Acta 1840:906–912

Prockop DJ (1997) Marrow stromal cells as stem cells for nonhematopoietic tissues. Science 276: 71–74

Reynolds JD, Bennett KM, Cina AJ et al (2013) S-nitrosylation therapy to improve oxygen delivery of banked blood. Proc Natl Acad Sci USA 110:11529–11534

Riahi S, Rowley CN (2014) Why can hydrogen sulfide permeate cell membranes? J Am Chem Soc 136:15111–15113

Ríos-González BB, Román-Morales EM, Pietri R, López-Garriga J (2014) Hydrogen sulfide activation in hemeproteins: the sulfheme scenario. J Inorg Biochem 133:78–86

Sen N, Paul BD, Gadalla MM et al (2012) Hydrogen sulfide-linked sulfhydration of NF-κB mediates its antiapoptotic actions. Mol Cell 45:13–24

Seth D, Stamler JS (2011) The SNO-proteome: causation and classifications. Curr Opin Chem Biol 15:129–136

Shulman JM, De Jager PL, Feany MB (2011) Parkinson's disease: genetics and pathogenesis. Annu Rev Pathol 6:193–222

Sparatore A, Perrino E, Tazzari V et al (2008) Pharmacological profile of a novel H_2S-releasing aspirin. Free Radic Biol Med 46:586–592

Steudel R, Drozdova Y, Miaskiewicz K, Hertwig RH, Koch W (1997) How unstable are thiosulfoxides? An ab initio MO study of various disulfanes RSSR (R = H, Me, Pr, All), their branched isomers R2SS, and the related transition states. J Am Chem Soc 119:1990–1996

Szabó C, Papapetropoulos A (2011) Hydrogen sulphide and angiogenesis: mechanisms and applications. Br J Pharmacol 164:853–865

Szczesny B, Módis K, Yanagi K et al (2014) AP39, a novel mitochondria-targeted hydrogen sulfide donor, stimulates cellular bioenergetics, exerts cytoprotective effects and protects against the loss of mitochondrial DNA integrity in oxidatively stressed endothelial cells in vitro. Nitric Oxide 41:120–130

Talipov MR, Timerghazin QK (2013) Protein control of S-nitrosothiol reactivity: interplay of antagonistic resonance structures. J Phys Chem B 117:1827–1837

Terzić V, Padovani D, Balland V, Artauda I, Galardon E (2014) Electrophilic sulfhydration of 8-nitro-cGMP involves sulfane sulfur. Org Biomol Chem 12:5360–5364

van Montfort RL, Congreve M, Tisi D, Carr R, Jhoti H (2003) Oxidation state of the active-site cysteine in protein tyrosine phosphatase 1B. Nature 423:773–777

Vandiver MS, Paul BD, Xu R et al (2013) Sulfhydration mediates neuroprotective actions of parkin. Nat Commun 4:1626

Vitvitsky V, Kabil O, Banerjee R (2012) High turnover rates for hydrogen sulfide allow for rapid regulation of its tissue concentrations. Antioxid Redox Signal 17:22–31

Wakabayashi N, Dinkova-Kostova AT, Holtzclaw WD, Kang MI, Kobayashi A, Yamamoto M, Kensler TW, Talalay P (2004) Protection against electrophile and oxidant stress by induction of the phase 2 response: fate of cysteines of the Keap1 sensor modified by inducers. Proc Natl Acad Sci U S A 101:2040–2045

Wang R (2002) Two's company, three's a crowd: can H_2S be the third endogenous gaseous transmitter? FASEB J 16:1792–1808
Wedmann R, Bertlein S, Macinkovic I, Boeltz S, Miljkovic J, Munoz L, Herrmann M, Filipovic MR (2014) Working with "H_2S": facts and apparent artifacts. Nitric Oxide 41:85–96
Whiteman M, Winyard PG (2011) Hydrogen sulfide and inflammation: the good, the bad, the ugly and the promising. Expert Rev Clin Pharmacol 4:13–32
Whiteman M, Li L, Kostetski I et al (2006) Evidence for the formation of a novel nitrosothiol from the gaseous mediators nitric oxide and hydrogen sulphide. Biochem Biophys Res Commun 343:303–310
Wood JL (1987) Sulfane sulfur. Methods Enzymol 143:25–29
Xie ZZ, Shi MM, Xie L et al (2014) Sulfhydration of p66Shc at cysteine59 mediates the antioxidant effect of hydrogen sulfide. Antioxid Redox Signal. doi:10.1089/ars.2013.5604
Xu L, Eu JP, Meissner G, Stamler JS (1998) Activation of the cardiac calcium release channel (ryanodine receptor) by poly-S-nitrosylation. Science 279:234–237
Xu ZS, Wang XY, Xiao DM, Hu LF, Lu M, Wu ZY, Bian JS (2011) Hydrogen sulfide protects MC3T3-E1 osteoblastic cells against H_2O_2-induced oxidative damage-implications for the treatment of osteoporosis. Free Radic Biol Med 50(10):1314–23
Yadav PK, Yamada K, Chiku T, Koutmos M, Banerjee R (2013) Structure and kinetic analysis of H2S production by human mercaptopyruvate sulfurtransferase. J Biol Chem 288: 20002–200013
Yang G, Wu L, Jiang B et al (2008) H_2S as a physiologic vasorelaxant: hypertension in mice with deletion of cystathionine gamma-lyase. Science 322:587–590
Yang G, Zhao K, Ju Y et al (2013) Hydrogen sulfide protects against cellular senescence via S-sulfhydration of Keap1 and activation of Nrf2. Antioxid Redox Signal 18:1906–1919
Yang J, Gupta V, Carroll KS, Liebler DC (2014) Site-specific mapping and quantification of protein S-sulphenylation in cells. Nat Commun 5:4776
Yong QC, Hu LF, Wang S, Huang D, Bian JS (2010) Hydrogen sulfide interacts with nitric oxide in the heart: possible involvement of nitroxyl. Cardiovasc Res 88:482–491
Yong QC, Cheong JL, Hua F et al (2011) Regulation of heart function by endogenous gaseous mediators-crosstalk between nitric oxide and hydrogen sulfide. Antioxid Redox Signal 14: 2081–2091
Yoshida T, Inoue R, Morii T et al (2006) Nitric oxide activates TRP channels by cysteine S-nitrosylation. Nat Chem Biol 2:596–607
Zhang D, Macinkovic I, Devarie-Baez NO et al (2014) Detection of protein S-sulfhydration by a tag-switch technique. Angew Chem Int Ed Engl 53:575–581
Zhao Y, Bhushan S, Yang C et al (2013) Controllable hydrogen sulfide donors and their activity against myocardial ischemia-reperfusion injury. ACS Chem Biol 8:1283–1290
Zhao K, Ju Y, Li S, Altaany Z, Wang R, Yang G (2014) S-sulfhydration of MEK1 leads to PARP-1 activation and DNA damage repair. EMBO Rep 15:792–800
Zhou Z, von Wantoch Rekowski M, Coletta C et al (2012) Thioglycine and L-thiovaline: biologically active H_2S-donors. Bioorg Med Chem 20:2675–2678

Physiological Roles of Hydrogen Sulfide and Polysulfides

Hideo Kimura

Contents

Abstract

Hydrogen sulfide (H_2S) has been recognized as a signaling molecule as well as a cytoprotective molecule. H_2S modulates neurotransmission, regulates vascular tone, protects various tissues and organs, regulates inflammation, induces angiogenesis, and detects cellular oxygen levels. H_2S is produced from L-cysteine by cystathionine β-synthase (CBS), cystathionine γ-lyase (CSE), and

H. Kimura (✉)
National Institute of Neuroscience, National Center of Neurology and Psychiatry,
4-1-1 Ogawahigashi, Kodaira, Tokyo 187-8502, Japan
e-mail: kimura@ncnp.go.jp

P.K. Moore, M. Whiteman (eds.), *Chemistry, Biochemistry and Pharmacology of Hydrogen Sulfide*, Handbook of Experimental Pharmacology 230,
DOI 10.1007/978-3-319-18144-8_3

3-mercaptopyruvate sulfurtransferase (3MST) together with cysteine aminotransferase (CAT). Recently, a novel pathway for the production of H_2S from D-cysteine was identified, involving D-amino acid oxidase (DAO) together with 3MST. Sulfuration (also called sulfhydration), which adds sulfur atoms to the cysteine residues of target proteins to modify protein activity, has been extensively studied as a mode of H_2S action. Recently, hydrogen polysulfides (H_2S_n, where $n = 3$–7; $n = 2$ is termed as persulfide) have been found to sulfurate target proteins in the brain, including transient receptor potential ankyrin 1 (TRPA1) channels, Kelch-like ECH-associating protein 1 (Keap1), and phosphatase and tensin homolog (PTEN), much more potently than H_2S. The physiological stimuli that trigger the production of H_2S and polysulfides, and the mechanisms maintaining their local levels, remain unknown. Understanding the regulation of H_2S_n (including H_2S) production, and the specific stimuli that induce their release, will provide new insight into the biology of H_2S and will provide novel avenues for therapeutic development in diseases involving H_2S-related substances.

Keywords
H_2S • H_2S_n • CBS • CSE • 3MST • Sulfuration • Sulfhydration • TRPA1 • Keap1 • Nrf2 • PTEN

Abbreviations

3MST	3-mercaptopyruvate sulfurtransferase
ATP	Adenosine triphosphate
CAT	Cysteine aminotransferase
CBS	Cystathionine β–synthase
CO	Carbon monoxide
CSE	Cystathionine γ–lyase
CysSSH	Cysteine persulfide
DAO	D-amino acid oxidase
DHLA	Dihydrolipoic acid
DTT	Dithiothreitol
eNOS	Endothelial nitric oxide synthetase
Fe	Iron
GAPDH	Glyceraldehydes-3-phosphate dehydrogenase
GCL	Glutamate cysteine ligase
GSH	Glutathione
GSNO	Nitrosoglutathione
GSSH	Glutathione persulfide
GSSSH	Glutathione trisulfide
H_2O_2	Hydrogen peroxide
H_2S	Hydrogen sulfide

H_2S_n	Polysulfide
HPLC	High-performance liquid chromatography
HSNO	Thionitrous acid
Keap1	Kelch ECH-associating protein 1
LTP	Long-term potentiation
NaHS	Sodium hydrosulfide
NF-κB	Nuclear factor-κB
NMDA	N-methyl D,L-aspartate
nNOS	Neuronal nitric oxide synthetase
NO	Nitric oxide
NOS	Nitric oxide synthetase
Nrf2	Nuclear factor-erythroid 2-related factor 2
O_2	Oxygen
PD	Parkinson's disease
PERK	Protein kinase RNA-like ER kinase
PLP	Pyridoxal 5'-phosphate
PTEN	Phosphatase and tensin homolog
PTP	Protein tyrosine phosphatase
ROS	Reactive oxygen species
SAM	S-adenosyl-L-methionine
SNAP	S-nitroso-N-acetyl-D,L-penicillamine
SNP	Sodium nitroprusside
$SSNO^-$	Nitrosopersulfide
TNFα	Tumore necrosis factor α
TRPA1	Transient receptor potential ankyrin 1
VEGFR2	Vascular endothelium growth factor receptor 2
γGCS	γ-glutamyl cysteine synthetase

1 Introduction

Since Ramazzini's first description of hydrogen sulfide (H_2S) as a toxic gas in 1713, many papers have been published on H_2S toxicity. Enzymes involved in the H_2S-producing pathways were intensively studied from the 1950s to the 1970s (Meister et al. 1954; Cavallini et al. 1962; Braunstein et al. 1971). However, H_2S was thought to be a mere byproduct of metabolic pathways or simply a marker of enzyme activity. In 1989, measurements of the endogenous levels of sulfide in mammalian brains suggested a physiological role for this molecule (Goodwin et al. 1989; Warenycia et al. 1989). These initial studies by Goodwin et al. (1989) and Warenycia et al. (1989) overestimated the H_2S concentration, due to contaminant H_2S released from another cellular form of sulfur. Recent reevaluation confirmed the existence of H_2S in tissues, albeit at much lower concentrations than originally reported (Furne et al. 2008; Ishigami et al. 2009; Wintner et al. 2010). We have previously demonstrated a role for H_2S produced by cystathionine β-synthase

(CBS) as a neuromodulator in the brain and another role for H_2S produced by cystathionine γ-lyase (CSE) as a smooth muscle relaxant (Abe and Kimura 1996; Hosoki et al. 1997). The enzyme 3-mercaptopyruvate sulfurtransferase (3MST) also produces H_2S when 3MST is associated with thioredoxin (Shibuya et al. 2009; Mikami et al. 2011a; Yadav et al. 2013). Recently, we identified a novel pathway for the production of H_2S from D-cysteine, which involves the enzymes 3MST and D-amino acid oxidase (DAO) (Shibuya et al. 2013).

Various other effects of H_2S have been identified, including cytoprotective effects against oxidative stress and ischemia-reperfusion injury in various tissues and organs. H_2S protects neurons from oxidative stress by increasing the production of glutathione, a major intracellular antioxidant, and by scavenging reactive oxygen species (ROS) (Kimura and Kimura 2004; Whiteman et al. 2004; Kimura et al. 2010). The protective effects of H_2S against oxidative stress and ischemia-reperfusion injury have been demonstrated in the heart, the kidney, and in pancreatic islet β-cells (Elrod et al. 2007; Tripatara et al. 2008; Kaneko et al. 2009). H_2S induces angiogenesis by enhancing the expression of vascular endothelial growth factor (VEGF) and facilitating the phosphorylation of VEGF receptor 2 (VEGFR2); the regulation of nitric oxide synthase (NOS) may also be involved in this effect (Cai et al. 2007; Papapetropoulos et al. 2009; Polhemus et al. 2013). H_2S induces conditions similar to those found in hypoxia. H_2S metabolism is coupled with that of O_2, and H_2S concentrations are inversely related to those of O_2 in tissues. O_2-dependent H_2S inactivation may be effectively used as an O_2 sensor to accurately determine O_2 availability (Olson et al. 2006; Peng et al. 2010). H_2S exerts regulatory effects on inflammatory processes, thereby promoting the resolution of inflammation (Bhatia et al. 2005; Zanardo et al. 2006). H_2S also inhibits cancer development at various stages, but in colon cancer, H_2S is involved in promoting cellular bioenergetics, proliferation, and migration (Chattopadhyay et al. 2012; Szabo et al. 2013).

The mechanism by which H_2S exerts its actions has been proposed as sulfhydration or sulfuration: the addition of sulfur atoms to the cysteine residues of target proteins, inducing conformational changes to modify protein activity. It has been suggested that adenosine triphosphate (ATP)-dependent K^+ channels, Kelch-like ECH-associated protein 1 (Keap1), nuclear factor kappa-B (NF-κB), transient receptor potential ankyrin 1 (TRPA1), and parkin can be activated by sulfhydration or sulfuration (Mustafa et al. 2011; Ogawa et al. 2012; Sen et al. 2012; Vandiver et al. 2013; Yang et al. 2013). H_2S is sequentially oxidized to polysulfides with a varying number of sulfur atoms, until the number of sulfur atoms reaches eight; at that point, the sulfur molecules cyclize and separate from polysulfides (Nagy and Winterbourn 2010; Toohey 2011).

$$2nH_2S + 1/2(2n-1)O_2 \rightarrow H_2S_{2n} + (2n-1)H_2O \tag{1}$$

$$HS^- \leftrightarrow HSS^- \leftrightarrow HSSS^- \leftrightarrow \ldots\ldots \leftrightarrow HS_7{}^- \rightarrow S_8 \tag{2}$$

Recently, polysulfides (H_2S_n, where $n = 3$–7; when $n = 2$, the compound is termed as persulfide) were found to be present in the mammalian brain. Polysulfides activate TRPA1 channels, facilitate the translocation of nuclear factor-erythroid 2-related factor 2 (Nrf2) to the nucleus via release of Nrf2 from sulfhydrated Keap1, and modulate the activity of PTEN; interestingly, polysulfides mediate these processes more efficiently than does H_2S (Greiner et al. 2013; Kimura et al. 2013; Koike et al. 2013). When considering whether these biological effects are mediated by H_2S or polysulfides, the efficiency of the effects and the chemical properties of the molecules suggest that polysulfides, rather than H_2S, sulfurate (or sulfhydrate) target proteins. H_2S may instead be involved in reducing reactions, such as the reduction of cysteine disulfide bonds.

2 Endogenous Free H_2S

Initial studies reported relatively high concentrations (50–160 μM) of H_2S in mammalian brains (Goodwin et al. 1989; Warenycia et al. 1989; Savage and Gould 1990). These studies used the methylene blue method, which is performed under strong acidic conditions. At pH values less than 5.4, H_2S is released from acid-labile sulfur, which consists mostly of the iron-sulfur complex at the active site of respiratory chain enzymes (Ishigami et al. 2009). Due to the localization of most enzymes containing an iron-sulfur complex to the mitochondrion, where the pH value is approximately 8, H_2S may not be released from iron-sulfur complexes under physiological conditions. Therefore, H_2S levels measured with the methylene blue method did not reflect free H_2S but instead reflected contaminant H_2S released from acid-labile sulfur.

The reevaluated endogenous H_2S concentrations are much lower than those previously measured, with recent studies suggesting a concentration range of 10 nM to 3 μM (Ishigami et al. 2009; Wintner et al. 2010; Levitt et al. 2011). These concentrations were measured during a steady state, and changes in H_2S concentration during cell activity are not well understood. Given that the K_m values of CSE and CBS for cysteine are 1.7 mM and 3.6 mM, respectively, and that the intracellular concentration of cysteine is approximately 100 μM (the exception being levels of approximately 1 mM in the kidney), the production of H_2S by CSE and CBS must occur relatively slowly, producing less amount of H_2S (Stipanuk et al. 2002; Chiku et al. 2009; Singh et al. 2009).

3 Regulation of H_2S Production

In order to function as a signaling molecule, H_2S concentrations must be properly regulated by physiological stimuli. There are at least two sources of H_2S: (1) H_2S-producing enzymes and (2) the intracellular H_2S stores.

3.1 H_2S-Producing Enzymes

Three enzymes are known to produce H_2S: the pyridoxal 5′-phosphate (PLP)-dependent CBS and CSE and the PLP-independent 3MST (Meister et al. 1954; Cavallini et al. 1962; Braunstein et al. 1971).

3.1.1 CBS

CBS produces H_2S via a β-replacement reaction involving L-cysteine. By catalyzing the β-replacement of L-cysteine with L-homocysteine, CBS produces H_2S with a 23-fold greater specific activity than that of H_2S produced from L-cysteine alone (Singh et al. 2009). Three sites on the CBS protein regulate its activity: (1) an S-adenosyl-L-methionine (SAM)-binding domain at the carboxy-terminus; (2) a heme group at the amino-terminus; and (3) cysteine 346 (Cys346) for S-glutathionylation. These three methods of CBS regulation are expanded upon in the following text (Fig. 1). (1) SAM enhances the H_2S-producing activity of CBS by approximately twofold (Abe and Kimura 1996). The SAM-binding domain covers the catalytic site of CBS in the absence of SAM, and binding to SAM thereby activates the enzyme (Shan et al. 2001). (2) CBS containing the Fe(II) form of heme is inhibited by the binding of carbon monoxide (CO) or nitric oxide (NO), while

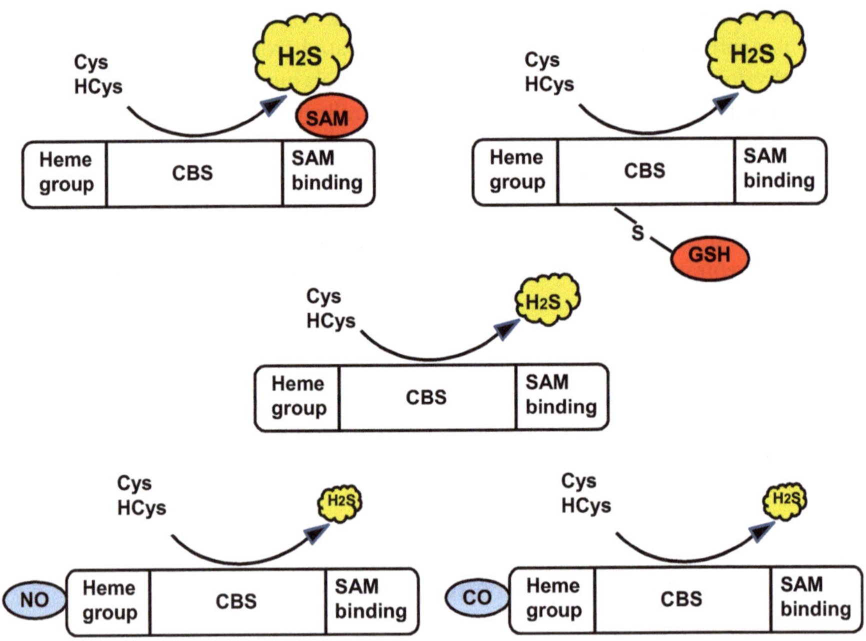

Fig. 1 The regulation of cystathionine β-synthase (CBS) activity. The activity of CBS is enhanced by the binding of S-adenosyl-L-methionine (SAM) to the SAM-binding domain located at the carboxy-terminus and by S-glutathionylation at Cys346. In contrast, the binding of nitric oxide (NO) or carbon monoxide (CO) to the heme group at the amino-terminus suppresses CBS activity

CBS activity is enhanced when Fe(II) is oxidized to Fe(III) (Taoka and Banerjee 2001). The regulation of H_2S production plays an important role in microcirculation in the brain (Morikawa et al. 2012). CO that is produced by heme oxygenase-2 in neurons suppresses the activity of CBS. Under hypoxic conditions, the production of CO is diminished, leading to the activation of CBS in astrocytes surrounding brain capillaries and the production of H_2S, which in turn vasodilates precapillary arterioles to restore blood flow. (3) Under conditions of oxidative stress, Cys346 of CBS is oxidized to a sulfenic acid, which binds to glutathione (Niu et al. 2014). CBS activity is enhanced by glutathionylation, increasing the production of cysteine and H_2S and in turn leading to increased glutathione production. Glutathionylation of CBS compensates for decreased levels of glutathione under oxidative stress.

The production of H_2S by CBS is enhanced by approximately 23-fold in the presence of homocysteine, by twofold when binding to SAM, and by threefold with glutathionylation. Therefore, the combination of these modifications to CBS may allow production of sufficient levels of H_2S to induce physiological responses despite low intracellular concentrations of cysteine (Abe and Kimura 1996; Singh et al. 2009; Niu et al. 2014).

3.1.2 CSE

CSE produces H_2S through an α,β-elimination reaction with L-cysteine (Chiku et al. 2009; Singh et al. 2009). It has been reported that CSE is regulated by Ca^{2+}/calmodulin (Yang et al. 2008). However, the study by Yang and colleagues (2008) examined CSE activity in the presence of 1–2 mM Ca^{2+}, which is the extracellular concentration of Ca^{2+}. As CSE is a cytosolic enzyme, its activity should be examined in the presence of intracellular concentrations of Ca^{2+} (approximately 100 nM). Our recent reevaluation of the relationship between Ca^{2+} levels and CSE activity showed that the H_2S-producing activity of CSE is maximal at low Ca^{2+} concentrations and is suppressed by Ca^{2+} in the presence of PLP (Mikami et al. 2013). H_2S may be constitutively produced by CSE in cells under steady-state conditions in which the intracellular concentration of Ca^{2+} is maintained at low levels, whereas H_2S production is suppressed once the intracellular concentration of Ca^{2+} increases upon cell stimulation. Because calmodulin and its selective inhibitor W-7 do not alter CSE activity, it appears that calmodulin is not involved in the regulation of CSE by Ca^{2+} (Mikami et al. 2013).

In addition to regulation of CSE activity, the expression level of CSE is also transcriptionally regulated. The CSE gene promoter region has an SP1-binding site, which is activated by the multifunctional, pro-inflammatory cytokine tumor necrosis factor α (TNFα) (Ishii et al. 2004; Sen et al. 2012).

3.1.3 3MST

3MST produces H_2S by metabolizing 3-mercaptopyruvate (3MP); 3MP is synthesized through the metabolism of L-cysteine and α-ketoglutarate by cysteine aminotransferase (CAT; identical to aspartate aminotransferase) (Meister et al. 1954; Taniguchi and Kimura 1974; Ubuka et al. 1978; Cooper 1983; Shibuya et al. 2009). 3MST is localized to both the mitochondria and the cytosol; CAT has

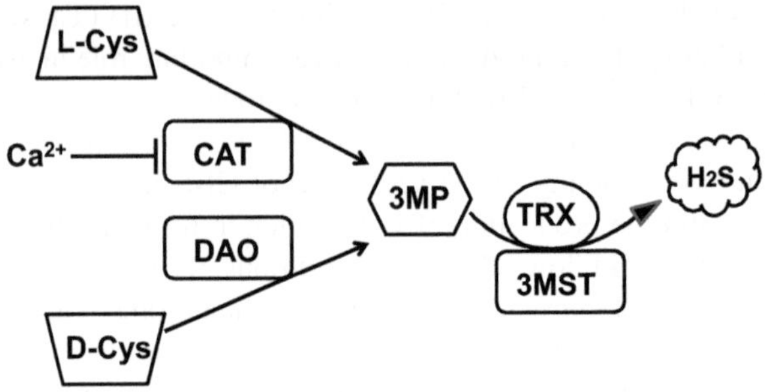

Fig. 2 Hydrogen sulfide (H_2S) production from L- and D-cysteine. L-Cysteine is metabolized to 3-mercaptopyruvate (3MP) by cysteine aminotransferase (CAT), which is regulated by Ca^{2+}. Achiral 3MP is also produced from D-cysteine by D-amino acid oxidase (DAO). 3MP is metabolized to H_2S by 3-mercaptopyruvate sulfurtransferase (3MST) accompanied by thioredoxin (TRX)

both a mitochondrial and a cytosolic form, with these two CAT forms sharing 48 % identity in their amino acid sequences (Ubuka et al. 1978; Akagi 1982; Doyle et al. 1990; Nagahara et al. 1998; Shibuya et al. 2009). Unlike the cytosol, mitochondria contain a sufficiently high cysteine concentration (approximately 1 mM) for the 3MST/CAT pathway to produce H_2S (Griffith 1999; Tateishi et al. 1977) (Fig. 2).

3MST requires a reducing agent to produce H_2S from 3MP, while neither CBS nor CSE requires such a reducing agent. The endogenous reducing agents, thioredoxin and dihydrolipoic acid (DHLA), associate with 3MST to produce H_2S (Mikami et al. 2011a; Yadav et al. 2013). Thioredoxin is present at approximately 20 μM in cells and is four times as potent as dithiothreitol (DTT), and DHLA is found at approximately 40 μM in mitochondria and is as potent as DTT. Other endogenous reducing agents do not associate with 3MST to produce H_2S. Dithiols such as thioredoxin (which has two cysteine residues at the active site), DHLA, and DTT associate with 3MST to release H_2S (Mikami et al. 2011a).

The activity of CAT is suppressed by Ca^{2+}, but as is the case with CSE, calmodulin is not involved in the regulation of CAT activity (Mikami et al. 2011b). Therefore, 3MP is supplied to 3MST under steady-state conditions, whereas the supply of 3MP is stopped once the intracellular concentrations of Ca^{2+} are increased.

3MST, together with DAO, produces H_2S from D-cysteine (Shibuya et al. 2013). DAO, which is located in peroxisomes, metabolizes D-cysteine to achiral 3MP (Fig. 2). Peroxisomes and mitochondria are in close vicinity to each other and can establish physical contact, and they exchange metabolites via a specific form of vesicular trafficking (Schumann and Subramani 2008). 3MP produced in peroxisomes is transported into mitochondria, where it is metabolized by 3MST to H_2S. This pathway is active in the brain and the kidney; this spatial restriction may be useful for the targeted production of H_2S in those specific organs following systemic delivery of D-cysteine. The 3MST/DAO pathway is highlighted in the kidney, where 60 times more H_2S is produced from D-cysteine than from L-cysteine (Shibuya et al. 2013).

3.2 The Intracellular H_2S Store

Several factors change the activity of H_2S-producing enzymes, and H_2S levels may change according to physiological stimuli. H_2S can be provided from bound sulfane sulfur, which releases H_2S under physiological reducing conditions. The levels of bound sulfane sulfur vary among tissues, and the amount of H_2S released from brain homogenates by DTT is approximately 1.5 μmol/g protein (Ishigami et al. 2009). Lysates of neurons and astrocytes release H_2S in the presence of 3 mM glutathione and 100 μM cysteine at pH 8.4. The pH of astrocytes can reach 8.4 when the extracellular concentration of K^+ is 10 mM, a level that can be induced by the excitation of nearby neurons. However, H_2S release from astrocytes has not been successfully detected; this is most likely due to the released H_2S being present at an insufficient concentration to be detected in perfusion medium (Ishigami et al. 2009).

4 Neuromodulation by H_2S and Polysulfides

H_2S facilitates the induction of hippocampal long-term potentiation (LTP) by enhancing the activity of N-methyl-D-aspartate (NMDA) receptors in the brain. LTP induction requires the activation of NMDA receptors, and reducing agents such as DTT reduce the cysteine disulfide bond at the hinge of the NMDA receptor ligand-binding domain to activate these receptors (Aizenman et al. 1989). In spite of having less reducing activity, the H_2S donor sodium hydrosulfide (NaHS) induces LTP more potently than does DTT, even when DTT is present at 10 times greater concentration than that of NaHS (Abe and Kimura 1996). It is therefore difficult to attribute the effects of H_2S on LTP to the reducing activity of H_2S alone.

Neurons are surrounded by astrocytes, which also express neurotransmitter receptors and are activated by neurotransmitters. The activated astrocytes, in turn, release gliotransmitters to modulate synaptic activity (Araque et al. 1999). Our previous research showed that NaHS activates transient receptor potential (TRP) channels to induce Ca^{2+} influx in astrocytes (Nagai et al. 2004). The involvement of this mechanism in the induction of LTP is discussed below.

In our previous research, we found that polysulfides activate TRP channels more potently than does H_2S (Nagai et al. 2006; Oosumi et al. 2010). High concentrations of NaHS (1–10 mM) activate TRPA1 channels in sensory neurons in the urinary bladder and dorsal ganglion (Streng et al. 2008; Ogawa et al. 2012). In these studies on sensory neurons, a portion of the NaHS-derived H_2S might be oxidized to polysulfides that activate TRPA1 channels. We found that polysulfides in the brain activated astrocytic TRPA1 channels to a much greater extent than did H_2S, inducing a maximal response (Kimura et al. 2013). A comparison of K_m values revealed that the affinity of TRPA1 channels for polysulfides was approximately 1000 times greater than for H_2S (Kimura et al. 2013). The level of endogenous polysulfides in the brain was found to be approximately 20 μM, using HPLC analysis with monobromobimane (Kimura et al. 2013).

Ogawa and colleagues determined that Cys422 and Cys622 in the TRPA1 channel amino-terminus are sensitive to H_2S (Ogawa et al. 2010). Given that DTT abolishes the effects of high concentrations of H_2S, these two cysteine residues must be sulfurated by polysulfides generated by oxidation of H_2S. The idea that polysulfide, rather than H_2S, sulfurates TRPA1 channels is based on the fact that substances with the same oxidation state do not react with each other. The oxidation state of sulfur in H_2S and in cysteine residues is −2. In contrast, the oxidation state of sulfur in polysulfides is 0, meaning that polysulfides easily react with cysteine residues to add sulfur for the generation of bound sulfane sulfur. It is also possible that one cysteine residue is sulfurated earlier than an adjacent cysteine, and the latter may react with the former to produce a cysteine disulfide bond (Fig. 3).

Shigetomi and colleagues recently demonstrated that the activated TRPA1 channels in astrocytes increase the intracellular concentrations of Ca^{2+}, which facilitate the release of the gliotransmitter D-serine from astrocytes; D-serine

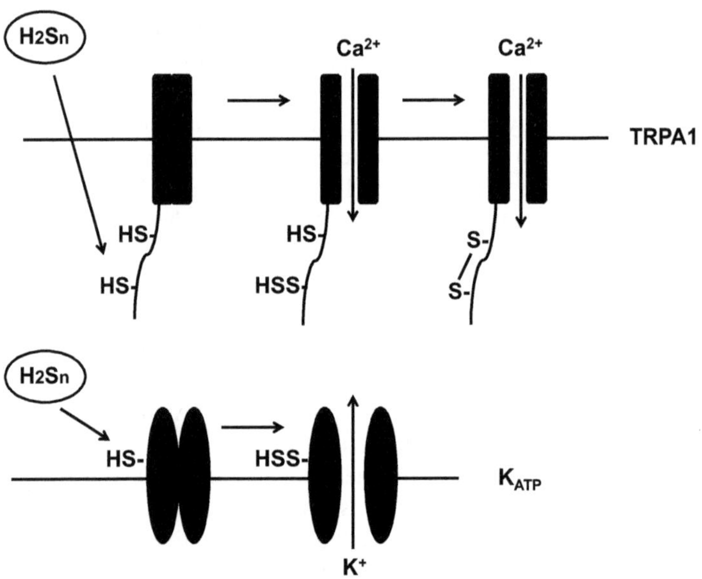

Fig. 3 The regulation of channel activity by sulfuration (sulfhydration). Transient receptor potential ankyrin 1 (TRPA1) channels and Adenosine triphosphate (ATP)-dependent K^+ (K_{ATP}) are activated by sulfuration (sulfhydration). H_2S has been proposed to sulfurate these channels, but polysulfides (H_2S_n) have much greater activity than H_2S. Thus, polysulfides have been proposed to be the molecular entity that induces sulfuration

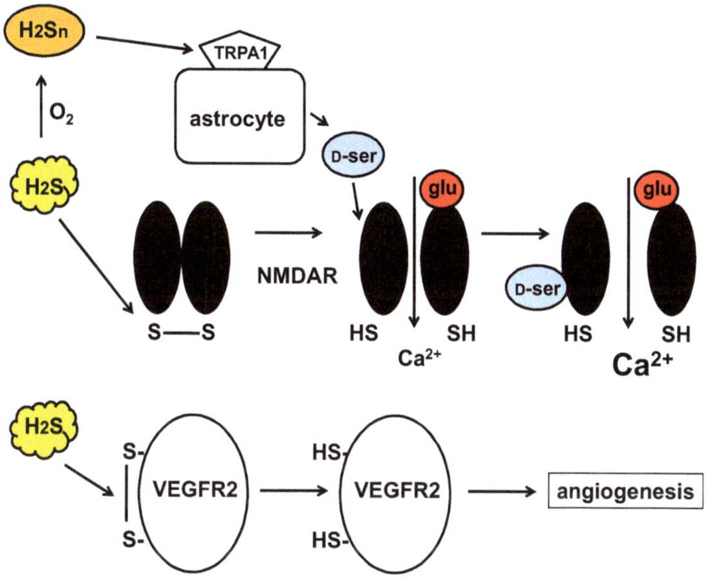

Fig. 4 H_2S enhances receptor activity by reducing a cysteine disulfide bond. H_2S enhances the activity of N-methyl-D-aspartate (NMDA) receptors by reducing the cysteine disulfide bond at the hinge of the ligand-binding domain. The activity of NMDA receptors is also enhanced by binding the gliotransmitter D-serine, released from astrocytes, which are activated by polysulfides (H_2S_n) through the activation of TRPA1 channels. H_2S also enhances the activity of vascular endothelial growth factor receptor 2 (VEGFR2) by reducing the cysteine disulfide bond between Cys1045 and Cys1024

activates NMDA receptors on postsynaptic neurons, leading to the induction of LTP (Shigetomi et al. 2012; 2013). A similar mechanism may be involved in the induction of LTP by H_2S and polysulfides. As described previously in this section, H_2S reduces the cysteine disulfide bond and enhances the activity of NMDA receptors, and polysulfides activate TRPA1 channels in astrocytes and induce D-serine release that enhances NMDA receptor activity. In contrast, DTT is not able to activate TRPA1 channels but reduces the cysteine disulfide bond in NMDA receptors. These findings explain the difference in LTP induction efficiency between H_2S/polysulfides and DTT (Abe and Kimura 1996; Kimura et al. 2013) (Fig. 4).

5 Cytoprotective Effects of H_2S and Polysulfides

H_2S protects various tissues/organs from oxidative stress or ischemia-reperfusion injury (Kimura and Kimura 2014; Elrod et al. 2007). Recently, polysulfides were found to have cytoprotective effect through a mechanism of sulfurating (sulfhydrating) target proteins (Kimura 2014).

5.1 Glutathione Production and ROS Scavenging

H_2S protects various tissues and organs from oxidative stress and ischemia-reperfusion injury. H_2S enhances the activity of the cysteine transporter and of the cystine/glutamate antiporter to increase the import of cystine even in the presence of high glutamate concentrations (Kimura and Kimura 2004). Cystine is reduced to cysteine inside cells and is used as a substrate for glutamate cysteine ligase (GCL) (previously known as γ-glutamylcysteine synthetase), which is a rate-limiting enzyme in glutathione production. H_2S enhances the activity of GCL. This enhancement only occurs when H_2S is applied extracellularly, not when GCL is directly exposed to H_2S (Kimura et al. 2010). This observation suggests that H_2S may activate a signaling pathway from the cell surface, which acts downstream to enhance the activity of GCL.

H_2S also scavenges ROS (Whiteman et al. 2004; Kimura et al. 2010). Considering the endogenous concentrations of H_2S (in the nM to μM range) and glutathione (in the mM range) and the fact that H_2S is not a strong reducing agent compared to glutathione, the suppression of oxidative stress by H_2S may be due to the enhancement of GCL activity and subsequent glutathione production, rather than the direct scavenging of ROS.

5.2 Translocation of Nrf2 to the Nucleus

H_2S also upregulates the transcription of antioxidant genes to exert its cytoprotective effect. Keap1 binds Nrf2 and retains it in the cytoplasm, and the Keap1/Nrf2 complex is ubiquitinated and degraded by the proteasome under steady-state conditions. The cysteine residues of Keap1 serve as sensors for oxidative stress, and their modification causes a conformational change in Keap1 that releases Nrf2, which is then translocated to the nucleus and upregulates the transcription of antioxidant genes (Wakabayashi et al. 2004). H_2S protects cardiac myocytes from ischemia-reperfusion injury by preserving the structure and function of mitochondria (Elrod et al. 2007). The translocation of Nrf2 to the nucleus has been suggested as a mechanism for the H_2S-mediated protection of myocytes (Calvert et al. 2009). Another study suggested that H_2S sulfurates (sulfhydrates) Keap1 to release Nrf2 for translocation to the nucleus (Yang et al. 2013). However, as described in section 4, it is difficult for H_2S to sulfurate (sulfhydrate) cysteine residues in Keap1 because H_2S and cysteine residues share the same oxidation state. Instead, polysulfide has been suggested as the molecular entity that efficiently sulfurates (sulfhydrates) Keap1 to release Nrf2 for nuclear translocation (Koike et al. 2013).

5.3 Modification of Proteins by Sulfuration (Sulfhydration)

Modification of enzyme activities by sulfuration was studied extensively from the 1960s to the 1980s (Kato et al. 1966; Branzoli and Massey 1974; Conner and Russell 1983). Mustafa et al. (2009) defined the process of sulfhydration (sulfuration) as the reaction in which sulfur provided by H_2S attaches to reactive cysteine residues in target proteins. Sulfuration of glyceraldehyde 3-phosphate dehydrogenase (GAPDH) by H_2S increases its catalytic activity by sevenfold, and DTT treatment removes the added sulfur to diminish GAPDH activity (Mustafa et al. 2009). In contrast, nitrosylation of the same GAPDH cysteine residue abolishes the catalytic activity. Thus, sulfuration has been proposed to have the opposite effect to nitrosylation. Sulfuration of GAPDH by H_2S is supported by the observation that the activity of GAPDH is reduced by 35 % in CSE-knockout mice. Actin polymerization is also enhanced by H_2S, and this effect is reversed by DTT (Mustafa et al. 2009). ATP-dependent K^+ channels provide a further example of the regulation of protein function by sulfuration. ATP-dependent K^+ channels are sulfurated by H_2S at Cys43 on the Kir6.1 subunit; this sulfuration prevents association of the subunit with ATP, leading to activation of the K^+ channel (Mustafa et al. 2011) (Fig. 3).

NF-κB, which plays a key role in regulating immune responses, is sulfurated by H_2S to increase its interaction with ribosomal protein S3, resulting in upregulation of antiapoptotic genes (Sen et al. 2012). Stress responses are also regulated by sulfuration. Sulfuration-mediated inhibition of protein tyrosine phosphatase (PTP) enables protein kinase RNA-like ER kinase (PERK) to retain its active state and phosphorylate eukaryotic translational initiation factor 2α (elF2α), inhibiting global protein synthesis (Krishnan et al. 2011). In Parkinson's disease, which affects the substantia nigra and presents as motor abnormalities, the E3 ubiquitin ligase parkin is affected and causes accumulation of α-synuclein, a major component of Lewy bodies. In the brains of patients with Parkinson's disease, parkin activity is decreased in line with reduced activity, sulfuration of Cys95 and Cys59 in parkin is diminished, and nitrosylation of parkin, which inactivates it, is increased (Vandiver et al. 2013).

Tao et al. (2013) proposed a role for H_2S in the reduction of cysteine disulfide bonds rather than sulfuration. H_2S induces angiogenesis by promoting the migration and tube formation of vascular endothelial cells, with the effects of H_2S being mediated by VEGFR2 (Cai et al. 2007; Papapetropoulos et al. 2009). H_2S activates VEGFR2 by reducing a disulfide bond between Cys1045 and Cys1024. Mass spectrometry analysis showed that H_2S reduced a cysteine disulfide bond contained within a synthesized hexapeptide but did not sulfurate any of the 20 free amino acids, including cysteine (Tao et al. 2013). The sulfuration of cysteine residues was only transiently observed as an intermediate step during the reduction of the disulfide bond by H_2S; the intermediate was immediately attacked by HS^- and reduced to cysteine (Fig. 4). These observations suggest that H_2S may be involved in reducing reactions such as the reduction of cysteine disulfide bonds, whereas polysulfides may mediate sulfuration. Because the reducing activity of H_2S is weaker than that of cysteine and glutathione, the regulation of local concentrations of H_2S and the accessibility to the targets need to be clarified.

5.4 Glutathione Polysulfide

Massey and colleagues initially reported that glutathione persulfide (GSSH or GSS^-) produced from glutathione trisulfide (GSSSG) reduces cytochrome c more efficiently than did glutathione (GSH) and that cysteine trisulfide has a similar effect (Massey et al. 1971). Francoleon and colleagues recently confirmed the reducing effect of GSSH on cytochrome c and found that GSSH was also able to suppress the activity of papain (Francoleon et al. 2011). GSSH was also found to scavenge H_2O_2 more efficiently than GSH (Ida et al. 2014).

Ida and colleagues (2014) proposed a mechanism of cysteine persulfide (CysSSH) and glutathione polysulfide production. CBS and CSE metabolize cystine to CysSSH, which produces GSSH in an exchange reaction with GSH. However, the physiological relevance of this pathway needs to be reevaluated. Although CSE has a high affinity for cystine (with a K_m value of 30–70 μM), CSE localizes to the cytosol where cystine is not present at a sufficient concentration to react with the enzyme. Cystine is an extracellular form of cysteine and is present at less than 0.2 μM in the cytosol, with only a few exceptions (such as the A549 cell line, containing approximately 12 μM cystine) (Stipanuk 1986; Ida et al. 2014). The extracellular concentration of cystine is only 40 μM in human blood (Brigham et al. 1960). The CSE- and CBS-catalyzed production of CysSSH and polysulfides was examined with cystine present at mM concentrations, which were far higher than the measured physiological concentration (Ida et al. 2014).

6 Cross Talk of H_2S with NO

The relaxation effect of H_2S on thoracic aorta smooth muscle was weak compared to the effects of H_2S on other smooth muscles found in the ileum and the portal vein; however, a synergistic effect was observed between H_2S and NO in thoracic aorta smooth muscle (Hosoki et al. 1997). A similar synergistic effect has been observed in the twitch response of the ileum (Teague et al. 2002). Cross-talk between H_2S and NO is also present in the production and transcription of an H_2S-producing enzyme. The NO donors nitroprusside (SNP) and S-nitroso-N-acetyl-DL-penicillamine (SNAP) enhance the production of H_2S and upregulate the transcription of CSE, respectively (Zhao et al. 2001).

Lefer and colleagues recently showed that H_2S activated endothelial NO synthetase (eNOS) to increase NO production, thereby providing a mechanism for the protection of the heart from ischemia-reperfusion injury (King et al. 2014) based on six observations outlined below. (1) The levels of H_2S and bound sulfane sulfur were reduced in CSE-knockout mice compared to wild-type mice. (2) The eNOS activation site was less phosphorylated in CSE-knockout mice than in wild-type mice, whereas the eNOS inhibitory site was more greatly phosphorylated in CSE-knockout mice. (3) The levels of NO metabolites in the blood and heart were lower in CSE-knockout mice than wild-type mice. (4) Administration of H_2S activated eNOS and increased NO bioavailability in CSE-knockout mice. (5) Ischemia-reperfusion injury was exacerbated in CSE-knockout mice. (6) The administration of H_2S failed to protect against myocardial ischemia-reperfusion injury in eNOS-deficient mice. H_2S also protected against pressure overload-induced heart failure by a similar mechanism (Kondo et al. 2013). Lefer and colleagues obtained interesting data relating to the levels of cyclic guanosine monophosphate (cGMP) in plasma and in the heart. Although plasma cGMP levels were significantly lower in CSE-knockout mice than in wild-type mice, cardiac cGMP levels were not significantly different between CSE-knockout and wild-type mice (King et al. 2014). This finding suggests that plasma cGMP may play a more significant role than cardiac cGMP in protecting the heart against ischemia-reperfusion injury.

In contrast, H_2S suppressed the activity of neuronal NOS (nNOS) in submucosal as well as myenteric-ganglion neurons innervated the colon (Sha et al. 2014). Endogenously produced H_2S acts in an autocrine fashion to inhibit NO generation from nNOS in colonic circular smooth muscle cells. This knowledge comes from the observation that the endogenous generation of NO was significantly greater in muscle preparations from CSE-knockout mice than in those from wild-type mice. As H_2S exerts different NO production outcomes depending on the type of NOS expressed, a more detailed understanding of the mechanism of H_2S is required (King et al. 2014; Sha et al. 2014).

The chemical interaction between H_2S and NO produces nitrosothiol, which releases NO in the presence of Cu^{2+} (Whiteman et al. 2006). H_2S (HS^-) releases NO from the NO intermediate or carrier, nitrosoglutathione (GSNO), potentiating vascular smooth muscle relaxation (Ondrias et al. 2008). Filipovic and colleagues reported that in the presence of human umbilical vein endothelial cells or Fe^{3+}-porphyrins, nitrite and H_2S interact with each other to produce an intermediate compound, thionitrous acid (HSNO), which produces NO and $HS^\bullet$, or produces nitroxyl (HNO) by a further reaction with H_2S (Filipovic et al. 2012). In contrast, Feelisch and colleagues reported that HSNO does not effectively release NO owing to a relatively strong S-N bond (Cortese-Krott et al. 2014). Feelisch and colleagues demonstrated that nitrosopersulfide ($SSNO^-$ or HSSNO), which is more stable than HSNO, efficiently releases NO and polysulfides, resulting in more effective activation of soluble guanylyl cyclase and relaxation of smooth muscle compared to that observed with the parent nitrosothiol (Cortese-Krott et al. 2014). However, these studies were performed in the presence of high concentrations of NaHS (in the mM range). Thus, the physiological relevance of the interaction between H_2S and NO needs to be further examined.

7 Concluding Remarks

Almost two decades have passed since the first demonstration of a physiological role for H_2S, and various aspects of this molecule have been extensively studied. Although the steady-state levels of H_2S have recently been reevaluated as much lower than initially reported, the range of change in H_2S concentrations due to cellular activity and the physiological stimuli to induce such changes are not well understood. One of the approaches to address these problems is to understand the regulation of H_2S metabolic turnover (Kimura 2012; Kabil and Banerjee 2014). It has been proposed that a balance between H_2S production and degradation regulates its physiological levels. In terms of H_2S production, CBS activity is enhanced by S-adenosyl-L-methionine and glutathionylation, and the activities of CSE and CAT are regulated by Ca^{2+}; H_2S degradation is performed by mitochondrial enzymes. However, the specific physiological stimuli required to effect changes in cellular H_2S levels have not been identified; identification of these specific stimuli is required to elucidate a role of H_2S as a signaling molecule.

Polysulfides have recently emerged as candidate molecules for the physiological mediators of H_2S signaling. Polysulfides activate channels, transcription factors, and enzymes by sulfuration (sulfhydration) of these target molecules; polysulfides also have more potent antioxidant effects than H_2S (Greiner et al. 2013; Kimura et al. 2013; Koike et al. 2013; Ida et al. 2014). Polysulfides may be produced by both enzymatic and nonenzymatic pathways. Although a pathway for the enzymatic production of polysulfides has been proposed, the concentrations of substrate required are well outside the physiologically relevant range (Ida et al. 2014). It is also necessary to clarify how nonenzymatic production of polysulfides is regulated.

The cross-talk between H_2S and NO has also been extensively studied. A synergistic effect between both substances was initially reported, and subsequently, the transcriptional regulation of H_2S-producing enzymes by NO was identified (Hosoki et al. 1997; Zhao et al. 2001; Teague et al. 2002). Recent studies demonstrated the regulation of NO production by H_2S through the activation or inhibition of NOS isoforms (Kondo et al. 2013; Sha et al. 2014). Further study is necessary to clarify the mechanisms underlying the varying effects of H_2S on different NOS isoforms and tissue types. The chemical interaction between H_2S and NO produces HSNO, HNO, or HSSNO, and the rate of production was changed in the presence of cells (Filipovic et al. 2012; Cortese-Krott et al. 2014). Because these studies were performed in the presence of high concentrations of substrates, the interaction between H_2S and NO must be reexamined under physiological conditions. It is also necessary to clarify whether H_2S and NO are present in cells at the same time and if enzymes producing these molecules are activated under the same conditions.

By understanding the biochemical nature of these molecules, as well as their mechanisms of action and regulation, the physiological function and therapeutic potential of H_2S and related molecules will be unveiled.

Acknowledgements This work was supported by a grant from the National Institute of Neuroscience and a KAKENHI (23659089) Grant-in-Aid for Challenging Exploratory Research and KAKENHI (26460115) Grant-in-Aid for Scientific Research to H.K.

Author Disclosure Statement There is no conflict of interest.

References

Abe K, Kimura H (1996) The possible role of hydrogen sulfide as an endogenous neuromodulator. J Neurosci 16:1066–1071

Aizenman E, Lipton DA, Loring RH (1989) Selective modulation of NMDA responses by reduction and oxidation. Neuron 2:1257–1263

Akagi R (1982) Purification and characterization of cysteine aminotransferase from rat liver cytosol. Acta Med Okayama 36:187–197

Araque A, Parpura V, Sanzgiri RP, Haydon PG (1999) Tripartite synapses: glia, the unacknowledged partner. Trends Neurosci 22:208–215

Bhatia M, Sidhapuriwala J, Moochhala SM, Moore PK (2005) Hydrogen sulphide is a mediator of carrageenan-induced hindpaw oedema in the rat. Br J Pharmacol 145:141–144

Branzoli U, Massey V (1974) Evidence for an active site persulfide residue in rabbit liver aldehyde oxidase. J Biol Chem 249:4346–4349

Braunstein AE, Goryachenkowa EV, Tolosa EA, Willhardt IH, Yefremova LL (1971) Specificity and some other properties of liver serine sulphhydrase:evidence for its identity with cystathionineβ-synthase. Biochim Biophys Acta 242:247–260

Brigham MP, Stein WH, Moore S (1960) The concentrations of cysteine and cystine in human blood plasma. J Clin Invest 39:1633–1638

Cai WJ, Wang MJ, Moore PK, Jin HM, Yao T, Zhu YC (2007) The novel proangiogenic effect of hydrogen sulfide is dependent on Akt phosphorylation. Cardiovasc Res 76:29–40

Calvert JW, Jha S, Gundewar S, Elrod JW, Ramachandran A, Pattillo CB, Kevil CG, Lefer DJ (2009) Hydrogen sulfide mediates cardioprotection through Nrf2 signaling. Circ Res 105:365–374

Cavallini D, Mondovi B, De Marco C, Scioscia-Santoro A (1962) The mechanism of desulphhydration of cysteine. Enzymologia 24:253–266

Chattopadhyay M, Kodela R, Nath N, Dastagirzqada YM, Velazquez-Martinez CA, Boring D, Kashfi K (2012) Hydrogen sulfide-reasing NSAIDs inhibit the growth of human cancer cells: a general property and evidence of a tissue type-independent effect. Biochem Pharmacol 83:715–722

Chiku T, Padovani D, Zhu W, Singh S, Vitvitsky V, Banerjee R (2009) H_2S biogenesis by human cystathionine γ–lyase leads to the novel sulfur metabolites lanthionine and homolanthionine and is responsive to the grade of hyperhomocysteinemia. J Biol Chem 284:11601–11612

Conner J, Russell PJ (1983) Elemental sulfur: a novel inhibitor of adenylate kinase. Biochem Biophys Res Commun 113:348–352

Cooper AJL (1983) Biochemistry of sulfur-containing amino acids. Annu Rev Biochem 52:187–222

Cortese-Krott MM, Fernandez BO, Santos JLT, Mergia E, Grman M, Nagy P, Kelm M, Butler A, Feelisch M (2014) Nitrosopersulfide ($SSNO^-$) accounts for sustained NO bioactivity of S-nitrosothiols following reaction with sulfide. Redox Biol 2:234–244

Doyle JM, Schinina ME, Bossa F, Doonan S (1990) The amino acid sequence of cytosolic aspartate aminotransferase from humahn liver. Biochem J 270:651–657

Elrod JW, Calvert JW, Morrison J, Doeller JE, Kraus DW, Tao L, Jiao X, Scalia R, Kiss L, Szabo C, Kimura H, Chow CW, Lefer DJ (2007) Hydrogen sulfide attenuates myocardial ischemia-reperfusion injury by preservation of mitochondrial function. Proc Natl Acad Sci USA 104:15560–15565

Filipovic MR, Miljkovic JL, Nauser T, Royzen M, Klos K, Shubina T, Koppenol WH, Lippard SJ, Ivanovic-Burmazovic I (2012) Chemical characterization of the smallest S-nitrosothiol, HSNO; cellular cross-talk of H_2S and S-nitrosothiols. J Am Chem Soc 134:12016–12027

Francoleon NE, Carrington SJ, Fukuto JM (2011) The reaction of H_2S with oxidized thiols: generation of persulfides and implications to H_2S biology. Arch Biochem Biophys 516:146–153

Furne J, Saeed A, Levitt MD (2008) Whole tissue hydrogen sulfide concentrations are orders of magnitude lower than presently accepted values. Am J Physiol Regul Integr Comp Physiol 295:R1479–R1498

Goodwin LR, Francom D, Dieken FP, Taylor JD, Warenycia MW, Reiffenstein RJ, Dowling G (1989) Determination of sulfide in brain tissue by gas dialysis/ion chromatography: postmortem studies and two case reports. J Anal Toxicol 13:105–109

Greiner R, Palinkas Z, Basell K, Becher D, Antelmann H, Nagy P, Dick TP (2013) Polysulfides link H_2S to protein thiol oxidation. Antioxid Redox Signal 19:1749–1765

Griffith OW (1999) Biologic and pharmacologic regulation of mammalian glutathione synthesis. Free Radic Biol Med 27:922–935

Hosoki R, Matsuki N, Kimura H (1997) The possible role of hydrogen sulfide as an endogenous smooth muscle relaxant in synergy with nitric oxide. Biochem Biophys Res Commun 237:527–531

Ida T, Sawa T, Ihara H, Tsuchiya Y, Watanabe Y, Kumagai Y, Suematsu M, Motohashi H, Fujii S, Matsunaga T, Yamamoto M, Ono K, Davarie-Baez NO, Xian M, Fukuto JM, Akaike T (2014) Reactive cysteine persulfides and S-polythiolation regulate oxidative stress and redox signaling. Proc Natl Acad Sci USA 111:7606–7611

Ishigami M, Hiraki K, Umemura K, Ogasawara Y, Ishii K, Kimura H (2009) A source of hydrogen sulfide and a mechanism of its release in the brain. Antioxid Redox Signal 11:205–214

Ishii I, Akahoshi N, Yu X-N, Kobayashi Y, Namekata K, Komaki G, Kimura H (2004) Murine cystathionine γ–lyase: complete cDNA and genomic sequences, promoter activity, tissue distribution and developmental expression. Biochem J 381:113–123

Kabil O, Banerjee R (2014) Enzymology of H_2S biogenesis, decay and signaling. Antioxid Redox Signal 20:770–782

Kaneko Y, Kimura T, Taniguchi S, Souma M, Kojima Y, Kimura Y, Kimura H, Niki I (2009) Glucose-induced production of hydrogen sulfide may protect the pancreatic beta-cells from apoptotic cell death by high glucose. FEBS Lett 583:377–382

Kato A, Ogura M, Suda M (1966) Control mechanism in the rat liver enzyme system converting L-methionine to L-cystine. 3. Noncompetitive inhibition of cystathionine synthetase-serine dehydratase by elemental sulfur and competitive inhibition of cystathionine-homoserine dehydratase by L-cysteine and L-cystine. J Biochem 59:40–48

Kimura H (2012) Metabolic turnover of hydrogen sulfide. Front Physiol 3, 101

Kimura H (2014) Signaling molecules: hydrogen sulfide and polysulfide. Antioxid Redox Signal 22(5):362–376

Kimura Y, Kimura H (2004) Hydrogen sulfide protects neurons from oxidative stress. FASEB J 18:1165–1167

Kimura Y, Goto Y-I, Kimura H (2010) Hydrogen sulfide increases glutathione production and suppresses oxidative stress in mitochondria. Antioxid Redox Signal 12:1–13

Kimura Y, Mikami Y, Osumi K, Tsugane M, Oka J-I, Kimura H (2013) Polysulfides are possible H_2S-derived signaling molecules in rat brain. FASEB J 27:2451–2457

King AL, Polhemus D, Bhushan S, Otsuka H, Kondo K, Nicholson CK, Bradley JM, Islam KN, Calvert JW, Tao Y-X, Dugas TR, Kelley EE, Elrod JW, Huang PL, Wang R, Lefer DJ (2014) Hydrogen sulfide cytoprotective signaling is endothelial nitric oxide synthase-nitric oxide dependent. Proc Natl Acad Sci USA 111:3182–3187

Koike S, Ogasawara Y, Shibuya N, Kimura H, Ishii K (2013) Polysulfide exerts a protective effect against cytotoxicity cuased by t-buthylhydroperoxide through Nrf2 signaling in neuroblastoma cells. FEBS Lett 587:3548–3555

Kondo K, Bhushan S, King AL, Prabhu SD, Hamid T, Koenig S, Murohara T, Predmore BL, Gojon GS, Gojon GJ, Wang R, Karusula N, Nicholson CK, Calvert JW, Lefer DJ (2013) H_2S protects against pressure overload-induced heart failure via upregulation of endothelial nitric oxide synthase. Circulation 127:1116–1127

Krishnan N, Fu C, Pappin DJ, Tonks NK (2011) H_2S-induced sulfhydration of the phosphatase PTP1B and its role in the endoplasmic reticulum stress response. Science Sig 4:ra86

Levitt MD, Abdel-Rehim MS, Furne J (2011) Free and acid-labile hydrogen sulfide concentrations in mouse tissues:Anomalously high free hydrogen sulfide in aortic tissue. Antioxid Redox Signal 15:373–378

Massey V, Williams CH, Palmer G (1971) The presence of S0-containing impurities in commercial samples of oxidized glutathione and their catalytic effect in the reduction of cytochrome c. Biochem Biophys Res Commun 42:730–738

Meister A, Fraser PE, Tice SV (1954) Enzymatic desulfuration of β-mercaptopyruvate to pyruvate. J Biol Chem 206:561–575

Mikami Y, Shibuya N, Kimura Y, Nagahara N, Ogasawara Y, Kimura H (2011a) Thioredoxin and dihydrolipoic acid are required for 3-mercaptopyruvate sulfurtransferase to produce hydrogen sulfide. Biochem J 439:479–485

Mikami Y, Shibuya N, Kimura Y, Nagahara N, Yamada M, Kimura H (2011b) Hydrogen sulfide protects the retina from light-induced degeneration by the modulation of Ca^{2+} influx. J Biol Chem 286:39379–39386

Mikami Y, Shibuya N, Ogasawara Y, Kimura H (2013) Hydrogen sulfide is produced by cystathionine γ–lyase at the steady-state low intracellular Ca^{2+} concentrations. Biochem Biophys Res Commun 431:131–135

Morikawa T, Kajimura M, Nakamura T, Hishiki T, Nakanishi T, Yukutake Y, Nagahata Y, Ishikawa M, Hattori K, Takenouchi T, Takahashi T, Ishii I, Matsubara K, Kabe Y, Uchiyama S, Nagata E, Gadalla MM, Snyder SH, Suematsu M (2012) Hypoxic regulation of the cerebral microcirculation is mediated by a carbon monoxide-sensitive hydrogen sulfide pathway. Proc Natl Acad Sci U S A 109:1293–1298

Mustafa AK, Gadalla MM, Sen N, Kim S, Mu W, Gazi SK, Barrow RK, Yang G, Wang R, Snyder SH (2009) H_2S signals through protein S-sulfhydration. Sci Signal 2:ra72

Mustafa AK, Sikka G, Gazi SK, Steppan J, Jung SM, Bhunia AK, Barodka VM, Gazi FK, Barrow RK, Wang R, Amzel LM, Berkowitz DE, Snyder SH (2011) Hydrogen sulfide as endothelium-derived hyperpolarizing factor sulfhydrates potassium channels. Circ Res 109:1259–1268

Nagahara N, Ito T, Kitamura H, Nishino T (1998) Tissue and subcellular distribution of mercapto-pyruvate sulfurtransferase in the rat: confocal laser fluorescence and immunoelectron microscopic studies combined with biochemical analysis. Histochem Cell Biol 110:243–250

Nagai Y, Tsugane M, Oka J, Kimura H (2004) Hydrogen sulfide induces calcium waves in astrocytes. FASEB J 18:557–559

Nagai Y, Tsugane M, Oka J-I, Kimura H (2006) Polysulfides induce calcium waves in rat hippocampal astrocytes. J Pharmacol Sci 100:200

Nagy P, Winterbourn CC (2010) Rapid reaction of hydrogen sulfide with the neutrophil oxidant hypochlorous acid to generate polysulfides. Chem Res Toxicol 23:1541–1543

Niu W-N, Yadav PK, Adamec J, Banerjee R (2014) S-Glutathionylation enhances human cystathionine β–synthase activity under oxidative stress conditions. Antioxid Red Signal 22 (5):350–356

Ogawa H, Takahashi K, Miura S, Imagawa T, Saito S, Tominaga M, Ohta T (2012) H_2S functions as a nociceptive messenger throughtransient receptor potential ankyrin 1 (TRPA1) activation. Neuroscience 218:335–343

Olson KR, Dombkowski RA, Russell MJ, Doellman MM, Head SK, Whitfield NL, Madden JA (2006) Hydrogen sulfide as an oxygen sensor/transducer in vertebrate hypoxic vasoconstriction and hypoxic vasodilation. J Exp Biol 209:4011–4023

Ondrias K, Stasko A, Cacanyiova S, Sulova Z, Krizanova O, Kristek F, Malekova L, Knezl V, Breier A (2008) H2S and HS- donor NaHS releases nitric oxide from nitrosothiols, metal nitrosyl complex, brain homogenate and murine L1210 leukaemia cells. Pflugers Arch Eur J Physiol 457:271–279

Oosumi K, Tsugane M, Ishigami M, Nagai Y, Iwai T, Oka J-I, Kimura H (2010) Polysulfide activates TRP channels and increases intracellular Ca^{2+} in astrocytes. Bull Jpn Soc Neurochem 49:517

Papapetropoulos A, Pyriochou A, Altaany Z, Yang G, Marazioti A, Zhou Z, Jeschke MG, Branski LK, Herndon DN, Wang R, Szabo C (2009) Hydrogen sulfide is an endogenous stimulator of angiogenesis. Proc Natl Acad Sci USA 106:21972–21977

Peng YJ, Nanduri J, Raghuraman G, Souvannakitti D, Gadalla MM, Kumar GK, Snyder SH, Prabhakar NR (2010) H_2S mediates O_2 sensing in the carotid body. Proc Natl Acad Sci USA 107:10719–10724

Polhemus DJ, Kondo K, Bhushan S, Bir SC, Kevil CG, Murohara T, Lefer DJ, Calvert JW (2013) Hydrogen sulfide attenuates cardiac dysfunction after heart failure via induction of angiogenesis. Circ Heart Fail 6:1077–1086

Savage JC, Gould DH (1990) Determination of sulfide in brain tissue and rumen fluid by ion-interaction reversed-phase high-performance liquid chromatography. J Chromatogr 526:540–545

Schumann U, Subramani S (2008) Special delivery from mitochondria to peroxisomes. Trends Cell Biol 18:253–256

Sen N, Paul BD, Gadalla MM, Mustafa AK, Sen T, Xu R, Kim S, Snyder SH (2012) Hydroen sulfide-linked sulfhydration of NF-κB mediates its intiapoptotic actions. Mol Cell 45:13–24

Sha L, Linden DR, Farrugia G, Szurszewski JH (2014) Effect of endogenous hydrogen sulfide on the transwall gradient of the mouse colon circular smooth muscle. J Physiol 592:1077–1089

Shan X, Dunbrack RLJ, Christopher SA, Kruger WD (2001) Mutation in the regulatory domain of cystathionine β–synthase can functionally suppress patient-derived mutations in cis. Hum Mol Genet 10:635–643

Shibuya N, Tanaka M, Yoshida M, Ogasawara Y, Togawa T, Ishii K, Kimura H (2009) 3-mercaptopyruvate sulfurtransferease produces hydrogen sulfide and bound sulfane sulfur in the brain. Antioxid Redox Signal 11:703–714

Shibuya N, Koike S, Tanaka M, Ishigami-Yuasa M, Kimura Y, Ogasawara Y, Fukui K, Nagahara N, Kimura H (2013) A novel pathway for the production of hydrogen sulfide from D-cysteine in mammalian cells. Nat Commun 4:1366

Shigetomi E, Tong X, Kwan KY, Corey DP, Khakh BS (2012) TRPA1 channels regulate astrocyte resting calcium and inhibitory synapse efficacy through GAT-3. Nat Neurosci 15:70–80

Shigetomi E, Jackson-Weaver O, Huckstepp RT, O'Dell TJ, Khakh BS (2013) TRPA1 channels are regulators of astrocyte basal calcium levels and long-term potentiation via constitutive D-serine release. J Neurosci 33:10143–10153

Singh S, Padovani D, Leslie RA, Chiku T, Banerjee R (2009) Relative contributions of cystathionine beta-synthase and gamma-cystathionase to H_2S biogenesis via alternative trans-sulfuration reactions. J Biol Chem 284:22457–22466

Stipanuk MH (1986) Metabolism of sulfur-containing amino acids. Ann Rev Nutr 6:179–209

Stipanuk MH, Londono M, Lee JI, Hu M, Yu AF (2002) Enzymes and metabolites of cysteine metabolism in nonhepatic tissues of rats show little response to changes in dietary protein or sulfur amino acid levels. J Nutr 132:3369–3378

Streng T, Axelsson HE, Hedlund P, Andersson DA, Jordt SE, Bevan S, Andersson KE, Hogestatt ED, Zygmunt PM (2008) Distribution and function of the hydrogen sulfide-sensitive TRPA1 ion channel in rat urinary bladder. Eur Urol 53:391–399

Szabo C, Coletta C, Chao C, Modis K, Szczesny B, Papapetropoulos A, Hellmich MR (2013) Tumor-derived hydrogen sulfide, produced by cystathionine –synthase, stimulates bioenergetics, cell proliferation, and angiogenesis in colon cancer. Proc Natl Acad Sci USA 110:12474–12479

Taniguchi T, Kimura T (1974) Role of 3-mercaptopyruvate sulfurtransferase in the formation of the iron-sulfur chromophore of adrenal ferredoxin. Biochim Biophys Acta 364:284–295

Taoka S, Banerjee R (2001) Characterization of NO binding to human cystathionine beta-synthase: possible implications of the effects of CO and NO binding to the human enzyme. J Inorg Biochem 87:245–251

Tateishi N, Higashi T, Naruse A, Nakashima K, Shiozaki H (1977) Rat liver glutathione: possible role as a reservoir of cysteine. J Nutr 107:51–60

Teague B, Asiedu S, Moore PK (2002) The smooth muscle relaxant effect of hydrogen sulphide in vitro: evidence for a physiological role to control intestinal contractility. Br J Pharmacol 137:139–145

Tao BB, Liu SY, Zhang CC, Fu W, Cai WJ, Wang Y, Shen Q, Wang MJ, Chen Y, Zhang LJ, Zhu YZ (2013) VEGFR2 functions as an H2S-targeting receptor protein kinase with its novel Cys1045-Cys1024 disulfide bond serving as a specific molecular switch for hydrogen sulfide actions in vascular endothelial cells. Antioxid Redox Signal 19:448–464

Toohey JI (2011) Sulfur signaling: is the agent sulfide or sulfane? Anal Biochem 413:1–7

Tripatara P, Patel NSA, Collino M, Gallicchio M, Kieswich J, Castiglia S, Benetti E, Stewart KN, Brown PAJ, Yaqoob MM, Fantozzi R, Thiemermann C (2008) Generation of endogenous hydrogen sulfide by cystathionine γ–lyase limits renal ischemia/reperfusion injury and dysfunction. Lab Invest 88:1038–1048

Ubuka T, Umemura S, Yuasa S, Kinuta M, Watanabe K (1978) Purification and characterization of mitochondrial cysteine aminotransferase from rat liver. Physiol Chem Phys 10:483–500

Vandiver MS, Paul BD, Xu R, Karuppagounder S, Rao F, Snowman AM, Ko HS, Lee YII, Dawson VL, Dawson TM, Sen N, Snyder SH (2013) Sulfhydration mediates neuroprotective actions of parkin. Nat Commun 4:1626

Wakabayashi N, Dinkova-Kostova AT, Holtzclaw WD, Kang MI, Kobayashi A, Yamamoto M, Kensler TW, Talalay P (2004) Protection against electrophile and oxidant stress by induction of the phase 2 response:fate of cysteines of the Keap1 sensor modified by inducers. Proc Natl Acad Sci USA 101:2040–2045

Warenycia MW, Goodwin LR, Benishin CG, Reiffenstein RJ, Grancom DM, Taylor JD, Dieken FP (1989) Acute hydrogen sulfide poisoning. Demonstration of selective uptake of sulfide by the brainstem by measurement of brain sulfide levels. Biochem Pharmacol 38:973–981

Whiteman M, Armstrong JS, Chu SH, Jia-Ling S, Wong BS, Hheung NS, Halliwell B, Moore PK (2004) The novel neuromodulator hydrogen sulfide: an endogenous peroxynitrite 'scavenger'? J Neurochem 90:765–768

Whiteman M, Li L, Kostetski I, Chu SH, Siau JL, Bhatia M, Moore PK (2006) Evidence for the formation of a novel nitrosothiol from the gaseous mediators nitric oxide and hydrogen sulphide. Biochem Biophys Res Commun 343:303–310

Wintner EA, Deckwerth TL, Langston W, Bengtsson A, Leviten D, Hill P, Insko MA, Dumpit R, VandenEkart E, Toombs CF, Szabo C (2010) A monobromobimane-based assay to measure the pharmacokinetic profile of reactive sulphide species in blood. Br J Pharmacol 160:941–957

Yadav PK, Yamada K, Chiku T, Koutmos M, Banerjee R (2013) Structure and kinetic analysis of H2S production by human mercaptopyruvate sulfurtransferase. J Biol Chem 288:20002–20013

Yang G, Wu L, Jiang B, Yang W, Qi J, Cao K, Meng Q, Mustafa AK, Mu W, Zhang S, Snyder SH, Wang R (2008) H_2S as a physiologic vasorelaxant: hypertension in mice with deletion of cystathionine γ-lyase. Science 322:587–590

Yang G, Zhao K, Ju Y, Mani S, Cao Q, Puukila S, Khaper N, Wu L, Wang R (2013) Hydrogen sulfide protects against cellular senescence via S-sulfhydration of Keap1 and activation of Nrf2. Antioxid Redox Signal 18:1906–1919

Zanardo RCO, Brancaleone V, Distrutti E, Fiorucci S, Cirino G, Wallace JL (2006) Hydrogen sulphide is an endogenous modulator of leukocyte-mediated inflammation. FASEB J 20:2118–2120

Zhao W, Zhang J, Lu Y, Wang R (2001) The vasorelaxant effect of H_2S as a novel endogenous gaseous K_{ATP} channel opener. EMBO J 20:6008–6016

Part III

Cardiovascular and Urogenital Systems

H_2S and Blood Vessels: An Overview

Guangdong Yang and Rui Wang

Contents

G. Yang
Cardiovascular and Metabolic Research Unit, Lakehead University, Thunder Bay, ON, Canada

School of Kinesiology, Lakehead University, Thunder Bay, ON, Canada

R. Wang (✉)
Cardiovascular and Metabolic Research Unit, Lakehead University, Thunder Bay, ON, Canada

Department of Biology, Lakehead University, Thunder Bay, ON, Canada
e-mail: rwang@lakeheadu.ca

P.K. Moore, M. Whiteman (eds.), *Chemistry, Biochemistry and Pharmacology of Hydrogen Sulfide*, Handbook of Experimental Pharmacology 230,
DOI 10.1007/978-3-319-18144-8_4

Abstract

The physiological and biomedical importance of hydrogen sulfide (H_2S) has been fully recognized in the cardiovascular system as well as in the rest of the body. In blood vessels, cystathionine γ-lyase (CSE) is a major H_2S-producing enzyme expressed in both smooth muscle and endothelium as well as periadventitial adipose tissues. Regulation of H_2S production from CSE is controlled by a complex integration of transcriptional, posttranscriptional, and posttranslational mechanisms in blood vessels. In smooth muscle cells, H_2S regulates cell apoptosis, phenotypic switch, relaxation and contraction, and calcification. In endothelial cells, H_2S controls cell proliferation, cellular senescence, oxidative stress, inflammation, etc. H_2S interacts with nitric oxide and acts as an endothelium-derived relaxing factor and an endothelium-derived hyperpolarizing factor. H_2S generated from periadventitial adipose tissues acts as an adipocyte-derived relaxing factor and modulates the vascular tone. Extensive evidence has demonstrated the beneficial roles of the CSE/H_2S system in various blood vessel diseases, such as hypertension, atherosclerosis, and aortic aneurysm. The important roles signaling in the cardiovascular system merit further intensive and extensive investigation. H_2S-releasing agents and CSE activators will find their great applications in the prevention and treatment of blood vessel-related disorders.

Keywords

Hydrogen sulfide • Cystathionine γ-lyase • Blood vessel • Smooth muscle cells • Endothelial cells • Periadventitial adipose tissues • Blood vessel-related disorders

1 Hydrogen Sulfide Is a Gasotransmitter

The physiological and biomedical importance of hydrogen sulfide (H_2S) has been fully recognized in the cardiovascular system as well as in the rest of the body. The gasotransmitter identity of H_2S is validated against six criteria of gasotransmitters (Wang 2002, 2014):

1. H_2S is a small gas molecule. Once produced inside eukaryotes, it is partially dissolved in biological milieu in its free or bounded forms and partially dissociated to HS^-. It is occasionally squabbled over whether H_2S or nitric oxide (NO) in eukaryotes is still *gas*. Oxygen in the blood is a gas and this does not need verification by seeing oxygen gas bubbles. This is the same case with gasotransmitters. NO, carbon monoxide (CO), H_2S, and NH_3 are gas molecules in our bodies.
2. H_2S is freely permeable to the plasma membrane and intracellular organelle membranes. The same signal relay sequence for neurotransmitters is not required for H_2S signal since there will be no synaptic barrier or cognate

membrane receptors for the transmembrane movement of this gas molecule. The idea of a "gas channel" has been around for a while. No experimental evidence exists to date, however, which shows the reliance of transmembrane movement of H_2S on such "channels". Furthermore, one has to distinguish the transmembrane movement of undissociated gas molecule from that of its dissociated ionic forms. Whereas ammonia gas (NH_3) freely permeates cell membranes, ammonium ion (NH_4^+) passes the membrane through ion channels.

3. H_2S is endogenously generated in mammalian cells under both physiological and pathophysiological conditions. L-cysteine and homocysteine are the most important substrates of enzymatic H_2S production. Yes. H_2S is a metabolic product of reverse-transsulfuration pathway. But the production of H_2S is more than a metabolic need, more than a sulfur transfer phenomenon, and more than the degradation of cysteine or homocysteine. It interacts with different protein targets, alters the directions of multiple signaling pathways, and forms signaling webs and nets intracellularly and intercellularly. And H_2S does all these in response to the functional regulation as well as the metabolic needs of the body.
4. H_2S has well-defined specific functions at physiologically relevant concentrations. The effects of H_2S on the cardiovascular system, neuronal system, respiratory system, and gastrointestinal system, to name a few, have been extensively and convincingly documented. These effects of H_2S are realized at its physiologically relevant concentrations because decreasing endogenous H_2S level (the knockout or knockdown of H_2S-generating enzymes from the whole animal to tissue to cell levels) or increasing it (overexpression or knock-in of these enzymes) has been clearly correlated to the correspondingly functional changes in different systems.
5. The application of H_2S donors (fast releasing or slow releasing) has shown the similar effects as endogenous H_2S on different biological systems.
6. H_2S is involved in signal transduction and has specific cellular and molecular targets. It induces *S*-sulfhydration of numerous proteins. It regulates the levels and activities of traditional second messengers, such as cGMP, cAMP, and intracellular calcium. Its action and production are sensitive to cellular oxygen levels. Mitochondrial bioenergetics, endoplasmic reticulum stress, and gene transcription and translation in the nucleus are within the impact radius of H_2S.

It is relevant here to comment on the contextual connection of neurotransmitter and gasotransmitter to the conventional denotation of "transmitter". A biological transmitter usually refers to a biological molecule that is generated in response to the homeostatic need and serves for "communication" to meet this need. Neurotransmitter and gasotransmitter both conduct the "communication" between the homeostatic needs and functional changes. The former does this via second messengers, and the latter directly interacts with its downstream signaling webs and nets.

2 Endogenous Production of H_2S in Blood Vessels

Different enzymes are involved in the production of H_2S from both vascular smooth muscle cells (SMCs) and endothelial cells (ECs). Cystathionine β-synthase (CBS) is critical for the transsulfuration of homocysteine to generate cystathionine and then to H_2S. Similar to CBS, cystathionine γ-lyase (CSE) is also a pyridoxal-5′-phosphate-dependent H_2S-generating enzyme. In the cardiovascular system, the transformation of L-cysteine to H_2S is mainly catalyzed by CSE with ammonium and pyruvate as two coproducts.

CSE gene in the cardiovascular system was cloned for the first time in 2001 (Zhao et al. 2001). This vascular CSE gene cloned from rat mesenteric artery tissues (GenBank #AB052882) shares the same sequence with that of rat liver CSE gene (GenBank #AY032875). Hosoki et al. (1997) detected CSE mRNA in rat thoracic aorta and portal vein (Hosoki et al. 1997). Zhao et al. showed CSE mRNA in rat mesenteric artery, tail artery, and pulmonary arteries (Zhao et al. 2001). The first Western blot study on CSE protein expression in the cardiovascular system was reported in 2006 (Yang et al. 2006) when CSE proteins were observed in human aorta vascular SMCs. In 2008, the expression of CSE proteins in vascular endothelium was reported (Yang et al. 2008). Following studies have demonstrated endothelial expression of CSE in mice, rats, and humans (Altaany et al. 2013; Papapetropoulosa et al. 2009). To date, CSE proteins and/or mRNA have been detected in newborn pig cerebral microvessels (Leffler et al. 2011), mouse renal artery SMCs (Sen et al. 2012b), rat pulmonary artery SMCs and ECs (Sun et al. 2011; Chen et al. 2009), mouse pulmonary artery SMCs and ECs (Wang et al. 2011), human pulmonary artery SMCs (Kiss et al. 2008), bovine pulmonary artery SMCs and sea lion resistance pulmonary arteries (Olson et al. 2010), rat hepatic artery and portal vein (Siebert et al. 2008), and human internal mammary artery (Webb et al. 2008).

The detection of CBS proteins in blood vessels has been reported in hepatic artery and pulmonary artery. CBS proteins were observed in rat pulmonary artery rings (Sun et al. 2011) and bovine pulmonary artery endothelial cells (Olson et al. 2010). In rat hepatic artery and portal vein, both CSE and CBS proteins were localized, but interestingly the terminal branches of the hepatic afferent vessels only expressed CSE (Siebert et al. 2008).

Cysteine aminotransferase (CAT) and 3-mercaptopyruvate sulfurtransferase (MST) are other two enzymes involved in H_2S production in the cardiovascular system. Whereas CAT uses PLP as its cofactor, zinc is the cofactor of MST. The sequential reactions catalyzed by CAT and MST lead to the transformation of cysteine to 3-mercaptopyruvate (3-MP) to sulfane sulfur. This bound sulfur will need to be released or reduced to free H_2S (Wang 2012b). CAT protein was found in vascular ECs, and MST protein was localized in both ECs and SMCs of rat thoracic aortae (Shibuya et al. 2009). MST protein was also localized in bovine pulmonary artery SMCs and sea lion resistance pulmonary arteries (Olson et al. 2011).

3 Regulation of CSE-Mediated Vascular Production of H_2S

H_2S production through CSE is both tonic under resting conditions and phasic upon specific stimulations. The activation of muscarinic cholinergic receptor by acetylcholine in vascular ECs leads to the elevation of intracellular calcium. Subsequently, calcium-activated calmodulin stimulates CSE to produce H_2S in ECs (Yang et al. 2008). Testosterone is another endogenous CSE stimulator in vascular system (Bucci et al. 2009). Its vasorelaxant effect on rat aortic rings in vitro was inhibited by DL-propargylglycine (PPG) and β-cyano-L-alanine (BCA), two inhibitors of CSE. As well, H_2S production was increased by testosterone but inhibited by PPG and BCA (Bucci et al. 2009).

It has been shown that vascular endothelial growth factor (VEGF) stimulated H_2S production in cultured human umbilical vein endothelial cells (HUVECs) (Papapetropoulos et al. 2009). This effect of VEGF can be attributed to increased intracellular calcium and calcium-activated calmodulin in ECs. Whether VEGF affects CSE gene expression is unknown. Hassan et al. showed that platelet-derived growth factor-BB (PDGF-BB) upregulated CSE mRNA and protein levels in mesangial cells (Hassan et al. 2012). Hypoxia is another factor that leads to increased H_2S production in human placenta and rat liver, uterus, and fetal membranes (Patel et al. 2009). CSE gene expression in the vascular system may also be inhibited by insulin (Wang 2004).

CSE activities in vascular SMCs were increased by NO (Zhao et al. 2001). After incubating rat aortic tissue homogenates with a NO donor for 90 min, H_2S generation from the homogenates was significantly increased. One of the underlying mechanisms for this effect of NO is the upregulation of CSE expression, which was confirmed 6 h after incubating cultured vascular SMCs with a NO donor (Zhao et al. 2001). Similar observations were made by Patel et al. that NO donors increased both the expression and activity of CSE proteins in rat fetal membranes (Patel et al. 2009).

The regulation of CSE gene expression mostly occurs at the CSE promotor site. The specific protein 1 (Sp1) transcription factor, nuclear factor erythroid-2-related factor-2 (Nrf2), and farnesoid X receptor (FXR) ligand can all bind to the CSE promotor, hence stimulating CSE transcription (Zhang et al. 2011; Yang et al. 2011; Hassan et al. 2012; Renga et al. 2009). This transcriptional regulation mechanism can explain the inhibitory effect of microRNA 21 (miR-21) on CSE expression since miR-21 directly repressed the expression of Sp1 (Yang et al. 2012a, b). In contrast, increased CSE expression by TNF-α resulted from TNF-α-stimulated Sp1 binding to the CSE promoter (Sen et al. 2012a). In the case of Nrf2, it binds to an antioxidant-responsive element (ARE) to mediate the targeted gene transcription after Nrf2 is translocated into the nucleus. The CSE promotor contains an ARE sequence and this provides an oxidative stress-sensitive mechanism for regulating CSE expression. Hassan et al. showed that PDGF-BB-induced CSE expression in mesangial cells is inhibited by co-treatment with antioxidants or by Nrf2 knockout. Furthermore, stabilization of Nrf2 protein upregulated CSE protein expression (Hassan et al. 2012). The evidence for the interaction of FXR with the CSE promotor was derived from HepG2 cells. The human CSE gene contains an FXR-responsive element in its 5′-flanking region. Treatment of HepG2 cells with an FXR ligand increased CSE expression and mutation of FXR blocks FXR ligand-induced CSE expression (Renga et al. 2009).

4 H_2S and SMCs

4.1 H_2S and SMC Apoptosis, Proliferation, and Migration

Abnormal SMC proliferation and apoptosis are among the causative factors for vascular remodeling. H_2S regulation of SMC proliferation and apoptosis has been extensively studied (Fig. 1). We previously demonstrated that exogenously applied H_2S or endogenous H_2S derived from overexpressed CSE gene inhibits proliferation and induces apoptosis of human aorta SMCs by activating ERK and caspase 3 (Yang et al. 2004, 2006). Du et al. also found that H_2S dose-dependently suppressed the proliferation of rat SMCs through the MAPK pathway (Du et al. 2004). Region-specific chromatin remodeling of MAPK signaling pathway-associated genes such as Ntf3, Pcna, and Pdgfα can be regulated by H_2S, and Brg1 acts as a switch to turn these genes "on" in a spatially and temporally specific manner to inhibit SMC proliferation (Li et al. 2013). The hypoxia-induced proliferation of pulmonary artery SMCs is the main cause of pulmonary arterial hypertension, and H_2S was demonstrated to inhibit $CoCl_2$-induced pulmonary arterial SMC proliferation by the upregulation of cyclooxygenase-2 and prostacyclin (Li et al. 2014). Indeed, SMCs from CSE knockout mice displayed an increased proliferation rate in vitro and in vivo, and these cells were more susceptible to apoptosis induced by an oxidative stress inducer (H_2O_2) or a high dose of H_2S (100 uM) (Yang et al. 2010). CSE knockout mice exhibited decreased endogenous H_2S level in the cardiovascular system and impaired endothelium-dependent vasorelaxation and age-dependent hypertension (Yang et al. 2008). The altered SMC proliferation in CSE knockout mice provides new insight into the pathogenesis of hypertension and underscores the importance of H_2S in homeostatic control of vascular integrity. Li et al. further found that the proliferation of cultured vascular SMCs isolated from wild-type mice was inhibited, but that from CSE gene knockout mice increased, by estrogen treatments, indicative of the interaction of H_2S and estrogen in regulating SMC proliferation (Li et al. 2012). In a rat model of pulmonary hypertension and pulmonary artery structural remodeling induced by high pulmonary blood flow, the inhibition of endogenous H_2S production by PPG markedly decreased the rate of pulmonary artery SMC apoptosis but supplements of H_2S donor increased pulmonary artery SMC apoptosis, as demonstrated by positive TUNEL staining (Li et al. 2009). Consistent with these in vivo findings, Baskar et al. observed that S-diclofenac, a novel molecule containing an H_2S-releasing dithiolthione moiety, stabilized p53 and induced the expressions of downstream proteins, such as p21, p53AIP1, and Bax, to repress SMC proliferation (Baskar et al. 2008).

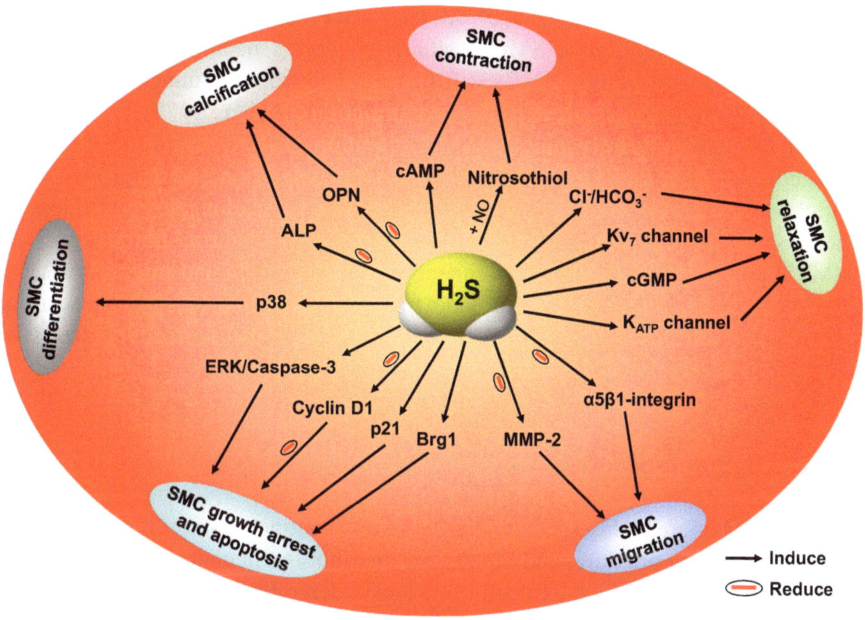

Fig. 1 H_2S signaling in SMC functions. H_2S maintains SMC differentiation by inducing p38 MAPK. H_2S stimulates SMC growth arrest and induces SMC apoptosis by activating ERK/caspase 3, p21, and/or Brg1 pathways but inactivating cyclin D1. H_2S also attenuates SMC migration by inhibiting the expressions of MMP-2 and α5β1-integrin. By activating Cl^-/HCO_3^-, Kv_7 channel, cGMP, and/or K_{ATP} channel, H_2S induces SMC relaxation, while H_2S mediates SMC contraction by enhancing cAMP production and interacting with NO to form nitrosothiol. H_2S also reduces SMC calcification by inhibiting the activities of ALP and OPN. Abbreviation used in this figure: ALP, alkaline phosphatase; Brg1, Brahma-related gene 1; cAMP, adenosine 3′,5′-cyclic monophosphate; cGMP, cyclic guanosine monophosphate; ERK, extracellular-signal-regulated kinase; H_2S, hydrogen sulfide; K_{ATP} channel, ATP-sensitive potassium channel; Kv_7 channel, voltage-gated potassium channel subunit K_v7; MMP-2, matrix metallopeptidase 2; NO, nitric oxide; OPN, osteopontin; SMC, smooth muscle cell

Altered SMC proliferation and apoptosis have been considered as key events in vascular injury in diseases such as atherosclerosis and restenosis after invasive intervention. CSE knockout mice fed with atherogenic diet developed early fatty streak lesions in the aortic root and increased aortic intimal proliferation (Mani et al. 2013). Treatment of CSE knockout mice with NaHS inhibited the accelerated atherosclerosis development. By using a rat model of vascular remodeling induced by balloon injury, Meng et al. found that CSE expression and H_2S production are significantly reduced in the blood vessel during the development of neointimal formation after balloon injury and the administration of NaHS attenuated the development of neointimal hyperplasia by inhibiting SMC proliferation (Meng et al. 2007). NaHS induced a significant reduction in cell proliferation in the neointima. We also noticed that CSE deficiency in mice led to increased neointima formation in carotid arteries 4 weeks after ligation, which were attenuated by NaHS administration (Yang et al. 2012a, b). All these data suggest that endogenous H_2S is critical for the inhibition of SMC proliferation during neointimal formation. In response to vascular injury, SMCs will first migrate from the tunica media to the intima, where they contribute to neointima formation (Thyberg 1998). Exogenously applied NaHS significantly inhibited SMC migration, and SMCs and aortic explants isolated from CSE knockout mice exhibited more migration and outgrowth compared with that from wild-type mice. SMCs became more elongated and spread in the absence of CSE. The interactions among $\alpha 5\beta 1$-integrin, fibronectin, and MMP-2 promoted CSE deficiency-enhanced SMC adhesion and migration (Yang et al. 2012a, b). These studies provide further insight into the pathogenesis of proliferative cardiovascular disorders and underscore the protective effects of H_2S in maintaining vascular homeostasis.

4.2 H_2S and SMC Phenotype Modulation

Vascular SMCs are highly specialized cells whose contractile status regulates blood vessel tone, blood pressure, and blood flow distribution. In contrast to cardiac and skeletal muscle cells which exit cell cycle and undergo final differentiation, vascular SMCs are highly plastic and can switch their phenotypes between contractile and synthetic phenotypes in response to extracellular stimuli or damage (Owens 2007). SMC phenotypic switching is believed to play a key role in many cardiovascular diseases, such as hypertension, atherosclerosis, coronary heart diseases, postangioplasty restenosis, and transplantation arteriopathy (Thyberg 1998). The H_2S-regulated SMC phenotypic switch has been explored (Fig. 1). CSE expression and H_2S production were reduced in proliferated SMCs compared with differentiated SMCs in a Sp1-dependent manner (Yang et al. 2011). In the presence of 10 % serum, H_2S strikingly induced the expressions of SMC differentiation maker genes in proliferative human aorta SMCs. In addition, H_2S-stimulated SMC differentiation maker gene expressions were reversed by co-treatment of the cells with SB203580, a p38 MAPK inhibitor. In the absence of serum, exogenously applied H_2S did not change the expressions of SM-MHC and calponin in SMCs. However, the inhibition of endogenous H_2S production by PPG significantly repressed the expressions of SM myosin heavy chain and calponin (Yang et al. 2011). Induction of SMC differentiation maker gene expressions by H_2S in the presence of, but not absence, serum suggests that H_2S may coordinate the expression of proliferative and contractile proteins to induce differentiated SMC phenotype. All these data indicate that CSE/H_2S system is essential for the maintenance of SMC differentiation.

4.3 H_2S and SMC Relaxation and Contraction

H_2S-induced vasorelaxation is a well-known vascular event, and the most widely characterized cellular target for H_2S in SMCs is the ATP-sensitive K^+ (K_{ATP})

channels (Fig. 1). In the vascular system, specifically in SMCs, K_{ATP} channels contribute significantly toward vasodilation in response to various vasoactive substances. The opening of K_{ATP} channels hyperpolarizes cell membrane and inactivates voltage-dependent L-type Ca^{2+} channels, leading to cell relaxation and blood vessel dilation by reducing intracellular free Ca^{2+} concentration (Wang 2012a, b). It has been previously shown that H_2S at physiologically relevant concentrations induced the relaxation of rat aortic tissue and transient reduction of blood pressure, and these vascular effects of H_2S were mediated by a direct stimulation of K_{ATP} channels and subsequent hyperpolarization of aortic SMCs (Zhao et al. 2001; Cheng et al. 2004; Wang 2014; Sun et al. 2011; Siebert et al. 2008). Using the whole-cell and single-channel patch-clamp technique, Tang et al. demonstrated that H_2S activates K_{ATP} channels and hyperpolarized cell membrane in rat mesenteric artery SMCs (Tang et al. 2005, 2010). H_2S enhanced the amplitude of whole-cell K_{ATP} currents and increased the open probability of single K_{ATP} channels. Furthermore, inhibition of endogenous H_2S production with PPG reduced whole-cell K_{ATP} currents. H_2S also causes the relaxation of human airway SMCs via stimulating sarcolemmal K_{ATP} channels (Fitzgerald et al. 2014). Other studies proved that H_2S interacts with the SUR subunits of the K_{ATP} channel complex to cause the channel to open, and the sulfhydryl groups located on the extracellular surface of the SUR subunits are potential targets for H_2S *S*-sulfhydration (Jiang et al. 2010). The deletion of extracellular cysteine 6 or 26 of SUR1 subunits caused the loss of channel sensitivity to H_2S. It appears that H_2S firstly breaks the disulfide bond between cysteine 6 and cysteine 26 and then caused their *S*-sulfhydrations. In addition to K_{ATP} channels, H_2S is reported to activate Kv7 voltage-gated potassium channels (particularly the Kv7.4 subtype) in SMCs, and the activation of Kv7 channel mediates a significant part of the vasorelaxing effects of H_2S (Martelli et al. 2013).

H_2S may induce SMC relaxation by altering intracellular pH. H_2S has been shown to decrease intracellular pH in a dose-dependent manner. Ionic exchangers, including Na^+/H^+ and Cl^-/HCO^{3-} and Ca^{2+} ATPase, maintain the resting pH in SMCs between 7.1 and 7.2. Preincubation of SMCs with a selective inhibitor of Cl^-/HCO_3^-, but not Na^+/H^+ exchanger inhibitor, prevented the drop of intracellular pH and vasorelaxation caused by H_2S, suggesting that Cl^-/HCO^{3-} exchanger is involved in H_2S-induced relaxation in SMCs (Lee et al. 2007). Intracellular acidosis could activate K_{ATP} channel in SMCs and decrease vascular tone. As such, H_2S-induced opening of K_{ATP} channels may be partially triggered by intracellular acidification (Liu et al. 2011).

H_2S may also cause vasorelaxation by increasing cGMP level in SMCs. Incubation of cultured rat aortic SMCs with NaHS led to a concentration-dependent increase in cGMP levels. The NaHS-induced rise in cGMP was evident as early as 1 min, reached a maximum at 3 min, and remained elevated for at least 10 min. Blockade of CSE activity by PPG or BCA or knockdown of CSE mRNA by siRNA resulted in a significant reduction of cGMP accumulation (Bucci et al. 2010). In contrast, overexpression of CSE elevated intracellular cGMP level. Vascular tissue levels of cGMP in CSE knockout mice were lower than those in wild-type control mice. Intracellular cGMP levels reflect the balance between the rate of cGMP synthesis via guanylyl cyclases and breakdown by phosphodiesterases. It appears that H_2S does not activate soluble guanylate cyclase, because H_2S-induced vasorelaxation is not inhibited by a soluble guanylate cyclase inhibitor (Coletta et al. 2012). Further studies showed that H_2S acts as an endogenous inhibitor of phosphodiesterase. In a cell-free assay, Bucci et al. demonstrated that NaHS at 10–30 nM significantly inhibits phosphodiesterase activity and causes a reduction in the breakdown of 5′-GMP (Bucci et al. 2012). H_2S also ameliorated the reduction in cGMP levels brought about by overexpression of phosphodiesterase 5A (Bucci et al. 2010, 2012). Because phosphodiesterases are involved in the degradation of both cAMP and cGMP, the researchers did not test the cellular effects of H_2S on cAMP level in SMCs. In contrast, another study by Lim et al. proved that NaHS significantly reverses forskolin-induced cAMP accumulation in SMCs (Lim et al. 2008). cAMP plays important roles in the regulation of mature contractile phenotype in SMCs. The effect of H_2S on SMC relaxation is actually biphasic depending on its concentration (Kubo et al. 2007; Liu et al. 2011). At higher level, H_2S produces vasorelaxation effect, while it induces vasoconstriction at lower concentration (Ali et al. 2006). To this end, the researchers observed that NaHS at a concentration range of 10–100 μM concentration-dependently reverses the vasodilation caused by isoprenaline and salbutamol (two β-adrenoceptor agonists) and forskolin (a selective adenylyl cyclase activator) in phenylephrine-precontracted rat aortic rings (Lim et al. 2008). Therefore, the authors proposed that the contractile effect of H_2S observed in isolated rat aorta is, at least partially, associated with reducing cAMP level in SMCs. In addition, H_2S may react with NO to form a compound, probably nitrosothiol, which leads to less NO bioavailability (Whiteman et al. 2006).

4.4 H_2S and SMC Calcification

Vascular calcification is implicated in the pathogenesis of various vascular diseases and resulted from passive precipitation of calcium and phosphate. Vascular calcification is now considered to be an active, regulative process similar to osteogenesis (McCarty and DiNicolantonio 2014). Calcified vessels have decreased capacity for vasodilatation and increased stiffness and promote a form of thrombus and atherosclerotic plaque rupture. Osteoblastic differentiation of SMCs is involved in the pathogenesis of vascular calcification. H_2S has been shown to ameliorate SMC calcification. In a rat vascular calcification model induced by the administration of vitamin D3 plus nicotine, aortic CSE expression and H_2S content were significantly reduced (Wu et al. 2006). Supplement of NaHS significantly reduced aortic calcium mineral deposits, OPN mRNA expression, and ALP activity, pointing to a regulatory role of CSE/H_2S pathway in the pathogenesis of vascular calcification. By using cultured SMCs, Zavaczki et al. explored the roles of H_2S in phosphate-induced osteoblastic transformation and mineralization (Zavaczki et al. 2011). H_2S inhibited calcium deposition in the extracellular matrix and suppressed the induction of the genes involved in osteoblastic transformation of SMCs, including alkaline phosphatase, osteocalcin, and Cbfa1. H_2S also prevented phosphate uptake and phosphate-triggered upregulation of the sodium-dependent phosphate cotransporter. H_2S, regardless of its exogenous or endogenous origin, is a potent inhibitor of phosphate-induced calcification and osteoblastic differentiation of SMCs. In contrast, silencing CSE by siRNA and inhibition of CSE activity by PPG attenuated receptor activator of nuclear factor κ-B ligand-induced tartrate-resistant acid phosphatase type 5 activities and pit formation in RAW264.7 cells. Moreover, knockdown of CSE suppressed the expression of osteoclast differentiation markers. A large-scale proteomics study also identified that CSE acts in early stages of osteoclastogenesis (Itou et al. 2014). These results suggest that CSE is a potent inducer of calcium resorption in inflammatory cells.

5 H_2S and ECs

5.1 H_2S and EC Proliferation and Angiogenesis

H_2S significantly stimulates endothelial cell growth and angiogenesis (Wang 2012a, b; Polhemus and Lefer 2014; Liu et al. 2011) (Fig. 2). Several groups have demonstrated that H_2S significantly promotes cell growth and capillary-like structure formation of cultured ECs (Altaany et al. 2013; Papapetropoulosa et al. 2009; Liu et al. 2011). Supplement of exogenous H_2S increased cellular infiltration and neovascularization in mouse Matrigel, enhanced the length of vascular network in the chick chorioallantoic membrane model, and promoted the formation of collateral vessels in ischemic hind limbs in rats (Papapetropoulosa et al. 2009; Köhn et al. 2012a; Hoefer 2007). A delayed wound healing was found in CSE knockout mice when compared with wild-type littermates (Papapetropoulosa et al. 2009). Endothelial progenitor cells are a population of rare cells that circulate in the blood with the ability to differentiate into ECs. H_2S was shown to improve endothelial progenitor cell function in diabetic wound healing of type 2 diabetic mice. Conversely, PPG treatment reduced progenitor cell function and delayed wound healing (Liu et al. 2014). Under hypoxic condition, H_2S induced endothelial proliferation and migration by promoting VEGF and HIF-1a expression and increasing HIF-1a-binding activity (Liu et al. 2010). In contrast, H_2S decreased cell proliferation and capillary tube formation of EA.hy926 cells under hypoxia by inhibiting the expression of VEGF and HIF-1a (Wu et al. 2012). The discrepancy of the aforementioned observations may be due to different hypoxic models, H_2S concentrations, and cell types. Pupo et al. showed that NaHS at 10 uM did not exert any effect on cell migration and proliferation of normal human microvascular ECs (HMVECs) (Pupo et al. 2011). This apparent discrepancy may be ascribed to tissue specificity. Different from HMVECs, H_2S donors failed to induce cell proliferation but promote their migration of ECs obtained from human breast carcinoma (B-TEC). B-TECs pretreated with PPG showed drastically reduced migration

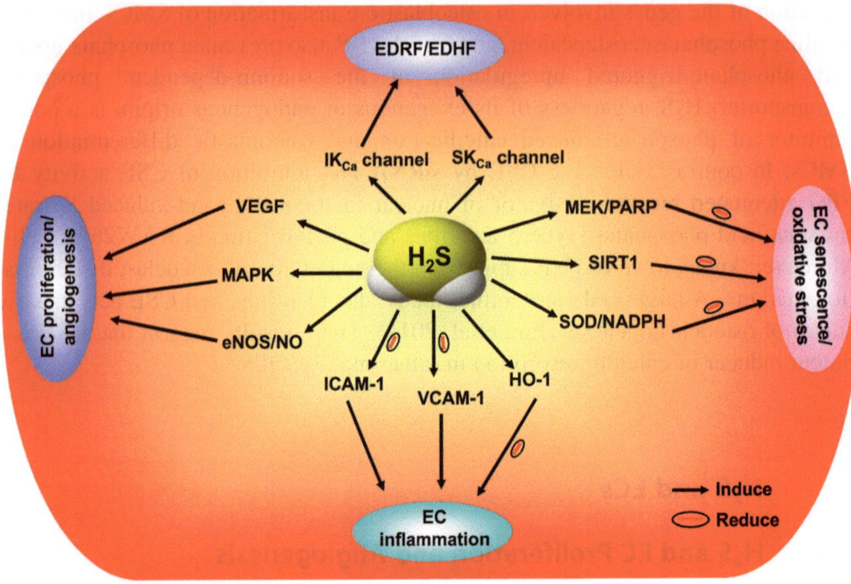

Fig. 2 H_2S signaling in EC functions. H_2S stimulates EC proliferation and angiogenesis by activating VEGF, MAPK, and/or eNOS/NO pathways. H_2S attenuates EC inflammation by inhibiting the expressions of ICAM-1 and VCAM-1 but stimulating HO-1. H_2S lowers EC senescence and oxidative stress by activating MEK/PARP, SIRT1, and/or SOD/NADPH pathways. H_2S acts as EDRF and EDHF at least through IK_{Ca} channel and SK_{Ca} channel. Abbreviation used in this figure: EDHF, endothelium-derived hyperpolarizing factor; EDRF, endothelium-derived relaxing factor; eNOS, endothelial nitric oxide synthase; HO-1, heme oxygenase-1; ICAM-1, intercellular adhesion molecule 1; IKCa channel, intermediate-conductance, calcium-activated potassium channel; MAPK, mitogen-activated protein kinase; NADPH, nicotinamide adenine dinucleotide phosphate-oxidase; NO, nitric oxide; PARP, poly (ADP-ribose) polymerase; SIRT1, sirtuin-1; SK_{Ca} channel, small-conductance calcium-activated potassium channel; SOD, superoxide dismutase; VCAM-1, vascular cell adhesion molecule 1; VEGF, vascular endothelial growth factor

induced by VEGF (Pupo et al. 2011). The authors concluded that H_2S plays a role in proangiogenic signaling of tumor-derived but not normal human ECs.

In addition to the direct stimulatory effect on EC growth and angiogenesis, H_2S was also reported to ameliorate stress-induced cell death of ECs. Exposure of primary human umbilical vein endothelium cells (HEVECs) to a high concentration of glucose (25 mM) resulted in the induction of apoptosis, but pretreatment with NaHS (50 μM) for 30 min attenuated the pro-apoptotic effect of 25 mM glucose. Further investigation of the apoptotic mechanisms in the cells demonstrated that high glucose upregulated the ratio of Bax/Bcl-2, activated caspase-3, increased the levels of reactive oxygen species and malondialdehyde, and suppressed superoxide dismutase activity (Guan et al. 2012). All these effects of glucose could be prevented by pretreatment with 50 μM NaHS. Pretreatment with NaHS (0.05-0.1 mM) attenuated methionine- or oxidized low-density lipoprotein-induced oxidative stress and cell death of ECs (Tyagi et al. 2009; Muellner et al. 2009).

Exposure of bEnd3 microvascular ECs to elevated extracellular glucose induced mitochondrial formation of ROS, and H_2S replacement protected against ROS formation, improved endothelial metabolic state, and maintained normal endothelial function (Suzuki et al. 2011). H_2S was also found to preserve the activities and protein levels of the antioxidant enzymes, superoxide dismutase, catalase, glutathione peroxidase, and glutathione-S-transferase in H_2O_2-exposed ECs (Wen et al. 2013). Zofenoprilat is a sulfhydryl-containing angiotensin-converting enzyme inhibitor, which can enhance CSE-dependent availability of H_2S. Zofenoprilat was shown to exert a protective effect on doxorubicin-induced endothelial damage without affecting its antitumor efficacy (Monti et al. 2013).

5.2 H_2S and EC Senescence and Oxidative Stress

Growing evidence shows that the progress of vascular aging alters cardiovascular function and subsequently increases the risk of cardiovascular diseases. Vascular aging has been largely associated with senescence of the vascular endothelium. Several lines of evidence point to the implication of H_2S signaling in the process of endothelial senescence (Fig. 2). Calorie restriction is reported to decelerate biological aging process, resulting in longer maintenance of youthful health and an increase in both median and maximum life span. Calorie restriction maintained normal H_2S level in vascular tissues from rats during aging, suggesting a protective role of H_2S in vascular aging (Predmore et al. 2010). Oxidative stress is a driving factor for vascular aging. H_2O_2 treatments of HUVECs lead to high rate of senescent cells, which was attenuated by NaHS incubation putatively through the modulation of SIRT1 activity (Suo et al. 2013). H_2S also improved the function of senescent HUVECs. Zhao et al. found that H_2S attenuates cellular senescence and DNA damage in HUVECs by MEK1 *S*-sulfhydration and PARP-1 activation (Zhao et al. 2014). In the presence of H_2S, activated PARP-1 recruits XRCC1 and DNA ligase III to DNA breaks to mediate DNA damage repair. AP39, a mitochondria-targeted H_2S donor, was shown to exert a concentration-dependent effect on mitochondrial activity in bEnd.3 murine microvascular ECs, as evidenced by the stimulation of mitochondrial electron transport and cellular bioenergetic function. Furthermore, AP39 pretreatment protected against glucose oxidase-induced mitochondrial DNA damage (Szczesny et al. 2014). A recent study showed that H_2S increases the life span of *Caenorhabditis elegans*. The life span-prolonging and health-promoting effects of H_2S in *C. elegans* are likely due to the antioxidant action (Qabazard et al. 2014). Similarly, compared with the lower passage of ECs, the higher passage of ECs had lower SOD activity and higher H_2O_2 level, whereas NaHS pretreatment reversed the changes of SOD activity and H_2O_2 level, indicating that H_2S delays senescence of HUVECs through lessening oxidative stress (Qi et al. 2012; Muellner et al. 2009). By using porcine pulmonary arterial ECs, Muzaffar et al. demonstrated that H_2S inhibited superoxide formation and upregulation of NADPH oxidase through the adenylyl cyclase-PKA pathway

(Muzaffar et al. 2008). In spite of these findings, the in vivo role of H_2S in regulating EC senescence and vascular aging remains unclear.

5.3 H_2S and EC Inflammation

The regulatory role of H_2S in inflammation involves the endothelium and its interaction with leukocytes (Whiteman and Winyard 2011; Zanardo et al. 2006). H_2S donors (NaHS and Na_2S) inhibited but PPG promoted aspirin-induced leukocyte adherence in mesenteric venules via the activation of K_{ATP} channels (Zanardo et al. 2006). H_2S also suppressed leukocyte infiltration in an air pouch model and carrageenan-induced paw edema, implicating a protective role of H_2S in acute inflammation by acting at the leukocyte–endothelium interface (Perna et al. 2013). Further study demonstrated that H_2S treatment of ECs decreased the expression of MCP-1, intercellular adhesion molecule-1 (ICAM-1), and vascular cell adhesion molecule-1 (VCAM-1) at the mRNA and protein levels. In an in vitro model entailing monocyte adhesion to an endothelial monolayer, H_2S prevented the increase in monocyte adhesion induced by tumor necrosis factor-α (TNF-α). Pan et al. also showed that H_2S dose-dependently suppressed TNF-α-induced mRNA and protein expressions of ICAM-1 and VCAM-1 in HUVECs, possibly through the upregulation of HO-1 (Pan et al. 2011). By using another H_2S donor, SPRC, the same group found that H_2S exerts anti-inflammatory effects on TNF-α-stimulated ECs through scavenging ROS, inhibiting JNK1/2/NF-κB pathways, and attenuating adhesion molecule expression (Pan et al. 2012). NaHS was shown to reduce oxLDL-induced foam cell formation in macrophages and TNF-α-stimulated ICAM-1 expression in HUVECs (Zhao et al. 2011; Wang et al. 2013b). In contrast, Choi et al. showed that Korean red ginseng extracts inhibit the expression of inflammatory mediators, including IL-8 and IL-6, via reduced CSE expression and H_2S production in ECs (Choi et al. 2012). Supplement of exogenously applied H_2S reversed the Korean red ginseng extract-improved inflammation status in ECs.

H_2S protects vascular tissues from atherogenic damage by inhibiting adhesion molecule expression and suppressing monocyte adhesion to the activated endothelium (Wang et al. 2009). In spontaneously hypertensive rat (SHR), the expression of ICAM-1 and NF-kappaB p65 protein in aortic ECs was significantly higher, and NaHS treatment reduced blood pressure in SHR rats and downregulated the expressions of ICAM-1 and NF-kappaB p65 in aortic ECs. On the other hand, inhibition of H_2S production enhanced the expressions of ICAM-1 and NF-kappaB p65 protein in aortic ECs (Jin et al. 2008), suggesting that H_2S might attenuate the development of hypertension by suppressing endothelial inflammation reactions.

5.4 Interplay Between H_2S and NO in ECs

Both H_2S and NO are major gasotransmitters produced in ECs, and growing evidence showed that the "cross talk" between NO and H_2S mediates the cardioprotective effect of H_2S (Jamroz-Wiśniewska et al. 2014). Predmore et al. first showed that H_2S stimulated eNOS phosphorylation and NO production in ECs, and pharmacological inhibition of Akt, the kinase responsible for eNOS Ser 1177 phosphorylation, attenuated the stimulatory effect of H_2S on NO production (Predmore et al. 2011). Al Tanny et al. confirmed that H_2S promotes NO production in ECs via the activation of a cascade of phosphorylation events, starting from p38 MAPK to Akt to eNOS (Altaany et al. 2013). Deficiencies in H_2S signaling can directly impact on processes regulated by NO (Coletta et al. 2012). H_2S promotes EC tube formation, proliferation, and angiogenesis by both NO-dependent and NO-independent mechanisms. H_2S also modulates eNOS via *S*-sulfhydration and prevents eNOS coupling collapse and thus increases NO and decreases $ONOO^-$ and O_2 levels in ECs (Al Taany et al. 2014). Exposure of ECs to H_2S increased intracellular cGMP in a PI3K/Akt and NO-dependent manner, and NO and H_2S are mutually required for the physiological control of vascular function (Coletta et al. 2012). H_2S may stimulate NO production by different mechanisms. Kida et al. showed that NaHS dose-dependently increased NO production in cultured ECs by releasing calcium from the intracellular store in endoplasmic reticulum. NaHS-induced eNOS phosphorylation and NO production were abolished by the ryanodine receptor inhibitor, inositol 1,4,5-triphosphate receptor inhibitor, and calcium chelator but not by the PI3K/Akt inhibitor and the absence of extracellular calcium (Kida et al. 2013). NaHS was also shown to increase eNOS expression and NO production in rat corpus, suggesting H_2S could exert its pro-erectile effects by augmenting NO pathway. In contrast, Na_2S, another H_2S donor, was shown to reduce the level of phospho-eNOS (serine 1177) and inhibit eNOS activity in cultured mouse aortic ECs (Chai et al. 2014). The discrepancy may be due to the difference of cell type and H_2S dose used. Further elucidation of the H_2S–NO relationship in the vascular biology would provide more insight into the vasodilator function of H_2S and improve our understanding of the pathogenic mechanisms for cardiovascular diseases.

5.5 H_2S Acts as an Endothelium-Derived Relaxing Factor (EDRF) and an Endothelium-Derived Hyperpolarizing Factor (EDHF)

In response to a variety of chemical and physical stimuli, ECs produce and release various vasoactive factors, including EDRF and EDHF. Both EDRF and EDHF relax vascular SMCs, causing blood vessel to expand in diameter (Wang 2012a, b). NO is a well-described EDRF, but other EDRFs are also produced and released from the endothelium. H_2S shares many features with NO. Accumulating evidence supports the concept that H_2S acts as both EDRF and EDHF (Baragatti et al. 2013; Han et al. 2013; Skovgaard et al. 2011; Wang 2009) (Fig. 2). Endogenously generated H_2S induces vasorelaxation in part through an endothelium-dependent mechanism (Yang et al. 2008; Zhao et al. 2001). The vasorelaxation elicited by H_2S is greater in small resistance arteries as compared with that in larger conduit arteries such as the aorta (Tang et al. 2013). Knocking out CSE expression attenuated acetylcholine-induced membrane hyperpolarization of the isolated ECs and led to membrane depolarization of the whole vascular tissues. Different from EDRFs, the unique property of EDHF is its sole effect on hyperpolarizing vascular SMCs so as to close voltage-dependent calcium channels (Wang 2012a, b). The effect of EDHF is mainly mediated by the opening of small-conductance and/or intermediate-conductance K_{Ca} channels (Mustafa et al. 2011; Wang 2012a, b). Using electrophysiological microelectrode technique, Tang et al. demonstrated that the unique EC dependence of H_2S-induced SMC hyperpolarization was linked to the opening of IK_{Ca} and/or SK_{Ca} channels (Tang et al. 2013). The identification of H_2S as an EDRF/EDHF will not only help better understand the mechanisms underlying endothelium-dependent vasorelaxation of different types of vascular tissues but also shed light on devising novel therapeutic agents to deal with specific cardiovascular diseases.

6 H_2S and Periadventitial Adipose Tissue (PAT)

Adventitia is the outermost connective tissue covering a blood vessel, also called the tunica adventitia or the tunica externa. PAT, the major part of adventitia and defined as the accumulation of adipocytes around vascular structures, can be found in the proximity of virtually all blood vessels (Gollasch 2012). PAT is composed of various cells such as adipocytes, fibroblasts, and macrophages and can release various active agents, e.g., adipocyte-derived relaxing factor (ADRF), which play important roles in modulating the vascular tone. Recent studies have shown that H_2S can be generated from PAT and act as an ADRF, contributing to vascular relaxation (Fang et al. 2009). Immunohistochemical staining revealed the presence of CSE protein in PAT. In isolated PAT, H_2S production was reduced in an age-dependent manner. The CSE/H_2S pathway was upregulated in PAT as a compensatory mechanism against the elevated blood pressure in hypertension (Bełtowski 2013). In consistent with this study, Köhn et al. showed that the inhibition of CSE activity by PPG reversed the anti-contractile effect of PAT in rat aorta (Köhn et al. 2012a, b). Statins are well-known drugs to reduce plasma LDL cholesterol, improve endothelial function, ameliorate oxidative stress, and maintain coagulation–fibrinolysis balance. It was recently found that atorvastatin, one kind of statins, increased H_2S production in PAT and H_2S mediated the protective role of statins in the cardiovascular system. Atorvastatin augmented the anti-contractile effect of PAT, most likely in an H_2S- and K_{ATP} channel-dependent manner, because its effect was abolished by CSE inhibitor PPG and K_{ATP} channel blocker glibenclamide (Wójcicka et al. 2011). A statin-induced increase in H_2S production may also contribute to the antiatherogenic effect of statins since H_2S inhibits platelet aggregation, vascular SMC proliferation, LDL oxidation, and local inflammatory reaction.

7 H_2S and Blood Vessel-Related Disorders

Blood vessel diseases are also called peripheral vascular diseases or artery diseases, including hypertension, atherosclerosis, aortic aneurysm, etc. H_2S plays an important protective role in various blood vessel diseases (Wang 2012a, b; Yang 2011).

7.1 H_2S and Hypertension

Hypertension or high blood pressure is a chronic medical condition in which the blood pressure is elevated. Hypertension results from a complex interaction of vasoactive factors with various types of cells in blood vessel walls. Recent studies have shown that CSE/H_2S system produces antihypertensive effects in different hypertensive models. An acute intravenous bolus injection of NaHS caused a transient fall or mean arterial blood pressure in anesthetized rats (Zhao et al. 2001). Exogenous H_2S administration lowers blood pressure and prevents the hypertrophy of intramyocardial arterioles and aortic thickening of spontaneously hypertensive rats (SHR) (Du et al. 2003; Zhao et al. 2008). Injection of a single dose of GYY4137 (a H_2S donor) alleviated L-NAME-induced hypertension in rats, and chronic treatment with GYY4137 successfully reduced blood pressure of SHR (Li et al. 2008a). ACS14, another H_2S-releasing donor, reduced blood pressure in buthionine sulfoximine-induced hypertensive rats (Rossoni et al. 2010). Inhibition of H_2S production by PPG worsened hypoxic pulmonary hypertension in rats. Direct evidence also demonstrated that complete deficiency of CSE in mice markedly reduced endogenous H_2S levels in the vascular system and led to age-dependent development of hypertension (Yang et al. 2008). Preeclampsia is a disorder of pregnancy characterized by high blood pressure and contributes to maternal and fetal morbidity and mortality worldwide. The cause and pathogenesis of preeclampsia has yet to be definitively uncovered. Wang et al. recently showed that plasma H_2S and CSE expression in the placenta were reduced in pregnancies complicated by preeclampsia in comparison with gestational age-matched controls (Wang et al. 2013a). Inhibition of CSE activity induces hypertension and causes placental abnormalities in pregnant mice owing to the inhibition of H_2S production. These discoveries suggest that a dysfunctional CSE/H_2S pathway may contribute to the pathogenesis of preeclampsia and targeting at the CSE/H_2S system would be an effective therapy for preeclampsia (Wang et al. 2013a, b).

7.2 H_2S and Atherosclerosis

Atherosclerosis is a chronic progressive pathological process in large- and medium-sized arteries, caused by the buildup of fatty/cholesterol plaques on the ECs of arteries, endothelial inflammation, and SMC proliferation. H_2S plays an anti-atherosclerotic role, and its deficiency leads to the development and progression of atherosclerosis (Zhang et al. 2012a, b; Mani et al. 2013, 2014; Xu et al. 2014;

Wang et al. 2009). Plasma H_2S level and aortic CSE activity was decreased in apolipoprotein E (ApoE) knockout mice with advanced atherosclerosis. Treatment of ApoE knockout mice with NaHS resulted in reduced atherosclerosis plaque, while inhibition of CSE activity enlarged plaque size (Wang et al. 2009). GYY4137, a H_2S donor, was also shown to decrease aortic atherosclerotic plaque formation and partially restored aortic endothelium-dependent relaxation in ApoE knockout mice (Li et al. 2008a; Liu et al. 2013; Qiao et al. 2010). Another study showed that onion extracts boosted endogenous production of H_2S and lessened atherosclerotic lesions in rats (Zhang et al. 2012a, b). In a high-fat and high-vitamin D1 diet-induced atherosclerotic rat model, H_2S was shown to slow down the development of atherosclerosis by improving the damage of vessels and inhibiting the expression of VEGF (Zhang et al. 2012a, b). Treatment with NaHS significantly inhibited arterial restenosis following balloon angioplasty in rabbits by reducing the intimal area and the intima/media ratio, while PPG treatment had a tendency to worsen the severe restenosis (Ma et al. 2012). Whether CSE deficiency impacts on the development of atherosclerosis has been directly addressed by knocking out the CSE gene in high-fat-fed mice. Decreased endogenous H_2S production in CSE knockout mice leads to the vascular remodeling and early development of atherosclerosis (Mani et al. 2013). The atherosclerotic plaque development is rescued by H_2S donor NaHS via reducing vessel intimal proliferation and inhibiting adhesion molecule expression. In contrast, antihypertensive (hydralazine), antioxidant (N-acetylcysteine), or lipid-lowering (ezetimibe) agents have no effect on high-fat-diet-induced plaque formation in CSE knockout mice, implying that hypertension, higher oxidative stress, and abnormal lipid profile do not play major roles in atherosclerosis development in CSE knockout mice. In addition, the knockout of CSE from ApoE knockout mice accelerated plaque formation even under normal diet, indicative of a potential therapeutic implication of endogenous H_2S (generated by CSE) in atherosclerosis. Interestingly, estrogen attenuates atherosclerosis development by stimulating H_2S production in female ovariectomized ApoE knockout mice fed with a high-fat diet, suggesting that H_2S mediates estrogen-induced vascular protection (Zhou et al. 2013; Fu et al. 2013).

7.3 H_2S and Aortic Aneurysms

Aortic aneurysms, including thoracic and abdominal aortic aneurysms, are the most life-threatening cardiovascular complication in Marfan syndrome, leading to aortic expansion, dissection, rupture, and sudden death. Enzymatic degradation of extracellular matrix (ECM) protein by matrix metalloproteinases (MMPs) leads to dilation of the aortic wall and constitutes the most prominent characters of aortic aneurysms. The regulatory role of H_2S in vascular remodeling during aortic aneurysms has not been explored yet. Early studies showed that H_2S is involved in vascular remolding during the development of hypertension and neointimal formation. Rats with high pulmonary blood flow for 11 weeks showed a significant pulmonary hypertension and pulmonary artery collagen remodeling in association with a decrease in lung tissue H_2S content (Li et al. 2008b). Supplement of exogenous H_2S lowered pulmonary artery collagen I and collagen III protein levels and normalized pulmonary hypertension, suggesting that the downregulation of H_2S is involved in the development of pulmonary artery collagen remodeling induced by high pulmonary blood flow. Vascular H_2S production was lower in SHR compared with Wistar Kyoto (WKY) rats, and the inhibitory role of H_2S on collagen generation was stronger in the SHR than in the WKY rats (Zhao et al. 2008). In a carotid artery-injured mouse model, H_2S was shown to mitigate vascular remodeling from endothelial damage by decreasing the expressions of TIMP-3 and MMP-9 (Vacek et al. 2010). In addition, we found that H_2S inhibited SMC migration and neointima formation by suppressing $\alpha5\beta1$-integrin-dependent MMP-2 expression. The inhibitory roles of H_2S on the MMP/TIMP system suggest that H_2S can block excess degradation of ECM and maintain the normal structure of the aorta (Yang et al. 2012a, b). Future studies need to determine the role of H_2S in vascular degeneration and aortic aneurysm formation. The molecular mechanisms involved in the pathogenesis of aortic aneurysm formation as well as the mechanisms underlying CSE/H_2S system-regulated MMP activation are also waiting to be explored. All these endeavors will provide a new therapeutic avenue for the prevention and treatment of aortic aneurysms.

8 Prospective

Undoubtedly, the research scope and depth on H_2S signaling in the cardiovascular system, especially in blood vessel-related disorders, will continue to expand and deepen. The identification of new cellular targets and development of novel agents to enhance endogenous H_2S generation are highly expected. A better understanding of the roles of H_2S in vascular remodeling and the regulatory mechanisms for endogenous production of H_2S in blood vessel can provide insight into potential therapeutic interventions against blood vessel-associated disorders. The interaction of H_2S with numerous biological molecules, such as NO and CO, needs to be better characterized. At the end of the day, all these fundamental and mechanistic studies will become meaningless if the related discoveries cannot be translated from bench side to bedside so that human health and welfare will be improved. That day will not be too far away.

Acknowledgment The study has been supported by an operating grant to RW from Canadian Institutes of Health Research and a New Investigator award to GY from the Heart and Stroke Foundation of Canada.

References

Ali MY, Ping CY, Mok YY, Ling L, Whiteman M, Bhatia M, Moore PK (2006) Regulation of vascular nitric oxide in vitro and in vivo; a new role for endogenous hydrogen sulphide? Br J Pharmacol 149:625–634

Altaany Z, Yang G, Wang R (2013) Crosstalk between hydrogen sulfide and nitric oxide in endothelial cells. J Cell Mol Med 17:879–888

Al Taany Z, Ju Y, Yang G, Wang R (2014) The coordination of S-sulfhydration, S-nitrosylation, and phosphorylation of endothelial nitric oxide synthase by hydrogen sulfide. Sci Signal 7:ra87

Baragatti B, Ciofini E, Sodini D, Luin S, Scebba F, Coceani F (2013) Hydrogen sulfide in the mouse ductus arteriosus: a naturally occurring relaxant with potential EDHF function. Am J Physiol Heart Circ Physiol 304:H927–H934

Baskar R, Sparatore A, Del Soldato P, Moore PK (2008) Effect of S-diclofenac, a novel hydrogen sulfide releasing derivative inhibit rat vascular smooth muscle cell proliferation. Eur J Pharmacol 594:1–8

Bełtowski J (2013) Endogenous hydrogen sulfide in perivascular adipose tissue: role in the regulation of vascular tone in physiology and pathology. Can J Physiol Pharmacol 91:889–898

Bucci M, Mirone V, Di Lorenzo A, Vellecco V, Roviezzo F, Brancaleone V, Ciro I, Cirino G (2009) Hydrogen sulfide is involved in testosterone vascular effect. Eur Urol 5:378–384

Bucci M, Papapetropoulos A, Vellecco V, Zhou Z, Pyriochou A, Roussos C, Roviezzo F, Brancaleone V, Cirino G (2010) Hydrogen sulfide is an endogenous inhibitor of phosphodiesterase activity. Arterioscler Thromb Vasc Biol 30:1998–2004

Bucci M, Papapetropoulos A, Vellecco V, Zhou Z, Zaid A, Giannogonas P, Cantalupo A, Dhayade S, Karalis KP, Wang R, Feil R, Cirino G (2012) cGMP-dependent protein kinase contributes to hydrogen sulfide-stimulated vasorelaxation. PLoS ONE 7:e53319

Chai Q, Lu T, Wang XL, Lee HC (2014) Hydrogen sulfide impairs shear stress-induced vasodilation in mouse coronary arteries. Pflugers Arch 467(2):329–340 [Epub ahead of print]

Chen YH, Wu R, Geng B, Qi YF, Wang PP, Yao WZ, Tang CS (2009) Endogenous hydrogen sulfide reduces airway inflammation and remodeling in a rat model of asthma. Cytokine 45:117–123

Cheng Y, Ndisang JF, Tang G, Cao K, Wang R (2004) Hydrogen sulfide-induced relaxation of resistance mesenteric artery beds of rats. Am J Physiol Heart Circ Physiol 287:H2316–H2323

Choi KS, Song H, Kim EH, Choi JH, Hong H, Han YM, Hahm KB (2012) Inhibition of hydrogen sulfide-induced angiogenesis and inflammation in vascular endothelial cells: potential mechanisms of gastric cancer prevention by korean red ginseng. J Ginseng Res 36:135–145

Coletta C, Papapetropoulos A, Erdelyi K, Olah G, Módis K, Panopoulos P, Asimakopoulou A, Gerö D, Sharina I, Martin E, Szabo C (2012) Hydrogen sulfide and nitric oxide are mutually dependent in the regulation of angiogenesis and endothelium-dependent vasorelaxation. Proc Natl Acad Sci USA 109:9161–9166

Du J, Yan H, Tang C (2003) Endogenous H2S is involved in the development of spontaneous hypertension. Beijing Da Xue Xue Bao 35:102

Du J, Hui Y, Cheung Y, Bin G, Jiang H, Chen X, Tang C (2004) The possible role of hydrogen sulfide as a smooth muscle cell proliferation inhibitor in rat cultured cells. Heart Vessels 19:75–80

Fang L, Zhao J, Chen Y, Ma T, Xu G, Tang C, Liu X, Geng B (2009) Hydrogen sulfide derived from periadventitial adipose tissue is a vasodilator. J Hypertens 27:2174–2185

Fitzgerald R, DeSantiago B, Lee DY, Yang G, Kim JY, Foster DB, Chan-Li Y, Horton MR, Panettieri RA, Wang R, An SS (2014) H2S relaxes isolated human airway smooth muscle cells via the sarcolemmal K(ATP) channel. Biochem Biophys Res Commun 446:393–398

Fu X, Zhou K, Gao Q, Zheng S, Chen H, Li P, Zhang Y, Suo K, Simoncini T, Wang T (2013) 17β-estradiol attenuates atherosclerosis development: the possible role of hydrogen sulfide. Int J Cardiol 167:1061–1063

Gollasch M (2012) Vasodilator signals from perivascular adipose tissue. Br J Pharmacol 165:633–642, Review
Guan Q, Zhang Y, Yu C, Liu Y, Gao L, Zhao J (2012) Hydrogen sulfide protects against high-glucose-induced apoptosis in endothelial cells. J Cardiovasc Pharmacol 59:188–193
Han J, Chen ZW, He GW (2013) Acetylcholine- and sodium hydrosulfide-induced endothelium-dependent relaxation and hyperpolarization in cerebral vessels of global cerebral ischemia-reperfusion rat. J Pharmacol Sci 121:318–326
Hassan MI, Boosen M, Schaefer L, Kozlowska J, Eisel F, von Knethen A, Beck M, Hemeida RA, El-Moselhy MA, Hamada FM, Beck KF, Pfeilschifter J (2012) Platelet-derived growth factor-BB induces cystathionine γ-lyase expression in rat mesangial cells via a redox-dependent mechanism. Br J Pharmacol 166:2231–2242
Hoefer IE (2007) Something is rotten in the state of angiogenesis – H2S as gaseous stimulator of angiogenesis. Cardiovasc Res 76:1–2
Hosoki R, Matsuki N, Kimura H (1997) The possible role of hydrogen sulfide as an endogenous smooth muscle relaxant in synergy with nitric oxide. Biochem Biophys Res Commun 237:527–531
Itou T, Maldonado N, Yamada I, Goettsch C, Matsumoto J, Aikawa M, Singh S, Aikawa E (2014) Cystathionine γ-lyase accelerates osteoclast differentiation: identification of a novel regulator of osteoclastogenesis by proteomic analysis. Arterioscler Thromb Vasc Biol 34:626–634
Jamroz-Wiśniewska A, Gertler A, Solomon G, Wood ME, Whiteman M, Bełtowski J (2014) Leptin-induced endothelium-dependent vasorelaxation of peripheral arteries in lean and obese rats: role of nitric oxide and hydrogen sulfide. PLoS One 9:e86744
Jiang B, Tang G, Cao K, Wu L, Wang R (2010) Molecular mechanism for H(2)S-induced activation of K(ATP) channels. Antioxid Redox Signal 12:1167–1178
Jin HF, Sun Y, Liang JM, Tang CS, DU JB (2008) Hypotensive effects of hydrogen sulfide via attenuating vascular inflammation in spontaneously hypertensive rats. Zhonghua Xin Xue Guan Bing Za Zhi 36:541–545
Kida M, Sugiyama T, Yoshimoto T, Ogawa Y (2013) Hydrogen sulfide increases nitric oxide production with calcium-dependent activation of endothelial nitric oxide synthase in endothelial cells. Eur J Pharm Sci 48:211–215
Kiss L, Deitch EA, Szabó C (2008) Hydrogen sulfide decreases adenosine triphosphate levels in aortic rings and leads to vasorelaxation via metabolic inhibition. Life Sci 83:589–594
Köhn C, Dubrovska G, Huang Y, Gollasch M (2012a) Hydrogen sulfide: potent regulator of vascular tone and stimulator of angiogenesis. Int J Biomed Sci 8:81–86
Köhn C, Schleifenbaum J, Szijártó IA, Markó L, Dubrovska G, Huang Y, Gollasch M (2012b) Differential effects of cystathionine-γ-lyase-dependent vasodilatory H2S in periadventitial vasoregulation of rat and mouse aortas. PLoS ONE 7:e41951
Kubo S, Kajiwara M, Kawabata A (2007) Dual modulation of the tension of isolated gastric artery and gastric mucosal circulation by hydrogen sulfide in rats. Inflammopharmacology 15:288–292
Lee SW, Cheng Y, Moore PK, Bian JS (2007) Hydrogen sulphide regulates intracellular pH in vascular smooth muscle cells. Biochem Biophys Res Commun 358:1142–1147
Leffler CW, Parfenova H, Basuroy S, Jaggar JH, Umstot ES, Fedinec AL (2011) Hydrogen sulfide and cerebral microvascular tone in newborn pigs. Am J Physiol Heart Circ Physiol 300:H440–H447
Li L, Liu D, Bu D, Chen S, Wu J, Tang C, Du J, Jin H (2013) Brg1-dependent epigenetic control of vascular smooth muscle cell proliferation by hydrogen sulfide. Biochim Biophys Acta 1833:1347–1355
Li L, Whiteman M, Guan YY, Neo KL, Cheng Y, Lee SW, Zhao Y, Baskar R, Tan CH, Moore PK (2008a) Characterization of a novel, water-soluble hydrogen sulfide-releasing molecule (GYY4137): new insights into the biology of hydrogen sulfide. Circulation 117:2351–2360
Li X, Du J, Jin H, Geng B, Tang C (2008b) Sodium hydrosulfide alleviates pulmonary artery collagen remodeling in rats with high pulmonary blood flow. Heart Vessels 23:409–419

Li W, Jin HF, Liu D, Sun JH, Jian PJ, Li XH, Tang CS, DU JB (2009) Hydrogen sulfide induces apoptosis of pulmonary artery smooth muscle cell in rats with pulmonary hypertension induced by high pulmonary blood flow. Chin Med J (Engl) 122:3032–3038

Li H, Mani S, Cao W, Yang G, Lai C, Wu L, Wang R (2012) Interaction of hydrogen sulfide and estrogen on the proliferation of vascular smooth muscle cells. PLoS ONE 7:e41614

Li Y, Liu G, Cai D, Pan B, Lin Y, Li X, Li S, Zhu L, Liao X, Wang H (2014) H_2S inhibition of chemical hypoxia-induced proliferation of HPASMCs is mediated by the upregulation of COX-2/PGI2. Int J Mol Med 33:359–366

Lim JJ, Liu YH, Khin ES, Bian JS (2008) Vasoconstrictive effect of hydrogen sulfide involves downregulation of cAMP in vascular smooth muscle cells. Am J Physiol Cell Physiol 295: C1261–C1270

Liu X, Pan L, Zhuo Y, Gong Q, Rose P, Zhu Y (2010) Hypoxia-inducible factor-1α is involved in the pro-angiogenic effect of hydrogen sulfide under hypoxic stress. Biol Pharm Bull 33:1550–1554

Liu YH, Yan CD, Bian JS (2011) Hydrogen sulfide: a novel signaling molecule in the vascular system. J Cardiovasc Pharmacol 58:560–569

Liu Z, Han Y, Li L, Lu H, Meng G, Li X, Shirhan M, Peh MT, Xie L, Zhou S, Wang X, Chen Q, Dai W, Tan CH, Pan S, Moore PK, Ji Y (2013) The hydrogen sulfide donor, GYY4137, exhibits antiatherosclerotic activity in high fat fed apolipoprotein E(−/−) mice. Br J Pharmacol 169:1795–1809

Liu F, Chen DD, Sun X, Xie HH, Yuan H, Jia WP, Chen AF (2014) Hydrogen sulfide improves wound healing via restoration of endothelial progenitor cell functions and activation of angiopoietin-1 in type 2 diabetes. Diabetes 63:1763–1778

Ma B, Liang G, Zhang F, Chen Y, Zhang H (2012) Effect of hydrogen sulfide on restenosis of peripheral arteries after angioplasty. Mol Med Rep 5:1497–1502

Mani S, Li H, Untereiner A, Wu L, Yang G, Austin RC, Dickhout JG, Lhoták Š, Meng QH, Wang R (2013) Decreased endogenous production of hydrogen sulfide accelerates atherosclerosis. Circulation 127:2523–2534

Mani S, Untereiner A, Wu L, Wang R (2014) Hydrogen sulfide and the pathogenesis of atherosclerosis. Antioxid Redox Signal 20:805–817

Martelli A, Testai L, Breschi MC, Lawson K, McKay NG, Miceli F, Taglialatela M, Calderone V (2013) Vasorelaxation by hydrogen sulphide involves activation of Kv7 potassium channels. Pharmacol Res 70:27–34

McCarty MF, DiNicolantonio JJ (2014) The molecular biology and pathophysiology of vascular calcification. Postgrad Med 126:54–64

Meng QH, Yang G, Yang W, Jiang B, Wu L, Wang R (2007) Protective effect of hydrogen sulphide on balloon injury-induced neointima hyperplasia in rat carotid arteries. Am J Pathol 170:1406–1414

Monti M, Terzuoli E, Ziche M, Morbidelli L (2013) The sulphydryl containing ACE inhibitor Zofenoprilat protects coronary endothelium from Doxorubicin-induced apoptosis. Pharmacol Res 76:171–181

Muellner MK, Schreier SM, Laggner H, Hermann M, Esterbauer H, Exner M, Gmeiner BM, Kapiotis S (2009) Hydrogen sulfide destroys lipid hydroperoxides in oxidized LDL. Biochem J 420:277–281

Mustafa AK, Sikka G, Gazi SK, Steppan J, Jung SM, Bhunia AK, Barodka VM, Gazi FK, Barrow RK, Wang R, Amzel LM, Berkowitz DE, Snyder SH (2011) Hydrogen sulfide as endothelium-derived hyperpolarizing factor sulfhydrates potassium channels. Circ Res 109:1259–1268

Muzaffar S, Jeremy JY, Sparatore A, Del Soldato P, Angelini GD, Shukla N (2008) H_2S-donating sildenafil (ACS6) inhibits superoxide formation and gp91phox expression in arterial endothelial cells: role of protein kinases A and G. Br J Pharmacol 155:984–994

Olson KR, Whitfield NL, Bearden SE, St LJ, Nilson E, Gao Y, Madden JA (2010) Hypoxic pulmonary vasodilation: a paradigm shift with a hydrogen sulfide mechanism. Am J Physiol Regul Integr Comp Physiol 298:R51–R60

Owens GK (2007) Molecular control of vascular smooth muscle cell differentiation and phenotypic plasticity. Novartis Found Symp 283:174–191

Pan LL, Liu XH, Gong QH, Wu D, Zhu YZ (2011) Hydrogen sulfide attenuated tumor necrosis factor-α-induced inflammatory signaling and dysfunction in vascular endothelial cells. PLoS ONE 6:e19766

Pan LL, Liu XH, Zheng HM, Yang HB, Gong QH, Zhu YZ (2012) S-propargyl-cysteine, a novel hydrogen sulfide-modulated agent, attenuated tumor necrosis factor-α-induced inflammatory signaling and dysfunction in endothelial cells. Int J Cardiol 155:327–332

Papapetropoulosa A, Pyriochoua A, Altaany Z, Yang G, Maraziotia A, Jeschkec MG, Branskic LK, Herndonc DN, Wang R, Szabo C (2009) Hydrogen sulfide is an endogenous stimulator of angiogenesis. Proc Natl Acad Sci USA 106:21972–21977

Patel P, Vatish M, Heptinstall J, Wang R, Carson RJ (2009) The endogenous production of hydrogen sulphide in intrauterine tissues. Reprod Biol Endocrinol 7:10

Perna AF, Sepe I, Lanza D, Capasso R, Zappavigna S, Capasso G, Caraglia M, Ingrosso D (2013) Hydrogen sulfide reduces cell adhesion and relevant inflammatory triggering by preventing ADAM17-dependent TNF-α activation. J Cell Biochem 114:1536–1548

Polhemus DJ, Lefer DJ (2014) Emergence of hydrogen sulfide as an endogenous gaseous signaling molecule in cardiovascular disease. Circ Res 114:730–737

Predmore BL, Alendy MJ, Ahmed KI, Leeuwenburgh C, Julian D (2010) The hydrogen sulfide signaling system: changes during aging and the benefits of caloric restriction. Age (Dordr) 32:467–481

Predmore BL, Julian D, Cardounel AJ (2011) Hydrogen sulfide increases nitric oxide production from endothelial cells by an akt-dependent mechanism. Front Physiol 2:104

Pupo E, Pla AF, Avanzato D, Moccia F, Cruz JE, Tanzi F, Merlino A, Mancardi D, Munaron L (2011) Hydrogen sulfide promotes calcium signals and migration in tumor-derived endothelial cells. Free Radic Biol Med 51:1765–1773

Qabazard B, Li L, Gruber J, Peh MT, Ng LF, Kumar SD, Rose P, Tan CH, Dymock BW, Wei F, Swain SC, Halliwell B, Stürzenbaum SR, Moore PK (2014) Hydrogen sulfide is an endogenous regulator of aging in caenorhabditis elegans. Antioxid Redox Signal 20:2621–2630

Qi HN, Cui J, Liu L, Lu FF, Song CJ, Shi Y, Yan CD (2012) Exogenous hydrogen sulfide delays the senescence of human umbilical vein endothelial cells by lessening oxidative stress. Sheng Li Xue Bao 64:425–432

Qiao W, Chaoshu T, Hongfang J, Junbao D (2010) Endogenous hydrogen sulfide is involved in the pathogenesis of atherosclerosis. Biochem Biophys Res Commun 396:182–186

Renga B, Mencarelli A, Migliorati M, Distrutti E, Fiorucci S (2009) Bile-acid-activated farnesoid X receptor regulates hydrogen sulfide production and hepatic microcirculation. World J Gastroenterol 15:2097–2108

Rossoni G, Manfredi B, Tazzari V, Sparatore A, Trivulzio S, Del Soldato P, Berti F (2010) Activity of a new hydrogen sulfide-releasing aspirin (ACS14) on pathological cardiovascular alterations induced by glutathione depletion in rats. Eur J Pharmacol 648:139–145

Sen N, Paul BD, Gadalla MM, Mustafa AK, Sen T, Xu R, Kim S, Snyder SH (2012a) Hydrogen sulfide-linked sulfhydration of NF-κB mediates its antiapoptotic actions. Mol Cell 45:13–24

Sen U, Sathnur PB, Kundu S, Givvimani S, Coley DM, Mishra PK, Qipshidze N, Tyagi N, Metreveli N, Tyagi SC (2012b) Increased endogenous H_2S generation by CBS, CSE, and 3MST gene therapy improves ex vivo renovascular relaxation in hyperhomocysteinemia. Am J Physiol Cell Physiol 303:C41–C51

Shibuya N, Mikami Y, Kimura Y, Nagahara N, Kimura H (2009) Vascular endothelium expresses 3-mercaptopyruvate sulfurtransferase and produces hydrogen sulfide. J Biochem 46:623–626

Siebert N, Cantré D, Eipel C, Vollmar B (2008) H_2S contributes to the hepatic arterial buffer response and mediates vasorelaxation of the hepatic artery via activation of K(ATP) channels. Am J Physiol Gastrointest Liver Physiol 295:G1266–G1273

Skovgaard N, Gouliaev A, Aalling M, Simonsen U (2011) The role of endogenous H_2S in cardiovascular physiology. Curr Pharm Biotechnol 12:1385–1393, Review

Sun Y, Tang CS, Jin HF, Du JB (2011) The vasorelaxing effect of hydrogen sulfide on isolated rat aortic rings versus pulmonary artery rings. Acta Pharmacol Sin 32:456–464

Suo R, Zhao ZZ, Tang ZH, Ren Z, Liu X, Liu LS, Wang Z, Tang CK, Wei DH, Jiang ZS (2013) Hydrogen sulfide prevents H_2O_2-induced senescence in human umbilical vein endothelial cells through SIRT1 activation. Mol Med Rep 7:1865–1870

Suzuki K, Olah G, Modis K, Coletta C, Kulp G, Gerö D, Szoleczky P, Chang T, Zhou Z, Wu L, Wang R, Papapetropoulos A, Szabo C (2011) Hydrogen sulfide replacement therapy protects the vascular endothelium in hyperglycemia by preserving mitochondrial function. Proc Natl Acad Sci USA 108:13829–13834

Szczesny B, Módis K, Yanagi K, Coletta C, Le Trionnaire S, Perry A, Wood ME, Whiteman M, Szabo C (2014) AP39, a novel mitochondria-targeted hydrogen sulfide donor, stimulates cellular bioenergetics, exerts cytoprotective effects and protects against the loss of mitochondrial DNA integrity in oxidatively stressed endothelial cells in vitro. Nitric Oxide S1089–8603 (14):00206–00207

Tang G, Wu L, Liang W, Wang R (2005) Direct stimulation of K(ATP) channels by exogenous and endogenous hydrogen sulfide in vascular smooth muscle cells. Mol Pharmacol 68:1757–1764

Tang G, Wu L, Wang R (2010) Interaction of hydrogen sulfide with ion channels. Clin Exp Pharmacol Physiol 37:753–763, Review

Tang G, Yang G, Jiang B, Ju Y, Wu L, Wang R (2013) H_2S is an endothelium-derived hyperpolarizing factor. Antioxid Redox Signal 19:1634–1646

Thyberg J (1998) Phenotypic modulation of smooth muscle cells during formation of neointimal thickenings following vascular injury. Histol Histopathol 13:871–891

Tyagi N, Moshal KS, Sen U, Vacek TP, Kumar M, Hughes WM Jr, Kundu S, Tyagi SC (2009) H_2S protects against methionine-induced oxidative stress in brain endothelial cells. Antioxid Redox Signal 11:25–33

Vacek TP, Gillespie W, Tyagi N, Vacek JC, Tyagi SC (2010) Hydrogen sulfide protects against vascular remodeling from endothelial damage. Amino Acids 39:1161–1169

Wang R (2002) Two's company, three's a crowd – can H_2S be the third endogenous gaseous transmitter? FASEB J 16:1792–1798

Wang R (ed) (2004) Signal transduction and the gasotransmitters: NO, CO, and H_2S in biology and medicine. Humana, Totowa

Wang R (2009) Hydrogen sulfide: a new EDRF. Kidney Int 76:700–704, Review

Wang R (2012a) Shared signaling pathways among gasotransmitters. Proc Natl Acad Sci USA 109:8801–8802

Wang R (2012b) Physiological implications of hydrogen sulfide – a whiff exploration that blossomed. Physiol Rev 92:791–896

Wang R (2014) Gasotransmitter: growing pains and joys. Trends Biochem Sci 39:227–232

Wang Y, Zhao X, Jin H, Wei H, Li W, Bu D, Tang X, Ren Y, Tang C, Du J (2009) Role of hydrogen sulfide in the development of atherosclerotic lesions in apolipoprotein E knockout mice. Arterioscler Thromb Vasc Biol 29:173–179

Wang P, Zhang G, Wondimu T, Ross B, Wang R (2011) Hydrogen sulfide and asthma. Exp Physiol 96:847–852

Wang K, Ahmad S, Cai M, Rennie J, Fujisawa T, Crispi F, Baily J, Miller MR, Cudmore M, Hadoke PW, Wang R, Gratacós E, Buhimschi IA, Buhimschi CS, Ahmed A (2013a) Dysregulation of hydrogen sulfide producing enzyme cystathionine γ-lyase contributes to maternal hypertension and placental abnormalities in preeclampsia. Circulation 127:2514–2522

Wang XH, Wang F, You SJ, Cao YJ, Cao LD, Han Q, Liu CF, Hu LF (2013b) Dysregulation of cystathionine γ-lyase (CSE)/hydrogen sulfide pathway contributes to ox-LDL-induced inflammation in macrophage. Cell Signal 25:2255–2262

Webb GD, Lim LH, Oh VM, Yeo SB, Cheong YP, Ali MY, El Oakley R, Lee CN, Wong PS, Caleb MG, Salto-Tellez M, Bhatia M, Chan ES, Taylor EA, Moore PK (2008) Contractile and vasorelaxant effects of hydrogen sulfide and its biosynthesis in the human internal mammary artery. J Pharmacol Exp Ther 324:876–882

Wen YD, Wang H, Kho SH, Rinkiko S, Sheng X, Shen HM, Zhu YZ (2013) Hydrogen sulfide protects HUVECs against hydrogen peroxide induced mitochondrial dysfunction and oxidative stress. PLoS ONE 8:e53147

Whiteman M, Winyard PG (2011) Hydrogen sulfide and inflammation: the good, the bad, the ugly and the promising. Expert Rev Clin Pharmacol 4:13–32

Whiteman M, Li L, Kostetski I, Chu SH, Siau JL, Bhatia M, Moore PK (2006) Evidence for the formation of a novel nitrosothiol from the gaseous mediators nitric oxide and hydrogen sulphide. Biochem Biophys Res Commun 343:303–310

Wójcicka G, Jamroz-Wiśniewska A, Atanasova P, Chaldakov GN, Chylińska-Kula B, Bełtowski J (2011) Differential effects of statins on endogenous H_2S formation in perivascular adipose tissue. Pharmacol Res 63:68–76

Wu SY, Pan CS, Geng B, Zhao J, Yu F, Pang YZ, Tang CS, Qi YF (2006) Hydrogen sulfide ameliorates vascular calcification induced by vitamin D3 plus nicotine in rats. Acta Pharmacol Sin 27:299–306

Wu B, Teng H, Yang G, Wu L, Wang R (2012) Hydrogen sulfide inhibits the translational expression of hypoxia-inducible factor-1α. Br J Pharmacol 167:1492–1505

Xu S, Liu Z, Liu P (2014) Targeting hydrogen sulfide as a promising therapeutic strategy for atherosclerosis. Int J Cardiol 172:313–317

Yang G (2011) Hydrogen sulfide in cell survival: a double-edged sword. Expert Rev Clin Pharmacol 4:33–47

Yang G, Sun X, Wang R (2004) Hydrogen sulfide-induced apoptosis in human aorta smooth muscle cells is associated with activation of MAP kinases and caspase-3. FASEB J 18:1782–1784

Yang G, Wu L, Wang R (2006) Pro-apoptotic effect of endogenous H_2S on human aorta smooth muscle cells. FASEB J 20:553–555

Yang G, Wu L, Jiang B, Yang W, Qi J, Cao K, Meng Q, Mustafa AK, Mu W, Zhang S, Snyder SH, Wang R (2008) H_2S as a physiologic vasorelaxant: hypertension in mice with deletion of cystathionine gamma-lyase. Science 322:587–590

Yang G, Wu L, Bryan S, Khaper N, Mani S, Wang R (2010) Cystathionine gamma-lyase deficiency and overproliferation of smooth muscle cells. Cardiovasc Res 86:487–495

Yang G, Pei Y, Teng H, Cao Q, Wang R (2011) Specificity protein-1 as a critical regulator of human cystathionine gamma-lyase expression in smooth muscle cells. J Biol Chem 286:26450–26460

Yang G, Li H, Tang G, Wu L, Cao Q, Xu C, Wang R (2012a) Increased neointimal formation in cystathionine gamma-lyase deficient mice: role of hydrogen sulfide in $\alpha5\beta1$-integrin and matrix metalloproteinase-2 expression in smooth muscle cells. J Mol Cell Cardiol 52:677–688

Yang G, Pei Y, Cao Q, Wang R (2012b) MicroRNA-21 regulates cystathionine gamma-lyase expression in smooth muscle cells by targeting at specificity protein-1. J Cell Physiol 227:3192–3200

Zanardo RC, Brancaleone V, Distrutti E, Fiorucci S, Cirino G, Wallace JL (2006) Hydrogen sulfide is an endogenous modulator of leukocyte-mediated inflammation. FASEB J 20:2118–2120

Zavaczki E, Jeney V, Agarwal A, Zarjou A, Oros M, Katkó M, Varga Z, Balla G, Balla J (2011) Hydrogen sulfide inhibits the calcification and osteoblastic differentiation of vascular smooth muscle cells. Kidney Int 80:731–739

Zhang L, Yang G, Tang G, Wu L, Wang R (2011) Rat pancreatic level of cystathionine γ-lyase is regulated by glucose level via specificity protein 1 (SP1) phosphorylation. Diabetologia 54:2615–2625

Zhang H, Guo C, Wu D, Zhang A, Gu T, Wang L, Wang C (2012a) Hydrogen sulfide inhibits the development of atherosclerosis with suppressing CX3CR1 and CX3CL1 expression. PLoS One 7:e41147

Zhang H, Guo C, Zhang A, Fan Y, Gu T, Wu D, Sparatore A, Wang C (2012b) Effect of S-aspirin, a novel hydrogen-sulfide-releasing aspirin (ACS14), on atherosclerosis in apoE-deficient mice. Eur J Pharmacol 697:106–116

Zhao W, Zhang J, Lu Y, Wang R (2001) The vasorelaxant effect of H_2S as a novel endogenous gaseous K_{ATP} channel opener. EMBO J 20:6008–6016

Zhao X, Zhang LK, Zhang CY, Zeng XJ, Yan H, Jin HF, Tang CS, DU JB (2008) Regulatory effect of hydrogen sulfide on vascular collagen content in spontaneously hypertensive rats. Hypertens Res 31:1619–1630

Zhao ZZ, Wang Z, Li GH, Wang R, Tan JM, Cao X, Suo R, Jiang ZS (2011) Hydrogen sulfide inhibits macrophage-derived foam cell formation. Exp Biol Med (Maywood) 236:169–176

Zhao K, Ju Y, Li S, Al Tanny Z, Wang R, Yang G (2014) S-sulfhydration of MEK1 leads to PARP-1 activation and DNA damage repair. EMBO Rep 15(7):792–800 [Epub ahead of print]

Zhou K, Gao Q, Zheng S, Pan S, Li P, Suo K, Simoncini T, Wang T, Fu X (2013) 17β-estradiol induces vasorelaxation by stimulating endothelial hydrogen sulfide release. Mol Hum Reprod 19:169–176

Hydrogen Sulfide and Urogenital Tract

Roberta d'Emmanuele di Villa Bianca, Giuseppe Cirino, and Raffaella Sorrentino

Contents

Abstract

In this chapter the role played by H_2S in the physiopathology of urogenital tract revising animal and human data available in the current relevant literature is discussed. H_2S pathway has been demonstrated to be involved in the mechanism

R. d'Emmanuele di Villa Bianca • G. Cirino (✉) • R. Sorrentino
Department of Pharmacy, University of Naples, Federico II, Via D. Montesano, 49, Naples 80131, Italy
e-mail: cirino@unina.it

P.K. Moore, M. Whiteman (eds.), *Chemistry, Biochemistry and Pharmacology of Hydrogen Sulfide*, Handbook of Experimental Pharmacology 230,
DOI 10.1007/978-3-319-18144-8_5

underlying penile erection in human and experimental animal. Both cystathionine-β synthase (CBS) and cystathionine-γ lyase (CSE) are expressed in the human corpus cavernosum and exogenous H_2S relaxes isolated human corpus cavernosum strips in an endothelium-independent manner. Hydrogen sulfide pathway also accounts for the direct vasodilatory effect operated by testosterone on isolated vessels. Convincing evidence suggests that H_2S can influence the cGMP pathway by inhibiting the phosphodiesterase 5 (PDE-5) activity. All these findings taken together suggest an important role for the H_2S pathway in human corpus cavernosum homeostasis. However, H_2S effect is not confined to human corpus cavernosum but also plays an important role in human bladder. Human bladder expresses mainly CBS and generates in vitro detectable amount of H_2S. In addition the bladder relaxant effect of the PDE-5 inhibitor sildenafil involves H_2S as mediator.

In conclusion the H_2S pathway is not only involved in penile erection but also plays a role in bladder homeostasis. In addition the finding that it involved in the mechanism of action of PDE-5 inhibitors strongly suggests that modulation of this pathway can represent a therapeutic target for the treatment of erectile dysfunction and bladder diseases.

Keywords
Bladder • Corpus cavernosum • Erectile dysfunction (ED) • Benign prostatic hyperplasia (BPH) • Lower urinary tract symptoms (LUTS) • PDE-5 inhibitor • CBS • CSE • 3MST

1 Introduction

Hydrogen sulfide (H_2S) presence in mammalian tissues has been known since the 1980s, but H_2S has been considered mainly as a metabolic waste product with no potential physiological activity. The first evidence, indicating H_2S as an endogenous mediator was published in 1996 by Abe and Kimura (1996) that described a role for this mediator in the brain. Some years, after this first evidence was published, many research groups have focused their interest on this new pathway as described in the other chapters. H_2S together with nitric oxide (NO) and carbon monoxide (CO) is now classified as gasotransmitter. Like the other two gasses, it has the ability to diffuse easily through the cellular membranes. Hydrogen sulfide can also interact with proteins such as ion channels or enzymes regulating their state either directly via chemical modification such as by sulfuration and sulfhydration or indirectly via second messengers.

The role of H_2S in the homeostatic control of our body is now consistently supported by the literature, and in the other chapters of the present book, its synthesis, measurement, chemistry, and interaction have been reported. This chapter will deal with the role played by H_2S in the physiopathology of urogenital tract revising animal and human data available in the current relevant literature.

2 Hydrogen Sulfide in Male Sexual Function

2.1 The Male Reproductive System

Male reproductive system consists of a number of sex organs that concur to the human reproductive process localized around the pelvic region. Briefly, the main male sex organs are the penis and the testicles important for producing semen and sperm. In particular, the penis and the scrotum represent the external genital organs. The penile erectile apparatus consists of two vascularized paired corpora cavernosa. The corpus spongiosum together with the urethra is related to the ventral area of the penis. The corpora cavernosa, the corpus spongiosum, and the glans penis are composed of septa of smooth muscle and erectile tissue that enclose vascular cavities. The tunica albuginea forms a thick fibrous coat to the spongy tissue of the corpora cavernosa and corpus spongiosum, and it consists of two layers. The muscles involved in penile function are the ischiocavernosum and bulbospongiosus. The ischiocavernous muscle inserts into the medial and inferior surface of the corpora and increases penile turgidity during erection. In the flaccid state, the smooth muscle fibers are tonically contracted by the sympathetic system. During erection, trabeculae smooth muscle is relaxed, and this allows a cascade of events. The arterioles of the penis are dilated while the veins are passively compressed between the tunica albuginea and the peripheral sinusoids reducing the venous outflow. This process is known as the venous-occlusive mechanism, and it leads to erection that is sustained by an increase of intra-cavernous blood pressure of about 100 mmHg. The contraction of the ischiocavernous muscle further increases the intra-cavernous pressure leading to the rigid erection phase.

The main role of the scrotum is to hold and protect the testes. It also contains numerous nerves and blood vessels. Between the male internal genital organs, there is the epididymis, a whitish mass of tightly coiled tubes cupped against the testicles. It acts as a maturation and storage for sperm before they pass into the vas deferens, which carry sperm to the ampullary gland and prostatic ducts. Testosterone is the most important sexual hormone in males and is released by the testes. This hormone has a crucial role in the development of sperm and is also responsible for the development of physical characteristics in men such as facial hair and deep voice. Moreover the accessory glands, such as the seminal vesicles and the prostate gland, provide fluids that lubricate the duct system and nourish the sperm cells. In particular, the prostate gland surrounds the ejaculatory ducts at the base of the male urethra, just below the bladder. The prostate gland is responsible for the proof semen, a liquid mixture of sperm cells, prostate fluid, and seminal fluid. This gland is also responsible for making the semen milky in appearance by mixing calcium to the semen coming from seminal vesicle. The semen remains cloudy and clumpy until the prostatic pro fibrinolysis is formed into fibrinolysis and lysis of the fibrinogen from the seminal vesicle fluids occurs.

2.2 Erectile Function and Dysfunction

Erection is a vascular event which relies upon interaction of neural and humoral mechanisms at various levels. Indeed, since the penis receives innervations from sacral parasympathetic (pelvic), thoraçolumbar sympathetic, and somatic (pudenda) nerves, the erection phenomenon requires participation of these three systems. Thus, erection is a consequence of a complex integration of several signals. In simple words, erection is essentially a spinal reflex that can be initiated by recruitment of penile afferents, but also by visual, olfactory, and imaginary stimuli. All these stimuli may contribute to the increase in intra-cavernous pressure (Cirino et al. 2006) and thus to penile erection. Erectile dysfunction (ED) is defined as the consistent or recurrent inability of a man to attain and/or maintain a penile erection sufficient for sexual activity (2nd International Consultation on Sexual Dysfunction-Paris, June 28th–July 1st, 2003). Of note, ED shares many of the risk factors that contribute to the development and the progression of cardiovascular diseases such as age, hypercholesterolemia, obesity, diabetes, and smoking (Brunner et al. 2005). Several clinical studies have shown that ED is often associated with cardiovascular disturbance and therefore it could be taken in account as an early sign of cardiovascular diseases (Dong et al. 2011; Nehra et al. 2012). This hypothesis relies on the fact that penile artery size is smaller as compared to coronary arteries. Therefore, it is feasible that the same level of endothelial dysfunction can cause a more significant reduction of blood flow in erectile tissues compared to that elicited in coronary circulation. Therefore, ED could be considered a prognostic factor for possible cardiovascular problems (Gandaglia et al. 2014).

2.3 Hydrogen Sulfide and Erectile Function and Dysfunction

The corpora cavernosa, as briefly reported above, are composed of sinusoids bearing a single layer of endothelial cells surrounded by multiple layers of smooth muscle cells. Thus, the corpora cavernosa are vascular organs. Among the vasodilator agents, it is undisputed that nitric oxide (NO) is considered one of the most important endogenous factors since it is released not only by endothelial cells but also from plexus nerve. Recently hydrogen sulfide (H_2S), as discussed in the other chapters, has been shown to possess an important role in the modulation of smooth muscle cell tone. In particular, it has been demonstrated that H_2S relaxes smooth muscle cells, and this finding together with others have lead to study the role of this mediator in the penile physiology. To date, the role of this gas in corpus cavernosum function has been addressed in animal models as well as in the human tissues (d'Emmanuele di Villa Bianca et al. 2011; Qiu et al. 2012). The first evidence was published in 2006 by Srilatha and coworkers. These authors showed that the intra-cavernous injection of sodium hydrogen sulfide (NaHS) resulted in a significant increase in penile length and cavernous pressure in primates measured by using laser Doppler flow meter and cutaneous probe. To acquire

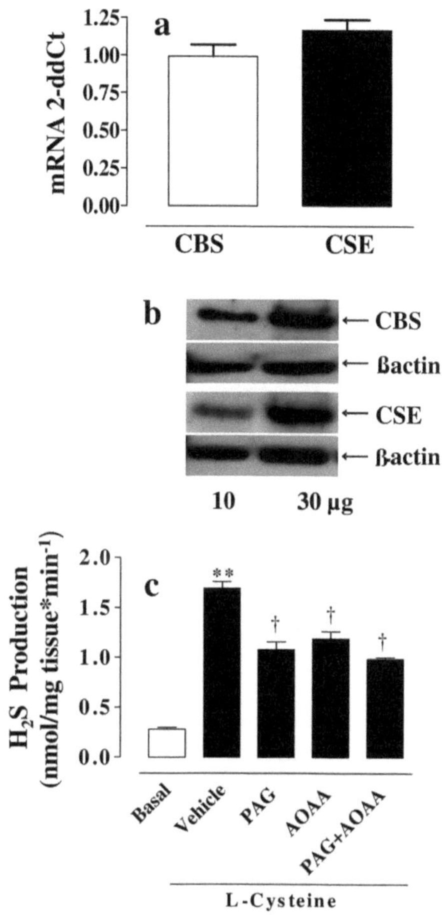

Fig. 1 CBS and CSE: activity, Western blot analysis, and qRT-PCR of human penile tissue. (**a**) HCC-expressed mRNA for both CBS and CSE as determined by qRT-PCR. (**b**) Representative Western blot analysis for CBS and CSE. (**c**) HCC homogenate produced H_2S under basal conditions (*open bar*). Incubation of HCC homogenate with 10 mM L-Cys caused a significant increase in the H_2S production compared with basal values (**$P < 0.001$). PAG (10 mM), 1 mM AOAA, or 10 mM PAG plus 1 mM AOAA significantly inhibited the L-Cys-induced increase in H_2S production ($^{\dagger}P < 0.01$). Data represent the mean SEM from 3 or 4 different human specimens (Reprinted with permission from d'Emmanuele di Villa Bianca et al. 2009)

further evidence on the role of H_2S in penile erection, a study on rats was also performed. Administration of dl-propargylglycine [PAG, a cysthatione-γ lyase (CSE) inhibitor] to rats resulted in a significant reduction in cavernous nerve stimulation-evoked perfusion pressure. This study suggested a possible role for endogenous H_2S in facilitating nerve-mediated penile tumescence (Srilatha et al. 2006). In 2009, d'Emmanuele di Villa Bianca and coworkers demonstrated the role and function of H_2S in human. The authors showed that both cysthatione-β synthase (CBS) and CSE, the main enzymes involved in H_2S synthesis, are expressed in the human corpus cavernosum (CC) (Fig. 1a, b). Moreover, by using human CC homogenates, as enzyme source, they confirmed that the enzymes present in human CC can efficiently convert L-Cys (the substrate) to H_2S (Fig. 1c). CBS and CSE are localized within muscular trabeculae and smooth muscle component of the penile artery (Fig. 2). Interestingly, the enzymes appear to be differently distributed within the penile structure. In particular, CSE but not

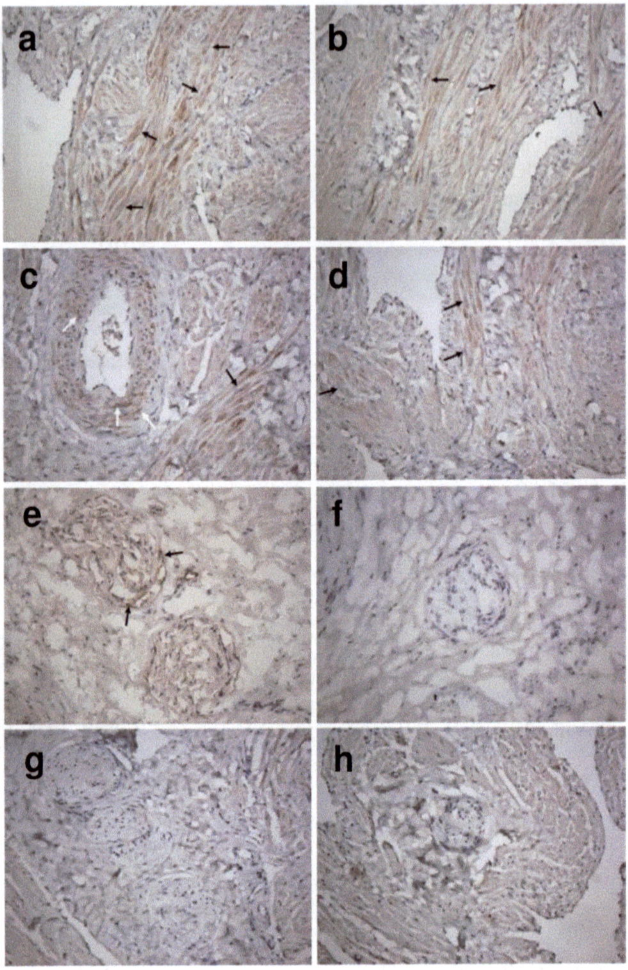

Fig. 2 Immunochemistry for CBS and CSE in HCC. Immunohistochemical detection of CSE and CBS in HCC tissue. Immunoreactivity and nuclear staining appear *brown* (DAB) and *blue* (hematoxylin counterstain), respectively. CSE was detected in trabecular muscular tissue (**a** and **b**, *black arrows*) and vascular smooth muscle cells (**c**, *white arrows*). Immunoreactivity for CBS was mostly observed in trabecular muscular tissue (**d**, *black arrows*). Results illustrated are from a single experiment and are representative of three different specimens. (Original magnification, 200.) Immunohistochemical detection of CSE and CBS in HCC nerve fibers. Immunoreactivity and nuclear staining appear *brown* (DAB) and *blue* (hematoxylin counterstain), respectively. CSE was detected in nerve fibers in cryostat (**e**, *arrows*) and not in paraffin (**g**) sections. Both cryostat (**f**) and paraffin (**h**) sections lacked immunoreactivity for CBS. Results illustrated are from a single experiment and are representative of three different specimens. (Original magnification, 200.) (Reprinted with permission from d'Emmanuele di Villa Bianca et al. 2009)

CBS is expressed in peripheral nerves (Fig. 2). Functional studies conducted by using human CC isolated strips demonstrated that both exogenous H_2S and L-Cys cause a concentration-dependent relaxation of human CC strips. This relaxant

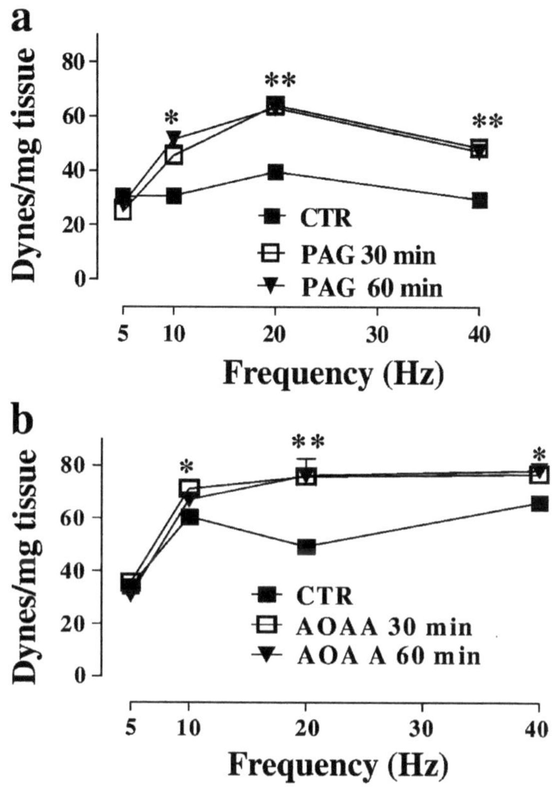

Fig. 3 EFS of HCC strips. EFS caused a frequency-related increase in basal tone. Incubation of HCC strips with 10 mM PAG for 30 or 60 min (**a**) or with 1 mM AOAA for 30 or 60 min (**b**) significantly increased the EFS-induced contraction [$*P < 0.05$; $**P < 0.01$ vs. control (CTR)]. Tissue responses to EFS are expressed as force in dynes per milligram of tissue. Experiments were performed on three different specimens (Reprinted with permission from d'Emmanuele di Villa Bianca et al. 2009)

effect was inhibited by the CBS inhibitor, aminoxyacetic acid (AOAA) and only slightly reduced by the endothelial NO-synthase inhibitor, L-NAME. Electrophysiological experiments performed using peripheral nerve electrical field stimulation of human penile tissue demonstrated that H_2S pathway is involved in penile homeostasis. Indeed, EFS of human CC strips, under resting conditions, caused an increase in tension that was significantly potentiated by inhibiting CSE and/or CBS with the selective inhibitors PAG and/or AOAA (Fig. 3). The role of H_2S pathway in erectile function was confirmed also in vivo by using an experimental animal model of penile erection. Using this model, it was demonstrated that either NaHS or L-Cys administration cause an increase in the intra-cavernous pressure in anesthetized rats (d'Emmanuele di Villa Bianca et al. 2009).

These studies have revealed that there are important differences between the human and rat tissue for what concerns the H_2S pathway. In fact, in rats, only CSE is expressed. This finding is also confirmed by the fact that CSE but not CBS inhibition significantly decreased H_2S production when rat cavernosal tissue homogenates were used as enzyme source to generate H_2S from L-cysteine in vitro. Concerning the role of CSE and H_2S in rat corpus cavernosum, it has been also shown that CSE inhibition by PAG causes a significant increase in non-adrenergic non-cholinergic-induced relaxation leading to an enhanced neurogenic relaxation of rat corpus cavernosum induced by EFS. The authors hypothesize that, since H_2S does not have a direct constrictor effect in corporal tissue, this effect might be due to either inhibition of NO synthesis by endogenous H_2S or to a direct chemical reaction between H_2S and NOS products (Ghasemi et al. 2012). Thus, while CSE in rats accounts for the production of H_2S, in human, the H_2S pathway is sustained by both CBS and CSE, which are also differently distributed in the human corpus cavernosum (Fig. 2) indicating that human and rat penile tissues are different for what concerns the H_2S pathway involvement in penile erectile mechanism(s).

2.4 Mechanism(s) of Action that Sustain the Effect of H_2S in Erection

2.4.1 Ion Channels and EDHF as H_2S Target

Ion channels are intimately involved in the biochemical events associated with smooth muscle function, and their activation/inactivation is tightly associated with tumescence/detumescence function during erection. As already reported in the previous chapters, H_2S activates within the vasculature the adenosine triphosphate (ATP)-activated potassium (K_{ATP}) channel. The mechanism of activation proposed involves S-sulfhydration by H_2S of the cysteine residues located on a specific subunit of the extracellular loop of the K_{ATP} channel complex leading to opening of this channel (Jiang et al. 2010; Mustafa et al. 2011). In particular, potassium channels are important in mediating physiologically relevant relaxation responses in human, rabbit, and rat corpus cavernosum strips (Christ et al. 1993; Mirone et al. 2000; Karicheti and Christ 2001; Spektor et al. 2002; Ruiz Rubio et al. 2004). In this context, it has been proposed that K_{ATP} channels could have an important role as modulator of corporal smooth muscle tone (Christ et al. 1993). Indeed, in diabetes, the K_{ATP} channels-mediated relaxation is significantly reduced in human CC strips (Venkateswarlu et al. 2002). Interestingly, in human CC strips glybenclamide, a selective K_{ATP} channel inhibitor, significantly impairs H_2S-induced relaxation (d'Emmanuele di Villa Bianca et al. 2009) indicating that these channels are involved in H_2S effect. K_{ATP} channels may also act by increasing cyclic adenosine monophosphate (cAMP) and thus cause relaxation. In fact, K_{ATP} channels are a physiologically important target of the adenylate cyclase/cAMP/PKA signaling pathway (Nelson et al. 2011), confirming indirectly the possible involvement of this channel as proposed by Srilatha et al. in 2007. In fact, the effect of NaHS in rabbit CC can be blocked by using MDL 12,330A, an adenylate cyclase inhibitor (Srilatha et al. 2007). On this basis, it is feasible to hypothesize that in diabetes, where there is a reduced activity of these channels, the H_2S pathways may be involved in ED associated with the diabetes. However, these issues need to be addressed more carefully in order to define the contribution of this mechanism to the physiology of erection as well as its involvement in ED.

Due to the ability of H_2S to directly interact with proteins, another channel that seems to be its target is the calcium-dependent potassium (K_{Ca}) channel.

The physiological role of this channel in human CCSM cells have been demonstrated (Christ et al. 1993). As mentioned before, NO is a key mediator in erectile function, and it derives from both peripheral nervous terminations and of course from endothelial cells. However, the existence of an unidentified endothelial factor that promotes smooth muscle hyperpolarization and relaxation, resistant to NO synthase (NOS) and cyclooxygenase (COX) inhibition, has been clearly established (Busse et al. 2002). This factor, named endothelial-derived hyperpolarizing factor (EDHF), has particular functional relevance in small arteries. This pathway involves the activation of two populations of endothelial potassium channels, the small/intermediate conductance and the large K_{Ca} channels (Grgic et al. 2009). It has been clearly demonstrated that in human penile resistance arteries, as opposite to human CC, there is a significant component in the relaxation response to acetylcholine that is resistant to NOS and COX inhibition attributed to EDHF-like activity (Angulo et al. 2003). This EDHF-like activity is impaired in human penile resistance arteries from diabetic patients. In this framework fits H_2S since there are significant data supporting H_2S as an EDHF (Wang 2003, 2009; d'Emmanuele di Villa Bianca et al. 2011; Tang et al. 2013).

Inhibitors of the phosphodiesterase-5 (PDE-5) are drugs largely used to treat men with ED to increase the half-life of cyclic guanosine monophosphate (cGMP) in the corpus cavernosum facilitating its relaxation and therefore penile erection. However, although sildenafil has been shown to be efficacious for treating ED, the percentage of efficacy in diabetic patients is clearly reduced when compared to nondiabetic men (Rendell et al. 1999; Price et al. 1998; Vickers and Satyanarayana 2002). Indeed, the presence of ED and diabetes in a patient represents a prognostic factor for a poor response to sildenafil but also with the other new PDE-5 inhibitors tadalafil (Saenz de Tejada et al. 2002) and vardenafil (Goldstein et al. 2003). The mechanism(s) responsible for this poor clinical outcome is not, as yet, understood. One hypothesis formulated is related to the impairment of EDHF in human penile resistance vessels of diabetes patients that is not restored by PDE-5 treatment. This hypothesis is supported by a study where it has been shown that dobesilate, a drug that specifically enhances endothelium-dependent relaxation attributed to EDHF, restores the endothelial function in penile arteries of diabetic patients (Angulo et al. 2003). Therefore, it is also feasible, and it needs to be explored the possibility, that the combination of an H_2S donor and a PDE-5 inhibitor may overcome the poor outcome of PDE-5 inhibitors in diabetic patients and could be beneficial in the oral treatment of diabetic ED.

Other H_2S target may be the calcium permeable channels. In fact, other authors have reported the ability of H_2S to interfere not only with potassium channels but also with calcium homeostasis particularly in neurons, cardiomyocytes, and endothelial cells. Although some effects of H_2S on calcium signals are secondary to K_{ATP} modulation, there is growing consensus about the existence of a direct effect of H_2S on Ca^{2+}-permeable channels (Munaron et al. 2013). The modulation/activation of these channels seems to be tightly related to the H_2S concentration. At the present stage, there are no data available in the current literature on CCSM cells in both animal models and human tissues.

2.4.2 Testosterone, Gender Difference, and Biosynthesis of H_2S

Testosterone (T), as already reported above, plays a critical role not only in human male sexual behavior, but several studies have indicated a relation between androgen levels and sexual interest, libido, and the frequencies of orgasm and nocturnal erections. Moreover, there is a growing body of evidence that low T levels are associated with an increased cardiovascular and cancer mortality (Hackett et al. 2014). Thus, T seems to have a protective effect in man as occurs for estrogen in woman. For example, T plasma levels negatively correlate with hypertension, diabetes, and severe coronary artery disease. While it is clear that T restoration to a physiological level is beneficial in hypogonadal subjects, the positive role of T treatment is questionable in men who are not clearly hypogonadal or eugonadal. Another important well-accepted concept is that T level correlates to ED (Isidori et al. 2014). In particular, patients with organic ED have lower free T levels than patients with psychogenic ED. Moreover, a strong positive correlation exists between free T levels and the degree of trabecular smooth muscle relaxation, as measured by resistive index at dynamic duplex ultrasound (Aversa et al. 2000). The role of T in regulating CCSM and penile arterial tone has been extensively explored in animal models. In this regards, a dual separate effect with or without the endothelium has been shown. The effect in presence of endothelium is related to NO; in particular, there are several evidences supporting a role for androgens in regulating the expression and activity of NOS isoforms in the corpus cavernosum in animal models (Traish et al. 2007). The increase in T plasma levels in response to sexual stimulation raised the question of its biological significance. To this aim, several indirect evidences support the presence of a local vascular effect of T in men. In fact, it has been demonstrated that penile erection in the healthy male is associated with a significant increase in T level in both circulating and in the cavernous plasma. Moreover, in the flaccid state, the T level in cavernous plasma is significantly lower when compared to the systemic concentration, and on this basis, the authors conclude that these differential evaluation could represent a diagnostic tool (Becker et al. 2000). Interestingly, during the penile tumescence and rigidity, T level in corpus cavernosum increases significantly in comparison to the flaccid state, and this increase is not observed in ED patients (Becker et al. 2001). These results further support the hypothesis that T, through the androgen receptor, has a direct effect on the cavernous smooth muscle. Androgen receptors are present in the human CC (Schultheiss et al. 2003), and T induces relaxation by activating smooth muscle K_{ATP} channels in human CC strips (Yildiz et al. 2009). This study was the first to report a direct non-genomic relaxant effect of T on human CC in vitro. The authors have demonstrated that T causes a rapid vasorelaxation partially mediated by an increase of potassium efflux through K_{ATP} channels, but not involving BK_{Ca}, voltage-dependent inward rectifier K channel, or voltage K channels. Thus, T-induced relaxation on smooth muscle CC seems to involve an endothelium-dependent (particularly trough NO pathway) and -independent (probably by K channels) mechanism (Aversa et al. 2000). It has been confirmed that these channels are also involved in T-induced relaxation within the vasculature and that T effect involves H_2S pathway. In fact, it has been shown that T causes an increase in H_2S level involving the L-Cys/H_2S pathway in rat aorta (Bucci et al. 2009). The increase of H_2S induced by T and its vasodilatory effect was prevented by the androgen receptor antagonist nilutamide in rat aorta rings indicating that the interaction with the androgen receptor is a key issue (Bucci et al. 2009). In this context, it has been recently shown the heat shock protein 90, which plays a role in the activation of androgen receptor, is also involved in CSE activation. In the same work, the presence of a similar mechanism for progesterone and 17-β-estradiol was ruled out. This latter observation well fit with the finding that H_2S levels in human blood, collected from male healthy volunteers, were higher than those in female samples (Brancaleone et al. 2014).

Thus, if T level is related to ED, it could well be that an impairment of T-mediated H_2S production plays a role in ED. On this specific subject, there is only one study available, performed using an in vivo rat model. In this work, the role of endogenous H_2S in ED induced by androgen deficiency has been investigated. Serum T levels were significantly reduced after 2 and 4 weeks from castration, and in this condition, the H_2S pathway was significantly impaired as well as there was a reduction in intra-cavernous pressure increase elicited by electrical stimulation. On the other hand, T replacement resulted efficacious to restore H_2S release and to improve intra-cavernous pressure in rats. These in vivo data support the hypothesis that H_2S pathway may be one of the mechanisms underlying androgen role in erection (Zuo et al. 2014). Moreover, it has been demonstrated that aging significantly impairs NO and H_2S level both in plasma and corpus cavernosum tissue. A reduction of the intra-cavernous pressure is countered by NaHS or sildenafil after 10 weeks of treatment. The link between T and H_2S was further confirmed by Srilatha and coauthors, who have shown a marked increase in T or estradiol levels after NaHS supplementation. These data support the idea that ED in aging may be also linked to a derangement in the H_2S pathway accompanied by low T levels (Srilatha et al. 2012).

All these findings support the importance of androgens in regulating smooth muscle function in the penis. A possible clinical application could be the use of a combination of PDE-5, T, and/or H_2S donors/substrate (such as L-Cysteine) to be used in patients nonresponder to PDE-5 therapy (Isidori et al. 2014).

2.4.3 Interaction Between H_2S and NO

There is a deep discussion in literature concerning the role of NO/cGMP pathway in H_2S effects, and this issue is under investigation in several anatomical district (see other chapters). For example, it has been demonstrated that CSE activity is upregulated by NO (Zhao et al. 2001), and it is partially inhibited by NOS blockade (Zhao and Wang 2002). The data available in the current literature demonstrated that H_2S (1) inhibits eNOS activity partly through inhibition of eNOS phosphorylation by reducing Akt phosphorylation (Geng et al. 2007), (2) stimulates the activity of several upstream kinase such as Akt and in turn activates eNOS by phosphorylation (Cai et al. 2007), (3) acts as PDE inhibitor (Bucci et al. 2010), (4) directly causes eNOS phosphorylation (Altaany et al. 2014), (5) is mutually required with NO in order to elicit angiogenesis and vasodilatation (Coletta et al. 2012). Therefore, there are several evidence implying a cross talk at vascular level between NO and H_2S. As opposite, few studies are available on human CC tissues. The relaxing effect of H_2S on human CC strips has been demonstrated to be endothelium independent (d'Emmanuele di Villa Bianca et al. 2009), but this result does not exclude a possible indirect effect of H_2S since, in presence of the endothelium blockade of NOS, L-NAME causes a significant inhibition of H_2S-induced relaxation at higher concentration. This latter result is in line with a recent study by Meng and coworkers (Meng et al. 2013) demonstrating that incubation of rat CC tissues with NaHS leads to an increase in eNOS but not nNOS mRNA. In this study, the authors show that the increased mRNA expression of eNOS correlates with protein expression as well as with NO production. However, caveolin-1 expression, a dominant inhibitory interaction partner of eNOS, was not modified by H_2S. On this basis, the authors conclude that H_2S could be particularly useful in improving the clinical outcome of ED patients, whose erectile impairment involves a weakened function of endothelial-derived NO. However, whether H_2S increases the soluble guanylil cyclase activity directly was not examined. The possible role of H_2S on modulating the NO/cGMP pathway has been, instead, suggested in vascular studies, where it has been shown that H_2S increases cGMP levels acting as a PDE inhibitor delaying cGMP degradation (Bucci et al. 2010; Coletta et al. 2012). Preliminary data indicate that an increase in cGMP level drive by PDE-5 can cause an increase in H_2S production in mice corpus cavernosum (Dikmen et al. 2013). Many aspects concerning the interaction between these two gasses are still matter of debate in the current relevant literature. A recent review addresses in depth the possible role of this cross talk in erectile function/dysfunction (Yetik-Anacak et al. 2014).

2.5 Hydrogen Sulfide and Vas Deferens

The vas deferens transports sperm from the epididymis to the ejaculatory ducts in anticipation of ejaculation. Due to the role of H_2S on erectile function, its presence and role in *vas deferens* have also been evaluated. Both CBS and CSE are functionally expressed in the vas deferens of rat, mice, and human. The endogenous H_2S causes a smooth muscle relaxation of vas deferens (Li et al. 2011). The same group has also demonstrated that NaHS-induced effect is mediated by BK_{Ca} channel. Indeed, they observed a consistent reversion of the relaxant effect of rat vas deferens-induced relaxation by performing a pharmacological modulation with iberiotoxin or tetraethylammonium. The authors also demonstrated that H_2S modulation of K_{Ca} channels requires a redox signaling. Indeed, N-ethylmaleimide, a sulfhydryl alkylation compound protecting thiols from oxidation, inhibited NaHS relaxation as opposite to DTT, a strong reducing agent, that did not affect the H_2S response of vas deferens. Besides, the presence of the BK_{Ca} channels in rat vas deferens smooth muscle cells was also confirmed. Finally, the involvement of the NO pathway, the transient receptor potential (TRP) channels and of the K_{ATP} channels were excluded (Li et al. 2012).

2.6 Hydrogen Sulfide and Human Prostate

Regarding the role of H_2S in the prostate, it has been reported by Guo and coworkers (Guo et al. 2012) that in human prostatic tissues (obtained from patients undergoing surgery for prostatic cancer) and cells, this pathway is physiologically present. The presence of H_2S pathway has been also confirmed by using biopsy from cancer-free human prostate (Gai et al. 2013). Moreover, Guo and coworkers have demonstrated that both cell activity and CBS/CSE protein levels are higher in the androgen-dependent prostate cancer cell LNCaP than in all the other cell lines evaluated and that dihydrotestosterone downregulates this activity. Varying H_2S levels, as well as CBS/CSE expression, in human prostate stromal and epithelial compartments have been also described. In particular, prostate epithelium expresses both CBS and CSE as opposite to stromal where only CSE is expressed. Overall, these data suggest that H_2S pathway can be involved in prostate cancer and benign prostatic hyperplasia. These results also further confirm that a link between H_2S and T exists. In physiological condition, T can modulate the H_2S synthesis contributing to erectile function. In pathologic condition, such as androgen-dependent prostate cancer, where H_2S pathway is over expressed, it may represent a potential therapeutic target (Guo et al. 2012). Indeed, H_2S does not alter significantly androgen receptor expression or its phosphorylation but inhibits androgen receptor transactivation probably at the DNA-binding level. This effect may involve a posttranslational regulation of androgen receptor by S-sulfhydration leading to conformational change and alteration of protein function. Thus, H_2S may decrease the genomic effect associated with androgen receptor activation exhibiting an anti-proliferating effect, data that are in agreement with the higher cell proliferating rate associated with aging in prostate tissues of CSE knock-out mice (Zhao et al. 2014).

3 Hydrogen Sulfide in Female Sexual Function

There is very little published on the possible involvement of the H_2S pathway in female sexual physiology. In 2009, a pilot study has been published suggesting that H_2S pathway plays a physiological role in female sexual apparatus. In particular, the authors have studied the effect of exogenous H_2S in vaginal and clitoral cavernosal smooth muscle strips from New Zealand white female rabbits. By using H_2S donors and several inhibitors towards different enzymes and channels, the authors have shown that H_2S vasodilatory effect involves cAMP, (cyclic adenosine 3′5′-monophosphate), NO-cGMP (cyclic guanosine monophosphate), and K_{ATP} channels. Of particular interest is the finding that inhibition of H_2S-induced relaxation was observed only when a combination of both adenylate- and guanylate- cyclase inhibitors was used indicating that both nucleotides may concur to H_2S effect. Thus, H_2S pathway appears to be involved also in female sexual responses. Moreover, these data further support the evidence that there exists an interplay between NO and H_2S pathways (Srilatha et al. 2009).

4 Hydrogen Sulfide and Bladder

4.1 Bladder Function

The urinary bladder has two important functions: storage of urine and emptying. Storage of urine occurs at low pressure, which implies that the bladder relaxes during the filling phase. The wall of the bladder is composed of 3 layers:

1. Outer layer of loose connective tissue, containing blood and lymphatic vessels and nerves.
2. Middle layer, consisting of a mass of interlacing smooth muscle fibers and elastic tissue—this is called the detrusor muscle.
3. Inner layer composed of transitional epithelium.

The storage and periodic elimination of urine are dependent upon the reciprocal activity of two functional units in the urinary tract: a reservoir (urinary bladder) and an outlet (bladder neck, smooth and striated muscle of the urethra). During urine storage, the outlet is closed, and the bladder is quiescent, allowing intravesical pressure to remain low over a wide range of bladder volumes. During voiding, the muscles of outlet relax and the bladder smooth muscles contract, raising intravesical pressure and inducing urine flow. These changes are coordinated by three sets of nerves (parasympathetic, sympathetic, and somatic) emerging from the sacral and thoracolumbar levels of spinal cord. Thus, when the bladder is filled, the relaxation of the wall stimulates the afferent fibers, and input is transmitted to the cortex eliciting the micturition reflex. During the micturition, the detrusor muscle and the longitudinal muscle of the neck and the urethral sphincter contract by activation of parasympathetic nerves. Contextually, an inhibitory input of the somatic nerves causes a relaxation of the external sphincter allowing the urine ejection. Different signaling molecules contribute to the fine tuning of the urinary bladder, acting both in autocrine and paracrine manner. The receptors involved are specifically distributed among different types of cells within the urinary bladder structure. Various neurotransmitters, including acetylcholine, norepinephrine, dopamine, serotonin, excitatory and inhibitory amino acids, adenosine thriphosphate, nitric oxide, and neuropeptides, are implicated in the neuronal regulation of micturition (de Groat and Yoshimura 2001). Several disorders can affect the bladder such as cystitis, urinary stones, bladder cancer, urinary incontinence, hematuria, urinary retention, cystocele, bed-wetting, dysuria, overactive bladder, and lower urinary tract symptoms (LUTS). Disturbances of storage function may result in LUTS, such as urgency, frequency, and urge incontinence, the components of the overactive bladder syndrome (Abrams et al. 2002). The overactive bladder syndrome, which may be due to involuntary contractions of the smooth muscle of the bladder (detrusor) during the storage phase, is a common problem (Milsom et al. 2001). H_2S has been shown to be involved in the bladder function in different species such as trout, rat, pig, and human (Fusco et al. 2012; Gai et al. 2013; Fernandes et al. 2013a, b; Patacchini et al. 2004; Dombkowski et al. 2006).

4.2 Animal Studies

H_2S is endogenously produced in urinary bladder of trout, mice, and rat implying that the contribution of this pathway to bladder function is conserved in different species. Similarly, to human tissues, homogenates of bladder can generate detectable amount of H_2S in basal or stimulated conditions (Dombkowski et al. 2006; Matsunami et al. 2012; Gai et al. 2013). The expression of the enzymes responsible of H_2S production has been reported in bladder of rat, pig, and mice by using different methods. CSE and CBS were detected by immunohistochemistry in nerve fibers and are widely distributed within the smooth muscle layer of the pig bladder neck (Fernandes et al. 2013a). CBS, CSE, and 3MST have been also shown to be present in rat bladder (Gai et al. 2013). In mouse bladder, only the presence of CSE has been reported so far, (Matsunami et al. 2012). However, the presence of CBS cannot be excluded since it has not been evaluated (Matsunami et al. 2012).

In the urinary tract, H_2S donors caused smooth muscle relaxation and contraction depending on the species and experimental conditions used. NaHS (exogenous source of H_2S) or L-cysteine (the substrate) relaxes rat bladder strips in a concentration-dependent manner (Gai et al. 2013). This effect can be inhibited by incubating tissues with glybenclamide, a K_{ATP} channel inhibitor, CSE, and CBS blockers (Gai et al. 2013). In trout bladder, NaHS, as well as Na_2S (exogenous donor of H_2S less used in the current literature), inhibits spontaneous contractions and relaxes precontracted strips (Dombkowski et al. 2006). In trout bladder, the H_2S response does not appear to be mediated by K_{ATP} channels as occurs in human and rat since glybenclamide was ineffective. Similarly, the other types of K^+ channels were not involved in H_2S-induced relaxation, as well. The rat urinary bladder can be indirectly contracted through H_2S stimulation of capsaicine-sensitive nerves as reported by Patacchini and coworkers (Patacchini et al. 2004, 2005). This discrepancy between rat and human could be due to the different derivation of the tissue (Kardong 2005). In rat, H_2S has a contractile effect that is mediated by the activation of sensory neurons in a ruthenium red-sensitive but not capsazepine-dependent manner. Thus, it is feasible that in this case, the molecular target of H_2S could be either a receptorial domain located on the transient receptor potential vanilloid receptor 1 (TRPV1) cation channel, independent from those bound by vanilloids and other TRPV1 activators, or another ruthenium red-sensitive TRP cation channels co-expressed on primary afferent neuron terminals (Patacchini et al. 2005). The super family of TRP includes the transient receptor potential ankyrin (TRPA) (Nilius et al. 2007), which has been found on capsaicine-sensitive primary sensory neurons (Story et al. 2003; Bautista et al. 2005). The TRPA1 activators allyl isothiocyanate and cinnamaldehyde contract rat urinary bladder through stimulation of the capsaicin-sensitive nerves (Andrade et al. 2006; Patacchini et al. 1990). Therefore, an action of H_2S also on TRPA1 in urinary bladder has been suggested. This hypothesis is based upon the following findings i) co-localization of TRPA1 and TRPV1 in rat bladder afferents ii) H_2S mimics the TRPA1 agonist cinnamaldehyde effect stimulating the micturition reflex after protamine sulfate pretreatment (Streng et al. 2008). Thus, TRPA1 represents a molecular target for H_2S in rat bladder.

H_2S-mediated O_2 sensing has been demonstrated in a variety of O_2-sensing tissues in vertebrate cardiovascular and respiratory systems, including smooth muscle in systemic and respiratory blood vessels and airways, carotid artery, adrenal medulla, and other peripheral as well as central chemoreceptors (Olson 2014). Concerning the possible involvement of this mechanism in urinary tract, it has been shown that H_2S is involved in O_2 sensing/signal transduction in the trout urinary bladder (Dombkowski et al. 2006) showing some similitude with mammalian tissue (Olson 2014).

An upregulation of CSE has been found in cyclophosphamide-induced cystitis in mouse, suggesting a role for the endogenous H_2S in the pathogenesis of this experimental model (Matsunami et al. 2012). In addition, a link between H_2S and $Ca_v3.2$ T-type channel has also been suggested in the cystitis-related nociceptive changes (Matsunami et al. 2012).

The control of the bladder neck tone is important during the voiding phase either in physiological or in pathological conditions. In this regard, CBS and CSE, as reported above, have been found expressed in nerve fibers in the pig bladder neck smooth muscle layer suggesting the involvement of H_2S as mediator released following muscle distension (Fernandes et al. 2013a). Both electrical field or GYY4137, an H_2S donor, relaxed the pig bladder neck strips through K_{ATP} channel activation as well as COX1-derived prostanoids (Fernandes et al. 2013a, b) indicating also a functional role at this level. These studies provided also evidence for a role of endogenous H_2S released from nerves involvement in the inhibitory transmission of the outflow region (Fernandes et al. 2013a). Additionally, H_2S promotes the release of inhibitory neuropeptide by the activation of TRPA1 and TRPV1 (Fernandes et al. 2013b). Therefore, at the present stage, K_{ATP} channels, TRPA1, and TRPV1 are considered feasible targets for H_2S in the bladder function.

4.3 Human Studies

Human bladder expresses all three enzymes responsible for the H_2S production namely CBS, CSE, and 3-MST (Fusco et al. 2012; Gai et al. 2013). Indeed, human bladder homogenate generates a basal level of production of H_2S that can be enhanced by incubation with L-cysteine (the substrate). This effect is reversed by either PAG (CSE inhibitor) or AOAA (CBS inhibitor) (Fig. 4). Functional studies performed using isolated human bladder strips have demonstrated that both sodium hydrogen sulfide (NaHS) or L-cysteine relax human bladder strips precontracted by carbachol. Thus, H_2S pathway is involved in the regulation of bladder homeostasis.

As discussed, K_{ATP} channels are activated by H_2S. It is known that urinary bladder smooth muscles express K_{ATP} channels, and they have been shown to be involved in the regulation of bladder contractility (Andersson 1992; Bonev and Nelson 1993). The presence of mRNA for sulfonylurea receptors has been demonstrated in both pig and human detrusor (Buckner et al. 2000). The detrusor smooth muscle expresses a variety of K^+ channels that are responsible for the detrusor smooth muscle excitability and contractility (Andersson and Arner 2004;

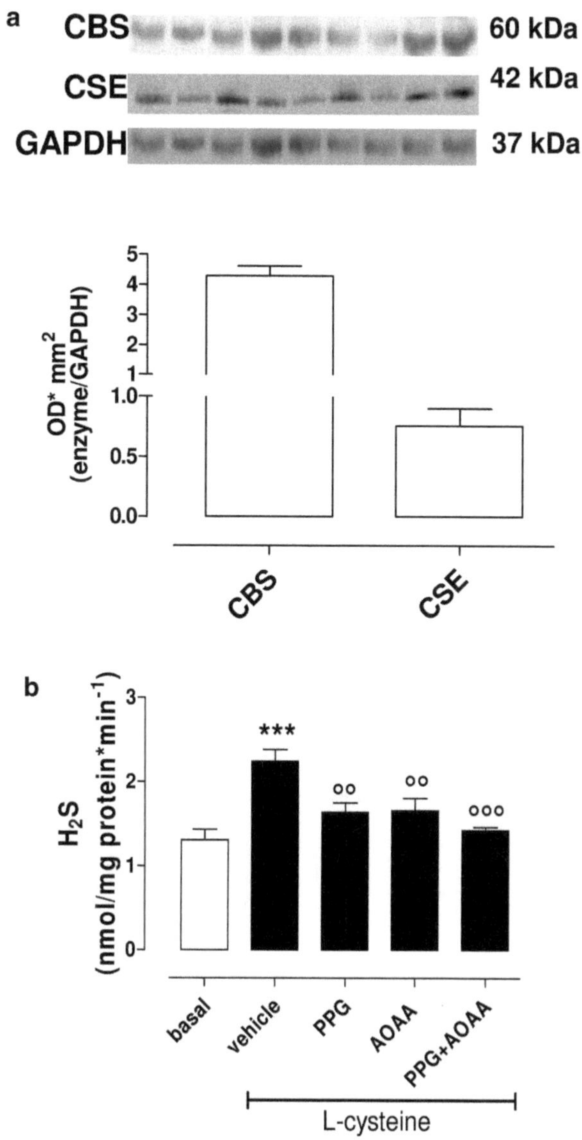

Fig. 4 Cystathionine b-synthase (CBS) and cystathionine g-lyase (CSE) western blot, and activity in human bladder dome. (**a**) Representative western blot for CBS and CSE (*upper*) and densitometric analysis for CBS and CSE normalized to glyceraldehyde-3-phosphate dehydrogenase (GAPDH; *lower*) from nine patients; data were expressed as the mean plus or minus standard error of the mean (SE). (**b**) Production of hydrogen sulfide (H_2S) in human tissue homogenates from five patients under basal condition (*open bar*) and after incubation with L-cysteine 10 mM (*black bars*). L-cysteine caused a significant increase in H_2S production (***$p < 0.001$ vs. basal) that was significantly reverted by CSE inhibitor (D,L-propargylglycine [PPG]; 10 mM) and/or CBS inhibitor (aminooxyacetic acid [AOAA]; 1 mM) (°°$p < 0.01$ and °°°$p < 0.001$, respectively, vs. L-cysteine). Data were calculated as nanomoles per milligram of protein per minute and expressed as mean ± SE (Reprinted with permission from Fusco et al. 2012)

Brading 2006; Christ and Hodges 2006; Gopalakrishnan and Shieh 2004). Interestingly, the density of K_{ATP} channels is significantly higher in detrusor smooth muscle (Petkov et al. 2001; Kajioka et al. 2008; Shieh et al. 2001). In this context, it has been shown that ~1 % of the functional K_{ATP} channels need to be activated in order to inhibit detrusor smooth muscle action potentials and related phasic contractions (Petkov et al. 2001). Studies on isolated detrusor muscle from human and several animal species have demonstrated that K_{ATP} channel openers reduce not only spontaneous contractions but also those induced by electrical stimulation, carbachol, and low external K^+ concentrations (Andersson 1993). Interestingly, the relaxant effect elicited by H_2S is inhibited by glybenclamide, a K_{ATP} channel inhibitor (Gai et al. 2013). Therefore, it is feasible to hypothesize that once released, H_2S causes relaxation of the detrusor muscle that involves K_{ATP} channel activation thereby contributing to the tonic regulation of the bladder tone. Bladder tone regulation is a key issue in LUTS. LUTS markedly increase with age in both males and females, and it represents a major problem in the elderly population. Recently, the PDE-5 inhibitor tadalafil has been approved by FDA for LUTS treatment. In addition, it has been reported that sildenafil, a well-known PDE-5 inhibitor, can directly relax human bladder strips and that this effect involves the H_2S pathway (Fusco et al. 2012). Indeed, sildenafil incubation of human bladder samples causes a time- and concentration-dependent increase in H_2S production (Fig. 5). In conclusion, the human data, available in the current

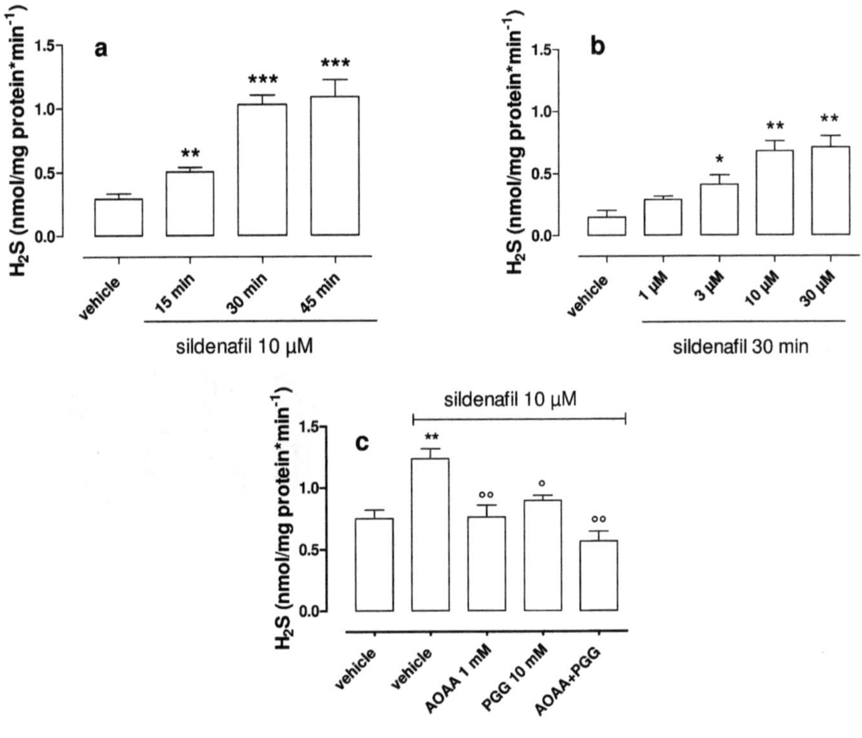

Fig. 5 Sildenafil-induced hydrogen sulfide (H_2S) production in human bladder dome. (**a**) Sildenafil 10 mM caused a time-dependent increase in H_2S production (***$p < 0.001$ vs vehicle; **$p <$ 0.01 vs. 30 min and 45 min). (**b**) Sildenafil incubation for 30 min caused a concentration-dependent increase in H_2S production compared to vehicle (*$p < 0.05$ and **$p < 0.001$ vs. vehicle). (**c**) Combination of CBS and/or CSE inhibitors (DL-propargylglycine [PPG] 10 mM and/or aminooxyacetic acid [AOAA] 1 mM) significantly inhibited sildenafil-induced H_2S production (°$p < 0.05$ and °°$p < 0.01$ vs sildenafil 10 mM; **$p < 0.01$ vs. vehicle). Data were calculated as nanomoles per milligram of protein per minute and expressed as mean ± SEM for five patients (Reprinted with permission from Fusco et al. 2012)

relevant literature, suggest that the H_2S pathway is involved not only in the physiology of human bladder function but also in LUTS as well as in overactive bladder. However, further studies are necessary in order to better define the role of this pathway.

5 Hydrogen Sulfide and Urethra

The urethra together with smooth muscles in the bladder controls the storage and the voiding of urine. The presence as well as the ability of CBS, CSE, and 3-MST to convert L-cysteine into H_2S has been reported both in human and rat urethra (Gai et al. 2013). On the functional side, H_2S has been shown to have no effect on the basal tone but to relax the human urethra strips (Gratzke et al. 2009). The relaxing effect on human urethra strips has been proposed to involve the TRPA1 receptors that are expressed on nerves fibers, urothelium, and interstitial cells (Gratzke et al. 2009).

6 Conclusions and Future Directions

On the basis of the findings present in the current literature, from non-mammalian and mammalian, including human, it is clear that H_2S is endogenously produced, and the enzymes responsible of its biosynthesis are constitutively present in the genitourinary tract. In addition, H_2S appears to be a phylogenetically ancient and versatile regulatory molecule. Nevertheless, the H_2S response in the genitourinary tract, as reported above, appears to be specie dependent. Therefore, more studies are needed in order to clearly define the role of this pathway in human since the translation of preclinical data to human is not always possible. For what concerns, drugs used in therapy that have been proposed to involve H_2S as mediator, there are some considerations that can be made. It is now well established that patients' response to a therapy is not equally effective, and who once were called nonresponders are now considered as a specific sub group. Indeed, it is now clear that genetics plays a great role in determining the nature of the drug responses. However, it is also true that, most likely, there are other therapeutic targets that have not as yet been defined within a specific pathology. In this regard, in both ED and in pathologies associated with bladder dysfunction, there is a need to find new therapies. Indeed, PDE-5 inhibitors often do not resolve the ED in diabetes patients or fail to act in a certain number of eligible patients with no comorbidities. In fact, the efficacy of PDE-5 inhibitor, which is a mainstay in the treatment of ED, is negatively associated with the nerve and endothelium damage. These latter features are associated with several pathologies such as cardiovascular disease, diabetes, obesity, and post-prostatectomy state and in turn lead to an impairment in the signaling of the NO/cGMP pathway. In this context, the H_2S pathway represents an attractive target since a link between PDE-5 inhibitors and H_2S pathway has been already shown. Indeed, tadalafil limits myocardial infarction through H_2S

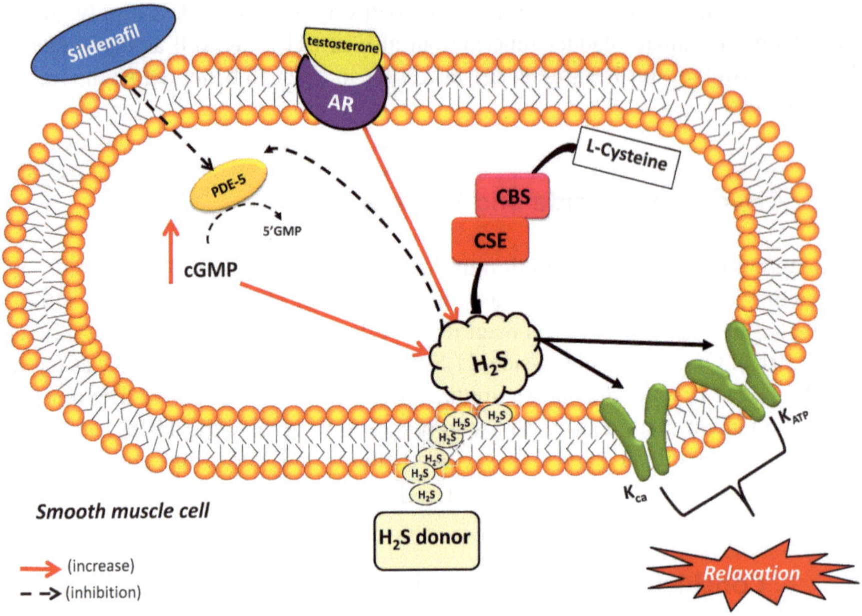

Fig. 6 Drugs proposed to involve hydrogen sulfide (H_2S) pathway in their mechanism of action. The cartoon reproduces a smooth muscle cell of the urogenital tract, where the involvement of H_2S in testosterone and sildenafil mechanism of action is depicted. Adenosine triphosphate-activated potassium channel (K_{ATP}); Androgen receptor AR; calcium-dependent potassium channel (K_{Ca}); cyclic guanosine monophosphate (cGMP); Cysthatione-β synthase (CBS); Cysthatione-γ lyase (CSE), phosphodiesterase-5 (PDE-5)

signaling (Salloum et al. 2009), and sildenafil causes an increase in H_2S production by CBS and CSE activities in human bladder (Fusco et al. 2012). A very preliminary attempt to develop a drug working on H_2S and cGMP pathways has been taken by Shukla and coworkers who have synthesized and characterized an H_2S-donating derivative of sildenafil (ACS6) (Shukla et al. 2009). For what concerns the bladder, it has to be stressed that the widely used antimuscarinic drugs often fails to alleviate the LUTS symptoms. Also in this case, by looking at the recent findings, highlighted in this chapter, the H_2S pathway represents an attractive therapeutic target that may allow to develop new drugs. In Fig. 6 are shown the drugs for which the involvement of H_2S in their mechanism of action has been proposed.

Another possible therapeutic approach that rises from what has been discussed is the possibility to modulate H_2S levels by exogenous supplementation. Intuitively, the easiest way should be to use H_2S donors that release the gas slowly, in order to reproduce as much as possible, the physiological conditions. Alternatively, the H_2S production could be enhanced by inducing H_2S synthesis by using L-cysteine or other substrates. However, this approach could fail in condition where a down-regulation of the enzyme CBS and CSE occurs. Several studies have addressed these issues in preclinical setting (Kashfi and Olson 2013). The activity of garlic-derived molecules, generally considered as H_2S releasers following metabolization, such as diallyl disulfide, diallyl sulfide, diallyl trisulfide, and diallyl tetrasulfide, has been profusely investigated (Jacob et al. 2008). Garlic and its bioactive component, the S-allyl cysteine, have been shown to restore erectile function in diabetic rats by preventing ROS formation through modulation of NADPH oxidase subunit expression (Yang et al. 2013). In particular, anti-inflammatory and anticancer effects have been demonstrated. However, no data, concerning their efficacy, on urogenital tract are available. The GYY4137 is actually the most attracting synthetic H_2S donor. It inhibits lipid accumulation exhibiting anti-atherosclerotic activity both in vitro and in vivo (Xu et al. 2014) and exerts anti-inflammatory (Li et al. 2013) and anticancer (Kashfi 2014) activity. Also in this case, there are no data available on GYY4137 effect in urogenital tract diseases.

References

Abe K, Kimura H (1996) The possible role of hydrogen sulfide as an endogenous neuromodulator. J Neurosci 16:1066–1071

Abrams P, Cardozo L, Fall M, Griffiths D, Rosier P, Ulmsten U, van Kerrebroeck P, Victor A, Wein A (2002) The standardization of terminology of lower urinary tract function: report from the Standardisation Sub-committee of the International Continence Society. Neurourol Urodyn 21:167–178

Altaany Z, Ju Y, Yang G, Wang R (2014) The coordination of S-sulfhydration, S-nitrosylation, and phosphorylation of endothelial nitric oxide synthase by hydrogen sulfide. Sci Signal 7:ra87

Andersson KE (1992) Clinical pharmacology of potassium channel openers. Pharmacol Toxicol 70:244–254

Andersson KE (1993) Pharmacology of lower urinary tract smooth muscles and penile erectile tissues. Pharmacol Rev 45:253–308

Andersson KE, Arner A (2004) Urinary bladder contraction and relaxation: physiology and pathophysiology. Physiol Rev 84:935–986

Andrade EL, Ferreira J, André E, Calixto JB (2006) Contractile mechanisms coupled to TRPA1 receptor activation in rat urinary bladder. Biochem Pharmacol 72:104–114

Angulo J, Cuevas P, Fernández A, Gabancho S, Allona A, Martín-Morales A, Moncada I, Videla S, Sáenz de Tejada I (2003) Diabetes impairs endothelium-dependent relaxation of human penile vascular tissues mediated by NO and EDHF. Biochem Biophys Res Commun 312:1202–1208

Aversa A, Isidori AM, De Martino MU, Caprio M, Fabbrini E, Rocchietti-March M, Frajese G, Fabbri A (2000) Androgens and penile erection: evidence for a direct relationship between free testosterone and cavernous vasodilation in men with erectile dysfunction. Clin Endocrinol (Oxf) 53:517–522

Bautista DM, Movahed P, Hinman A, Axelsson HE, Sterner O, Högestätt ED, Julius D, Jordt SE, Zygmunt PM (2005) Pungent products from garlic activate the sensory ion channel TRPA1. Proc Natl Acad Sci USA 102:12248–12252

Becker AJ, Uckert S, Stief CG, Truss MC, Machtens S, Scheller F, Knapp WH, Hartmann U, Jonas U (2000) Cavernous and systemic testosterone levels in different phases of human penile erection. Urology 56:125–129

Becker AJ, Uckert S, Stief CG, Scheller F, Knapp WH, Hartmann U, Jonas U (2001) Cavernous and systemic testosterone plasma levels during different penile conditions in healthy males and patients with erectile dysfunction. Urology 58:435–440

Bonev AD, Nelson MT (1993) ATP-sensitive potassium channels in smooth muscle cells from guinea pig urinary bladder. Am J Physiol Cell Physiol 264:C1190–C1200

Brading AF (2006) Spontaneous activity of lower urinary tract smooth muscles: correlation between ion channels and tissue function. J Physiol 570:13–22

Brancaleone V, Vellecco V, Matassa DS, d'Emmanuele di Villa Bianca R, Sorrentino R, Ianaro A, Bucci M, Esposito F, Cirino G (2014) Crucial role of androgen receptor in vascular H_2S biosynthesis induced by testosterone. Br J Pharmacol 172(6):1505–1515

Brunner H, Cockcroft JR, Deanfield J, Donald A, Ferrannini E, Halcox J, Kiowski W, Lüscher TF, Mancia G, Natali A, Oliver JJ, Pessina AC, Rizzoni D, Rossi GP, Salvetti A, Spieker LE, Taddei S, Webb DJ, Working Group on Endothelins and Endothelial Factors of the European Society of Hypertension (2005) Endothelial function and dysfunction. Part II: association with cardiovascular risk factors and diseases. A statement by the working group on endothelins and endothelial factors of the european society of hypertension. J Hypertens 23:233–246

Bucci M, Mirone V, Di Lorenzo A, Vellecco V, Roviezzo F, Brancaleone V, Imbimbo C, Cirino G (2009) Hydrogen sulphide is involved in testosterone vascular effect. Eur Urol 56:378–383

Bucci M, Papapetropoulos A, Vellecco V, Zhou Z, Pyriochou A, Roussos C, Roviezzo F, Brancaleone V, Cirino G (2010) Hydrogen sulfide is an endogenous inhibitor of phosphodiesterase activity. Arterioscler Thromb Vasc Biol 30:1998–2004

Buckner SA, Milicic I, Daza A, Davis-Taber R, Scott VE, Sullivan JP, Brioni JD (2000) Pharmacological and molecular analysis of ATP-sensitive K channels in the pig and human detrusor. Eur J Pharmacol 400:287–295

Busse R, Edwards G, Félétou M, Fleming I, Vanhoutte PM, Weston AH (2002) EDHF: bringing the concepts together. Trends Pharmacol Sci 23:374–380

Cai WJ, Wang MJ, Moore PK, Jin HM, Yao T, Zhu YC (2007) The novel proangiogenic effect of hydrogen sulfide is dependent on Akt phosphorylation. Cardiovasc Res 76:29–40

Christ GJ, Hodges S (2006) Molecular mechanisms of detrusor and corporal myocyte contraction: identifying targets for pharmacotherapy of bladder and erectile dysfunction. Br J Pharmacol 147(Suppl 2):S41–S55

Christ GJ, Brink PR, Melman A, Spray DC (1993) The role of gap junctions and ion channels in the modulation of electrical and chemical signals in human corpus cavernosum smooth muscle. Int J Impot Res 5:77–96

Cirino G, Fusco F, Imbimbo C, Mirone V (2006) Pharmacology of erectile dysfunction in man. Pharmacol Ther 111:400–423

Coletta C, Papapetropoulos A, Erdelyi K, Olah G, Módis K, Panopoulos P, Asimakopoulou A, Gerö D, Sharina I, Martin E, Szabo C (2012) Hydrogen sulfide and nitric oxide are mutually dependent in the regulation of angiogenesis and endothelium-dependent vasorelaxation. Proc Natl Acad Sci USA 109:9161–9166

d'Emmanuele di Villa Bianca R, Sorrentino R, Maffia P, Mirone V, Imbimbo C, Fusco F, De Palma R, Ignarro LJ, Cirino G (2009) Hydrogen sulfide as a mediator of human corpus cavernosum smooth-muscle relaxation. Proc Natl Acad Sci USA 106:4513–4518

d'Emmanuele di Villa Bianca R, Sorrentino R, Coletta C, Mitidieri E, Rossi A, Vellecco V et al (2011) Hydrogen sulfide-induced dual vascular effect involves arachidonic acid cascade in rat mesenteric arterial bed. J Pharmacol Exp Ther 337:59–64

de Groat WC, Yoshimura N (2001) Pharmacology of the lower urinary tract. Annu Rev Pharmacol Toxicol 41:691–721

Dikmen A, d'Emmanuele di Villa Bianca R, Mitidieri E, Donnarumma E, Sevin G, Cirino G, Sorrentino R, Yetik-Anacak G (2013) New mechanism for the beneficial effect of sildenafil on erectile function: H2S. Nitric Oxide 31:S38

Dombkowski RA, Doellman MM, Head SK, Olson KR (2006) Hydrogen sulfide mediates hypoxia-induced relaxation of trout urinary bladder smooth muscle. J Exp Biol 209:3234–3340

Dong JY, Zhang YH, Qin LQ (2011) Erectile dysfunction and risk of cardiovascular disease: meta-analysis of prospective cohort studies. J Am Coll Cardiol 58:1378–1385

Fernandes VS, Ribeiro AS, Barahona MV, Orensanz LM, Martínez-Sáenz A, Recio P, Martínez AC, Bustamante S, Carballido J, García-Sacristán A, Prieto D, Hernández M (2013a) Hydrogen sulfide mediated inhibitory neurotransmission to the pig bladder neck: role of K_{ATP} channels, sensory nerves and calcium signaling. J Urol 190:746–756
Fernandes VS, Ribeiro AS, Martínez MP, Orensanz LM, Barahona MV, Martínez-Sáenz A, Recio P, Benedito S, Bustamante S, Carballido J, García-Sacristán A, Prieto D, Hernández M (2013b) Endogenous hydrogen sulfide has a powerful role in inhibitory neurotransmission to the pig bladder neck. J Urol 189:1567–1573
Fusco F, d'Emmanuele di Villa Bianca RD, Mitidieri E, Cirino G, Sorrentino R, Mirone V (2012) Sildenafil effect on the human bladder involves the L-cysteine/hydrogen sulfide pathway: a novel mechanism of action of phosphodiesterase type 5 inhibitors. Eur Urol 62:1174–1180
Gai JW, Wahafu W, Guo H, Liu M, Wang XC, Xiao YX, Zhang L, Xin ZC, Jin J (2013) Further evidence of endogenous hydrogen sulphide as a mediator of relaxation in human and rat bladder. Asian J Androl 15:692–696
Gandaglia G, Briganti A, Jackson G, Kloner RA, Montorsi F, Montorsi P, Vlachopoulos C (2014) A systematic review of the association between erectile dysfunction and cardiovascular disease. Eur Urol 65:968–978
Geng B, Cui Y, Zhao J, Yu F, Zhu Y, Xu G, Zhang Z, Tang C, Du J (2007) Hydrogen sulfide downregulates the aortic L-arginine/nitric oxide pathway in rats. Am J Physiol Regul Integr Comp Physiol 293:R1608–R1618
Ghasemi M, Dehpour AR, Moore KP, Mani AR (2012) Role of endogenous hydrogen sulfide in neurogenic relaxation of rat corpus cavernosum. Biochem Pharmacol 83:1261–1268
Goldstein I, Young JM, Fischer J, Bangerter K, Segerson T, Taylor T (2003) Vardenafil a new phosphodiesterase type 5 inhibitor, in the treatment of erectile dysfunction in men with diabetes: a multicenter double-blind placebo-controlled fixed-dose study. Diabetes Care 26: 777–783
Gopalakrishnan M, Shieh CC (2004) Potassium channel subtypes as molecular targets for overactive bladder and other urological disorders. Expert Opin Ther Targets 8:437–458
Gratzke C, Streng T, Waldkirch E, Sigl K, Stief C, Andersson KE, Hedlund P (2009) Transient receptor potential A1 (TRPA1) activity in the human urethra—evidence for a functional role for TRPA1 in the outflow region. Eur Urol 55:696–704
Grgic I, Kaistha BP, Hoyer J, Köhler R (2009) Endothelial Ca+ −activated K+ channels in normal and impaired EDHF-dilator responses–relevance to cardiovascular pathologies and drug discovery. Br J Pharmacol 157:509–526
Guo H, Gai JW, Wang Y, Jin HF, Du JB, Jin J (2012) Characterization of hydrogen sulfide and its synthases, cystathionine β-synthase and cystathionine γ-lyase, in human prostatic tissue and cells. Urology 79:483, e1-5
Hackett G, Kirby M, Sinclair AJ (2014) Testosterone deficiency, cardiac health, and older men. Int J Endocrinol 2014:143763
Isidori AM, Buvat J, Corona G, Goldstein I, Jannini EA, Lenzi A, Porst H, Salonia A, Traish AM, Maggi M (2014) A critical analysis of the role of testosterone in erectile function: from pathophysiology to treatment-a systematic review. Eur Urol 65:99–112
Jacob C, Anwar A, Burkholz T (2008) Perspective on recent developments on sulfur-containing agents and hydrogen sulfide signaling. Planta Med 74:1580–1592
Jiang B, Tang G, Cao K, Wu L, Wang R (2010) Molecular mechanism for H(2)S-induced activation of K(ATP) channels. Antioxid Redox Signal 12:1167–1178
Kajioka S, Nakayama S, Asano H, Seki N, Naito S, Brading AF (2008) Levcromakalim and MgGDP activate small conductance ATP-sensitive K+channels of K+ channel pore 6.1/ sulfonylurea receptor 2A in pig detrusor smooth muscle cells: uncoupling of cAMP signal pathways. J Pharmacol Exp Ther 327:114–123
Kardong KV (2005) The urogenital system. In: Vertebrates: compartive anatomy, function, evolution. McGraw Hill Higher Education, New York, pp 577–578 (Chapter 14)

Karicheti V, Christ GJ (2001) Physiological roles for K+ channels and gap junctions in urogenital smooth muscle: implications for improved understanding of urogenital function, disease and therapy. Curr Drug Targets 2:1–20

Kashfi K (2014) Anti-cancer activity of new designer hydrogen sulfide-donating hybrids. Antioxid Redox Signal 20:831–846

Kashfi K, Olson KR (2013) Biology and therapeutic potential of hydrogen sulfide and hydrogen sulfide-releasing chimeras. Biochem Pharmacol 85:689–703

Li J, Li Y, Du Y, Mou K, Sun H, Zang Y, Liu C (2011) Endogenous hydrogen sulfide as a mediator of vas deferens smooth muscle relaxation. Fertil Steril 95:1833–1835

Li Y, Zang Y, Fu S, Zhang H, Gao L, Li J (2012) H_2S relaxes vas deferens smooth muscle by modulating the large conductance Ca2 + –activated K+ (BKCa) channels via a redox mechanism. J Sex Med 9:2806–2813

Li L, Fox B, Keeble J, Salto-Tellez M, Winyard PG, Wood ME, Moore PK, Whiteman M (2013) The complex effects of the slow-releasing hydrogen sulfide donor GYY4137 in a model of acute joint inflammation and in human cartilage cells. J Cell Mol Med 17:365–376

Matsunami M, Miki T, Nishiura K, Hayashi Y, Okawa Y, Nishikawa H, Sekiguchi F, Kubo L, Ozaki T, Tsujiuchi T, Kawabata A (2012) Involvement of the endogenous hydrogen sulfide/Ca (v) 3.2 T-type Ca2+ channel pathway in cystitis-related bladder pain in mice. Br J Pharmacol 167:917–928

Meng J, Ganesan Adaikan P, Srilatha B (2013) Hydrogen sulfide promotes nitric oxide production in corpus cavernosum by enhancing expression of endothelial nitric oxide synthase. Int J Impot Res 25:86–90

Milsom I, Abrams P, Cardozo L, Roberts RG, Thuroff J, Wein AJ (2001) How widespread are the symptoms of an overactive bladder and how are they managed? A population-based prevalence study. BJU Int 87:760–766

Mirone V, Sorrentino R, di Villa BR, Imbimbo C, Palmieri A, Fusco F, Tajana G, Cirino G (2000) A standardized procedure for using human corpus cavernosum strips to evaluate drug activity. J Pharmacol Toxicol Methods 44:477–482

Munaron L, Avanzato D, Moccia F, Mancardi D (2013) Hydrogen sulfide as a regulator of calcium channels. Cell Calcium 53:77–84

Mustafa AK, Sikka G, Gazi SK, Steppan J, Jung SM, Bhunia AK, Barodka VM, Gazi FK, Barrow RK, Wang R, Amzel LM, Berkowitz DE, Snyder SH (2011) Hydrogen sulfide as endothelium-derived hyperpolarizing factor sulfhydrates potassium channels. Circ Res 109:1259–1268

Nehra A, Jackson G, Miner M, Billups KL, Burnett AL, Buvat J, Carson CC, Cunningham GR, Ganz P, Goldstein I, Guay AT, Hackett G, Kloner RA, Kostis J, Montorsi P, Ramsey M, Rosen R, Sadovsky R, Seftel AD, Shabsigh R, Vlachopoulos C, Wu FC (2012) The Princeton III Consensus recommendations for the management of erectile dysfunction and cardiovascular disease. Mayo Clin Proc 87:766–778

Nelson CP, Rainbow RD, Brignell JL, Perry MD, Willets JM, Davies NW, Standen NB, Challiss RA (2011) Principal role of adenylyl cyclase 6 in K^+ channel regulation and vasodilator signalling in vascular smooth muscle cells. Cardiovasc Res 91:694–702

Nilius B, Owsianik G, Voets T, Peters JA (2007) Transient receptor potential cation channels in disease. Physiol Rev 87:165–217

Olson KR (2014) Hydrogen sulfide as an oxygen sensor. Antioxid Redox Signal 22(5):377–397

Patacchini R, Maggi CA, Meli A (1990) Capsaicin-like activity of some natural pungent substances on peripheral endings of visceral primary afferents. Naunyn Schmiedebergs Arch Pharmacol 342:72–77

Patacchini R, Santicioli P, Giuliani S, Maggi CA (2004) Hydrogen sulfide (H_2S) stimulates capsaicin-sensitive primary afferent neurons in the rat urinary bladder. Br J Pharmacol 142:31–34

Patacchini R, Santicioli P, Giuliani S, Maggi CA (2005) Pharmacological investigation of hydrogen sulfide (H_2S) contractile activity in rat detrusor muscle. Eur J Pharmacol 509:171–177

Petkov GV, Heppner TJ, Bonev AD, Herrera GM, Nelson MT (2001) Low levels of K_{ATP} channel activation decrease excitability and contractility of urinary bladder. Am J Physiol Regul Integr Comp Physiol 280:R1427–R1433

Price DE, Gingell JC, Gepi-Attee S, Warcham K, Yates P, Boolell M (1998) Sildenafil: study of a novel oral treatment for erectile dysfunction in diabetic men. Diabet Med 15:821–825

Qiu X, Villalta J, Lin G, Lue TF (2012) Role of hydrogen sulfide in the physiology of penile erection. J Androl 33:529–535

Rendell MS, Rajfer J, Wicker PA, Smith MD (1999) Sildenafil for treatment of erectile dysfunction in men with diabetes. A randomized controlled trial. JAMA 281:421–426

Ruiz Rubio JL, Hernández M, Arcos L, Benedito S, Recio P, García P, García-Sacristán A, Prieto D (2004) Role of ATP-sensitive K+ channels in relaxation of penile resistance arteries. Urology 63:800–805

Saenz de Tejada I, Anglin G, Knight JR, Emmick JT (2002) Effects of tadalafil on erectile dysfunction in men with diabetes. Diabetes Care 25:2159–2164

Salloum FN, Chau VQ, Hoke NN, Abbate A, Varma A, Ockaili RA, Toldo S, Kukreja RC (2009) Phosphodiesterase-5 inhibitor, tadalafil, protects against myocardial ischemia/reperfusion through protein-kinase g-dependent generation of hydrogen sulfide. Circulation 120:S31–S36

Schultheiss D, Badalyan R, Pilatz A, Gabouev AI, Schlote N, Wefer J, von Wasielewski R, Mertsching H, Sohn M, Stief CG, Jonas U (2003) Androgen and estrogen receptors in the human corpus cavernosum penis: immunohistochemical and cell culture results. World J Urol 21:320–324

Shieh CC, Feng J, Buckner SA, Brioni JD, Coghlan MJ, Sullivan JP, Gopalakrishnan M (2001) Functional implication of spare ATP-sensitive K+ channels in bladder smooth muscle cells. J Pharmacol Exp Ther 296:669–675

Shukla N, Rossoni G, Hotston M, Sparatore A, Del Soldato P, Tazzari V, Persad R, Angelini GD, Jeremy JY (2009) Effect of hydrogen sulphide-donating sildenafil (ACS6) on erectile function and oxidative stress in rabbit isolated corpus cavernosum and in hypertensive rats. BJU Int 103:1522–1529

Spektor M, Rodriguez R, Rosenbaum RS, Wang HZ, Melman A, Christ GJ (2002) Potassium channels and human corporeal smooth muscle cell tone: further evidence of the physiological relevance of the Maxi-K channel subtype to the regulation of human corporeal smooth muscle tone in vitro. J Urol 167:2628–2635

Srilatha B, Adaikan PG, Moore PK (2006) Possible role for the novel gasotransmitter hydrogen sulphide in erectile dysfunction–a pilot study. Eur J Pharmacol 535:280–282

Srilatha B, Adaikan PG, Li L, Moore PK (2007) Hydrogen sulphide: a novel endogenous gasotransmitter facilitates erectile function. J Sex Med 4:1304–1311

Srilatha B, Hu L, Adaikan GP, Moore PK (2009) Initial characterization of hydrogen sulfide effects in female sexual function. J Sex Med 6:1875–1884

Srilatha B, Muthulakshmi P, Adaikan PG, Moore PK (2012) Endogenous hydrogen sulfide insufficiency as a predictor of sexual dysfunction in aging rats. Aging Male 15:153–158

Story GM, Peier AM, Reeve AJ, Eid SR, Mosbacher J, Hricik TR, Earley TJ, Hergarden AC, Andersson DA, Hwang SW, McIntyre P, Jegla T, Bevan S, Patapoutian A (2003) ANKTM1, a TRP-like channel expressed in nociceptive neurons, is activated by cold temperatures. Cell 112:819–829

Streng T, Axelsson HE, Hedlund P, Andersson DA, Jordt SE, Bevan S, Andersson KE, Högestätt ED, Zygmunt PM (2008) Distribution and function of the hydrogen sulfide-sensitive TRPA1 ion channel in rat urinary bladder. Eur Urol 53:391–399

Tang G, Yang G, Jiang B, Ju Y, Wu L, Wang R (2013) H_2S is an endothelium-derived hyperpolarizing factor. Antioxid Redox Signal 19:1634–1646

Traish AM, Goldstein I, Kim NN (2007) Testosterone and erectile function: from basic research to a new clinical paradigm for managing men with androgen insufficiency and erectile dysfunction. Eur Urol 52:54–70

Venkateswarlu K, Giraldi A, Zhao W, Wang HZ, Melman A, Spektor M, Christ GJ (2002) Potassium channels and human corporeal smooth muscle cell tone: diabetes and relaxation of human corpus cavernosum smooth muscle by adenosine triphosphate sensitive potassium channel openers. J Urol 168:355–361

Vickers MA, Satyanarayana R (2002) Phosphodiesterase type 5 inhibitors for the treatment of erectile dysfunction in patients with diabetes mellitus. Int J Impot Res 14:466–471

Wang R (2003) The gasotransmitter role of hydrogen sulfide. Antioxid Redox Signal 5:493–501

Wang R (2009) Hydrogen sulfide: a new EDRF. Kidney Int 76:700–704

Xu S, Liu Z, Liu P (2014) Targeting hydrogen sulfide as a promising therapeutic strategy for atherosclerosis. Int J Cardiol 172:313–317

Yang J, Wang T, Yang J, Rao K, Zhan Y, Chen RB, Liu Z, Li MC, Zhuan L, Zang GH, Guo SM, Xu H, Wang SG, Liu JH, Ye ZQ (2013) S-allyl cysteine restores erectile function through inhibition of reactive oxygen species generation in diabetic rats. Andrology 1:487–494

Yetik-Anacak G, Sorrentino R, Linder AE, Murat N (2014) Gas what: NO is not the only answer to sexual function. Br J Pharmacol 172(6):1434–1454

Yildiz O, Seyrek M, Irkilata HC, Yildirim I, Tahmaz L, Dayanc M (2009) Testosterone might cause relaxation of human corpus cavernosum by potassium channel opening action. Urology 74:229–232

Zhao W, Wang R (2002) H_2S-induced vasorelaxation and underlying cellular and molecular mechanisms. Am J Physiol Heart Circ Physiol 283:H474–H480

Zhao W, Zhang J, Lu Y, Wang R (2001) The vasorelaxant effect of H_2S as a novel endogenous gaseous K_{ATP} channel opener. EMBO J 20:6008–6016

Zhao K, Li S, Wu L, Lai C, Yang G (2014) Hydrogen sulfide represses androgen receptor transactivation by targeting at the second zinc finger module. J Biol Chem 289:20824–20835

Zuo C, Huang YM, Jiang R, Yang HF, Cheng B, Chen F (2014) Endogenous hydrogen sulfide and androgen deficiency-induced erectile dysfunction in rats. Zhonghua Nan Ke Xue 20:605–612

H_2S Is a Promoter of Angiogenesis: Identification of H_2S "Receptors" and Its Molecular Switches in Vascular Endothelial Cells

Bei-Bei Tao, Wen-Jie Cai, and Yi-Chun Zhu

Contents

Abstract

Angiogenesis is a physiological process in organ development and also a compensatory response in ischemia. When ischemia occurs, oxygen sensors in vascular endothelial cells sense the decrease in oxygen, thus activating downstream signaling pathways to promote the proliferation, migration, and tube formation of the endothelial cells. The new vasculatures are formed by sprouting from preexisting vessels, in order to maintain oxygen homeostasis in ischemic tissues (Folkman and Shing 1992). Collateral circulation is sometimes established under chronic ischemic conditions such as chronic myocardial

B.-B. Tao • Y.-C. Zhu (✉)
Key Laboratory of Bioactive Small Molecules and Research Center on Aging and Medicine, Department of Physiology and Pathophysiology, Fudan University Shanghai Medical College, 138 Yi Xue Yuan Road, Shanghai 200032, China
e-mail: yczhu@shmu.edu.cn

W.-J. Cai
Department of Basic Medicine, University of Shanghai for Science and Technology, Shanghai, China

P.K. Moore, M. Whiteman (eds.), *Chemistry, Biochemistry and Pharmacology of Hydrogen Sulfide*, Handbook of Experimental Pharmacology 230,
DOI 10.1007/978-3-319-18144-8_6

ischemia (Banai et al. 1994). However, naturally occurring angiogenesis is usually not sufficient to compensate for ischemia in ischemic tissues. Proangiogenic drugs may be useful to promote angiogenesis in these diseases.

Keywords
Angiogenesis • Myocardial ischemia • Endothelial cells • Hydrogen sulfide • Molecular switch

1 Introduction

Angiogenesis is a physiological process in organ development and also a compensatory response in ischemia. When ischemia occurs, oxygen sensors in vascular endothelial cells sense the decrease in oxygen, thus activate downstream signaling pathways to promote the proliferation, migration, and tube formation of the endothelial cells. The new vasculatures are formed by sprouting from preexisting vessels, in order to maintain oxygen homeostasis in ischemic tissues (Folkman and Shing 1992). Collateral circulation is sometimes established under chronic ischemic conditions such as chronic myocardial ischemia (Banai et al. 1994). However, naturally occurring angiogenesis is usually not sufficient to compensate for ischemia in ischemic tissues. Proangiogenic drugs may be useful to promote angiogenesis in these diseases.

Accumulating evidences have revealed some protective effects of hydrogen sulfide in the cardiovascular system. H_2S reduces blood pressure (Zhao et al. 2001) and ameliorates hypertension-induced myocardial remodeling in spontaneous hypertensive rats (Yan et al. 2004; Shi et al. 2007). H_2S provides protection in models of myocardial injury (Geng et al. 2004) and myocardial infarction (Chuah et al. 2007). However, the mechanisms underlying numerous biological effects of H_2S remain largely unknown. One of the most challenging questions in this new field is the identification of the "receptors" for H_2S to mediate its numerous biological effects. Here we review the studies reporting the proangiogenic role of H_2S and the recent work which reports the first H_2S "receptor" and its molecular switch identified in this proangiogenic model.

2 H_2S Promotes Angiogenesis

We report the proangiogenic role of in vitro and in vivo models of angiogenesis in 2007 (Cai et al. 2007). H_2S facilitates migration of vascular endothelial cells in both the wound healing model and the transwell model (Fig. 1A–D). H_2S also increases tube length and the number of branching points in an in vitro of tube formation (Fig. 1G–I). In addition, H_2S promotes angiogenesis in an in vivo model of Matrigel plug assay in mice which is evidenced by an increase in vascular density (Fig. 1L) and hemoglobin content (Fig. 1M) in the Matrigel plug. In 2009, Papapetropoulos

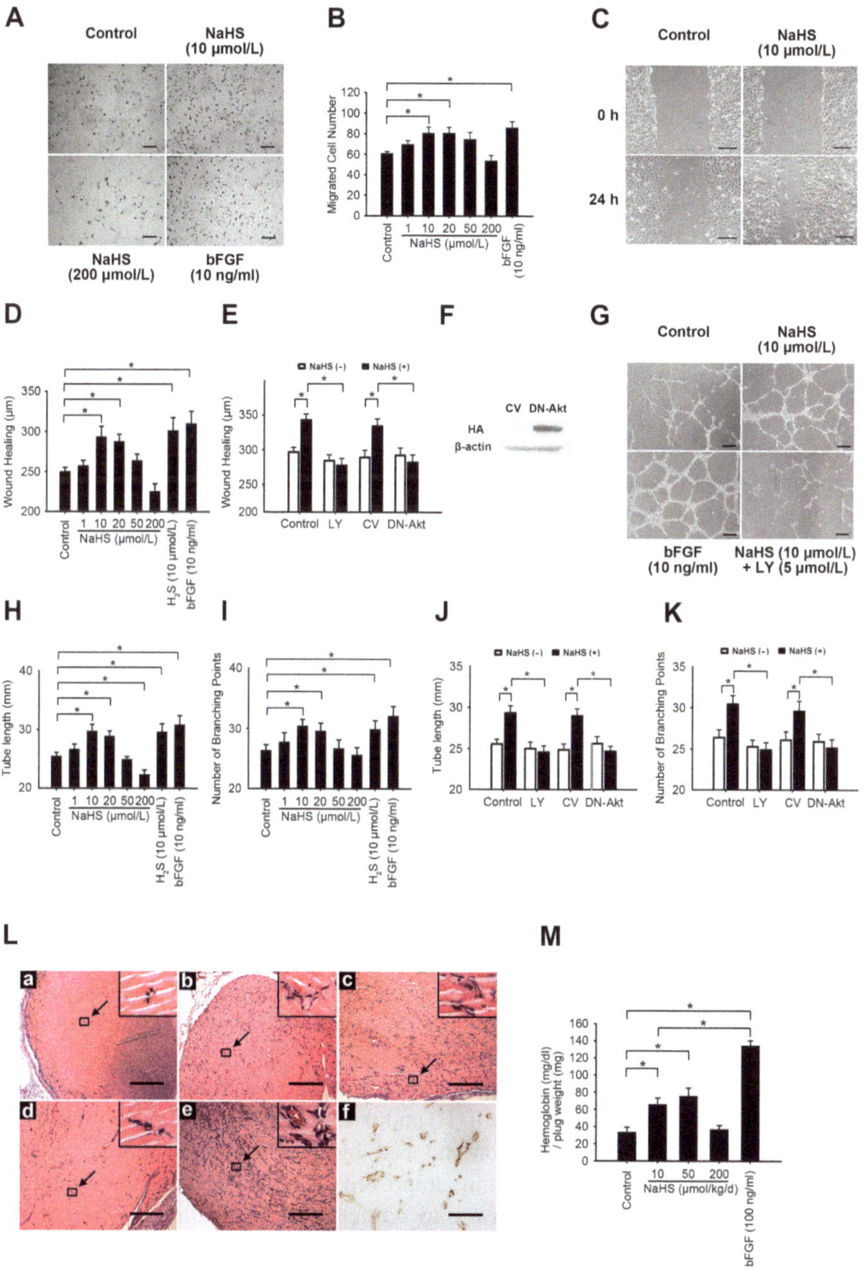

Fig. 1 H_2S promotes angiogenesis both in vitro and in vivo. (**A** and **B**) Cell migration was assessed by transwell migration assay. (**C** and **D**) Representative micrographs and statistical analysis of scratch wound healing assay of RF/6A endothelial cells. (**E–K**) NaHS-induced promotion of RF/6A endothelial cell migration was prevented by either LY 294002 (5 μmol/L) or transfection of the DN-Akt. The effects of H_2S on in vivo angiogenesis were assessed using

et al. further reported the proangiogenic role of endogenous H_2S in cystathionine γ-lyase (CSE) knockout mice. VEGF administration caused less angiogenesis in CSE knockout mice as compared with the wild-type mice. The authors suggested that endogenous H_2S has an important role in inducing angiogenesis (Papapetropoulos et al. 2009). This report also confirmed our previous results by using the chicken chorioallantoic membrane (CAM) model (Papapetropoulos et al. 2009). They found H_2S administration increased vascular length in chicken chorioallantoic membranes. In addition, the effects of H_2S in promoting the migration and tube formation of the vascular endothelial cells were also validated by Papapetropoulos et al. (2009).

3 The Proangiogenic Role of H_2S in Ischemic Disease Models

Though the proangiogenic role of H_2S had been reported by two independent groups using various in vitro and in vivo models, the role of H_2S in the angiogenesis in diseases was not examined. In 2010, we reported that chronic treatment with exogenous H_2S using an H_2S donor NaHS improved regional blood flow in ischemic tissues in a rat model of femoral artery ligation (Wang et al. 2010). H_2S significantly promoted collateral vessel growth in ischemic hind limbs of rats (Fig. 2A, B) and causes an increase in capillary density (Fig. 2D). In addition, regional tissue blood flow was also increased by H_2S treatment in ischemic tissue (Fig. 2C). These findings indicate that H_2S is a proangiogenic factor under both physiological and pathophysiological conditions. In the following years, the proangiogenic role of H_2S in various disease models has been reported by several independent groups. In 2012, Bir et al. reported a proangiogenic role of H_2S in a mouse model of hind limb ischemia (Bir et al. 2012). More recently in 2012, Qipshidze et al. examined the role of H_2S in chronic heart failure model of myocardial infarction induced by ligation of the left anterior descending coronary artery (Qipshidze et al. 2012). They found that the expression level of both H_2S-producing enzymes (CSE and CBS) was decreased in mice with myocardial infarction. Chronic treatment with NaHS for 1 month promoted angiogenesis in the myocardium using X-ray angiography and blood flow measurement with Doppler.

In addition to myocardial ischemia, H_2S also protects against cerebral ischemia by inducing angiogenesis in the brain (Hyunduk et al. 2014). In a model of transient focal cerebral ischemia/reperfusion induced by suturing endovascular internal carotid artery, H_2S administration improved significantly the neurological outcomes on the limb placing test and corner turn test without reducing the

Fig. 1 (continued) Matrigel plug assay in mice (**L** and **M**). (**L**) Representative photomicrographs of hematoxylin-eosin-stained Matrigel sections of mice treated with vehicle (**a**), various doses of NaHS (**b**, **c**, and **d** for 10, 50, and 200 $\mu mol \cdot kg^{-1} \cdot day^{-1}$ NaHS, respectively) and bFGF (**e**, 100 ng/ml in Matrigel). (**M**) Neovascularization in the Matrigel plugs was quantified by measuring hemoglobin content using the tetramethylbenzidine method

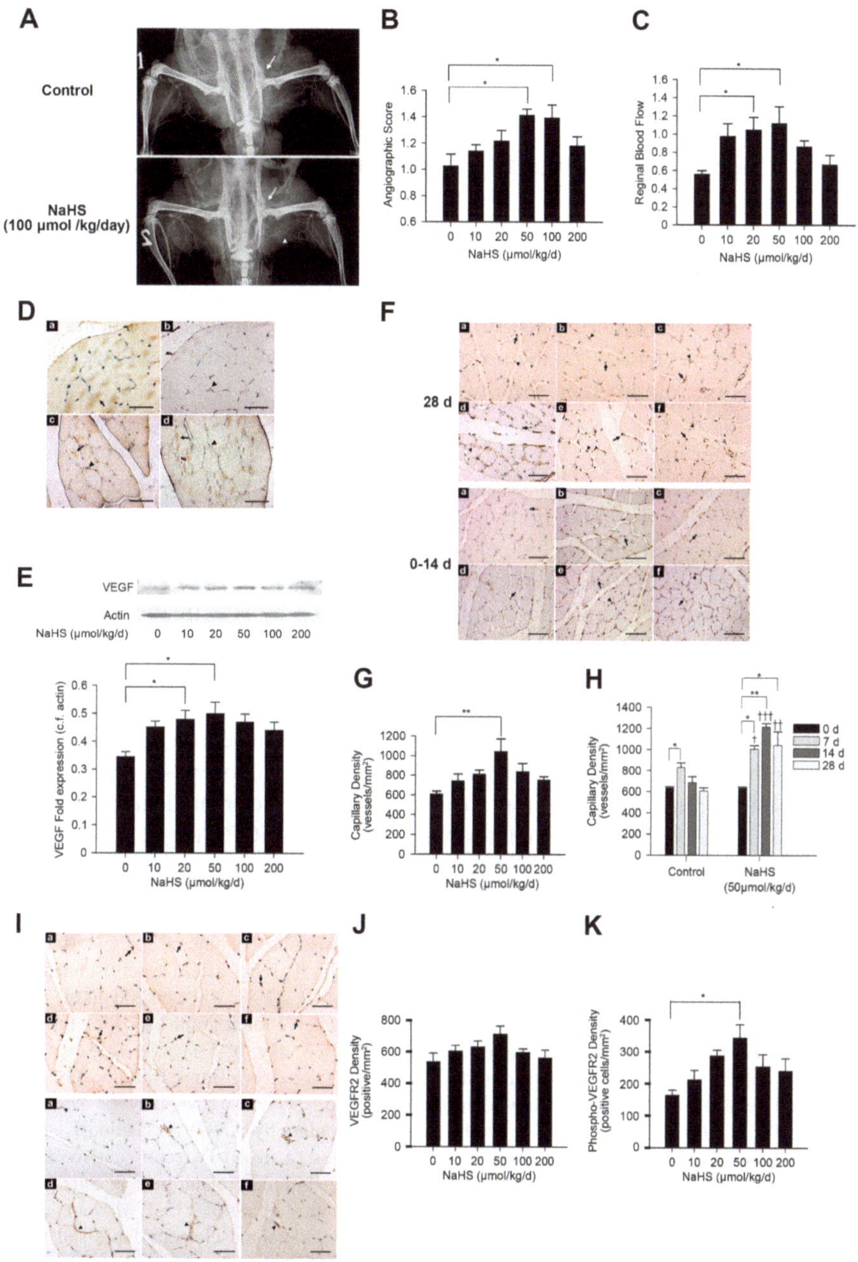

Fig. 2 NaHS treatment promoted collateral vessel formation and regional blood flow after femoral artery occlusion in the rat hindlimb ischemia model. (**A**) Representative postmortem angiograms obtained 4 weeks after surgery. *Arrow denotes* the site of ligation at the femoral artery. *Arrow head* indicates the typical "corkscrew" appearance of collateral vessels. (**B**) Quantitative analysis of collateral vessel development was performed by measuring the total length of the

infarction volume. In addition, H_2S enhanced the synthesis of endothelial cells in the ischemic region.

4 Signaling Pathways to Mediate the H_2S Effects

When H_2S was first reported as a proangiogenic factor, Cai et al. found that the proangiogenic effects of H_2S in vascular endothelial cells were dependent on the PI3K/Akt pathway (Fig. 1E, F, G, J, K) (Cai et al. 2007). However, the direct target molecule for H_2S remains unknown. It could be one of the signaling elements in this pathway or some elements in the upstream of the pathway in the vascular endothelial cells. This prompts us to explore for the "receptor" of H_2S in the subsequent studies.

On the other hand, the H_2S signals may be transduced with more complicated mechanisms where cell-cell interaction is involved. For example, the proangiogenic effects of H_2S in the hind limb ischemia model might involve an interaction between the vascular endothelial cells and the skeletal muscle cells. H_2S treatment caused an increase in VEGF expression in the skeletal muscle cells in the ischemic hind limb where collateral circulation was promoted by H_2S treatment (Fig. 2E) (Wang et al. 2010). An increase in the phosphorylation of Akt and VEGFR2 was observed in the vascular endothelial cells neighboring the skeletal muscle cells in the ischemic tissues (Fig. 2I, J, K). The data suggest that exogenous H_2S might stimulate the skeletal muscle cells to release VEGF which acts on VEGFR2 on the cell membrane of vascular endothelial cells to promote angiogenesis. In addition to VEGF, other proangiogenic factors are also involved in the proangiogenic effects of H_2S. It is reported that H_2S promotes biosynthesis of VEGF, Ang-1, and Ang-2 in

Fig. 2 (continued) contrast-opacified vessels. (**C**) Blood flow measured with microsphere assay. **F** (*upper part*) Representative micrographs showing the vascular endothelial cells (*arrow*) and capillary vessels (*arrow head*) in the gastrocnemius muscles 4 weeks after surgery stained with anti-rat CD34 antibody (**a** for the vehicle; **b, c, d, e,** and **f** for 10, 20, 50, 100, and 200 $\mu mol \cdot kg^{-1} \cdot day^{-1}$ NaHS groups, respectively). **F** (*lower part*) Representative micrographs showing the time-course study of CD34-positive cells. (**a–c**) Control group (**a**, 0 day; **b**, 7 days; **c**, 14 days); (**d–f**) 50 $\mu mol \cdot kg^{-1} \cdot day^{-1}$ NaHS group (**d**, 0 day; **e**, 7 days; **f**, 14 days). Bar = 100 μm. (**G** and **H**) Quantitative analysis of the capillary density. $*P < 0.05$, $**P < 0.01$ vs. day 0; $\dagger P < 0.05$, $\dagger\dagger P < 0.01$, $\dagger\dagger\dagger P < 0.001$ vs. control (at the same time point) by ANOVA. (**D** and **E**) NaHS increased VEGF expression in the ischemic hindlimb muscles. (**a** and **b**) Two consecutive sections showing that VEGF expression (*arrow*) was mainly localized in the skeletal muscle cells but not the vascular endothelial cells (*arrow head*). (**c** and **d**) Double staining of VEGF (*arrow*) and the vascular endothelial cells (*arrow head*) further clarified the skeletal muscle cells expressed VEGF in the groups treated with 50 $\mu mol \cdot kg^{-1} \cdot day^{-1}$ NaHS administration (**d**) and the vehicle (**c**). Bar = 100 μm. I (**a–f** in the *upper part*) Representative micrographs of VEGFR2 (*arrow*) in the gastrocnemius muscles. (**I**) (**a–f** in the *lower part*) Representative micrographs of phospho-VEGFR2 in the gastrocnemius muscles. **a** for the vehicle; **b, c, d, e,** and **f** for 10, 20, 50, 100, and 200 $\mu mol \cdot kg^{-1} \cdot day^{-1}$ NaHS groups, respectively. (**J** and **K**) Quantitative analysis of VEGFR2 and phospho-VEGFR2 density, respectively

astrocytes, increases mouse brain endothelial cells (MBECs) in the in vitro OGD model, and increases VEGF and Ang-1 in ischemic brain (Hyunduk et al. 2014). The expression of VEGF, VEGFR1, and VEGFR2 is increased in ischemic heart with H_2S treatment. Whereas, some factors with anti-angiogenic property such as endostatin, angiostatin, and parstatin were decreased (Qipshidze et al. 2012). These findings support the idea that H_2S is also a proangiogenic factor in disease status and some paracrine mechanisms might be involved in in vivo models.

On the other hand, more signaling elements may be involved in addition to the PI3K/Akt pathway found earlier in single cells. For example, ERK and HIF-1α might also be involved in mediating the H_2S effects (Hyunduk et al. 2014). Hyunduk et al. report that H_2S increases p-Akt, p-ERK, and HIF-1α in cultured astrocytes and MBECs.

Some chemical inhibitors such as glibenclamide (K_{ATP} channel blocker) and p38 inhibitors have been shown to block the H_2S effects in inducing the migration of vascular endothelial cells (Papapetropoulos et al. 2009). These data all suggest that the H_2S signals are mediated by some signaling pathways in a single cell. However, the signaling mechanisms underlying the H_2S effects remain largely unknown.

On the other hand, other gaseous signaling molecules including nitric oxide (NO) and carbon monoxide (CO) have been suggested to interact with H_2S in their proangiogenic effects. Coletta et al. found that H_2S and NO were mutually dependent in promoting angiogenesis (Coletta et al. 2012). An inhibitor of NO biosynthesis, L-NAME, blocks the H_2S effects in promoting endothelial cell proliferation. H_2S-induced angiogenesis in aortic rings is abolished in cultured vessels isolated from eNOS knockout mice. In the in vivo Matrigel plug model, H_2S-induced angiogenesis is also abrogated completely in eNOS knockout mice. These results indicate that H_2S-induced angiogenesis is dependent on the NO pathway. Moreover, Bir et al. show that H_2S increases eNOS activity by stimulating the phosphorylation of Akt and thereby increases NO production (Bir et al. 2012). This may explain how H_2S can increase NO biosynthesis. Indeed, future works are required to clarify the signaling mechanisms underlying the H_2S effects in angiogenesis.

5 H_2S Donors Developed to Treat Ischemic Diseases

Kan et al. report that S-propargyl-cysteine (SPRC), a novel endogenous hydrogen sulfide modulator, promotes proliferation, migration, and in vitro tube formation of the vascular endothelial cells (Kan et al. 2014). This proangiogenic effect of SPRC is also validated in rat aortic ring angiogenic model and the in vivo Matrigel plug assay. SPRC also increases angiogenesis in the mouse hind limb ischemia model and the rat myocardial ischemia model. SPRC treatment significantly increased STAT3 phosphorylation, and the proangiogenic effect of SPRC is blunted by siRNA-mediated knockdown of STAT3. The data indicate the involvement of STAT3 in this H_2S effect. However, co-crystallization shows that STAT3 does not directly interact with SPRC. SPRC enhances the interaction between VEGFR2 and STAT3 and increases nuclear translocation of STAT3, leading to an activation of VEGF transcription.

Diallyl trisulfide, a stable H_2S donor, is also reported as a proangiogenic factor. It has been found to have therapeutic potential in heart failure and to increase vascular density in the heart (Polhemus et al. 2013). Chronic treatment with diallyl trisulfide for 12 weeks significantly increased VEGF expression and decreased angiostatin expression in the myocardium in a pressure overload model of mice. In addition, diallyl trisulfide increased eNOS phosphorylation and enhanced NO bioavailability in the heart.

6 VEGFR2 Is the "Receptor" for H_2S to Induce Angiogenesis

All the works reviewed above support the proangiogenic role of H_2S which is first reported by Cai et al. in 2007. Since then, numerous efforts have been endowed to clarify the mechanisms underlying the H_2S effects. A series of signaling elements such as VEGF, PI3K, Akt, NO, and STAT3 have been suggested to mediate the H_2S effects. However, all these elements may reflect some signaling event secondary to the initial H_2S signals. That means there must be some target molecules in the vascular endothelial cells to directly interact with H_2S and transduce the initial H_2S signals to downstream signaling pathways to induce the H_2S effects in the vascular endothelial cells. We name the direct target molecule for H_2S as the "receptor" for H_2S. A typical ligand has a conformation to match with the structure of its receptor. However, H_2S is too small a molecule to have a conformation sufficient to dock with its "receptors." In fact, H_2S does cause significant biological effects in the vascular endothelial cells. This leads to a hypothesis that there is some "receptor" for H_2S in the vascular endothelial cells. This "receptor" must have a direct interaction with H_2S and initiate a cascade of signaling events leading to increased migration of the cells. With this idea, we searched for the H_2S "receptor" using the in vitro model of H_2S-induced migration of vascular endothelial cells. We focused on the PI3K/Akt pathway identified in the previous studies (Cai et al. 2007). Purified recombinant proteins of PI3K and Akt were reacted with H_2S in a cell-free system and the activity of these two kinases was not changed (Fig. 3b, c). Moreover, no chemical modification was found in these two kinases reacted with H_2S using tandem mass spectrometry. Since blockade of these two elements abolished the H_2S effects, we tried to search for the H_2S "receptor" in the pathways upstream of PI3K and Akt and finally found that VEGFR2 functioned as the direct target molecule for H_2S. H_2S directly increased the activity of recombinant VEGFR2 in a cell-free system (Fig. 3a). In vascular endothelial cells, blockade of VEGFR2 using either the chemical inhibitor SU5416 or siRNA-mediated knockdown both significantly prevented the proangiogenic effect of H_2S. Therefore, we believed that VEGFR2 functioned as a "receptor" for H_2S in the vascular endothelial cells. We proposed that there might be some chemical modifications in VEGFR2 induced by H_2S treatment. At glance of the data, we did not find any modifications in any of the free amino acid residues in VEGFR2 using tandem mass

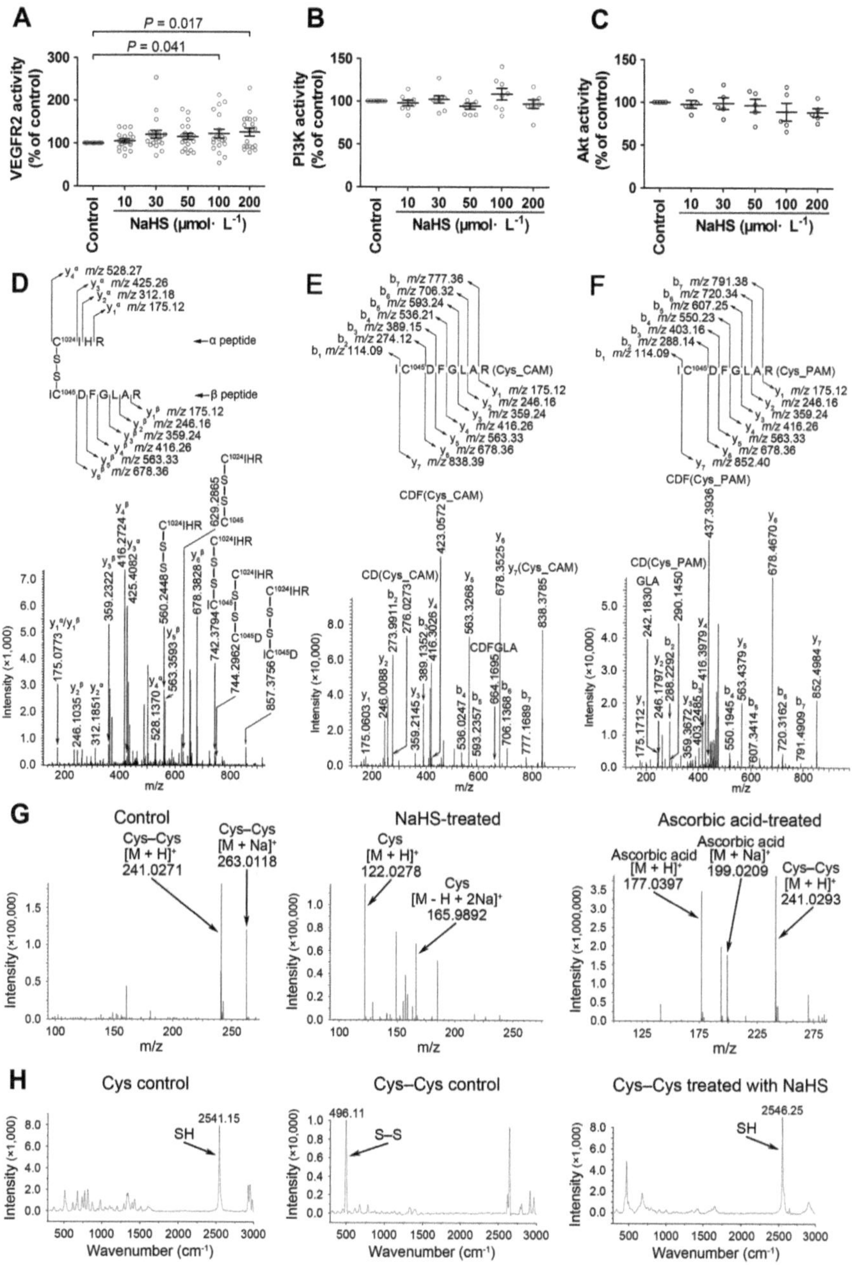

Fig. 3 (**a–c**) NaHS directly activated VEGFR2 without affecting the activity of PI3K and Akt in a cell-free system. (**d**) CID spectra of $[M+3H]^{3+}$ m/z 473.90 from a tryptic digest of VEGFR2 in the absence of NaHS showing an S–S bond between Cys1045 and Cys1024. (**e**) CID spectra of $[M+2H]^{2+}$ m/z 476.24 from a tryptic digest of VEGFR2 in the presence of DTT showing the β peptide containing Cys1045. (**f**) CID spectra of $[M+2H]^{2+}$ m/z 483.25 from a tryptic digest of VEGFR2 in the presence of NaHS showing the β peptide containing Cys1045. Cys_CAM,

spectrometry. However, a new disulfide bond between Cys1045 and Cys1024 was surprisingly discovered in the intracellular kinase core of VEGFR2, and this disulfide bond was cleaved by H_2S (Fig. 3d, e, f). More mass spectrometry and Raman spectroscopy experiments were performed to testify the effects of H_2S on each individual amino acid contained in protein molecules and the disulfide bonds contained in model chemicals. The results further validate our conclusion that H_2S does not modify any free amino acid residues in VEGFR2, however, cleaves the disulfide bonds (Fig. 3g, h). In order to further investigate the role of the Cys1045-Cys1024 disulfide bonds in VEGFR2, we mutated Cys1045 to alanine and acquired the mutant VEGFR2-C1045A. The activity of the recombinant VEGFR2-C1045A was significantly higher than the wild-type VEGFR2. The data suggest that Cys1045-Cys1024 is an inhibitory intrinsic motif in VEGFR2. Molecular dynamic calculations further showed that the breaking of Cys1045-Cys1024 disulfide bond restored the activity of VEGFR2 by shifting it to an active "Phe-in" conformation which is necessary for ATP binding (Fig. 4a, b). The data indicate that the Cys1045-Cys1024 disulfide bond actually serves as a molecular switch for H_2S to regulate the structure and function of its "receptor", VEGFR2. The next question is how H_2S specifically targets this molecular switch? We further performed quantum chemical studies and mass spectrometry at various pH values and different time points. The data revealed that it was HS^- (the aqueous solution of H_2S is actually a mixture of H_2S and HS^-), but not H_2S, that breaks the disulfide bond molecular switch (Fig. 4c, d, e, f). Quantum chemical calculations showed that HS^- specifically recognized the sulfur atoms of the disulfide bond and attacked one of the sulfur atoms of the disulfide bond. This was a nucleophilic attack where HS^- attacked the S–S bond via interaction with its frontier molecular orbitals: the highest occupied molecular orbital (HOMO) of the nucleophile HS^- and the lowest unoccupied molecular orbital (LUMO) of the electrophile under attack. Two steps of reactions and two HS^- anions are required to cleave one S–S bond (Tao et al. 2013):

$$\begin{aligned}\text{Reaction 1}: \ & HS^- + C_3H_6NO_2 - S - S - C_3H_6NO_2 \rightarrow C_3H_6NO_2 - S^- \\ & + HS - S - C_3H_6NO_2 \\ \text{Reaction 2}: \ & HS^- + HS - S - C_3H_6NO_2 \rightarrow C_3H_6NO_2 - S^- + HS - SH\end{aligned}$$

With these data we conclude that H_2S/HS^- targets its "receptor" with a mechanism beyond the typical ligand-receptor docking. The targeting is based on an interaction of the electron orbitals of the nucleophile sulfur atom of H_2S/HS^- and the electrophile sulfur atom of the S–S bond. This is actually a new atomic biological mechanism based on the morphology and energy of the frontier molecular orbitals involved in the reaction.

Fig. 3 (continued) Carboxyamidomethyl cysteine; Cys_PAM, propionamide cysteine. (**g** and **h**) ESI-MS spectra and Raman bands showing hydrogen sulfide-induced cleavage of the S–S bond in the model chemical of Cys–Cys

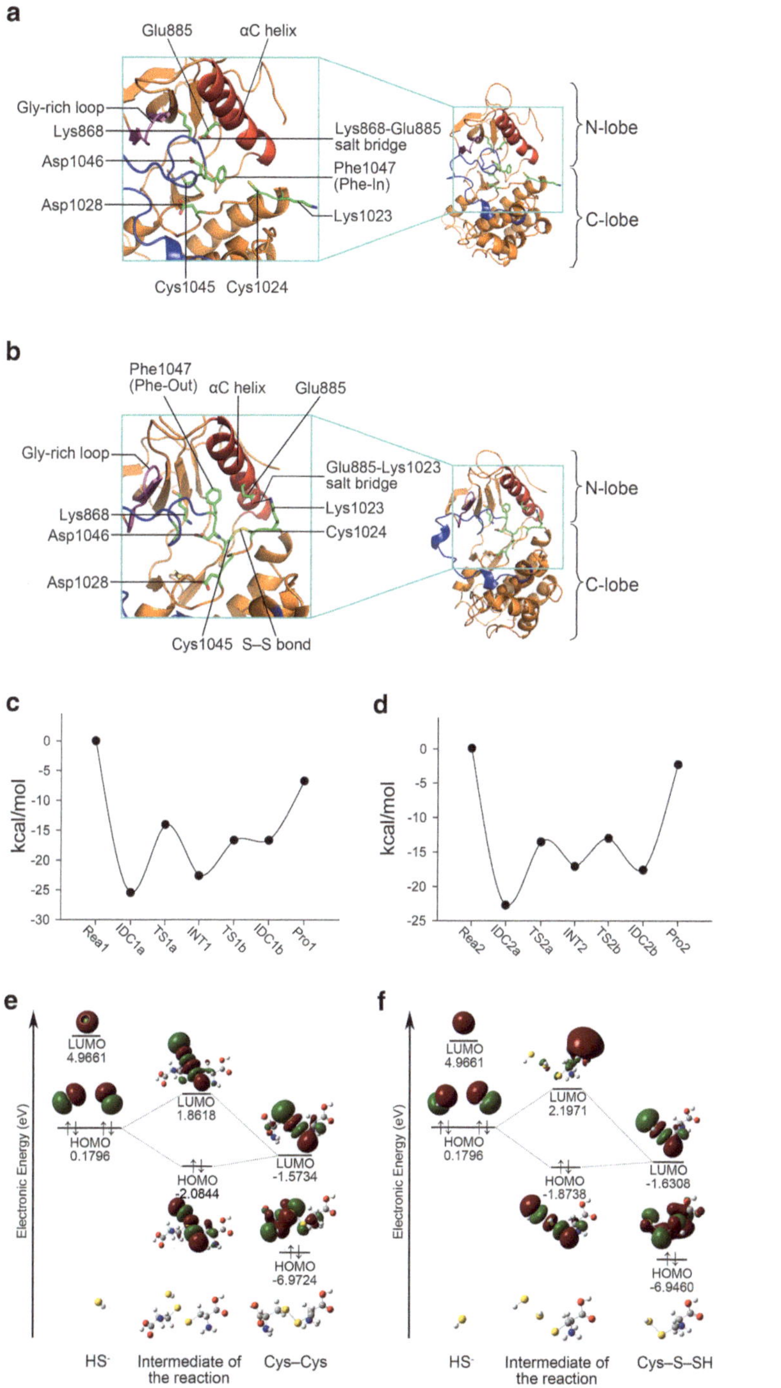

Fig. 4 (**a** and **b**) Molecular dynamics simulation showing conformation of the kinase core of VEGFR2. (**a**) There is a conserved salt bridge between Lys868 and Glu885 in VEGFR2 without the Cys1045-Cys1024 S–S bond. The DFG motif is in the "Phe-In" conformation. (**b**) The Lys868-

7 The Receptor Tyrosine Kinase Family Are Candidate "Receptors" for H_2S

Since the direct target of H_2S is actually the disulfide bond, it is reasonable that there might be other "receptors" for H_2S. If a protein contains a functional disulfide bond, this protein might be a candidate "receptor" for H_2S. It is known that VEGFR2 belongs to the family of receptor tyrosine where various members with different extracellular ligand-binding domain have a similar intracellular kinase domain. The Cys1045-Cys1024 molecular switch is located in the intracellular kinase domain of VEGFR2. H_2S may penetrate through cell membrane and directly activate the intracellular kinase domain without interacting with the extracellular ligand-binding domain of VEGFR2. This gives rise to an idea that other members of the receptor tyrosine kinase family may also contain such a molecular switch and thereby serving as potential "receptors" for H_2S. Interestingly, we have recently found that H_2S also directly activates EGFR (epidermal growth factor receptor) (Ge et al. 2014) and the insulin receptor (Xue et al. 2013), both of which belong to the receptor tyrosine kinase family (Fig. 5a, h). Some disulfide bond molecular switch was detected in EGFR which was labile to H_2S-induced cleavage (Fig. 5b, c). To find the precise location of the molecular switch in EGFR, mutation of some potential cysteine residue which is necessary to form a disulfide bond molecular switch was performed. The results showed that the Cys798 residue located in the intracellular kinase domain of EGFR is required to form the disulfide bond molecular switch for H_2S though the other cysteine residue required for the switch remains to be identified (Fig. 5d, e, f, g). Moreover, we have also shown that endogenous H_2S was required for VEGF to activate VEGFR2 in the vascular endothelial cells. This was evidenced by experiments using siRNA-mediated knockdown of CSE where the proangiogenic effect of VEGF was blunted in the vascular endothelial cells with CSE knockdown (Tao et al. 2013). VEGFR2 does not respond to its extracellular ligand without endogenous H_2S. The receptor tyrosine kinase family may be renamed as H_2S-dependent receptor tyrosine kinase family. In addition to the receptor tyrosine kinase family, some phosphatases, including protein tyrosine phosphatases (PTPs), the lipid phosphatase (PTEN), and low-molecular-weight PTPs, have functional disulfide bonds which are sensitive to redox regulation (Salmeen and Barford 2005). Those phosphatases all contain cysteine residues that are prone to form disulfide bonds or other oxidized forms such as a sulfonamide bond. Cho et al. found a redox regulation of disulfide

Fig. 4 (continued) Glu885 salt bridge is disrupted and the DFG motif rotates anticlockwise from the "Phe-In" conformation in VEGFR2 with the Cys1045-Cys1024 S–S bond. (**c**) Potential energy surface for the nucleophilic attack of HS^- on Cys–S–S–Cys (reaction 1) (IDC1a, the ion dipole complex 1a; TS1a, transition state 1a; INT1, intermediate 1; TS1b, transition state 1b; IDC1b, ion dipole complex 1b). (**d**) Potential energy surface for the nucleophilic attack of HS^- on Cys–S–SH (reaction 2) (IDC2a, the ion dipole complex 2a; TS2a, transition state 2a; INT2, intermediate 2; TS2b, transition state 2b; IDC2b, ion dipole complex 2b). (**e** and **f**) Quantum chemical mechanisms underlying hydrogen sulfide-induced breaking of the S–S bond with the model chemical Cys–Cys

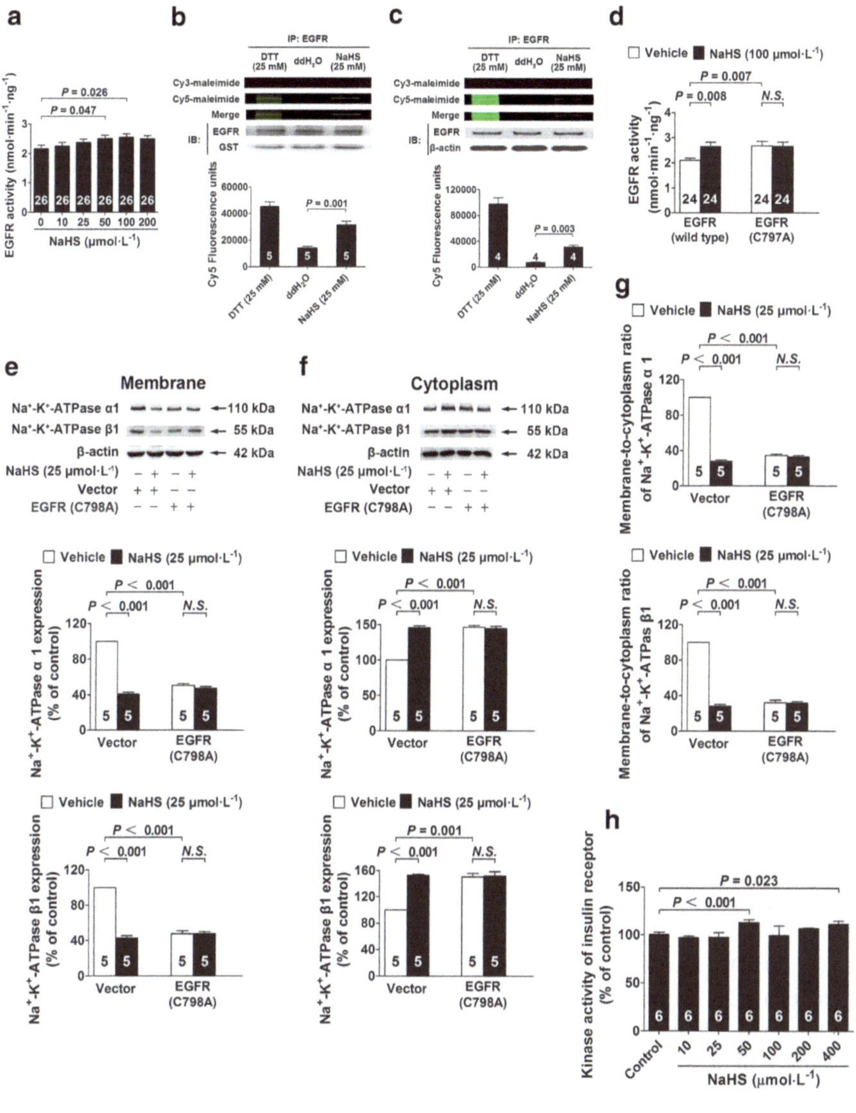

Fig. 5 H_2S directly activates EGFR (**a**) and breaks its disulfide bond 9 (**b** and **c**). NEM was used to cover all –SH groups of Cys residues in the intracellular kinase domain of recombinant EGFR proteins (**b**) or EGFR immunoprecipitated from cell lysate (**c**). Free thiols in EGFR were labeled firstly by Cy3-conjugated maleimide and then by Cy5-conjugated maleimide after DTT or NaHS treatment. (**d–g**) EGFR Cys797 (human) or Cys798 (rat) is required for H_2S to activate EGFR and induce Na^+/K^+-ATPase endocytosis. (**d**) Mutation at Cys797 in human recombinant EGFR proteins. (**e–g**) Rat renal tubular epithelial cells transfected with the vectors expressing the mutant EGFR (C798A) or control vectors. (**h**) H_2S directly increases activity of insulin receptor

bonds in PTEN (C71-C124), Cdc25A (C384-C430), and LMW-PTP (C12-C17) (Cho et al. 2004). Therefore, it is possible that these phosphatases might also function as "receptors" for H_2S.

Moreover, it is reported that a disulfide bond could be formed between two protein molecules to regulate their function. For example, a disulfide bond is formed between Cys32 in the active site of Trx-1 and Cys212 at the C-terminal of PTEN and this binding inactivates PTEN (Song et al. 2007; Meuillet et al. 2004). The data suggest that a protein complex may also function as potential "receptor" for H_2S with some disulfide bond between two protein molecules serving as molecular switch for H_2S.

8 Two Theories: Disulfide Bond Molecular Switch Versus Protein S-sulfhydration

Mustafa et al. proposed an S-sulfhydration theory to explain the interaction between H_2S and its target molecules (Mustafa et al. 2009). According to the S-sulfhydration theory, H_2S oxidizes the –SH group of the cysteine residues to yield an –S–SH group and thereby change the conformation and function of a protein molecule. Future works are required to clarify how H_2S could oxidize the –SH group and how such a modification at the cysteine residues would change the conformation and function of a protein molecule. Another interesting question is does H_2S-induced oxidization of the cysteine residues (S-sulfhydration) have selectivity for different cysteine residues in a protein molecule? In case of VEGFR2, there are 11 cysteine residues in the intracellular kinase domain. Does H_2S oxidize all the cysteine residues or some selective cysteine residues in VEGFR2 and how such a modification would activate this kinase? We have examined this idea using mass spectrometry and did not find any S-sulfhydration in the kinase domain of VEGFR2 (Fig. 3f) (Tao et al. 2013). Actually H_2S did not cause chemical modification on any of the 20 amino acids that constitute protein molecules in a cell-free system (Tao et al. 2013). However, H_2S cleaved disulfide bonds in various model chemicals (Tao et al. 2013). It is possible that H_2S might be oxidized to sulfane sulfur (Bailey et al. 2014). It is the oxidants of H_2S to oxidize the free thiols of the cysteine residues in a protein molecule to yield S-sulfhydration of a protein (Greiner et al. 2013). However, this is actually the effects of the oxidants of H_2S. Indeed, it remains to be clarified if the H_2S effects are mediated by the oxidants of H_2S but not H_2S itself.

A more interesting field about the role of element sulfur is now arising. We believe that element sulfur can be transformed to a reductive form like H_2S or an oxidative form like sulfane sulfur depending on the microenvironment in the body. For instance, H_2S can be easily oxidized under normoxic conditions but will remain in a reductive form during hypoxia. The role of endogenously generated H_2S may be tissue/cell type specific.

All these hypotheses including the disulfide bond molecular switch proposed by us remain to be validated in future studies. This will surely prompts the field of H_2S biology to a new landmark.

References

Bailey TS, Zakharov LN, Pluth MD (2014) Understanding hydrogen sulfide storage: probing conditions for sulfide release from hydrodisulfides. J Am Chem Soc 136(30):10573–10576

Banai S, Shweiki D, Pinson A, Chandra M, Lazarovici G, Keshet E (1994) Upregulation of vascular endothelial growth factor expression induced by myocardial ischaemia: implications for coronary angiogenesis. Cardiovasc Res 28:1176–1179

Bir SC, Kolluru GK, McCarthy P et al (2012) Hydrogen sulfide stimulates ischemic vascular remodeling through nitric oxide synthase and nitrite reduction activity regulating hypoxia-inducible factor-1α and vascular endothelial growth factor-dependent angiogenesis. J Am Heart Assoc 5, e004093

Cai WJ, Wang MJ, Moore PK, Jin HM, Yao T, Zhu YC (2007) The novel proangiogenic effect of hydrogen sulfide is dependent on Akt phosphorylation. Cardiovasc Res 76:29–40

Cho SH, Lee CH, Ahn Y, Kim H, Kim H, Ahn CY, Yang KS, Lee SR (2004) Redox regulation of PTEN and protein tyrosine phosphatases in H(2)O(2) mediated cell signaling. FEBS Lett 560 (1–3):7–13

Chuah SC, Moore PK, Zhu YZ et al (2007) S-allylcysteine mediates cardioprotection in an acute myocardial infarction rat model via a hydrogen sulfide-mediated pathway. Am J Physiol Heart Circ Physiol 293:H2693–H2701

Coletta C, Papapetropoulos A, Erdelyi K, Olah G, Módis K, Panopoulos P, Asimakopoulou A, Gerö D, Sharina I, Martin E, Szabo C (2012) Hydrogen sulfide and nitric oxide are mutually dependent in the regulation of angiogenesis and endothelium-dependent vasorelaxation. Proc Natl Acad Sci U S A 109(23):9161–9166

Folkman J, Shing Y (1992) Angiogenesis. J Biol Chem 267:10931–10934

Ge SN, Zhao MM, Wu DD, Chen Y, Wang Y, Zhu JH, Cai WJ, Zhu YZ, Zhu YC (2014) Hydrogen sulfide targets EGFR Cys797/Cys798 residues to induce Na(+)/K(+) –ATPase endocytosis and inhibition in renal tubular epithelial cells and increase sodium excretion in chronic salt-loaded rats. Antioxid Redox Signal 21:2061–2082

Geng B, Chang L, Pan CS, Qi YF, Zhao J, Pang YZ, Du JB, Tang CS (2004) Endogenous hydrogen sulfide regulation of myocardial injury induced by isoproterenol. Biochem Biophys Res Commun 318:756–763

Greiner R, Pálinkás Z, Bäsell K, Becher D, Antelmann H, Nagy P, Dick TP (2013) Polysulfides link H_2S to protein thiol oxidation. Antioxid Redox Signal 19(15):1749–1765

Hyunduk J, Mi-Young O, Young-Ju K, In-Young C, Yang H s, Wi-Sun R, Seung-Hoon L, Byung-Woo Y (2014) Hydrogen sulfide treatment induces angiogenesis after cerebral ischemia. J Neurosci Res 92(11):1520–1528

Kan J, Wei G, Chengrong H, Guangzhi B, Yichun Z, Yi Zhun Z (2014) S-propargyl-cysteine, a novel water-soluble modulator of endogenous hydrogen sulfide, promotes angiogenesis through activation of signal transducer and activator of transcription 3. Antioxid Redox Signal 20(15):2303–2316

Meuillet EJ, Mahadevan D, Berggren M, Coon A, Powis G (2004) Thioredoxin-1 binds to the C2 domain of PTEN inhibiting PTEN's lipid phosphatase activity and membrane binding: a mechanism for the functional loss of PTEN's tumor suppressor activity. Arch Biochem Biophys 429(2):123–133

Mustafa AK, Gadalla MM, Sen N, Kim S, Mu W, Gazi SK, Barrow RK, Yang G, Wang R, Snyder SH (2009) H_2S signals through protein S-sulfhydration. Sci Signal 2(96):ra72

Qipshidze N, Metreveli N, Mishra PK et al (2012) Hydrogen sulfide mitigates cardiac remodeling during myocardial infarction via improvement of angiogenesis. Int J Biol Sci 8(4):430–441

Papapetropoulos A, Pyriochou A, Altaany Z et al (2009) Hydrogen sulfide is an endogenous stimulator of angiogenesis. Proc Natl Acad Sci USA 106(51):21972–21977

Polhemus DJ, Kondo K, Bhushan S, Bir SC, Kevil CG, Murohara T, Lefer DJ, Calvert JW (2013) Hydrogen sulfide attenuates cardiac dysfunction after heart failure via induction of angiogenesis. Circ Heart Fail 6(5):1077–1086

Salmeen A, Barford D (2005) Functions and mechanisms of redox regulation of cysteine-based phosphatases. Antioxid Redox Signal 7(5–6):560–577

Shi YX, Chen Y, Zhu YZ et al (2007) Chronic sodium hydrosulfide treatment decreases medial thickening of intramyocardial coronary arterioles, interstitial fibrosis, and ROS production in spontaneously hypertensive rats. Am J Physiol Heart Circ Physiol 293(4):H2093–H2100

Song Z, Saghafi N, Gokhale V, Brabant M, Meuillet EJ (2007) Regulation of the activity of the tumor suppressor PTEN by thioredoxin in Drosophila melanogaster. Exp Cell Res 313 (6):1161–1171

Tao BB, Liu SY, Zhang CC, Fu W, Cai WJ, Wang Y, Shen Q, Wang MJ, Chen Y, Zhang LJ, Zhu YZ, Zhu YC (2013) VEGFR2 functions as an H_2S-targeting receptor protein kinase with its novel Cys1045-Cys1024 disulfide bond serving as a specific molecular switch for hydrogen sulfide actions in vascular endothelial cells. Antioxid Redox Signal 19(5):448–464

Wang MJ, Cai WJ, Li N et al (2010) The hydrogen sulfide donor NaHS promotes angiogenesis in a rat model of hind limb ischemia. Antioxid Redox Signal 12(9):1065–1077

Xue R, Hao DD, Sun JP, Li WW, Zhao MM, Li XH, Chen Y, Zhu JH, Ding YJ, Liu J, Zhu YC (2013) Hydrogen sulfide treatment promotes glucose uptake by increasing insulin receptor sensitivity and ameliorates kidney lesions in type 2 diabetes. Antioxid Redox Signal 19 (1):5–23

Yan H, Du J, Tang C (2004) The possible role of hydrogen sulfide on the pathogenesis of spontaneous hypertension in rats. Biochem Biophys Res Commun 313:22–27

Zhao WM, Zhang J, Lu YJ, Wang R (2001) The vasorelaxant effect of H2S as a novel endogenous gaseous KATP channel opener. EMBO J 20:6008–6016

Hydrogen Sulfide and Platelets: A Possible Role in Thrombosis

Michael Emerson

Contents

Abstract

Platelets are circulating blood elements with key roles in haemostasis and thrombosis. Platelets are activated by a range of stimuli including exposed subendothelial components. Haemostasis also depends upon the effects of inhibitory substances, including the gasotransmitter nitric oxide whose effects on platelets are well documented. Evidence is also emerging to suggest that H_2S is generated enzymatically by platelets and can impact their function. Exposure of platelets to H_2S from slow-release compounds inhibits aggregation and exerted anti-thrombotic effects in vivo. The mechanisms by which H_2S impacts platelet function and the importance of interactions between H_2S and other gasotransmitters remain unclear. H_2S is therefore emerging as a potentially important regulator of platelet activation and thrombosis. Further study is required to evaluate its importance as a regulator of platelet physiology and associated pathological conditions such as myocardial infarction and stroke.

M. Emerson (✉)
National Heart and Lung Institute, Imperial College London, London SW7 2AZ, UK
e-mail: m.emerson@imperial.ac.uk

P.K. Moore, M. Whiteman (eds.), *Chemistry, Biochemistry and Pharmacology of Hydrogen Sulfide*, Handbook of Experimental Pharmacology 230,
DOI 10.1007/978-3-319-18144-8_7

Keywords
Cardiovascular • Haemostasis • Hydrogen sulfide • Nitric oxide • Platelet • Thrombosis

1 Platelet Development and Structure

Platelets are anucleate circulating blood elements derived from pluripotent megakaryocytes in bone marrow. Platelets are 2–3 μm in diameter and circulate for 5–9 days before undergoing phagocytosis in the spleen and liver.

Platelet structure and morphology are maintained by the cytoskeleton which consists of three discrete parts, the spectrin-rich membrane skeleton, which includes the open canalicular system and maintains membrane integrity. The other cytoskeletal components are the actin-rich cytoskeleton, which generates contractile forces, and the microtubule ring, which maintains the discoid shape of the resting platelet. Platelets are also distinctly rich in intracellular vesicles which, upon activation, release a range of mediators that drive adhesion, aggregation and immune responses. Platelets also contain a specialised endoplasmic reticulum called the dense tubular system which stores calcium. A distinct feature of the platelet external membrane is the presence of a range of receptors which upon occupation drive the process of platelet activation. Platelets express receptors for subendothelial proteins such as collagen; the product of the coagulation cascade, thrombin; substances released from granules such as ADP and products of enzyme activity such as thromboxane A_2 derived from cyclooxygenase. The basic platelet intracellular structure is shown in Fig. 1.

2 Platelet Activation and Role in Haemostasis

The primary role of platelets is to maintain vascular integrity and cease blood loss following injury. Platelets circulate in a quiescent state, but upon vascular injury and exposure of blood to the subendothelial matrix, platelet surface receptors bind to exposed collagen and other macromolecules leading to adhesion of platelets to the vessel wall. Adhesion initiates a cascade of signalling events culminating in calcium release. The rise in intracellular calcium leads to rearrangement of the cytoskeleton and extension of filopodia so that the platelet changes from a spherical disc to a stellate shape with an increased surface area. In addition, intracellular granules release their contents to propagate the platelet activation process by activating adjacent platelets. Platelet aggregation, the linking of adjacent platelets by fibrinogen, follows calcium-mediated activation of the surface integrin αIIbβIII leading to the formation of a platelet plug which seals the breached vessel.

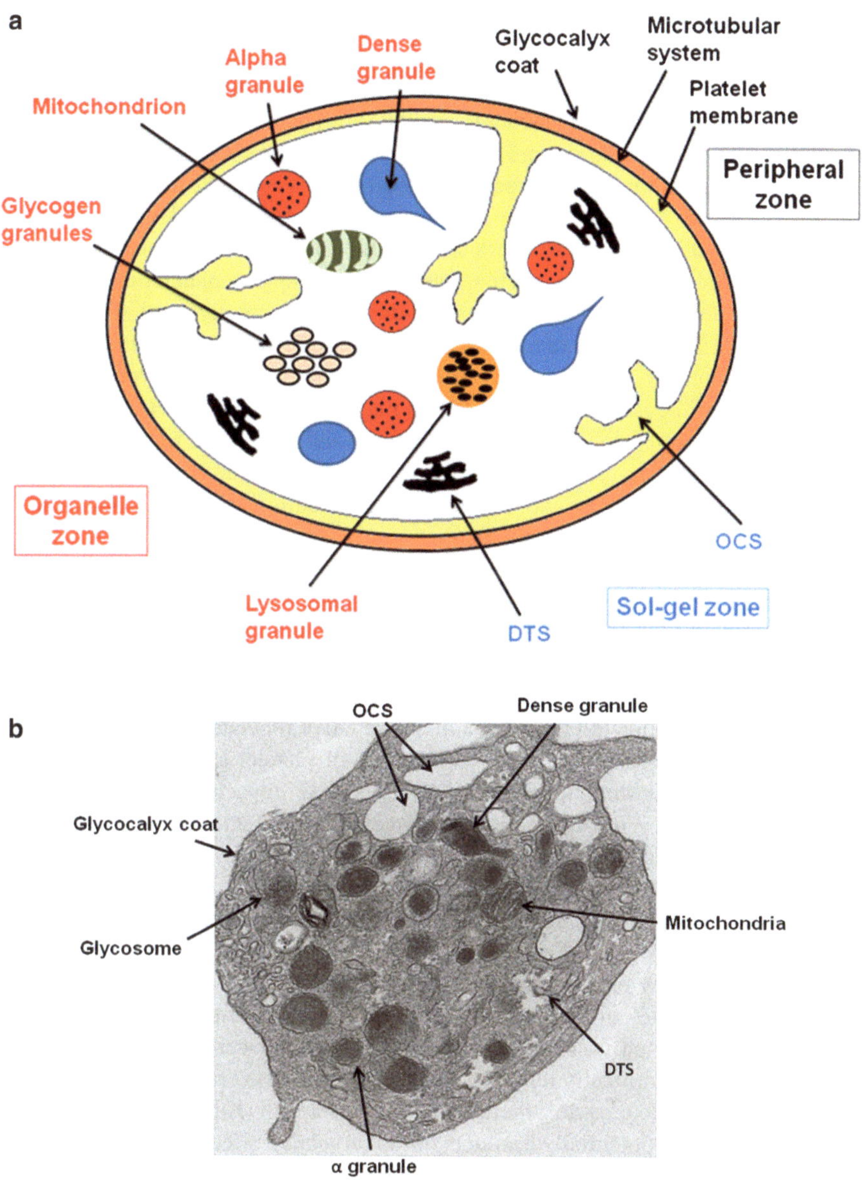

Fig. 1 Platelet intracellular morphology. (**a**) Basic diagram demonstrating the structure and organisation of platelets. DTS-dense tubular system; OCS-open canalicular system. (**b**) Electron microscopy image of a platelet from author's laboratory (Emerson, unpublished)

3 Platelets and Thrombosis

Thrombosis is characterised by the formation of a deposit of blood constituents (the thrombus) in a blood vessel leading to partial or full occlusion and results in events including myocardial infarction or stroke. Arterial thrombosis is a high-shear, platelet-dependent process that occurs when blood is exposed to sites of arterial injury such as atherosclerotic plaque rupture. Surgical devices used in revascularisation can also provide a surface that initiates thrombosis. Despite the use of anti-thrombotic drugs, such as aspirin, which targets platelet cyclooxygenase, arterial thrombosis and resulting conditions such as myocardial infarction, unstable angina and stroke remain the most common end points in atherosclerotic disease. In summary, inappropriate activation of platelets is a key driver of common cardiovascular diseases, and platelets are a major therapeutic target in the prevention and treatment of conditions such as myocardial infarction. Limitations in the efficacy of platelet-targeted antithrombotics are due to drug resistance in a high proportion of individuals and incidences of bleeding associated with use of antithrombotics (Schroeder et al. 2006).

4 Nitric Oxide and Platelets

Of the main gasotransmitters, the best characterised in terms of impact upon platelet function is nitric oxide (NO). NO is a major negative regulator of platelet function and acts to counterbalance the positive drivers of platelet activation during both haemostatic and thrombotic platelet responses. NO impacts a range of platelet activation processes including shape change, secretion and calcium signalling. NO acts in platelets predominantly via activation of soluble guanylyl cyclase and subsequent PKG activation resulting in interaction with a wide range of target proteins which explains its diverse inhibitory activity. For review, see Naseem and Roberts (2011).

NO generated by endothelial nitric oxide synthase (eNOS) within endothelial cells is released into surrounding tissues such as vascular smooth muscle and the blood. NO activity on vascular smooth muscle cells induces vasodilation (Huang et al. 1995) and so critically regulates vessel tone and blood pressure. NO has been reported to reduce leukocyte adherence to the vessel wall (Kubes et al. 1991; Lefer et al. 1999), which potentially accounts for its protective effect against atherogenesis (Kuhlencordt et al. 2001). NO can inhibit platelet aggregation in vitro in whole blood (Yoshimoto et al. 1999) and in vivo (Emerson et al. 1999; Freedman et al. 1999; Moore et al. 2010).

Determining the role of endogenous NO and eNOS in the regulation of platelets and thrombosis has been somewhat controversial, with conflicting reports published. A lack of thrombotic phenotype in eNOS$^{-/-}$ mice has been reported by a number of groups (Dayal et al. 2006; Ozuyaman et al. 2005), whereas others contradict these reports by describing both enhanced and blunted thrombotic responses (Freedman et al. 1999; Heeringa et al. 2000; Iafrati et al. 2005). Some clarity in the field eventually emerged when real-time monitoring of the duration of the platelet aggregation response in vivo showed that loss of eNOS led to an effect primarily on the duration rather than the amplitude of the platelet aggregation response (Moore et al. 2010).

A second area of controversy has surrounded the issue of the expression and functional relevance of eNOS within platelets. Researchers have reported the presence of the L-arginine/NO pathway in platelets (Freedman et al. 1997; Radomski et al. 1990) and NO release from platelets in vitro (Malinski et al. 1993; Zhou et al. 1995). Furthermore, the hypercoagulability seen in $eNOS^{-/-}$ mice was restored by the transfusion of wild-type platelets suggesting that platelets generated NO from eNOS (Freedman et al. 1999). eNOS protein and mRNA expression have been reported in platelets (Aytekin et al. 2012; Berkels et al. 1997; Mehta et al. 1995; Patel et al. 2006); however, the presence of eNOS in platelets is a subject of fierce debate. Contrary to the above studies, other research groups have reported the lack of eNOS mRNA and protein expression in platelets (Gambaryan et al. 2008; Ozuyaman et al. 2005; Tymvios et al. 2009). In fact, Gambaryan et al. (Gambaryan et al. 2008) have shown that some commercially available eNOS antibodies apparently detect eNOS expression in $eNOS^{-/-}$ mice. Gambaryan et al. (2008) emphasise the need to have appropriate positive and negative controls when detecting protein expression, a method adopted by Tymvios et al. (Tymvios et al. 2009), who showed a lack of eNOS protein in human and mouse platelets and perhaps more importantly, the lack of a functional role of platelet NOS in regulating aggregation in vitro regardless of any argument regarding expression of NOS protein.

5 Hydrogen Sulfide and Platelets

The study of the biological relevance of H_2S in the regulation of platelets and thrombosis is in its infancy, and there is currently no consensus on the regulation of platelets by H_2S. Gaining an understanding of the biological roles of NO has been dependent upon studies with NO donors and NO synthase inhibitors. The roles of H_2S are, in contrast, unclear due to limited biological tools with which to explore its effects. As improved tools such as slow-release H_2S donors that mimic physiological exposure to H_2S and selective inhibitors of H_2S generating enzymes are developed and applied, the roles of H_2S in regulating platelets are likely to become more clear.

H_2S generating salts have been shown to inhibit platelet aggregation induced by a range of agonists including ADP, collagen, adrenaline, arachidonic acid and thrombin (Zagli et al. 2007). It is tempting therefore to speculate an inhibitory role of H_2S similar to that established for NO. These data cannot, however, be used to infer an inhibitory role of H_2S physiologically since the concentrations used (100 μM to 10 mM) were considerably higher than those likely to be found physiologically. A mechanism by which H_2S may inhibit platelets was not proposed so that a lack of clarity remains. A more recent study similarly found H_2S

generating salts in the μM to mM range to inhibit collagen-induced human platelet aggregation (Zhong et al. 2014). Inhibition of aggregation was associated with reduced secretion, impaired mobilisation of calcium from intracellular stores and reduction of a range of collagen-induced signalling events including PLCγ2, Akt and MAPK phosphorylation. In this study, H_2S did not affect platelet viability, suggesting an inhibitory effect mediated via suppression of key signalling events downstream of collagen activation rather than a toxicologic effect on the platelet.

In contrast to the above studies, H_2S generating salts at lower concentrations (0.1–100 μM) have been reported to enhance human platelet aggregation induced by the thrombin receptor activator peptide (TRAP-6) and to contribute to the enhanced platelet aggregation occurring during hyperhomocysteinemia (d'Emmanuele di Villa Bianca R et al. 2013). A full causative mechanism was not described to explain the ability of H_2S to enhance platelet aggregation although the arachidonic acid cascade was shown to be implicated. The studies reported to date therefore report both inhibitory and stimulatory effects of H_2S-generating salts upon platelet aggregation responses. The reports cannot be described as contradictory due to differences in the concentrations of H_2S-generating salts used in the studies and the varying choices of platelet agonists. Nonetheless, further study is warranted to clarify the role of H_2S in regulating platelet activation and aggregation.

Given the slow manner of synthesis of H_2S from enzymes (Li et al. 2008), experiments with rapidly and transiently generating sulfide salts such as those described above may be of limited value in determining the physiological role of H_2S in regulating platelets. A limited study with the slow-release compound GYY4137 showed an inhibitory effect of H_2S upon human platelet aggregation at equivalent H_2S concentrations in the nM range (Ilkan et al. 2013). In addition, GYY4137 reduced activation-induced platelet adhesion molecule expression and attenuated the morphological signs of platelet activation (Grambow et al. 2013). In vivo, GYY4137 prolonged venular thrombus formation and tail-vein bleeding (Grambow et al. 2013) suggesting anti-thrombotic roles that may be in part platelet mediated. Further study is warranted, however, in order to clarify any physiological role of H_2S in the regulation of platelets at physiologically relevant concentrations.

As well as determining the functional roles of H_2S in regulating platelets, it is important to consider whether H_2S-generating enzymes are expressed by platelets and whether H_2S is generated endogenously by platelets. H_2S is synthesised by CBS and CSE in vascular endothelial cells (Shibuya et al. 2009; Whiteman et al. 2011) and has been shown to act as a physiologic vasodilator and regulator of blood pressure (Yang et al. 2008). Human platelets have been reported to possess both CBS and CSE by western blotting and to generate detectable amounts of H_2S from L-cysteine (d'Emmanuele di Villa Bianca R et al. 2013). In contrast, we have detected robust CBS expression by western blotting but very little or no CSE protein in human platelet lysates using three commercially available antibodies (unpublished data). There, therefore, appears to be a current consensus that platelets generate H_2S endogenously although the enzymatic source remains in dispute. In addition, application of L-cysteine, the substrate of H_2S generating enzymes, to platelets has been shown to modulate their function (d'Emmanuele di Villa Bianca R et al. 2013) suggesting functional relevance of endogenous H_2S in the regulation of platelet activation and potentially thrombosis.

The fact that platelets express H_2S generating enzymes and generate H_2S suggests that further study of the role of H_2S from both exogenous and endogenous sources should be pursued. In addition, it is important to highlight that platelets are highly sensitive to gasotransmitters released from the vasculature. In the NO field, the major source of enzymatic NO impacting platelets is that generated by the vascular endothelium rather than the platelet itself (Tymvios et al. 2009). H_2S from the vasculature should therefore be investigated as a modulator, not only of vascular tone and blood pressure but as a regulator of platelet function and thrombosis. This will require in vivo studies, including work with models of thrombosis and in vivo platelet aggregation.

6 H_2S and Thrombosis/Disease

H_2S from GYY4137 has been investigated in a mouse model of microvascular thrombus formation following vascular injury (Grambow et al. 2013). In this chapter, H_2S delayed venular thrombus formation suggesting a potential anti-thrombotic effect. The current goal of anti-thrombotic research is to develop compounds with anti-thrombotic efficacy but limited impact upon haemostatic function. In this way, effective anti-thrombotic therapy can be achieved without the bleeding complications associated with current therapeutics such as aspirin. In this study, GYY4137 increased tail bleeding time threefold. Since tail bleeding is used as an index of haemostatic function, further study is required to evaluate the comparative efficacy of H_2S releasing compounds relative to established therapeutics.

In addition to exerting direct effects on thrombus generation, H_2S has also been suggested to exert antiatherogenic effects in studies involving the development of atherosclerotic lesions in ApoE knock-out mice (Wang et al. 2009). A number of mechanisms have been suggested by which H_2S may attenuate atherogenesis including inhibition of superoxide formation (Muzaffar et al. 2008) and modification of the atherogenic modification of LDL by hypochlorite (Laggner et al. 2007). If confirmed in human studies, H_2S may prove to be an important driver not only of platelet function and thrombus formation but a determinant of the atherogenic process driving platelet dysfunction during atherothrombotic diseases.

7 Current State of Play and Future Directions

Although limited in terms of quantity and depth of study, there is a reasonable body of evidence suggesting that H_2S is both generated by platelets and impacts platelet function. The enzymatic source of H_2S in platelets remains unclear but given that these issues remain unresolved in the NO field, this is not a surprising state of affairs. Greater clarity of the role of H_2S in regulating platelet function and the mechanisms of action remain unresolved but are likely to become more clear as recently developed pharmacological tools are applied to the field.

One area of major interest to the platelet field is the interaction between NO and H_2S. Recently, NO and H_2S have been shown to interact in the regulation of angiogenesis, whereby H_2S amplifies NO signalling by inhibiting downstream phosphodiesterase (PDE) (Bucci et al. 2010; Coletta et al. 2012). Platelets are rich in PDE, and the importance of NO in the regulation of platelets as well as the emerging importance of H_2S as a modulator of platelet function suggests that interaction of gasotransmitters such as NO in the regulation of platelets and thrombosis should be considered. Evidence of the likely importance of the interaction between H_2S and NO recently emerged when the anti-thrombotic effect of H_2S was shown to be mediated in part by upregulation of NOS in a mouse model of thrombosis (Kram et al. 2013).

Currently, gasotransmitters are not direct targets for anti-thrombotic therapeutics, but their importance is highlighted not only by studies linking both NO and H_2S with platelet function and thrombosis but by their potential interactions with existing therapeutics, particularly prescribed antithrombotics. This is highlighted by the demonstration that blockade of the platelet P2Y12 receptor, a standard therapeutic intervention, enhances the inhibitory effects of NO upon platelet function (Kirkby et al. 2013). Thus, it is essential to comprehensively analyse the effects of H_2S upon platelets, thrombosis and its interaction with anti-thrombotic reagents in order to fully understand the roles of H_2S in haemostasis and thrombosis.

References

Aytekin M, Aulak KS, Haserodt S, Chakravarti R, Cody J, Minai OA et al (2012) Abnormal platelet aggregation in idiopathic pulmonary arterial hypertension: role of nitric oxide. Am J Physiol Lung Cell Mol Physiol 302(6):L512–520

Berkels R, Bertsch A, Zuther T, Dhein S, Stockklauser K, Rosen P et al (1997) Evidence for a NO synthase in porcine platelets which is stimulated during activation/aggregation. Eur J Haematol 58(5):307–313

Bucci M, Papapetropoulos A, Vellecco V, Zhou Z, Pyriochou A, Roussos C et al (2010) Hydrogen sulfide is an endogenous inhibitor of phosphodiesterase activity. Arterioscler Thromb Vasc Biol 30(10):1998–2004

Coletta C, Papapetropoulos A, Erdelyi K, Olah G, Modis K, Panopoulos P et al (2012) Hydrogen sulfide and nitric oxide are mutually dependent in the regulation of angiogenesis and endothelium-dependent vasorelaxation. Proc Natl Acad Sci USA 109(23):9161–9166

d'Emmanuele di Villa Bianca R, Mitidieri E, Di Minno MN, Kirkby NS, Warner TD, Di Minno G et al (2013) Hydrogen sulphide pathway contributes to the enhanced human platelet aggregation in hyperhomocysteinemia. Proc Natl Acad Sci USA 110(39):15812–15817

Dayal S, Wilson KM, Leo L, Arning E, Bottiglieri T, Lentz SR (2006) Enhanced susceptibility to arterial thrombosis in a murine model of hyperhomocysteinemia. Blood 108(7):2237–2243

Emerson M, Momi S, Paul W, Alberti PF, Page C, Gresele P (1999) Endogenous nitric oxide acts as a natural antithrombotic agent in vivo by inhibiting platelet aggregation in the pulmonary vasculature. Thromb Haemost 81(6):961–966

Freedman JE, Loscalzo J, Barnard MR, Alpert C, Keaney JF, Michelson AD (1997). Nitric oxide released from activated platelets inhibits platelet recruitment. The Journal of clinical investigation 100(2): 350-356.
Freedman JE, Sauter R, Battinelli EM, Ault K, Knowles C, Huang PL et al (1999) Deficient platelet-derived nitric oxide and enhanced hemostasis in mice lacking the NOSIII gene. Circ Res 84(12):1416–1421
Gambaryan S, Kobsar A, Hartmann S, Birschmann I, Kuhlencordt PJ, Muller-Esterl W et al (2008) NO-synthase-/NO-independent regulation of human and murine platelet soluble guanylyl cyclase activity. J Thromb Haemost 6(8):1376–1384
Grambow E, Mueller-Graf F, Delyagina E, Frank M, Kuhla A, Vollmar B (2013) Effect of the hydrogen sulfide donor GYY4137 on platelet activation and microvascular thrombus formation in mice. Platelets 25:166–174
Heeringa P, van Goor H, Itoh-Lindstrom Y, Maeda N, Falk RJ, Assmann KJ et al (2000) Lack of endothelial nitric oxide synthase aggravates murine accelerated anti-glomerular basement membrane glomerulonephritis. Am J Pathol 156(3):879–888
Huang PL, Huang Z, Mashimo H, Bloch KD, Moskowitz MA, Bevan JA et al (1995) Hypertension in mice lacking the gene for endothelial nitric oxide synthase. Nature 377(6546):239–242
Iafrati MD, Vitseva O, Tanriverdi K, Blair P, Rex S, Chakrabarti S et al (2005) Compensatory mechanisms influence hemostasis in setting of eNOS deficiency. Am J Physiol Heart Circ Physiol 288(4):H1627–H1632
Ilkan Z, Mustafa F, Apostoli G, Solomon A, Whiteman M, Emerson M (2013) Hydrogen sulfide inhibits human platelet aggregation. Nitric Oxide 31(Supp 2):S26
Kirkby NS, Lundberg MH, Chan MV, Vojnovic I, Solomon AB, Emerson M et al (2013) Blockade of the purinergic P2Y12 receptor greatly increases the platelet inhibitory actions of nitric oxide. Proc Natl Acad Sci USA 110(39):15782–15787
Kram L, Grambow E, Mueller-Graf F, Sorg H, Vollmar B (2013) The anti-thrombotic effect of hydrogen sulfide is partly mediated by an upregulation of nitric oxide synthases. Thromb Res 132(2):e112–e117
Kubes P, Suzuki M, Granger DN (1991) Nitric oxide: an endogenous modulator of leukocyte adhesion. Proc Natl Acad Sci U S A 88(11):4651–4655
Kuhlencordt PJ, Gyurko R, Han F, Scherrer-Crosbie M, Aretz TH, Hajjar R et al (2001) Accelerated atherosclerosis, aortic aneurysm formation, and ischemic heart disease in apolipoprotein E/endothelial nitric oxide synthase double-knockout mice. Circulation 104(4):448–454
Laggner H, Muellner MK, Schreier S, Sturm B, Hermann M, Exner M et al (2007) Hydrogen sulphide: a novel physiological inhibitor of LDL atherogenic modification by HOCl. Free Radic Res 41(7):741–747
Lefer DJ, Jones SP, Girod WG, Baines A, Grisham MB, Cockrell AS et al (1999) Leukocyte-endothelial cell interactions in nitric oxide synthase-deficient mice. Am J Physiol 276(6 Pt 2): H1943–1950
Li L, Whiteman M, Guan YY, Neo KL, Cheng Y, Lee SW et al (2008) Characterization of a novel, water-soluble hydrogen sulfide-releasing molecule (GYY4137): new insights into the biology of hydrogen sulfide. Circulation 117(18):2351–2360
Mehta JL, Chen LY, Kone BC, Mehta P, Turner P (1995) Identification of constitutive and inducible forms of nitric oxide synthase in human platelets. J Lab Clin Med 125(3):370–377
Moore C, Tymvios C, Emerson M (2010) Functional regulation of vascular and platelet activity during thrombosis by nitric oxide and endothelial nitric oxide synthase. Thromb Haemost 104 (2):342–349
Muzaffar S, Shukla N, Bond M, Newby AC, Angelini GD, Sparatore A et al (2008) Exogenous hydrogen sulfide inhibits superoxide formation, NOX-1 expression and Rac1 activity in human vascular smooth muscle cells. J Vasc Res 45(6):521–528
Naseem KM, Roberts W (2011) Nitric oxide at a glance. Platelets 22(2):148–152

Ozuyaman B, Godecke A, Kusters S, Kirchhoff E, Scharf RE, Schrader J (2005) Endothelial nitric oxide synthase plays a minor role in inhibition of arterial thrombus formation. Thromb Haemost 93(6):1161–1167

Patel B, Sharifi M, Milward AD, Oberprieler NG, Gibbins JM, Parkin S et al (2006) Platelet nitric oxide synthase is activated by tyrosine dephosphorylation: possible role for SHP-1 phosphatase. J Thromb Haemost 4(11):2423–2432

Radomski MW, Palmer RM, Moncada S (1990) An L-arginine/nitric oxide pathway present in human platelets regulates aggregation. Proc Natl Acad Sci U S A 87(13):5193–5197

Schroeder WS, Ghobrial L, Gandhi PJ (2006) Possible mechanisms of drug-induced aspirin and clopidogrel resistance. J Thromb Thrombolysis 22(2):139–150

Shibuya N, Mikami Y, Kimura Y, Nagahara N, Kimura H (2009) Vascular endothelium expresses 3-mercaptopyruvate sulfurtransferase and produces hydrogen sulfide. J Biochem 146 (5):623–626

Tymvios C, Moore C, Jones S, Solomon A, Sanz-Rosa D, Emerson M (2009) Platelet aggregation responses are critically regulated in vivo by endogenous nitric oxide but not by endothelial nitric oxide synthase. Br J Pharmacol 158(7):1735–1742

Wang Y, Zhao X, Jin H, Wei H, Li W, Bu D et al (2009) Role of hydrogen sulfide in the development of atherosclerotic lesions in apolipoprotein E knockout mice. Arterioscler Thromb Vasc Biol 29(2):173–179

Whiteman M, Le Trionnaire S, Chopra M, Fox B, Whatmore J (2011) Emerging role of hydrogen sulfide in health and disease: critical appraisal of biomarkers and pharmacological tools. Clin Sci (Lond) 121(11):459–488

Yang G, Wu L, Jiang B, Yang W, Qi J, Cao K et al (2008) H2S as a physiologic vasorelaxant: hypertension in mice with deletion of cystathionine gamma-lyase. Science 322(5901):587–590

Yoshimoto H, Suehiro A, Kakishita E (1999) Exogenous nitric oxide inhibits platelet activation in whole blood. J Cardiovasc Pharmacol 33(1):109–115

Zagli G, Patacchini R, Trevisani M, Abbate R, Cinotti S, Gensini GF et al (2007) Hydrogen sulfide inhibits human platelet aggregation. Eur J Pharmacol 559(1):65–68

Zhong L, Lv L, Yang J, Liao X, Yu J, Wang R et al (2014) The inhibitory effect of hydrogen sulfide on platelet aggregation and the underlying mechanisms. J Cardiovasc Pharmacol 64:481–487

Zhou Q, Hellermann GR, Solomonson LP (1995) Nitric oxide release from resting human platelets. Thromb Res 77(1):87–96

Part IV

Inflammation and Inflammatory Disease

H_2S and Inflammation: An Overview

Madhav Bhatia

Contents

Abstract

Inflammation is a response to traumatic, infectious, post-ischemic, toxic, or autoimmune injury. However, uncontrolled inflammation can lead to disease, and inflammation is now believed to be responsible for several disease conditions. Research in our laboratory has shown that hydrogen sulfide (H_2S) acts as a novel mediator of inflammation. At present, work in several research

M. Bhatia (✉)
Department of Pathology, University of Otago, Christchurch, 2 Riccarton Avenue
P.O. Box 4345, Christchurch 8140, New Zealand
e-mail: madhav.bhatia@otago.ac.nz

P.K. Moore, M. Whiteman (eds.), *Chemistry, Biochemistry and Pharmacology of Hydrogen Sulfide*, Handbook of Experimental Pharmacology 230,
DOI 10.1007/978-3-319-18144-8_8

groups worldwide is focused on determining the role of H_2S in inflammation. H_2S has been implicated in different inflammatory conditions. Most of this research involved working with animal models of disease and in vitro systems. Recent research, however, points to a role of H_2S in clinical inflammatory disease as well. This chapter describes our current understanding of the role of H_2S in inflammation.

Keywords
Hydrogen sulfide • Inflammation • Systemic inflammatory response syndrome • Leukocytes

1 Introduction

Inflammation is a response to traumatic, infectious, post-ischemic, toxic, or autoimmune injury. It is a highly orchestrated, tissue-based process, characterized by "rubor" (redness), "calor" (heat), "dolor" (pain), and "tumor" (swelling). Inflammation is a normal response to injury and a useful physiological event. However, uncontrolled inflammation can lead to disease. Recent research has shown a key role of hydrogen sulfide (H_2S) in inflammation. This has been possible with the use of experimental approaches, such as cell culture systems and in vivo disease models. At the time of writing this chapter, evidence has started emerging on the role of H_2S in clinical inflammatory disease.

In this chapter, our current understanding of the role of H_2S in inflammation is discussed.

2 Inflammation

Currently, inflammation is an important research question in systems biology in the academia, as well as a multibillion dollar market for the pharmaceutical industry. In a disease in which primary pathogenic events are unknown, control of inflammation is the next best option.

In response to the initial insult/injury, leukocytes are activated. A key component of the inflammatory process is the trafficking of inflammatory cells to the site of injury/infection. Cytokine/receptor interactions on the surface of these cells lead to the expression of gene products that bring about the inflammatory response. However, uncontrolled production of inflammatory products is injurious to host cells and, in some cases, can lead to cancer. Therefore, endogenous mechanisms have evolved to limit the production of inflammatory molecules and permit the resolution of the inflammatory response. An understanding of these mechanisms is important because defects in the pathway may contribute to inflammatory disorders, and the pathway itself may present targets for novel anti-inflammatory therapeutic strategies (Bhatia and Moochhala 2004; Bhatia 2010, 2012; Hegde and Bhatia 2011; Nathan 2002; Mazumder et al. 2010).

3 H_2S and Inflammation

Over the years, various studies have indicated a role of H_2S in the inflammatory process. In mammals, including humans, H_2S is produced by the action of the enzymes, cystathionine-β-synthase (CBS), cystathionine-γ-lyase (CSE), cysteine aminotransferase (CAT, EC 2.6.1.3), (followed by 3-mercaptopyruvate sulfurtransferase (MPST, EC 2.8.1.2), and cysteine lyase (EC 4.4.1.10), (Bhatia 2012; Li et al. 2011; Moore et al. 2003; Wang 2012). As the end product of CBS- and CSE-catalyzed cysteine metabolism, H_2S exerts a negative feedback effect on the activity of these enzymes (Bhatia 2012; Moore et al. 2003). Endogenous H_2S synthesized by CSE has been shown to be primarily responsible for its inflammatory action.

Recent work in our laboratory and others has shown a key role of H_2S as a mediator of inflammation in different clinical conditions. Current understanding of the role of H_2S in different inflammatory conditions is summarized in Table 1.

4 Disease Conditions with Role of H_2S in Inflammation

4.1 Acute Pancreatitis

Acute pancreatitis is a common clinical condition, the incidence of which has been increasing worldwide over recent years (Bhatia et al. 2000, 2005c; Bhatia 2012). For example, in the United States alone, acute pancreatitis is the most common reason for hospitalization (274,119 discharges in the year 2009, a 30 % increase over 2000) amongst all gastrointestinal diseases (Peery et al. 2012). It also inflicts a heavy economic burden; the direct cost in the United States alone in the year 2009 was US$2,599,686,000 (Peery et al. 2012). Acute pancreatitis was the cause of death in 3065 cases and a contributing cause in an additional 5500 deaths in the year 2009 (Peery et al. 2012).

Most cases of acute pancreatitis are secondary to biliary disease or excess alcohol consumption. The exact mechanisms by which diverse etiological factors induce an attack are still unclear, but once the disease process is initiated, common inflammatory and repair pathways are invoked. There is a local inflammatory reaction at the site of injury, which if marked leads to systemic inflammatory response syndrome (SIRS), and it is this systemic response that is believed to be ultimately responsible for the majority of the morbidity and mortality (Bhatia et al. 2000, 2005c; Bhatia 2012). Lung injury, which is clinically manifested as acute respiratory distress syndrome (ARDS), is a major component of the multiple organ dysfunction syndrome (MODS) that results from SIRS in acute pancreatitis.

CBS and CSE, the two major H_2S synthesizing enzymes, are highly expressed in the pancreas. Endogenously produced H_2S has been shown as a mediator of inflammation in acute pancreatitis (Bhatia et al. 2005a). mRNA for CSE is expressed in mouse pancreas and that pancreas homogenates convert L-cysteine to H_2S ex vivo. Also, circulating levels of H_2S are increased in mice upon induction

Table 1 Role of H_2S in inflammatory disease

Disease model	Effect on inflammation	Reference
Cerulein-induced acute Pancreatitis in the mouse	Plasma H_2S is increased in inflammation. Treatment with the cystathione-gamma-lyase (CSE) inhibitor propargylglycine (PAG) protects against acute pancreatitis and associated lung injury	Bhatia et al. (2005a)
	Treatment with slow H_2S-releasing diclofenac protects mice against acute pancreatitis-associated lung injury	Bhatia et al. (2008a)
	Treatment with s-propargyl-cysteine (SPRC), a slow H_2S-releasing donor, protects against acute pancreatitis and associated lung injury	Sidhapuriwala et al. (2012)
	Knockout mice genetically deficient in CSE ($CSE^{-/-}$) and protected against acute pancreatitis and associated lung injury	Ang et al. (2013)
Lipopolysaccharide (LPS)-induced endotoxemia in the mouse	Plasma H_2S levels, H_2S-synthesizing activity, and CSE expression are increased in inflammation. PAG treatment protects against inflammation	Li et al. (2005)
	Treatment with slow H_2S-releasing diclofenac protects mice against LPS-induced endotoxemia	Li et al. (2007)
	Treatment with morpholin-4-ium-4-methoxyphenyl(morpholino) phosphinodithioate (GYY4137), a slow H_2S-releasing donor, protects mice against LPS-induced endotoxemia	Li et al. (2009)
Cecal ligation and puncture (CLP)-induced sepsis	Plasma H_2S levels, H_2S-synthesizing activity, and CSE expression are increased in inflammation. PAG treatment protects against sepsis. The H_2S donor sodium hydrogen sulfide (NaHS) further aggravates inflammation in sepsis	Zhang et al. (2006)
Carrageenan-induced hindpaw edema in the mouse	H_2S-synthesizing activity increased in inflammation. PAG treatment protects against inflammation in this model	Bhatia et al. (2005b)
	Treatment with slow H_2S-releasing diclofenac protects mice against carrageenan-induced hindpaw edema	Sidhapuriwala et al. (2007)
Burn injury-induced inflammation	Plasma H_2S levels, H_2S-synthesizing activity, and CSE expression are increased in inflammation. PAG treatment protects against burn injury-induced inflammation	Zhang et al. (2010)
Trinitrobenzene sulfonic acid-induced colitis	Slow H_2S-releasing mesalamine protects against inflammation in colitis	Fiorucci et al. (2007)
Direct administration of H_2S as NaHS to mouse	Direct administration of H_2S by intraperitoneal (i.p) administration of NaHS causes lung inflammation characterized by an increase in lung myeloperoxidase (MPO) activity and histological evidence of lung injury	Bhatia et al. (2006)

of acute pancreatitis. The conversion of L-cysteine to H_2S in pancreas homogenates is significantly reduced in mice pretreated with DL-propargylglycine (PAG) (Bhatia et al. 2005a). Furthermore, treatment of animals with PAG (either prophylactic or therapeutic) reduces the severity of pancreatitis as evidenced by a significant attenuation of hyperamylasemia, acinar cell injury/necrosis, and pancreatic myeloperoxidase (MPO) activity and by histological evidence of diminished pancreatic injury. Lung injury in severe acute pancreatitis is characterized by sequestration of neutrophils within the lung (i.e., increased lung MPO activity) and histological evidence of lung injury. Prophylactic or therapeutic administration of PAG additionally protected mice against acute pancreatitis-associated lung injury as evidenced by a significant attenuation of lung MPO activity and by histological evidence of diminished lung injury (alveolar thickening and leukocyte infiltration) (Bhatia et al. 2005a). The pro-inflammatory role of H_2S synthesized by CSE has been confirmed recently using knockout mice deficient in CSE (Ang et al. 2013). CSE knockout mice are protected against acute pancreatitis and associated lung injury, when compared to wild-type controls (Ang et al. 2013). These effects of CSE blockade/gene deletion suggested an important pro-inflammatory role of H_2S in regulating the severity of pancreatitis and associated lung injury and raised the possibility that H_2S may exert similar activity in other forms of inflammation. Following this study, an important role of CBS in the pathogenesis of acute pancreatitis and associated lung injury has been shown (Shanmugam et al. 2009). Acute pancreatitis was associated with increased plasma and tissue H_2S and ammonia (NH_3). Prophylactic or therapeutic administration of aminooxyacetate (AOA), a reversible inhibitor of CBS, directly inhibited CBS in the pancreas thereby reducing H_2S and NH_3 production, and protected against acute pancreatitis, showing a role of H_2S synthesized by CBS in inflammation in acute pancreatitis (Shanmugam et al. 2009).

4.2 Sepsis

Sepsis is defined as the presence of infectious organisms, such as bacteria, viruses, protozoa, or fungi, and/or their toxins in blood or tissue and the systemic response that follows. Sepsis leading to organ failure characterizes severe sepsis, and septic shock is defined by severe sepsis accompanied by hypotension unresponsive to fluid resuscitation. Severe sepsis and septic shock are one of the leading causes of mortality among intensive care unit and postoperative care patients (Bhatia 2012; Bhatia et al. 2009; Levy et al. 2010; Martin et al. 2003; Ramnath et al. 2006). Sepsis is a major health problem worldwide. The incidence of sepsis in North America, for example, is 3.0 per 1000 population, which transforms into an annual number of 750,000 cases, with 210,000 of them being fatal and a large socioeconomic burden (Bhatia 2012; Bhatia et al. 2009; Levy et al. 2010; Martin et al. 2003; Ramnath et al. 2006). The incidence of mortality due to sepsis is increasing, and the most likely causes are the increased incidence of resistant pathogens and the advances of medical and surgical procedures that save the lives of many patients but at the cost of leaving them immunocompromized and in a state susceptible to death from severe sepsis and septic shock (Bhatia 2012; Bhatia et al. 2009; Levy et al. 2010; Martin et al. 2003; Ramnath et al. 2006).

H_2S has been shown to act as a mediator of inflammation in sepsis. In a clinically relevant mouse model of cecal ligation and puncture (CLP)-induced polymicrobial sepsis, liver CSE expression, H_2S synthesis, and plasma H_2S levels are significantly elevated (Zhang et al. 2006). Prophylactic, as well as therapeutic, administration of PAG significantly reduced sepsis-associated systemic inflammation, as evidenced by decreased MPO activity and histological changes in lung and liver and attenuated the mortality in CLP-induced sepsis (Zhang et al. 2006). On the other hand, administration of sodium hydrogen sulfide (NaHS), an H_2S donor, significantly aggravated sepsis-associated systemic inflammation (Zhang et al. 2006). Similar to CLP-induced sepsis, a pro-inflammatory action of H_2S has also been shown in lipopolysaccharide (LPS)-induced endotoxemia (Li et al. 2005; Collin et al. 2005). These studies show that H_2S plays a pro-inflammatory role in regulating the severity of sepsis and associated organ injury. H_2S has also been shown to modulate sinusoidal constriction in the liver and contribute to hepatic micorcirculatory dysfunction during endotoxemia (Norris et al. 2013), and endogenous H_2S formation mediates the liver damage in endotoxemia in the rat (Yan et al. 2013).

4.3 Burn Injuries

Burn injuries are ranked among the leading causes of morbidity and mortality worldwide. In severe cases, burn injuries lead to SIRS, and associated MODS, which is a major contributor to death following burns (Bhatia 2012; Church et al. 2006; Endorf and Ahrenholz 2011). H_2S has been shown to act as a critical mediator of severe burn injury (to 25 % total body surface area full thickness burn)-induced inflammation in mice (Zhang et al. 2010). The result of this study shows that burn injury in mice resulted in a significant increase in plasma H_2S levels, liver and lung CSE mRNA expression, and liver H_2S-synthesizing activity. The enhanced H_2S/CSE pathway correlates with increased burn-associated systemic inflammation, as evidenced by increased MPO activity and histological evidence of lung and liver injury. There was protection against systemic inflammation and multiple organ damage by prophylactic or therapeutic treatment of PAG. Administration of NaHS at the same time of burn injury resulted in a further increase in MPO activity and more severe tissue injury in the lung and liver. These findings show that H_2S plays a key pro-inflammatory role in burn injury (Zhang et al. 2010).

4.4 Joint Inflammation/Arthritis

Joint inflammation/arthritis is a major health problem worldwide. In the United States, for example, back pain and arthritis (osteoarthritis, rheumatoid arthritis) are amongst the most common and costly conditions, affecting more than 100 million individuals and costing more than US$200 billion per year (Ma et al. 2014). The possible role of endogenous H_2S in the development of joint inflammation induced by intraplantar administration of carrageenan to the rat hindpaw has been investigated. An increase in H_2S synthesis in inflamed hindpaws was seen, suggesting a localized overproduction of H_2S during inflammation (Bhatia et al. 2005c). Pretreatment with PAG resulted in a dose-dependent inhibition of hindpaw edema as well as hindpaw MPO activity. These findings suggest that H_2S is an endogenous mediator of the development of hindpaw local inflammation. Recent reports in the literature in human patients point to a role of H_2S in rheumatoid arthritis, but there is a difference of opinion as to the pro- versus anti-inflammatory action of H_2S (Fox et al. 2012; Kloesch et al. 2012).

4.5 Colitis

Colitis, or inflammation of the colon, presents itself in several clinical forms, such as inflammatory bowel disease (IBD), ulcerative colitis (UC), and Crohn's disease (CD). Mucosal changes in colitis are characterized by ulcerative lesions accompanied by prominent inflammatory infiltrates in the bowel wall (Polytarchou et al. 2014). Mesalamine (5-aminosalicylic acid) is the first-line therapy for colitis. An H_2S-releasing derivative of mesalamine has been reported to protect against inflammation in trinitrobenzene sulfonic-induced colitis (Fiorucci et al. 2007) and against nociception in hapten-induced colitis (Coletta et al. 2012).

4.6 Chronic Obstructive Pulmonary Disease

Chronic obstructive pulmonary disease (COPD) is a common clinical condition characterized by progressive airflow obstruction that is only partly reversible, inflammation in the airways, and systemic effects or comorbities. The main cause is tobacco smoking (Decramer et al. 2012). In patients with stable COPD, serum H_2S levels have been found to be significantly increased as compared to age-matched control subjects or those with acute exacerbation of COPD (AECOPD) (Chen et al. 2005). In this study, serum H_2S levels were negatively correlated with the severity of airway obstruction in patients with stable COPD, whereas they were positively correlated with the lung function in all patients with COPD and healthy controls. This study also reported that serum H_2S levels were lower in smokers than nonsmokers regardless of their health status (COPD or healthy controls). In addition, patients with AECOPD, whose serum H_2S levels were decreased, had greater neutrophil proportion but lower lymphocyte proportion in sputum than patients with stable COPD, suggesting a potential role of H_2S in regulating inflammatory response at different types or stages of COPD. This study (Chen et al. 2005) demonstrated that endogenous H_2S may participate in the development of airway obstruction in COPD and that the level of endogenous

H_2S may be correlated with the progression and severity of COPD. Also, there is a correlation between exhaled H_2S and exhaled nitric oxide (NO) in COPD (Sun et al. 2013).

5 Mechanisms of Action of H_2S in Inflammation

H_2S has been shown to contribute to inflammation via cytokines, chemokines, adhesion molecules (and leukocyte recruitment), and transient receptor potential vanilloid type 1 (TRPV1) and substance P.

5.1 Cytokines and Chemokines

In isolated pancreatic acinar cells stimulated by cerulein (an in vitro system that resembles acinar changes in acute pancreatitis), inhibition of H_2S formation by PAG has been shown to decrease mRNA expression and production of the chemokines monocyte chemoatractant protein (MCP)-1, macrophage inflammatory protein (MIP)-1α, and MIP-2 (Tamizhselvi et al. 2007, 2008). Cerulein-induced acute pancreatitis is associated with a significant increase in mRNA for MCP-1, MIP-1α, and MIP-2 in both the pancreas and lungs, suggesting that they are important early mediators in both local as well as distant inflammatory response (Tamizhselvi et al. 2008). Blockade of H_2S biosynthesis with PAG ameliorates the development of inflammatory process in cerulein-induced acute pancreatitis, acting through downregulation of chemokine expression (Tamizhselvi et al. 2008). Also, results indicate a key role of the phosphatidylinositol 3-kinase-protein kinase B pathway in relation to the action of H_2S on cerulein-induced cytokine production in isolated mouse pancreatic acinar cells (Tamizhselvi et al. 2009).

Both prophylactic and therapeutic administration of PAG significantly reduces the mRNA and protein levels of interleukin (IL)-1β, IL-6, tumor necrosis factor (TNF)-α, MCP-1, and MIP-2 in lungs and liver, coupled with decreased nuclear translocation and activation of NF-κB in lungs and liver following sepsis (Zhang et al. 2007a). In contrast, injection of NaHS significantly aggravates sepsis-associated systemic inflammation and increases nuclear factor (NF)-κB activation. In addition, H_2S-induced lung inflammation is blocked by the NF-κB inhibitor BAY 11-7082. Therefore, H_2S upregulates the production of pro-inflammatory mediators and exacerbates the systemic inflammation in sepsis through a mechanism involving NF-κB activation (Zhang et al. 2007a).

In human monocyte cell line U937, treatment with H_2S donor NaHS results in significant increase in the mRNA expression and protein production of TNF-α, IL-1β, and IL-6. This effect is mediated via NF-κB and extracellular signal-related kinase (ERK) pathway (Zhi et al. 2007). Recent research, using mouse macrophage RAW264.7 cells, has shown that the activation of macrophages by lipopolysaccharide (LPS) results in higher levels of CSE mRNA and protein as well as the increased production of pro-inflammatory cytokines and chemokines IL-1 β,

IL-6, TNF-α, and MCP-1. Silencing of the CSE gene by small interference ribonucleic acid (siRNA) results in decreased levels of pro-inflammatory cytokines (Badiei et al. 2013).

H_2S regulates inflammatory response by activating the extracellular signal-related kinase ERK pathway in polymicrobial sepsis (Zhang et al. 2008). Maximum phosphorylation of ERK1/2 and degradation of IκBα in lung and liver is observed 4 h after CLP. Inhibition of H_2S formation by PAG significantly reduces the phosphorylation of ERK1/2 in lung and liver, coupled with decreased degradation of IκBα and activation of NF-κB. In contrast, injection of NaHS significantly enhances the activation of ERK1/2 in lung and liver, therefore leading to a further rise in tissue NF-κB activity. In sepsis, pretreatment with PAG significantly reduces the production of cytokines and chemokines via NF-κB and ERK1/2, whereas exogenous H_2S greatly increased it. In addition, pretreatment with PD98059, an inhibitor of MEK-1, significantly prevented NaHS from aggravating systemic inflammation in sepsis (Zhang et al. 2008). Binding site for NF-κB has been shown on the CSE gene promoter and is critical for LPS-induced CSE expression (Wang et al. 2014). Therefore, H_2S may regulate systemic inflammatory response in sepsis via the NF-κB-ERK pathway.

5.2 Adhesion Molecules and Leukocyte Recruitment

H_2S induces intercellular adhesion molecule (ICAM)-1 expression and neutrophil adhesion to cerulein-treated pancreatic acinar cells through nuclear factor (NF)-κB and Src-family kinases (SFK) pathway (Tamizhselvi et al. 2010). H_2S activates SFKs in acinar cells, and inhibition of SFKs impairs H_2S-induced ICAM-1 expression secondary to the inhibition of NF-κB activation. The effect of SFK inhibition on NF-κB activation occurs together with IκBα degradation. The results further demonstrate that neutrophil attachment onto H_2S-treated acinar cells is increased and that inhibition of SFK function inhibits H_2S-induced neutrophil attachment onto acinar cells. Taken together, these data indicate that H_2S engages SFKs in order to signal ICAM-1 expression by a mechanism involving induction of NF-κB (Tamizhselvi et al. 2010).

Using intravital microscopy, it has been shown that in sepsis, prophylactic and therapeutic administration of PAG reduces leukocyte rolling and adherence significantly in mesenteric venules coupled with decreased mRNA and protein levels of adhesion molecules (ICAM-1, P-selectin, and E-selectin) in lung and liver. In contrast, administration of NaHS upregulates leukocyte rolling and attachment significantly, as well as tissue levels of adhesion molecules in sepsis. Conversely, in normal mice given NaHS to induce lung inflammation, NaHS treatment enhanced the level of adhesion molecules and neutrophil infiltration in lung. These alterations can be reversed by pretreatment with BAY 11-7082 (Zhang et al. 2007b). Therefore, H_2S acts as an important endogenous regulator of leukocyte activation and trafficking during an inflammatory response.

5.3 Transient Receptor Potential Vanilloid Type 1 and Substance P

Substance P is a mediator of inflammation and plays an important role in the pathogenesis of several inflammatory conditions. Intraperitoneal administration of the H_2S donor NaHS to mice causes an increase in circulating levels of substance P (Bhatia et al. 2006). NaHS, by itself, causes lung inflammation, as evidenced by a significant increase in lung MPO activity and histological evidence of lung injury. This is associated with a significant increase in lung levels of tumor necrosis factor (TNF)-α and interleukin (IL)-1β. In preprotachykinin (PPT)-$A^{-/-}$ mice, genetically deficient in substance P, H_2S does not cause any lung inflammation. Also, pretreatment of mice with CP-96345, an antagonist of the neurokinin-1 receptor (NK-1R), the receptor for substance P, protects mice against lung inflammation caused by H_2S. Depleting neuropeptide from sensory neurons by capsaicin significantly reduces the lung inflammation caused by H_2S. In addition, pretreatment of mice with capsazepine, an antagonist of the transient receptor potential vanilloid-1 (TRPV-1), protects mice against H_2S-induced lung inflammation. These results demonstrate an important role of substance P and neurogenic inflammation in H_2S-induced lung injury in mice (Bhatia et al. 2006).

In acute pancreatitis, PAG treatment significantly attenuates the increases in substance P concentrations in plasma, pancreas, and lung (Bhatia et al. 2008b). Moreover, administration of PAG significantly reduces PPT-A mRNA expression and NK-1R mRNA expression in both pancreas and lung when compared with cerulein-induced acute pancreatitis. This reduction in PPT-A mRNA expression and NK-1R mRNA expression is associated with a protection against acute pancreatitis and associated lung injury (Bhatia et al. 2008b). These results suggested that the pro-inflammatory effects of H_2S may be mediated by SP-NK-1R pathway in acute pancreatitis (Bhatia et al. 2008b). Furthermore, PPT-A deficiency and blockage of H_2S synthesis can regulate the toll-like receptor 4 (TLR4) pathway and subsequent innate immune response in acute pancreatitis, implying an interaction between SP/H_2S occurs via TLR4 and NF-kB pathway. PPT-A gene deletion regulates H_2S-induced TLR4 signaling pathway in cerulein-treated pancreatic acinar cells, suggesting that in acute pancreatitis, H_2S may upregulate the TLR4 pathway and NF-κB via substance P (Tamizhselvi et al. 2011).

In sepsis, PAG treatment significantly decreases the PPT-A gene expression and the production of substance P in lung, whereas administration of NaHS results in a further rise in the pulmonary level of substance P (Zhang et al. 2007c). PPT-A gene deletion and pretreatment with the NK-1R antagonist L703606 prevent H_2S from aggravating lung inflammation. In addition, septic mice genetically deficient in PPT-A gene or pretreated with L703606 do not exhibit further increase in lung permeability after injection of NaHS (Zhang et al. 2007c). These findings show that in sepsis, H_2S upregulates the generation of substance P that contributes to lung inflammation and lung injury mainly via activation of the NK-1R. Also, H_2S induces systemic inflammation and multiple organ damage characteristic of sepsis via transient receptor potential vanilloid type 1 (TRPV1)-mediated neurogenic inflammation (Ang et al. 2010). Pretreatment with capsazepine, a TRPV1 antagonist, significantly attenuates systemic inflammation and multiple organ damage caused by sepsis. Moreover, capsazepine delays the onset of lethality and protects against sepsis-associated mortality. As expected, administration of NaHS exacerbates sepsis, but capsazepine reverses these deleterious effects. In the presence of PAG, capsazepine causes no significant changes to the PAG-mediated attenuation of systemic inflammation, multiple organ damage, and mortality. More importantly, capsazepine has no effect on endogenous generation of H_2S, suggesting that H_2S is located upstream of TRPV1 activation, and may play a critical role in regulating the production and release of sensory neuropeptides in sepsis. This study showed for the first time that H_2S induces systemic inflammation and multiple organ damage in sepsis via TRPV1-mediated neurogenic inflammation (Ang et al. 2010). Capsazepine treatment results in an attenuation of circulating and pulmonary levels of substance P in septic mice. Capsazepine also inhibits NaHS-augmented substance P production but has no effect on PAG-mediated abrogation of substance P levels in both plasma and lung. Capsazepine significantly reduces H_2S-induced inflammatory cytokines, chemokines, and adhesion molecules expression and protects against lung and liver dysfunction in sepsis. In the absence of H_2S, capsazepine caused no significant changes to the PAG-mediated attenuation of sepsis-associated systemic inflammatory response and multiple organ dysfunction. Capsazepine inhibits phosphorylation of ERK1/2 and IκBα, concurrent with suppression of NF-κB activation. Results in this study showed that H_2S regulates TRPV1-mediated neurogenic inflammation in polymicrobial sepsis through enhancement of SP production and activation of the ERK-NF-κB pathway (Ang et al. 2011a). Also, a recent study has shown that H_2S upregulates cyclooxygenase-2 (COX-2) and prostaglandin E metabolite (PGEM) in sepsis-evoked acute lung injury via TRPV-1 channel activation (Ang et al. 2011b).

6 H_2S Donors

NaHS and sodium sulfide (Na_2S) are the two "classical" H_2S donors that release the gas rapidly in solution. In most of the studies, they have been substantiated a pro-inflammatory action of H_2S. In recent years, slow/controlled release donors have emerged, and results using these compounds have been quite interesting. For example, S-diclofenac (ACS 15) is H_2S-releasing diclofenac, which comprises an H_2S-releasing dithiol-thione moiety attached by an ester linkage to diclofenac (a nonsteroidal anti-inflammatory drug—NSAID). The effect of treatment with the NSAID diclofenac and its H_2S-releasing derivative ACS15 on cerulein-induced acute pancreatitis and the associated lung injury has been investigated (Bhatia et al. 2008a). Although these two drugs do not have any significant effect on the local pancreatic injury, ACS15 affords significant protection against acute pancreatitis-associated lung injury (Bhatia et al. 2008a). ACS 15 also more effectively inhibits hindpaw swelling and neutrophil infiltration after carrageenan injection as compared to its parent NSAID (Sidhapuriwala et al. 2007). Furthermore, in a rat model of LPS-induced endotoxemia, ACS 15 exhibits enhanced anti-inflammatory effect as compared to the parent drug (Li et al. 2007). Although these results suggest the potential for the use of controlled release H_2S donor compounds against inflammation, the protective action of ACS 15 in LPS-induced endotoxemia was associated with an inhibition of endogenous H_2S synthesis (Li et al. 2007). Therefore, protective actions of H_2S-releasing compounds may be caused by an inhibition of endogenous H_2S formation, possibly by a negative feedback mechanism caused by very low local levels of H_2S. Another slow H_2S-releasing drug is S-propargyl-cysteine (SPRC), a structural analog of S-allyl cysteine (SAC) with a common cysteine-containing structure. Pretreatment with SPRC has been shown to protect mice against acute pancreatitis and associated lung injury (Sidhapuriwala et al. 2012). Another slow H_2S-releasing compound is GYY4137 (morpholin-4-ium-4-methoxyphenyl(morpholino) phosphinodithioate), which has been reported to have an anti-inflammatory action (Li et al. 2009, 2013).

7 Conclusion

Figure 1 summarizes our current understanding of the mechanism of action of H_2S in inflammation. Several gaps in knowledge still remain on this subject. H_2S synthesis inhibitors, such as PAG, have proved useful in the proof of principle

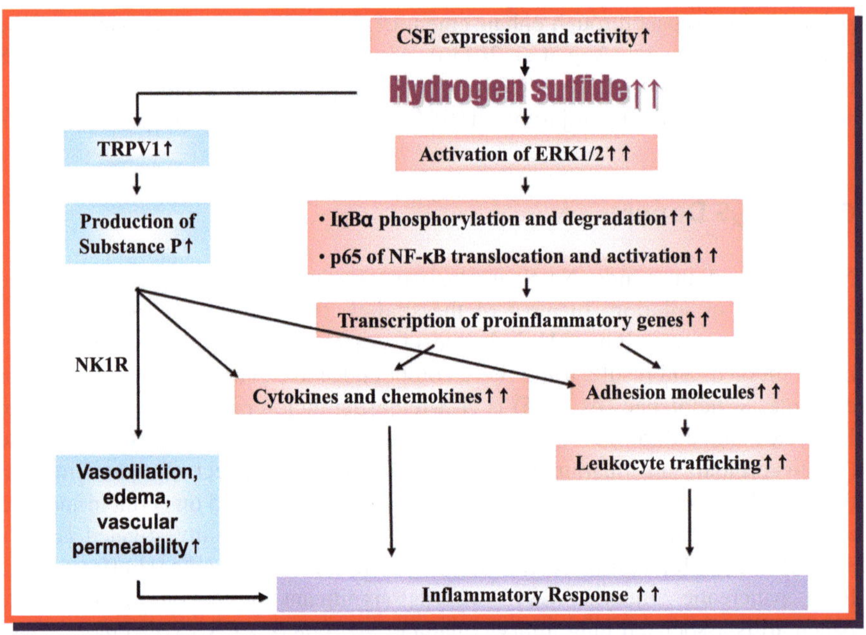

Fig. 1 A summary of our current understanding of the mechanisms of action of H_2S in inflammation

studies, but they are not entirely specific. Studies using CSE knockout mice and siRNA have contributed greatly to research in this area, but there is need to develop novel synthesis inhibitors that are more selective and have a better safety profile than the ones currently available. Also, following up on the progress made in basic/ preclinical research, now is the time to investigate the role of H_2S in inflammation in clinical disease. As described in this article, some groups have already started working on this, and early results have been promising.

Acknowledgements The author's laboratory is supported by research grants from the Lottery Health, Arthritis New Zealand, Maurice and Phyllis Paykel Trust, and University of Otago (Establishment Grant and University of Otago Research Grant).

References

Ang SF, Moochhala S, Bhatia M (2010) Hydrogen sulfide promotes transient receptor potential vanilloid 1-mediated neurogenic inflammation in polymicrobial sepsis. Crit Care Med 38:619–628

Ang SF, Moochhala SM, MacAry PA, Bhatia M (2011a) Hydrogen sulfide and neurogenic inflammation in polymicrobial sepsis: involvement of substance P and ERK-NF-κB signaling. PLoS ONE 6, e24535

Ang SF, Sio SW, Moochhala SM, MacAry PA, Bhatia M (2011b) Hydrogen sulfide upregulates cyclooxygenase-2 and prostaglandin E metabolite in sepsis-evoked acute lung injury *via* transient receptor potential vanilloid type 1 channel activation. J Immunol 187:4778–4787

Ang AD, Rivers-Auty J, Hegde A, Ishii I, Bhatia M (2013) The effect of CSE gene deletion in caerulein-induced acute pancreatitis in the mouse. Am J Physiol Gastrointest Liver Physiol 305:G712–G721

Badiei A, Rivers-Auty J, Ang AD, Bhatia M (2013) Inhibition of hydrogen sulfide production by gene silencing attenuates inflammatory activity of LPS-activated RAW 264.7 cells. Appl Microbiol Biotechnol 97:7845–7852

Bhatia M (2010) Hydrogen sulfide and substance P in inflammation. Antioxid Redox Signal 12:1191–1202

Bhatia M (2012) Role of hydrogen sulfide in the pathology of inflammation. Scientifica 2012, 159680

Bhatia M, Moochhala S (2004) Role of inflammatory mediators in the pathophysiology of acute respiratory distress syndrome. J Pathol 202:145–156

Bhatia M, Brady M, Shokuhi S, Christmas SE, Neoptolemos JP, Slavin J (2000) Inflammatory mediators in acute pancreatitis. J Pathol 190:117–125

Bhatia M, Wong FL, Fu D, Lau HY, Moochhala S, Moore PK (2005a) Role of hydrogen sulfide in acute pancreatitis and associated lung injury. FASEB J 19:623–625

Bhatia M, Sidhapuriwala J, Moochhala S, Moore PK (2005b) Hydrogen sulphide is a mediator of carrageenan-induced hindpaw oedema in the rat. Br J Pharmacol 145:141–144

Bhatia M, Wong FL, Cao Y, Lau HY, Huang J, Puneet P, Chevali L (2005c) Pathophysiology of acute pancreatitis. Pancreatology 5:132–144

Bhatia M, Zhi L, Zhang H, Ng SW, Moore PK (2006) Role of substance P in hydrogen sulfide-induced pulmonary inflammation in mice. Am J Physiol Lung Cell Mol Physiol 291:L896–L904

Bhatia M, Sidhapuriwala J, Sparatore A, Moore PK (2008a) Treatment with H_2S-releasing diclofenac protects mice against acute pancreatitis-associated lung injury. Shock 29:84–88

Bhatia M, Sidhapuriwala J, Ng SW, Tamizhselvi R, Moochhala S (2008b) Proinflammatory effects of hydrogen sulfide on substance P in caerulein-induced acute pancreatitis. J Cell Mol Med 12:580–590

Bhatia M, He M, Zhang H, Moochhala S (2009) Sepsis as a model of SIRS. Front Biosci [Animals in SIRS Research editor Bhatia M (a Special issue of encyclopedia of bioscience, published by frontiers in bioscience)] 14:4703–4711

Chen YH, Yao WZ, Geng B, Ding YL, Lu M, Zhao MW, Tang CS (2005) Endogenous hydrogen sulfide in patients with COPD. Chest 128:3205–3211

Church D, Elsayed S, Reid O, Winston B, Lindsay R (2006) Burn wound infections. Clin Microbiol Rev 19:403–434

Coletta C, Papapetropoulos A, Erdelyi K, Olah G, Módis K, Panopoulos P, Asimakopoulou A, Gerö D, Sharina I, Martin E, Szabo C (2012) Proc Natl Acad Sci U S A 109:9161–9166

Collin M, Anuar F, Murch O, Bhatia M, Moore PK, Thiemermann C (2005) Inhibition of endogenous hydrogen sulfide formation reduces the organ injury caused by endotoxemia. Br J Pharmacol 146:498–505

Decramer M, Janssens W, Miravitlles M (2012) Chronic obstructive pulmonary disease. Lancet 379:1341–1351

Endorf FW, Ahrenholz D (2011) Burn management. Curr Opin Crit Care 17:601–605

Fiorucci S, Orlandi S, Mencarelli A, Caliendo G, Santagada V, Distrutti E, Santucci L, Cirino G, Wallace JL (2007) Enhanced activity of a hydrogen sulphide-releasing derivative of mesalamine (ATB-429) in a mouse model of colitis. Br J Pharmacol 150:996–1002

Fox B, Schantz JT, Haigh R, Wood ME, Moore PK, Viner N, Spencer JP, Winyard PG, Whiteman M (2012) Inducible hydrogen sulfide synthesis in chondrocytes and mesenchymal progenitor cells: is H_2S a novel cytoprotective mediator in the inflamed joint? J Cell Mol Med 16:896–910

Hegde A, Bhatia M (2011) Hydrogen sulfide in inflammation: friend or foe? Inflamm Allergy Drug Targets 10:118–122

Kloesch B, Liszt M, Krehan D, Broell J, Kiener H, Steiner G (2012) High concentrations of hydrogen sulphide elevate the expression of a series of pro-inflammatory genes in fibroblast-like synoviocytes derived from rheumatoid and osteoarthritis patients. Immunol Lett 141:197–203

Levy MM, Dellinger RP, Townsend SR, Linde-Zwirble WT, Marshall JC, Bion J, Schorr C, Artigas A, Ramsay G, Beale R, Parker MM, Gerlach H, Reinhart K, Silva E, Harvey M, Regan S, Angus DC (2010) The surviving sepsis campaign: results of an international guideline-based performance improvement program targeting severe sepsis. Crit Care Med 38:367–374

Li L, Bhatia M, Zhu YZ, Ramnath RD, Wang ZJ, Anuar F, Whiteman M, Salto-Tellez M, Moore PK (2005) Hydrogen sulfide is a novel mediator of lipopolysaccharide-induced inflammation in the mouse. FASEB J 19:1196–1198

Li L, Rossoni G, Sparatore A, Lee LC, Del Soldato P, Moore PK (2007) Anti-inflammatory and gastrointestinal effects of a novel diclofenac derivative. Free Radic Biol Med 42:706–719

Li L, Salto-Tellez M, Tan CH, Whiteman M, Moore PK (2009) GYY4137, a novel hydrogen sulfide-releasing molecule, protects against endotoxic shock in the rat. Free Radic Biol Med 47:103–113

Li L, Rose P, Moore PK (2011) Hydrogen sulfide and cell signaling. Annu Rev Pharmacol Toxicol 51:169–187

Li L, Fox B, Keeble J, Salto-Tellez M, Winyard PG, Wood ME, Moore PK, Whiteman M (2013) The complex effects of the slow-releasing hydrogen sulfide donor GYY4137 in a model of acute joint inflammation and in human cartilage cells. J Cell Mol Med 17:365–376

Ma VY, Chan L, Carruthers KJ (2014) Incidence, prevalence, costs, and impact on disability of common conditions requiring rehabilitation in the United States: stroke, spinal cord injury, traumatic brain injury, multiple sclerosis, osteoarthritis, rheumatoid arthritis, limb loss, and back pain. Arch Phys Med Rehabil 95:986–995

Martin GS, Mannino DM, Eaton S, Moss M (2003) The epidemiology of sepsis in the United States from 1979 through 2000. N Engl J Med 348:1546–1554

Mazumder B, Li X, Barik S (2010) Translation control: a multifaceted regulator of inflammatory response. J Immunol 184:3311–3319

Moore PK, Bhatia M, Moochhala S (2003) Hydrogen sulphide: from the smell of the past to the mediator of the future? Trends Pharmacol Sci 24:609–611

Nathan C (2002) Points of control in inflammation. Nature 420:846–852

Norris EJ, Feilen N, Nguyen NH, Culberson CR, Shin MC, Fish M, Clemens MG (2013) Hydrogen sulfide modulates sinusoidal constriction and contributes to hepatic microcirculatory dysfunction during endotoxemia. Am J Physiol Gastrointest Liver Physiol 304:G1070–G1078

Peery AF, Dellon ES, Lund J, Crockett SD, McGowan CE, Bulsiewicz WJ, Gangarosa LM, Thiny MT, Stizenberg K, Morgan DR, Ringel Y, Kim HP, Dibonaventura MD, Carroll CF, Allen JK, Cook SF, Sandler RS, Kappelman MD, Shaheen NJ (2012) Burden of gastrointestinal disease in the United States: 2012 update. Gastroenterology 143:1179–1187

Polytarchou C, Koukos G, Iliopoulos D (2014) Systems biology in inflammatory bowel diseases: ready for prime time. Curr Opin Gastroenterol 30(4):339–346

Ramnath RD, Ng SW, He M, Sun J, Zhang H, Bawa MS, Bhatia M (2006) Inflammatory mediators in sepsis: cytokines, chemokines, adhesion molecules, and gases. J Organ Dysfunct 2:80–92

Shanmugam MK, Jing Z, Bhatia M (2009) Aminooxyacetate inhibits hydrogen sulfide and ammonium synthesis and protects mice in acute pancreatitis. Int J Integr Biol 8:7–14

Sidhapuriwala J, Li L, Sparatore A, Bhatia M, Moore PK (2007) Effect of S-diclofenac, a novel hydrogen sulfide releasing derivative, on carrageenan-induced hindpaw oedema formation in the rat. Eur J Pharmacol 569:149–154

Sidhapuriwala J, Hegde A, Ang AD, Zhu YZ, Bhatia M (2012) Effects of S-propargyl-cysteine (SPRC) in caerulein-induced acute pancreatitis in mice. PLoS ONE 7, e32574

Sun Y, Wang XM, Chen YH, Zhu RX, Liao CC (2013) Exhaled hydrogen sulfide in patients with chronic obstructive pulmonary disease and its correlation with exhaled nitric oxide. Chin Med J (Engl) 126:3240–3244

Tamizhselvi R, Moore PK, Bhatia M (2007) Hydrogen sulfide acts as a mediator of inflammation in acute pancreatitis: in vitro studies using isolated mouse pancreatic acinar cells. J Cell Mol Med 11:315–326

Tamizhselvi R, Moore PK, Bhatia M (2008) Inhibition of hydrogen sulfide synthesis attenuates chemokine production and protects mice against acute pancreatitis and associated lung injury. Pancreas 36:e24–e31

Tamizhselvi R, Sun J, Hua KY, Bhatia M (2009) Effect of hydrogen sulfide on PI3K-AKT pathway and on caerulein- induced cytokine production in isolated mouse pancreatic acinar cells. J Pharmacol Exp Ther 329:1166–1177

Tamizhselvi R, Koh YH, Sun J, Zhang H, Bhatia M (2010) Hydrogen sulfide-induces ICAM-1 expression and neutrophil adhesion to caerulein treated pancreatic acinar cells through NF-κB and Src-family kinases pathway. Exp Cell Res 316:1625–1636

Tamizhselvi R, Shrivastava P, Koh YH, Zhang H, Bhatia M (2011) Preprotachykinin-A gene deletion regulates hydrogen sulfide induced toll-like receptor 4 signaling pathway in cerulein-treated pancreatic acinar cells. Pancreas 40:444–452

Wang R (2012) Physiological implications of hydrogen sulfide: a whiff exploration that blossomed. Physiol Rev 92:791–896

Wang M, Guo Z, Wang S (2014) The binding site for the transcription factor, NF-κB, on the cystathionine γ-lyase promoter is critical for LPS-induced cystathionine γ-lyase expression. Int J Mol Med 34(2):639–645

Yan Y, Chen C, Zhou H, Gao H, Chen L, Chen L, Gao L, Zhao R, Sun Y (2013) Endogenous hydrogen sulfide formation mediates the liver damage in endotoxemic rats. Res Vet Sci 94:590–595

Zhang H, Zhi L, Moore PK, Bhatia M (2006) The role of hydrogen sulfide in cecal ligation and puncture induced sepsis in the mouse. Am J Physiol Lung Cell Mol Physiol 290:L1193–L1201

Zhang H, Zhi L, Moochhala S, Moore PK, Bhatia M (2007a) Hydrogen sulfide acts as an inflammatory mediator in cecal ligation and puncture induced sepsis in mice by up-regulating the production of cytokines and chemokines via NF-κB. Am J Physiol Lung Cell Mol Physiol 292:L960–L971

Zhang H, Zhi L, Moochhala S, Moore PK, Bhatia M (2007b) Endogenous hydrogen sulfide regulates leukocyte trafficking in cecal ligation and puncture induced sepsis. J Leukoc Biol 82:894–905

Zhang H, Hegde A, Ng SW, Adhikari S, Moochhala SM, Bhatia M (2007c) Hydrogen sulfide up-regulates substance P in polymicrobial sepsis associated Lung Injury. J Immunol 179:4153–4160

Zhang H, Moochhala S, Bhatia M (2008) Endogenous hydrogen sulfide regulates inflammatory response by activating the extracellular signal-regulated kinase (ERK) pathway in polymicrobial sepsis. J Immunol 181:4320–4331

Zhang J, Sio SW, Moochhala S, Bhatia M (2010) Role of hydrogen sulfide in severe burn injury-induced inflammation in the mouse. Mol Med 16:417–424

Zhi L, Ang AD, Zhang H, Moore PK, Bhatia M (2007) Hydrogen sulfide induces the synthesis of pro-inflammatory cytokines in human monocyte cell line U937 via ERK-NFκB pathway. J Leukoc Biol 81:1322–1332

Hydrogen Sulfide and Neuroinflammation

Kotaro Kida and Fumito Ichinose

Contents

Abstract

The innate and adaptive immune system plays an important role in diverse forms of central nervous system (CNS) pathologies including neurodegenerative diseases and peripheral nerve injury. Evidence for an innate inflammatory response in Alzheimer's disease (AD) was described 20 years ago, and subsequent studies have documented roles of inflammation in Parkinson's disease (PD), amyotrophic lateral sclerosis (ALS), multiple sclerosis (MS), and a growing number of other CNS pathologies. Although inflammation may not be the initiating factor for neurodegenerative pathologies, experimental data suggests that persistent inflammatory responses involving microglia and astrocytes, as well as blood monocyte-derived macrophages, clearly contribute to disease progression.

High levels of hydrogen sulfide exert toxic effects to CNS. On the other hand, low and physiological levels of H_2S may have beneficial effects on number of tissues including CNS. For example, a number of studies have reported that H_2S exerts anti-inflammatory and anti-apoptotic effects in CNS. In this chapter, studies related to the role of H_2S in neuroinflammation and neurodegeneration

K. Kida • F. Ichinose (✉)
Anesthesia Center for Critical Care Research, Department of Anesthesia, Critical Care, and Pain Medicine, Massachusetts General Hospital, Harvard Medical School, Boston, MA 02114, USA
e-mail: fichinose@mgh.harvard.edu

P.K. Moore, M. Whiteman (eds.), *Chemistry, Biochemistry and Pharmacology of Hydrogen Sulfide*, Handbook of Experimental Pharmacology 230,
DOI 10.1007/978-3-319-18144-8_9

will be reviewed and discussed. In particular, we will focus on the role of H_2S in neuroinflammation associated with PD.

Keywords
Neurodegenerative disease • Parkinson's disease • Microglia • Gliosis • Inflammatory cytokine

Parkinson's disease (PD) is one of the most common neurodegenerative diseases (Twelves et al. 2003). It is characterized by a slow and progressive degeneration of dopaminergic neurons in the substantia nigra (Hirsch et al. 1988). Although the etiology of PD is not fully understood, several mechanisms responsible for the neurodegeneration in PD have been suggested, including abnormal protein handling, oxidative stress, mitochondrial dysfunction, excitotoxicity, neuroinflammation, and apoptosis (Hirsch and Hunot 2009).

A number of animal models of PD, both toxin-induced and genetically engineered, have been created. While the toxin-induced PD models have provided considerable therapeutic insight into basal ganglia physiology and response to drug therapy, genetic models provide a powerful new set of molecular tools to study the etiology of PD (Melrose et al. 2006). Although none of the animal models accurately recapitulate the pathophysiological features of PD, 1-methyl-4-phenyl-1,2,3,6-tetrahydropyridine (MPTP) currently represents the most important and most frequently used parkinsonian toxin applied in animal models (Beal 2001; Przedborski et al. 2001; Schober 2004). MPTP is the only known dopaminergic neurotoxin capable of causing a clinical picture indistinguishable from idiopathic PD in humans (Langston et al. 1983). Moreover, MPTP produces a reliable and reproducible lesion of the nigrostriatal dopaminergic pathway after its systemic administration, which is often not the case for other neurotoxins (Bove et al. 2006).

Hydrogen sulfide appears to confer cytoprotection via multiple mechanisms including antioxidant and anti-apoptotic effects. For example, Kimura and colleagues demonstrated using a model of glutamate-induced oxidative stress that H_2S protects neurons from cell death by increasing the levels of the antioxidant, glutathione (Kimura and Kimura 2004). A recent study revealed that NaHS, an H_2S donor, protects PC12 cells from cytotoxicity and apoptosis induced by 1-methy-4-phenylpyridinium ion (MPP+), the active metabolite of MPTP (Yin et al. 2009).

1 Effects of H_2S in Animal Models of PD

Neuroprotective effects of H_2S donors (i.e., NaHS and Na_2S) have been reported in vitro and in vivo in a variety of animal models (Kimura et al. 2010; Kimura and Kimura 2004). In particular, a recent report by Hu and colleagues demonstrated neuroprotective effects of NaHS in rat models of PD induced by 6-OHDA or rotenone (Hu et al. 2009). While these studies suggest neuroprotective effects of H_2S or related compounds, solution of NaHS has been found to include polysulfides

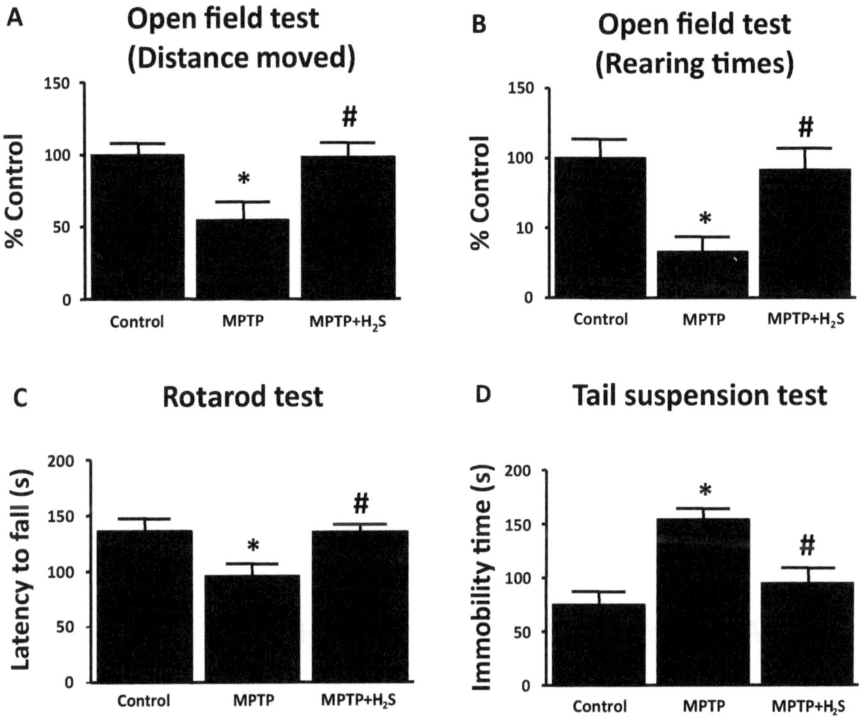

Fig. 1 Results of behavioral tests performed at 7 days after administration of MPTP or saline. Results of distance moved (**a**) and rearing times (**b**) in the open field test are shown as percent of the values in saline-treated control mice. (**c**) Latency to fall in the rotorod test is shown in seconds. (**d**) Immobility times in the tail-suspension test are shown in seconds. *Control* saline-treated mice, *MPTP* mice treated with MPTP that breathed air, *MPTP + H_2S* mice treated with MPTP that breathed H_2S. $N = 7$–9 in each group in all tests. *$P < 0.05$ vs. control. #$P < 0.05$ vs. MPTP. Reproduced from Kida et al. *Antioxid Redox Signal* 2011 Jul 15;15(2):343-52 with permission

and elemental sulfur in addition to H_2S (Doeller et al. 2005). To examine the neuroprotective effects of authentic H_2S, we sought to examine effects of H_2S inhalation in the clinically relevant MPTP-induced PD model in mice (Kida et al. 2010). Inhalation of H_2S at 40 ppm for 8 h a day for 7 days prevented MPTP-induced movement disorder in mouse as measured by open field test, rotarod test, and tail-suspension test (Fig. 1). The neuroprotective effects of H_2S inhalation were associated with marked attenuation of the MPTP-induced loss of TH-containing neurons in substantia nigra and striatum (Fig. 2).

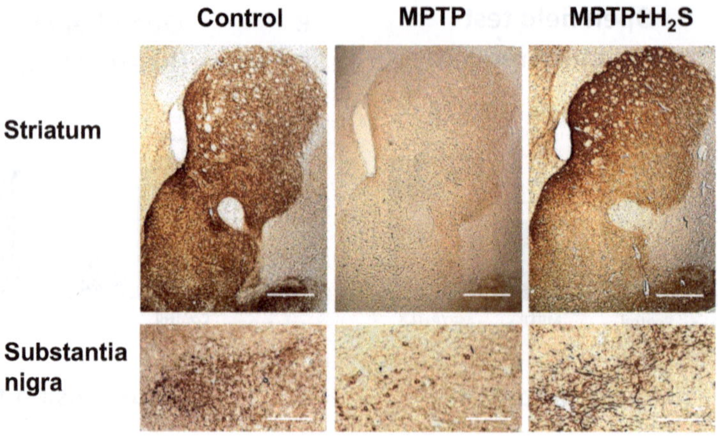

Fig. 2 Representative immunohistochemical staining of TH-positive neurons in mice 7 days after administration of MPTP or saline in striatum and substantia nigra. Reproduced from Kida et al. *Antioxid Redox Signal* 2011 Jul 15;15(2):343-52 with permission

2 Anti-apoptotic Effects of Inhaled H_2S

To determine the mechanisms responsible for the protective effects of H_2S inhalation after MPTP, we conducted extensive histological analysis of brain sections. Administration of MPTP increased the number of apoptotic cells in substantia nigra 1 day after MPTP administration as indicated by the presence of TUNEL-positive nuclei (Fig. 3a–d). In contrast, H_2S inhalation prevented the MPTP-induced apoptosis (Fig. 3e–h). Immunofluorescence double labeling demonstrated that the TUNEL-positive nuclei were found exclusively in substantia nigra especially in the area where TH-positive neurons reside (Fig. 3d, h, i). While majority of the TUNEL-positive nuclei did not overlay with TH-positive neurons (Fig. 3d, h, i, j), triple labeling with TUNEL, TH, and DAPI showed that a small number of TH-positive neurons contained a TUNEL-positive nucleus (Fig. 3j, k). These observations suggest that at least a part of TH-positive neurons died via apoptosis.

To further identify the cell types undergoing apoptosis in substantia nigra after MPTP administration, we stained brain sections with antibodies against GFAP (a marker of activated astrocytes) or Iba-1 (a marker of activated microglia) after TUNEL assay. Double labeling with TUNEL and GFAP revealed that GFAP-positive activated astrocytes did not overlay with any TUNEL-positive nuclei (Fig. 4a). On the other hand, double labeling with TUNEL and Iba-1 revealed that many TUNEL-positive nuclei were enveloped by Iba-1-positive activated microglia (Fig. 4b). In fact, triple labeling with TUNEL, Iba-1, and DAPI showed that majority of TUNEL-positive nuclei were distinct from nuclei of microglia (Fig. 4c), suggesting phagocytosis of apoptotic neurons by activated microglia. Taken together, these observations suggest that MPTP-induced apoptosis of

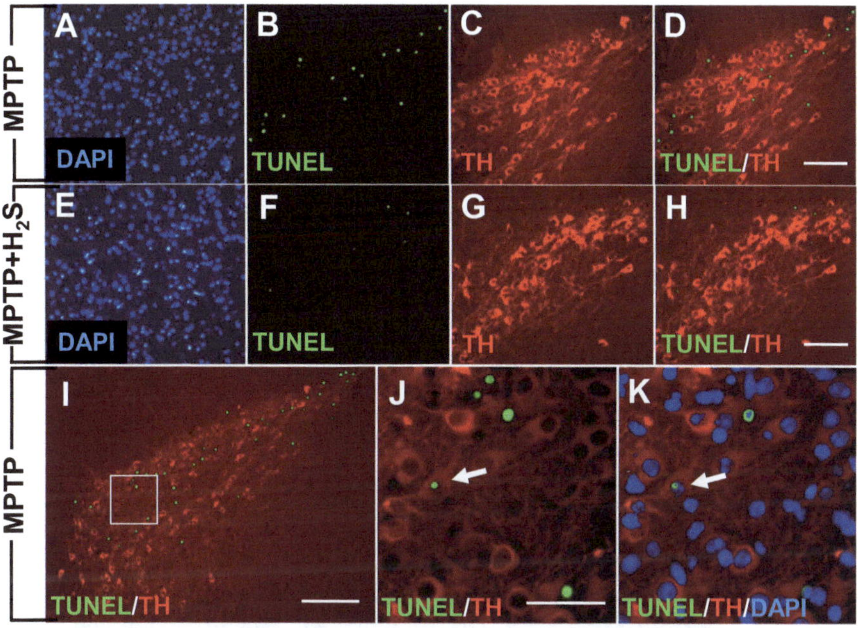

Fig. 3 Representative photomicrographs of brain sections containing substantia nigra obtained from mice 1 day after administration of MPTP without H_2S (MPTP, **a–d**, **i–k**) or MPTP with H_2S breathing (MPTP + H_2S, **e–h**). Sections were subjected to the TUNEL assay and stained with DAPI, anti-TH, anti-GFAP, or anti-Iba-1 antibody. Size bar = 100 μm for (**a–h**), 200 μm for (**i**), and 5 μm for (**j**) and (**k**). *White arrows* in (**j**) and (**k**) indicate a TH-positive neuron containing TUNEL-positive nuclei. Reproduced from Kida et al. *Antioxid Redox Signal* 2011 Jul 15;15 (2):343-52 with permission

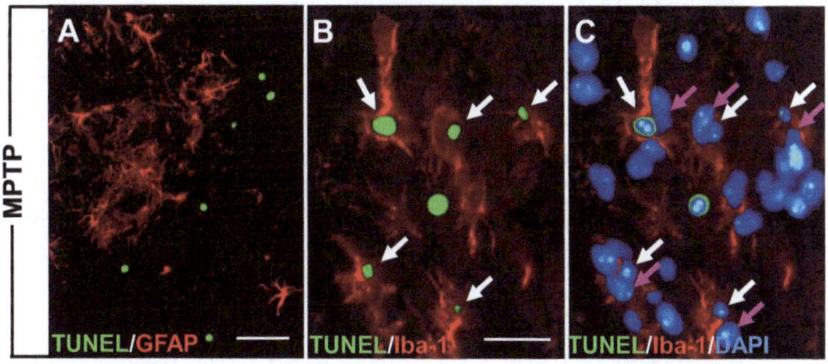

Fig. 4 Representative photomicrographs of brain sections 1 day after MPTP administration without H_2S breathing. (**a**) Double labeling with TUNEL (*green*) and GFAP (*red*). Size bar = 5 μm. (**b**) Double labeling with TUNEL (*green*) and Iba-1 (*red*). (**c**) Triple labeling with TUNEL (*green*), Iba-1 (*red*), and DAPI (*blue*). Size bar = 2.5 μm for (**b**) and (**c**). *White arrows* indicate TUNEL-positive nuclei and *pink arrows* indicate nuclei of microglia distinct from the TUNEL-positive nuclei. Reproduced from Kida et al. *Antioxid Redox Signal* 2011 Jul 15;15(2):343-52 with permission

TH-positive neurons in substantia nigra and apoptotic neurons were phagocytosed by activated microglia.

3 Effects of H_2S on Glial Activation

Immunofluorescence staining demonstrated few activated microglia and astrocytes in substantia nigra and striatum of saline-treated mice (Fig. 5a, d, g, j). In contrast, MPTP increased the number of activated microglia and astrocytes in substantia

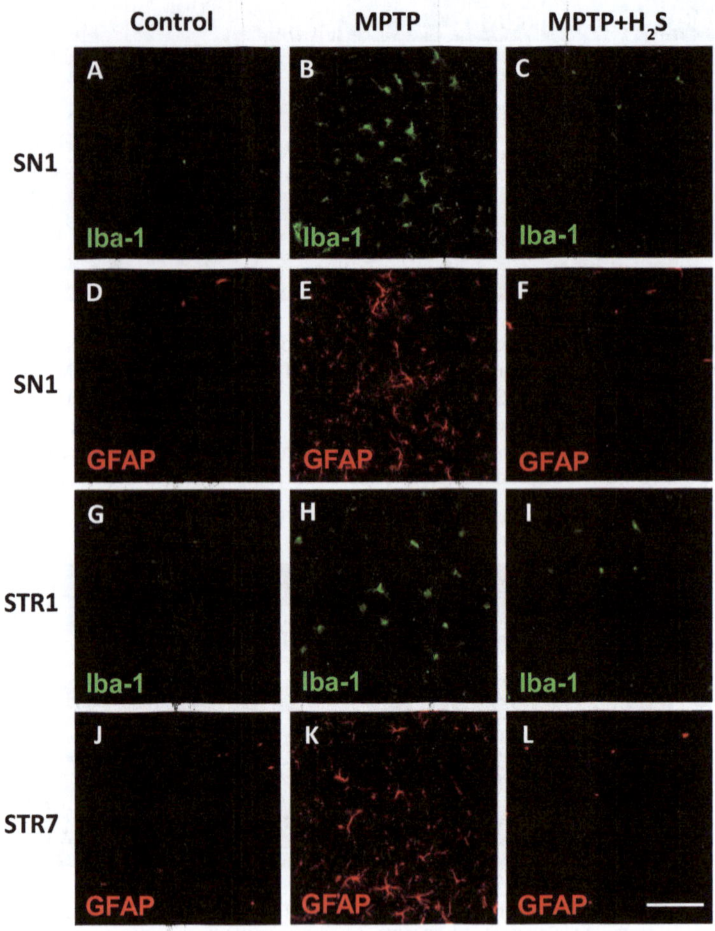

Fig. 5 Representative immunofluorescence staining for Iba-1 (microglia, *green*) and GFAP (astrocytes, *red*) in substantia nigra (SN) and striatum (STR) of saline-treated control mice (Control, **a**, **d**, **g**, and **j**), mice treated with MPTP that breathed air (MPTP, **b**, **e**, **h**, and **k**), and mice treated with MPTP that breathed H_2S (MPTP + H_2S, **c**, **f**, **i**, and **l**). These sections were obtained from mice 1 day (SN1 and STR1) or 7 days (STR7) after administration of MPTP or saline. Reproduced from Kida et al. *Antioxid Redox Signal* 2011 Jul 15;15(2):343-52 with permission

nigra on Day 1 (Fig. 5b, e). In striatum, the number of activated microglia increased on Day 1 (Fig. 5h) and the number of activated astrocytes increased on Day 7 after MPTP injection (Fig. 5k). Inhaled H_2S prevented the activation of both microglia and astrocytes in substantia nigra and striatum after MPTP administration (Fig. 5c, f, i, l).

4 Role of Endogenous H_2S in Parkinson's Disease

Parkin is an E3 ubiquitin ligase that ubiquitinates diverse substrates (Martin et al. 2011). Mutations in parkin that disrupt its catalytic activity are the most common cause of autosomal recessive form of PD, implying that loss of parkin is neurotoxic, while its enhancement is neuroprotective. Parkin may also participate in the pathophysiology of the common sporadic form of PD based on interactions with nitric oxide (NO). It has been reported that parkin is *S*-nitrosylated in brains of patients with PD (Chung et al. 2004; Yao et al. 2004). Nitrosylation of parkin inhibits its E3 ligase and neuroprotective activities, implying that the increased nitrosylation of parkin in PD is pathogenic. H_2S signals primarily by attaching to SH groups of cysteines in proteins, a process termed sulfhydration (Mustafa et al. 2009). Sulfhydration generally occurs on the same cysteines as nitrosylation so that the two processes may be reciprocal.

Vandiver and colleagues recently reported that parkin is physiologically sulfhydrated and that, whereas nitrosylation inactivates parkin, sulfhydration stimulates its activity (Vandiver et al. 2013). These authors identified major declines of parkin sulfhydration in the corpus striatum of PD patients. Their experiments also showed cytoprotective actions of H_2S donors, which appear to reflect sulfhydration of parkin. Mutation of C95, a principal site of parkin sulfhydration, largely prevented the protective influences of H_2S donors indicating that the donors act via parkin sulfhydration to enhance its ubiquitination activity. In addition, H_2S donors were cytoprotective in parkin-related models of neurotoxicity. These studies suggest that diminished sulfhydration of parkin may be pathogenic in PD and selective H_2S donors may be therapeutic.

5 Conclusion

In the last several years, a number of studies reported that H_2S donors exert beneficial effects in rodent models of PD (Hu et al. 2010; Kida et al. 2010; Lu et al. 2012), and an H_2S donating variant of L-DOPA has shown promising effects in cellular models of PD (Lee et al. 2010). In addition, H_2S donors have been shown to exert potent anti-inflammatory effects in other forms of inflammatory pathologies (Li et al. 2009; Shirozu et al. 2013; Tokuda et al. 2012; Whiteman et al. 2009). While these beneficial effects had been speculated to reflect general antioxidant and anti-inflammatory actions of H_2S, numerous studies have failed to reveal a beneficial effect of antioxidants in PD. For this reason, recent findings by

Vandiver and colleagues are particularly notable because they may provide a molecular mechanism whereby H_2S therapy may prevent one of the initiating pathogenetic mechanisms of PD, as well as to mitigate propagation of the pathological process by chronic inflammation. These recent observations suggest a unique potential of H_2S-based approach to the treatment of PD and other CNS pathologies characterized with progressive neurodegeneration.

References

Beal MF (2001) Experimental models of Parkinson's disease. Nat Rev Neurosci 2:325–334. doi:10.1038/35072550

Bove J, Zhou C, Jackson-Lewis V, Taylor J, Chu Y, Rideout HJ, Wu DC, Kordower JH, Petrucelli L, Przedborski S (2006) Proteasome inhibition and Parkinson's disease modeling. Ann Neurol 60:260–264. doi:10.1002/ana.20937

Chung KK, Thomas B, Li X, Pletnikova O, Troncoso JC, Marsh L, Dawson VL, Dawson TM (2004) S-nitrosylation of parkin regulates ubiquitination and compromises Parkin's protective function. Science 304:1328–1331. doi:10.1126/science.1093891

Doeller JE, Isbell TS, Benavides G, Koenitzer J, Patel H, Patel RP, Lancaster JR Jr, Darley-Usmar VM, Kraus DW (2005) Polarographic measurement of hydrogen sulfide production and consumption by mammalian tissues. Anal Biochem 341:40–51

Hirsch EC, Hunot S (2009) Neuroinflammation in Parkinson's disease: a target for neuroprotection? Lancet Neurol 8:382–397. doi:10.1016/s1474-4422(09)70062-6

Hirsch E, Graybiel AM, Agid YA (1988) Melanized dopaminergic neurons are differentially susceptible to degeneration in Parkinson's disease. Nature 334:345–348. doi:10.1038/334345a0

Hu LF, Lu M, Tiong CX, Dawe GS, Hu G, Bian JS (2010) Neuroprotective effects of hydrogen sulfide on Parkinson's disease rat models. Aging Cell 9:135–146. doi:10.1111/j.1474-9726.2009.00543.x

Hu LF, Zheng G, Zweier JL, Deshpande S, Irani K, Ziegelstein RC (2009) NADPH oxidase activation increases the sensitivity of intracellular Ca^{2+} stores to inositol 1,4,5-trisphosphate in human endothelial cells. J Biol Chem 275:15749–15757

Kida K, Yamada M, Tokuda K, Marutani E, Kakinohana M, Kaneki M, Ichinose F (2010) Inhaled hydrogen sulfide prevents neurodegeneration and movement disorder in a mouse model of Parkinson's disease. Antioxid Redox Signal 15:343–352. doi:10.1089/ars.2010.3671

Kimura Y, Kimura H (2004) Hydrogen sulfide protects neurons from oxidative stress. FASEB J 18(10):1165–1167

Kimura Y, Goto Y, Kimura H (2010) Hydrogen sulfide increases glutathione production and suppresses oxidative stress in mitochondria. Antioxid Redox Signal 12:1–13

Langston JW, Ballard P, Tetrud JW, Irwin I (1983) Chronic Parkinsonism in humans due to a product of meperidine-analog synthesis. Science 219:979–980

Lee M, Tazzari V, Giustarini D, Rossi R, Sparatore A, Del Soldato P, McGeer E, McGeer PL (2010) Effects of hydrogen sulfide-releasing L-DOPA derivatives on glial activation: potential for treating Parkinson disease. J Biol Chem 285:17318–17328. doi:10.1074/jbc.M110.115261

Li L, Salto-Tellez M, Tan CH, Whiteman M, Moore PK (2009) GYY4137, a novel hydrogen sulfide-releasing molecule, protects against endotoxic shock in the rat. Free Radic Biol Med 47:103–113

Lu M, Zhao FF, Tang JJ, Su CJ, Fan Y, Ding JH, Bian JS, Hu G (2012) The neuroprotection of hydrogen sulfide against MPTP-induced dopaminergic neuron degeneration involves uncoupling protein 2 rather than ATP-sensitive potassium channels. Antioxid Redox Signal 17:849–859. doi:10.1089/ars.2011.4507

Martin I, Dawson VL, Dawson TM (2011) Recent advances in the genetics of Parkinson's disease. Annu Rev Genomics Hum Genet 12:301–325. doi:10.1146/annurev-genom-082410-101440

Melrose HL, Lincoln SJ, Tyndall GM, Farrer MJ (2006) Parkinson's disease: a rethink of rodent models. Exp Brain Res 173:196–204. doi:10.1007/s00221-006-0461-3

Mustafa AK, Gadalla MM, Sen N, Kim S, Mu W, Gazi SK, Barrow RK, Yang G, Wang R, Snyder SH (2009) H2S signals through protein S-sulfhydration. Sci Signal 2:ra72. doi:10.1126/scisignal.2000464

Przedborski S, Jackson-Lewis V, Naini AB, Jakowec M, Petzinger G, Miller R, Akram M (2001) The Parkinsonian toxin 1-methyl-4-phenyl-1,2,3,6-tetrahydropyridine (MPTP): a technical review of its utility and safety. J Neurochem 76:1265–1274

Schober A (2004) Classic toxin-induced animal models of Parkinson's disease: 6-OHDA and MPTP. Cell Tissue Res 318:215–224. doi:10.1007/s00441-004-0938-y

Shirozu K, Tokuda K, Marutani E, Lefer D, Wang R, Ichinose F (2013) Cystathionine gamma-lyase deficiency protects mice from galactosamine/lipopolysaccharide-induced acute liver failure. Antioxid Redox Signal. doi:10.1089/ars.2013.5354

Tokuda K, Kida K, Marutani E, Crimi E, Bougaki M, Khatri A, Kimura H, Ichinose F (2012) Inhaled hydrogen sulfide prevents endotoxin-induced systemic inflammation and improves survival by altering sulfide metabolism in mice. Antioxid Redox Signal 17:11–21. doi:10.1089/ars.2011.4363

Twelves D, Perkins KS, Counsell C (2003) Systematic review of incidence studies of Parkinson's disease. Mov Disord 18:19–31. doi:10.1002/mds.10305

Vandiver MS, Paul BD, Xu R, Karuppagounder S, Rao F, Snowman AM, Ko HS, Lee YI, Dawson VL, Dawson TM, Sen N, Snyder SH (2013) Sulfhydration mediates neuroprotective actions of parkin. Nat Commun 4:1626. doi:10.1038/ncomms2623

Whiteman M, Li L, Rose P, Tan CH, Parkinson DB, Moore PK (2009) The effect of hydrogen sulfide donors on lipopolysaccharide-induced formation of inflammatory mediators in macrophages. Antioxid Redox Signal 12:1147–1154. doi:10.1089/ars.2009.2899

Yao D, Gu Z, Nakamura T, Shi ZQ, Ma Y, Gaston B, Palmer LA, Rockenstein EM, Zhang Z, Masliah E, Uehara T, Lipton SA (2004) Nitrosative stress linked to sporadic Parkinson's disease: S-nitrosylation of parkin regulates its E3 ubiquitin ligase activity. Proc Natl Acad Sci U S A 101:10810–10814. doi:10.1073/pnas.0404161101

Yin WL, He JQ, Hu B, Jiang ZS, Tang XQ (2009) Hydrogen sulfide inhibits MPP(+)-induced apoptosis in PC12 cells. Life Sci 85:269–275. doi:10.1016/j.lfs.2009.05.023

Part V

H_2S and the Nervous System

Brain, Learning, and Memory: Role of H_2S in Neurodegenerative Diseases

B.V. Nagpure and Jin-Song Bian

Contents

Abstract

For more than 300 years, the toxicity of hydrogen sulfide (H_2S) has been known to mankind. However, this point of view is changing as an increased interest was

B.V. Nagpure • J.-S. Bian (✉)
Department of Pharmacology, Yong Loo Lin School of Medicine, National University of Singapore, Singapore 117597, Singapore
e-mail: phcbjs@nus.edu.sg

P.K. Moore, M. Whiteman (eds.), *Chemistry, Biochemistry and Pharmacology of Hydrogen Sulfide*, Handbook of Experimental Pharmacology 230,
DOI 10.1007/978-3-319-18144-8_10

observed in H_2S biology in the last two decades. The scientific community has succeeded to unravel many important physiological and pathological effects of H_2S on mammalian body systems. Thus, H_2S is now referred to as a third endogenous gaseous mediator along with nitric oxide and carbon monoxide. Acting as a neuromodulator, H_2S facilitates long-term potentiation and regulates intracellular calcium levels, which are important processes in learning and memory. Aberrant endogenous production and metabolism of H_2S are implicated in pathogenesis of neurodegenerative diseases including Alzheimer's disease (AD) and Parkinson's disease (PD). Various H_2S donors have shown beneficial therapeutic effects in neurodegenerative disease models by targeting hallmark pathological events (e.g., amyloid-β production in AD and neuroinflammation in PD). The results obtained from many in vivo studies clearly show that H_2S not only prevents neuronal and synaptic deterioration but also improves deficits in memory, cognition, and learning. The anti-inflammatory, antioxidant, and anti-apoptotic effects of H_2S underlie its neuroprotective properties. In this chapter, we will overview the current understanding of H_2S in context of neurodegenerative diseases, with special emphasis on its corrective effects on impaired learning, memory, and cognition.

Keywords
Hydrogen sulfide • Neurodegeneration • Brain • Memory • Learning • Neuroinflammation

Abbreviation

3-MST	3-Mercaptopyruvate sulfurtransferase
Aβ	Amyloid-β
AD	Alzheimer's disease
APP	Amyloid precursor protein
CAT	Cysteine aminotransferase
CBS	Cystathionine-β-synthase
CNS	Central nervous system
CSE	Cystathionine-γ-lyase
H_2S	Hydrogen sulfide
K_{ATP}	ATP-sensitive potassium channel
LTP	Long-term potentiation
mitoK_{ATP}	Mitochondrial K_{ATP} channel
NaHS	Sodium hydrogen sulfide
NF-κB	Nuclear factor kappa-light-chain-enhancer of activated B cells
NMDA	*N*-methyl-D-aspartic acid
NSAIDs	Non-steroidal anti-inflammatory drugs

PD Parkinson's disease
ROS Reactive oxygen species

1 Introduction

Alzheimer's disease (AD), Parkinson's disease (PD), and other neurodegenerative diseases share a number of common aspects—old-age onset, hereditary or sporadic appearance, and the aggregation and deposition of misfolded proteins. These diseases show overlapping spectrum of clinical signs and symptoms in their advance stages. In 2005, 24.2 million people worldwide had dementia, which is one of the most common sign of AD. The developed world i.e., North America and Western Europe showed the highest prevalence of dementia. It is estimated that these parts of the world along with China will be home to around 55 % of total affected population worldwide by 2040 (Reitz et al. 2011). The prevalence of dementia grows with age. The extensive Delphi consensus study done by Ferri et al. revealed the growth from 1 % in 60–64 years age group to about 30 % in those of >85 years age group (Ferri et al. 2005). The incidence rate of AD and other dementias also increase exponentially with the age, mirroring the prevalence rate. The worst affected is the seventh and eighth decade of life (Reitz et al. 2011). PD is the second most common neurodegenerative disorder, with incidence rate rising sharply after the age of 60. The prevalence of PD in people over age of 50 is feared to be doubled to approximately nine million by 2030 in the world's most populated and industrialized countries. Placing a great burden on caregivers, AD, PD, and other neurodegenerative diseases are costing dearly to today's society

In recent decades, the scientific community has witnessed the rise of a whole new class of gaseous biological mediators in mammalian cells. These are simple gas molecules, which are lipid soluble and hence freely membrane permeable reaching intracellular organelle. After the seminal discovery of physiological effects of nitric oxide (NO) on blood vessels, carbon monoxide (CO) and hydrogen sulfide (H_2S) have been recently recognized as two more "gasotransmitters." The physiological role of H_2S was discovered by a Japanese group of scientists led by Kimura in 1996. In their pioneering study, the novel neuromodulator role of H_2S was transpired (Abe and Kimura 1996). Since then its possible roles in other body systems have been investigated worldwide. In mammalian central nervous system (CNS), its prominent effects include modulation of neurotransmission and long-term potentiation (LTP) (Abe and Kimura 1996) and induction of neuroprotection (Hu et al. 2011) from myriad of pathogenic agents. In mammalian cardiovascular system (CVS), its protective effects are deeply studied (Pan et al. 2006; Hu et al. 2008; Liu et al. 2012; Polhemus et al. 2014). The induction of relaxation (Lee et al. 2007; Yang et al. 2008) and constriction (Lim et al. 2008; Kohn et al. 2012) in various types of blood vessels is also documented (d'Emmanuele di Villa Bianca et al. 2011). The opposite effects of H_2S on systemic and localized inflammation have been observed in various mammalian tissues (Hegde and Bhatia

2011; Whiteman and Winyard 2011). In this book chapter, we will discuss the well-accepted and latest findings pertaining to neurophysiological and neuropharmacological effects of H_2S.

2 H_2S

2.1 Physical and Chemical Properties

At room temperature and ambient pressure, H_2S exists in a colorless gaseous form. The smell is pungent with distinctive rotten-egg odor. It is readily water soluble due to its weak acidic nature. Its solubility was measured to be 80 mM at 37 °C as equilibrium between H_2S, HS^-, and S^{2-}. The acid dissociation constant (pK_a) values of the first and second dissociation steps are recorded as 7.0 and >12.0, respectively (Vorobets et al. 2002; Kabil and Banerjee 2010; Mark et al. 2011). Thus, at physiological pH of 7.4 H_2S exists majorly as HS^- moiety along with minor presence of free H_2S in its dissociated form. The minute amounts of sulfide anions (S^{2-}) can also be detected. Even with the advent of various methods of H_2S measurement, it is almost impossible to determine the active form of H_2S (H_2S, HS^-, or S^{2-}) present in the biological system. Hence the all-encompassing term of H_2S is now used to refer the total sulfide content present in the solution (i.e., $H_2S + HS^- + S^{2-}$) (Zhao et al. 2014).

2.2 Biosynthesis in Mammalian CNS

In mammalian tissues, H_2S is biosynthesized from amino acid cysteine (Cys) and homocysteine (Hcy), which are recognized as the principle substrates for its endogenous production. They are acted upon by three different enzymes, namely cystathionine β-synthase (CBS), cystathionine γ-lyase (CSE), and 3-mercaptopyruvate sulfur transferase (3-MST) (Hu et al. 2011). Expressions of these enzymes are variable in different tissues. The study of this variation is important as the modulation of endogenous production of H_2S can be achieved by targeting each enzyme separately or concurrently.

CBS, a pyridoxal-5′-phosphate (PLP)-dependent enzyme, is abundantly expressed in brain and is the key source of H_2S there (Abe and Kimura 1996). The levels of CBS expression vary according to the stages of development. While the levels are low during embryonic brain, they increase notably from the late prenatal to the early postnatal period. After adulthood is reached, its level declines again (Enokido et al. 2005). Initially, the neuronal localization of CBS was demonstrated in major areas of brain including the hippocampus and the cerebellum (Robert et al. 2003). However, further study by Enokido et al. and Lee et al. revealed that H_2S is preferentially expressed in radial glia/astrocyte lineage of developing mouse brain (Enokido et al. 2005; Lee et al. 2009). In support of these findings, CBS has been found to be elevated in reactive astrocytes (Kimura and Kimura 2004). CBS initiates the *trans*-sulfuration pathway by catalyzing β-replacement of serine by Hcy to generate cystathionine and water. Furthermore, serine replacement by Cys as a substrate results into production of cystathionine and H_2S. Besides above-mentioned *trans*-sulfuration pathway, CBS also catalyzes condensation reactions between two molecules of Cys and β-replacement of Cys by water to produces H_2S (Kabil and Banerjee 2014). The reaction replacing Hcy by Cys yields maximum generation (~96 %) of H_2S in vitro. Whereas the minor share of 1–3 % is contributed by condensation reactions between two molecules of Cys (Singh et al. 2009).

CSE is yet another PLP-dependent enzyme, which mediates a reaction between thiocysteine and a thiol compound R-SH to generate H_2S (Kimura 2011). The substrate thiocysteine is generated from L-cystine which in turn is produced by two L-cysteine molecules (Yamanishi and Tuboi 1981). Expression of CSE is rather widely distributed among peripheral tissues including liver, pancreas, uterus, and intestine (Kimura 2011). The activity of CSE in the human brain is 100 folds higher than mouse brain and it plays an important role in CNS physiology (Diwakar and Ravindranath 2007). CSE is a rate-limiting enzyme during the generation of Cys from methionine in *trans*-sulfuration pathway. The availability of Cys is an essential factor in glutathione (GSH) synthesis, which in turn is needed for mitochondrial function preservation to maintain redox homeostasis in cells. Thus CSE offers neuroprotection against oxidative stress by maintaining GSH and protein thiol homeostasis (Diwakar and Ravindranath 2007). Under physiologic conditions, CSE generates H_2S mainly from Cys. However under hyperhomocysteinemic (abnormally high level of homocysteine in the blood) conditions, Hcy (γ-elimination) becomes its major substrate taking over Cys. This alteration might be responsible for elevated production of H_2S in severe hyperhomocysteinemic conditions (Singh et al. 2009). Hence, it is probable that CSE becomes a major enzyme to produce H_2S in conditions like hyperhomocysteinemia (Hu et al. 2011), which incidentally is a common feature of neurodegenerative diseases including AD and PD.

The third enzyme, 3-mercaptopyruvate sulfotransferase (3-MST), was identified in the neurons. The research group detected the significant presence of H_2S in the brain homogenate preparation of $CBS^{-/-}$ mice (Shibuya et al. 2009). Kimura further observed that 3-MST acts together with cysteine aminotransferase (CAT) to generate H_2S from Cys in the presence of α-ketoglutarate (Kimura et al. 2010). However, it is suggested that 3-MST is unable to produce H_2S in normal physiological conditions as they exert their activities at higher alkaline pH level (Shibuya et al. 2009). Furthermore, it requires endogenous reducing substances such as thioredoxin and dihydrolipoic acid (DHLA) for the production of H_2S (Kimura 2014). Aspartate can also act as a substrate for CAT, competitively binding to it and attenuating H_2S synthesis (Guo et al. 2012).

Recently, Shibuya et al. discovered the additional pathway for H_2S biosynthesis in mammalian cells. 3-MST along with D-amino acid oxidase (DAO) produces H_2S from D-cysteine by the interaction of mitochondria and peroxisomes. It was evident that this D-cysteine-dependent pathway operates predominantly in the cerebellum and the kidney. The protective effects of D-cysteine were observed against oxidative stress in cerebellar neurons and against ischemia-reperfusion injury in the kidney (Shibuya and Kimura 2013).

2.3 Storage and Metabolism

Although endogenous H_2S can be synthesized and released immediately, the storage forms of H_2S are also known. Acid-labile sulfur is primarily contained in iron-sulfur center of mitochondrial enzymes and can release H_2S only in acidic pH of 5.4. Due to higher instability of iron-sulfur complexes, the release of H_2S is readily achieved. Bound sulfane sulfur, which is localized in cytoplasm, consists of divalent sulfur bond (e.g., persulfide form). It releases H_2S under reducing conditions of pH 8.4 (Ishigami et al. 2009). It is possible that H_2S produced by 3-MST/CAT enzymatic pathway is stored in the bound sulfane sulfur form. The decreased amount of bound sulfane sulfur has been detected in cells without 3-MST/CAT compared to the cells with 3-MST/CAT (Shibuya et al. 2009).

H_2S is catabolized in mammalian cells though various pathways. The major mechanism is through its mitochondrial oxidation in different tissues (Hildebrandt and Grieshaber 2008). In a reaction catalyzed by quinone oxidoreductase enzyme, H_2S is converted into persulfides. Persulfides are, in turn, oxidized in sulfite and thiosulfite. In physiological normoxic conditions, the thiosulfite is further metabolized into excretable form of sulfate. H_2S catabolism by quinone oxidoreductase enzyme seems to be universal in mammalian tissues, with possible exception of the brain (Mikami et al. 2011). H_2S can also be methylated to produce methane thiol by the action of enzyme thiol-*S*-methyltransferase. Non-mitochondrial heme proteins such as hemoglobin and myoglobin also catabolize intracellular H_2S by oxidation (Berzofsky et al. 1971; Stein and Bailey 2013). To a weaker extent, H_2S can also interact with reactive oxygen and nitrogen species. It is interesting to know that the presence of oxygen (O_2) is a very influential factor in deciding the fate of cellular H_2S as O_2 is capable of spontaneous oxidization of H_2S (van Kampen and Zijlstra 1983; Stein and Bailey 2013). The intracellular concentration of H_2S is kept firmly in low range, owing to the highly efficient nature of above-mentioned mechanisms.

3 Protective Roles of H_2S in CNS

The molecular basis of protective effects of H_2S in CNS during neurodegenerative diseases has received a considerable attention in recent years. This section summarizes the major protective effects of H_2S in CNS with brief discussion about underlying molecular mechanism (Fig. 1).

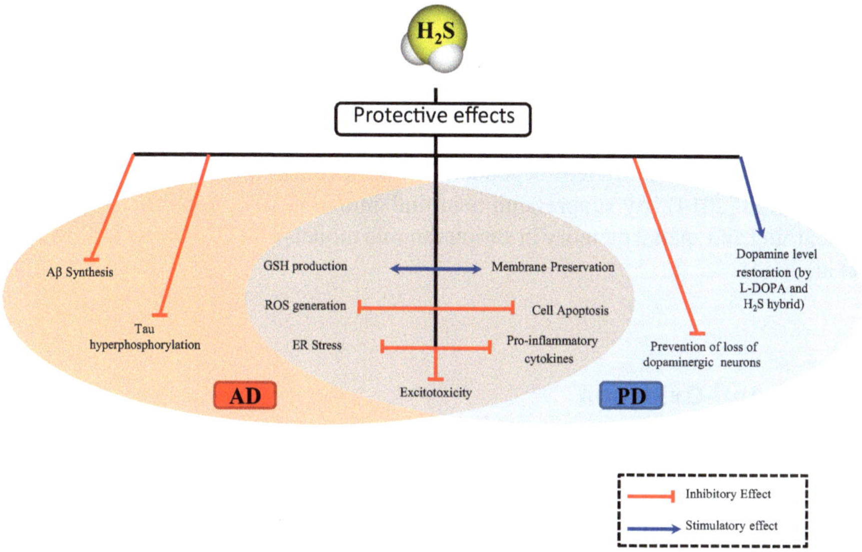

Fig. 1 The various targets of H_2S for its neuroprotective effects in the neurodegenerative diseases. The hallmark events in AD pathology i.e., Aβ generation and tau hyperphosphorylation are downregulated by H_2S. H_2S donors and hybrids produce positive therapeutic effects in PD pathology as well. Apart from reducing loss of neurons in substantia nigra of striatum, it also restores dopamine levels. H_2S elevates the intracellular levels of glutathione (GSH) while preserving membrane integrity, thus making neurons less susceptible to cellular injury. H_2S is also shown to prevent excessive cellular apoptosis, lower oxidative stress, reduce neuroinflammation by several pro-inflammatory cytokines, and prevent cells from excitotoxicity caused by aberrant neurotransmitter release from neurons and astrocytes

3.1 Anti-Inflammation

Neuroinflammation is strongly implicated in the pathologies of all major neurodegenerative diseases. The non-neuronal cells like microglia, astrocytes, and oligodendrocytes constitute the microenvironment under which normal neuronal functions are kept smoothly running. However, in case of any pathological or toxic insult, their over/chronic activation can initiate many unwanted and potentially neurotoxic cascades.

The pioneering work of Hu et al. revealed the anti-inflammatory properties of H_2S in lipopolysaccharide (LPS)-induced neuroinflammation in both primary cultured microglia and immortalized murine BV-2 microglial cells (Hu et al. 2007). In support of above findings, other studies also demonstrated the anti-inflammatory effects of H_2S-releasing NSAIDs in various neurodegenerative disease models. Although there is a lot to be done in order to completely reveal the underlying mechanisms of anti-inflammatory effect of H_2S, above-mentioned and other studies implicated inhibition of nuclear factor-kappa B (NF-κB) and MAPK signaling cascades in the observed effects. Additionally, in an interesting study analyzing the neuroprotective effects of *S*-propargyl-cysteine (a novel H_2S-modulating agent), its inhibitory effects on tumor necrosis factor (TNF)-α and TNF-α receptor 1 (TNFR1) were demonstrated (Gong et al. 2011). Earlier this year, Zhao et al. shed some light upon a previously unrecognized mechanism underlying H_2S suppression of neuroinflammation. They found that H_2S polarized microglia to an anti-inflammatory (M2) phenotype by activating calmodulin-dependent protein kinase kinase β (CaMKKβ)-dependent AMP-activated protein kinase (AMPK) (Zhou et al. 2014). By suppressing neuroinflammation, H_2S imparts improvement in learning and spatial memory in various animal models of AD (Xuan et al. 2012; He et al. 2014).

3.2 Anti-Oxidation

The brain is highly prone to oxidative stress than other organs in mammalian body (Chance et al. 1979). Accounting for relatively small fraction of body weight (2 %), it is constantly supplied with large amount of total body oxygen consumption (20 %) and total body glucose (25 %) in order to maintain high metabolic turnover (Uttara et al. 2009). During the mitochondrial generation of ATP via oxidative pathway, huge amount of ROS and hydrogen peroxide (H_2O_2) are generated. Furthermore, the biochemical structure of neurons makes them particularly vulnerable to ROS. The double bonds present in abundant unsaturated fatty acids are easily susceptible to peroxidation (Butterfield et al. 2002). To worsen the conditions, brain is not equipped with high antioxidant activity. Additionally, the higher concentration of iron and vitamin C in various regions of brain leads to increased interaction between O_2 and metal ions, culminating in ROS generation (Floyd and Carney 1992).

Oxidative stress caused by overproduction of ROS is detrimental and one of the etiological factors of many neurodegenerative diseases. Kimura et al. found that H_2S protects primary neurons from oxidative glutamate toxicity (oxytosis) caused by glutamate. H_2S elevated antioxidant GSH levels by enhancing the activity of gamma-glutamylcysteine synthetase and upregulating cystine transport. The upregulation of gamma-glutamylcysteine synthetase activity facilitates the redistribution of GSH into mitochondria, thus protecting cells against oxidative stress damage (Kimura and Kimura 2004). Later, the same group also discovered that H_2S protects immortalized mouse hippocampal cells from oxytosis by activating ATP-dependent K^+ (K_{ATP}) and Cl^- channels (Kimura et al. 2006). A study conducted by Lu et al. demonstrated that H_2S protects astrocytes via enhancing glutamate uptake function of glutamate transporter-1 and elevating GSH production. This phenomenon prevents excessive accumulation of glutamate in synaptic clefts protecting neurons from excitotoxicity (Lu et al. 2008). Besides these, H_2S has been show to downregulate peroxynitrite-mediated tyrosine nitration and inactivation of alpha1-antiproteinase inhibiting peroxynitrite-induced cytotoxicity, intracellular protein nitration, and protein oxidation in human neuroblastoma SH-SY5Y cells (Whiteman et al. 2004).

3.3 Anti-Apoptosis

Apart from antioxidant effects, H_2S is also known to possess anti-apoptotic properties conferring neuroprotection. Hu and colleagues discovered that H_2S inhibits apoptosis induced by rotenone (a toxin used to establish PD model) by preserving mitochondrial functions in human neuroblastoma cell line (SH-SY5Y). They observed that H_2S regulated the mitoK_{ATP} channel and thus impeded the apoptosis cascade (prevention of mitochondrial membrane potential (MMP) dissipation, cytochrome c release, and caspase-9/3 activation) (Hu et al. 2009). The anti-apoptotic effect was supported by other studies as well. Zhang et al. found out that H_2S attenuated neuronal injury induced by vascular dementia via inhibiting apoptosis in rats (Zhang et al. 2009). In yet another study, H_2S imparted the cytoprotective effect to PC12 cells against amyloid β (25–35)-induced apoptosis (Tang et al. 2008).

4 H_2S in Neurodegenerative Diseases

4.1 AD

AD is the most common neurodegenerative disease in today's society (Reitz et al. 2011). Classically, AD pathology is characterized by aggregation and deposition of Aβ plaques along with hyperphosphorylated tau in the brain (Goedert and Spillantini 2006).

More and more evidences are coming into light depicting a certain relevance of endogenous H_2S generation and AD pathology. In the very first study of its kind, Morrison et al. showed that concentration of *S*-adenosylmethionine, a CBS activator, was notably depleted in brains of AD patients (Morrison et al. 1996). It was later found out that any interference in *trans*-sulfuration pathway (please refer to Sect. 2.3 of this chapter) results in both elevation in total serum Hcy (Clarke et al. 1998) and depletion in H_2S production (Dwyer et al. 2004) in AD patients. It should also be noted that Hcy neurotoxicity itself inhibits CBS, reducing H_2S production (Tang et al. 2011). Furthermore, an extensive clinical study conducted by Liu et al. has proven the direct correlation between severity of AD and H_2S (Liu et al. 2008).

Recently, many studies have shown that H_2S affects amyloidogenesis in the brain. The downregulated generation of Aβ was observed in primary neuron culture (Zhu et al. 2014), SH-SY5Y neuroblastoma cell line (Nagpure and Bian 2014), and APP/PS1 transgenic mice (He et al. 2014). Aβ is synthesized by sequential enzymatic cleavage of amyloid precursor protein (APP). Besides suppressing the expression of APP (Nagpure and Bian 2014), H_2S has also been reported to inhibit the activities and expressions of the cleaving enzymes, BACE1 (Zhang et al. 2011) and γ-secretase (Nagpure and Bian 2014). In an interesting study investigating the effects of H_2S-rich Tabiano's spa-water on three experimental models of AD, Giuliani et al. found the lowered phosphorylation of tau protein at Thr181,

Ser396, and Ser202. In this particular study they also observed anti-amyloidogenic effects of H_2S in AD mouse model harboring human transgenes APPSwe, PS1M146V, and tauP301L (Giuliani et al. 2013). The therapeutic strategy of tau manipulation is of great interest. Many studies have revealed that heat-shock proteins (Hsps), the molecular chaperone families, are involved in prevention of tau aggregation and tau degradation (Thompson et al. 2012; Voss et al. 2012). As H_2S is known to elevate the expression of hsp90 (Xie et al. 2012), it is possible that H_2S could be used as a tau aggregation inhibitor to maintain intracellular microtubule infrastructure

H_2S targets multiple processes underlying AD pathology (Fig. 2) while exerting its neuroprotective effect. Hence H_2S can be of potential therapeutic value in treatment of AD.

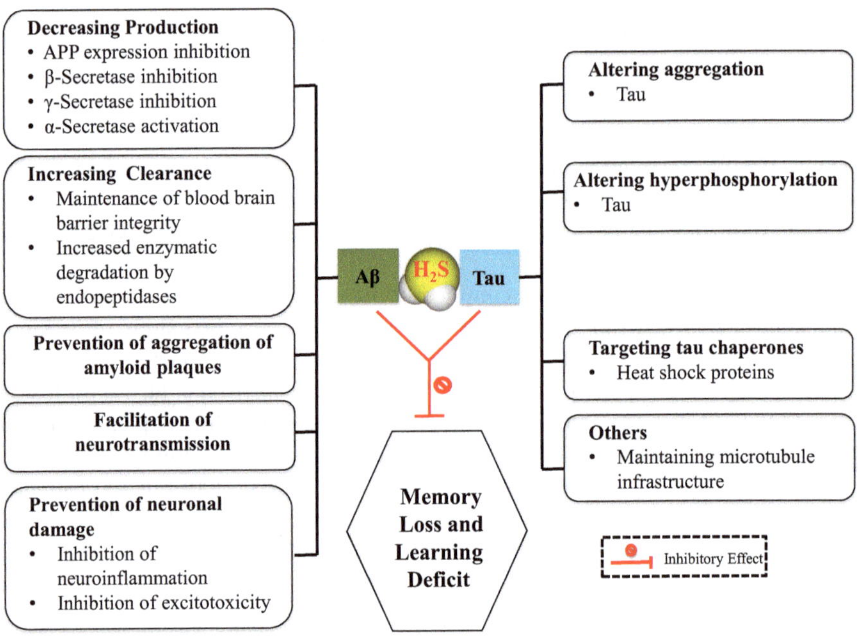

Fig. 2 Effect of H_2S on Aβ and tau aggregation and its potential role in prevention of memory loss and learning disabilities. H_2S inhibits aggregation of amyloid protein by decreasing the amyloid burden and enhancing its clearance. H_2S facilitates neurotransmission by acting as a neuromodulator. It also inhibits the production of several pro-inflammatory cytokines, maintaining the microenvironment in which neurons can function properly. By preventing hyperphosphorylation of tau proteins and inhibiting tau chaperons (Hsps), H_2S can inhibit its aggregation into neurofibrillary tangles and preserve intracellular microtubule infrastructure

4.2 Vascular Dementia

Vascular dementia (VD), a heterogeneous group of brain disorders in which cognitive impairment is attributable to cerebrovascular pathologies, is responsible for at least 20 % of cases of dementia, being second only to AD (Gorelick et al. 2011; Iadecola 2013). Zhang et al. suggest that H_2S could protect the brain against VD injury induced by cerebral ischemia reperfusion through inhibiting the apoptosis in the hippocampus. They found that NaHS-treated rats had a greater ratio of Bcl-2 (anti-apoptotic) over Bax (pro-apoptotic) with increased Bcl-2 expression and decreased Bax expression in the hippocampus (Zhang et al. 2009). It is generally believed that inflammation (Malaguarnera et al. 2006; Liu et al. 2007), oxidative stress (Liu et al. 2007), and vascular factors (Brown et al. 2007; Stephan and Brayne 2008) play important roles in the VD pathology. As discussed earlier in this chapter, H_2S possesses potent anti-inflammatory (Hu et al. 2007) and anti-oxidative action (Kimura and Kimura 2004). Hence, it appears that H_2S may protect against VD injury by targeting multiple signaling pathways and events.

4.3 PD

PD is the second most common neurodegenerative disease histopathologically characterized by progressive degeneration of dopaminergic neurons in substantia nigra of midbrain.

Many studies indicate that hyperhomocysteinemia is common in the patients of PD (O'Suilleabhain et al. 2004; Zoccolella et al. 2010). Various in vivo experiments in PD animal models detected reduced levels of H_2S in substantia nigra and striatum regions of the brain. These findings suggest that impaired endogenous production of H_2S has a substantial effect on pathogenesis and progression of PD. Furthermore, the administration of exogenous H_2S has shown protective effects against underlying pathologic mechanisms. The study conducted by Yin et al. demonstrated that NaHS, a fast H_2S donor, protected PC12 cells from cytotoxicity and apoptosis induced by MPP+, the active metabolite of MPTP. They found that H_2S inhibited the loss of MMP and the accumulation of intracellular ROS (Yin et al. 2009). Recently, Xie et al. evaluated the therapeutic effects of ACS 84, a H_2S-releasing L-DOPA derivative, in a 6-OHDA-induced PD model (Xie et al. 2013). In this study, they observed that ACS84 not only attenuated 6-OHDA-induced cell injury and ROS production but also downregulated antioxidant enzyme expression. These results bolster the initial findings of Calvert and colleagues (Calvert et al. 2009) about upregulation of endogenous antioxidants by H_2S via stimulation of nuclear-factor-E2-related (Nrf2)-dependent signaling pathway. Previously, it was shown that pretreatment with NaHS could suppress rotenone-induced cellular injury and apoptotic cell death in human neuroblastoma SH-SY5Y cells via regulation of mitoK_{ATP} channel/p38- and JNK-MAPK pathway (Hu et al. 2009). In yet another study using neurotoxin 6-OHDA to induce PD model, H_2S was shown to inhibit cell injury by stimulating prosurvival PKC/PI3K/Akt pathway (Tiong et al. 2010).

These findings in vitro studies were supported by the observations done in animal models of PD. Hu et al. found out that the systemic administration of NaHS dramatically reversed the progression of movement dysfunction, loss of tyrosine-hydroxylase (TH)-positive neurons in the striatum, and the elevated malondialdehyde level in injured striatum caused by 6-OHDA or rotenone (Hu et al. 2010). Inhaled H_2S also prevented the MPTP-induced movement disorder and the degeneration of TH-containing neurons by upregulating heme oxygenase-1 and glutamate-cysteine ligase (Kida et al. 2011). The anti-inflammatory, antioxidant, and neuroprotective properties shown by new H_2S releasing hybrids are encouraging, thus making them ideal candidates for PD treatment. ACS 84, a well-known L-DOPA hybrid, has been effective in reducing the release of pro-inflammatory cytokines and NO from stimulated microglia and astrocytes (Lee et al. 2010). Besides relieving from inflammation, H_2S-releasing L-DOPA hybrids also restore the depleted dopamine levels by inhibiting mono-amine oxidase B activity (Sparatore et al. 2011).

These findings highlight the potential therapeutic benefit of H_2S in PD which can be achieved either by the administration of exogenous H_2S or the modulation of endogenous H_2S production.

5 Learning and Memory

The life-long thoughts, behavior, and actions of humans are mostly controlled by two processes occurring in CNS: learning and memory. The intrinsic cellular mechanisms are different for learning (procurement), consolidation, and evocation of memories (Kandel 2001). Notably, the modifications of synaptic connectivity, in terms of number and structures of synapses, are probably the underlying mechanisms for learning and memory. The ability of synaptic plasticity of neurons relies heavily on many intracellular secondary signaling pathways. Therefore, modulation of synthesis of proteins involved in activation and/or deactivation of signal transduction pathways in CNS plays a key role in the processes of learning and memory (Milner et al. 1998).

5.1 Endogenous H_2S Levels: Relation with Learning and Memory

As mentioned earlier, H_2S is generated endogenously in brain, including hippocampus. It is a small but an important region primarily associated with learning and memory functions. The intracellular concentrations of a CBS activator, *S*-adenosylmethionine (SAM), were found to be decreased in AD brains than those in the normal brains (Morrison et al. 1996). In an extensive clinical study conducted in many patients of AD and VD, the plasma H_2S levels were demonstrated to be significantly lower than the normal controls (Liu et al. 2008). In accordance, Eto et al. showed that the levels of H_2S are drastically declined in the brains of AD patients compared with those of the age-matched normal individuals (Eto et al. 2002). A recent work characterized the levels of H_2S at various time points through the development and progression of AD in double-transgenic APP/PS1 mice. This research group observed the decreased levels of H_2S and protein expression of CBS in the cortex and hippocampus of 9- and 12-month-old AD mice (He et al. 2014). Thus, the plasma H_2S concentration is directly correlated with the severity of memory-related symptoms of AD

5.2 Effect of H_2S on Glutamatergic Neurotransmission

The interaction of H_2S with the glutamatergic neurotransmission is well established. Kimura et al. found that H_2S enhances the responses of NMDA receptors to the neurotransmitter glutamate in neurons (Abe and Kimura 1996). According to the previously published studies, H_2S protects brain cells from death due to elevated glutamate levels in synapses. The underlying mechanism is via enhancing glutamate uptake function by glutamate transporters (Lu et al. 2008). It is well known that H_2S attenuates the development of opioid dependence and alleviates heroin withdrawal symptoms in animals. The neural mechanism involved in these effects is closely related to inhibition of cAMP/PKA pathway (Jiang et al. 2012; Yang et al. 2013). Although this is the case, the modulation of the glutamatergic system can also be involved since presynaptic glutamate synaptic transmission by NMDA receptors is regulated directly by μ-opioid receptors (Garzon et al. 2012). Thus, these results signify the importance of H_2S in modulating adaptations in the brain glutamatergic system involved in synaptic plasticity.

5.3 Role of H_2S in Intracellular Calcium ($[Ca^{2+}]i$) Regulation

Many research groups have evaluated possible mechanisms by which H_2S modulates learning and memory processes. The effect of H_2S on $[Ca^{2+}]i$ homeostasis deserves a special mention here. $[Ca^{2+}]i$ is critical for normal neuron-glia communication and regulation of synaptic plasticity. It has been found that H_2S is capable of regulating $[Ca^{2+}]i$ in all important brain cell types, namely neurons (Yong et al. 2010), microglia (Lee et al. 2006), and astrocytes (Nagai et al. 2004). LTP is considered as a major cellular mechanism underlying the learning and memory functions. Kimura et al. discovered that the physiological concentrations of H_2S selectively enhance NMDA receptor-mediated currents and facilitate the induction of hippocampal LTP (Abe and Kimura 1996). L-type calcium channels contribute to LTP and fear memory formation (Bauer et al. 2002). Furthermore, they are essential in both the acquisition and retrieval of long-term recognition memory (Seoane et al. 2009). H_2S activates L-type calcium channels and NMDA receptors on plasma membrane (Lee et al. 2006). Yong et al. discovered that the action of H_2S on $[Ca^{2+}]i$ can be suppressed by using inhibitors of PKA, phospholipase C (PLC), and protein kinase C (PKC), suggesting the role of PKA and PLC/PKC pathways in the regulatory effect of H_2S on $[Ca^{2+}]i$ (Yong et al. 2010) and possibly on functions of learning and memory. Voltage-gated $Ca_V2.1$ (P/Q-type) Ca^{2+} channels are essential in modulation of processes such as neurotransmitter release and synaptic plasticity. It was found that context-associated memory retrieval was dependent on these particular Ca^{2+} channels (Chen et al. 2012). Mallmann et al. observed deficits in spatial learning and reference memory, reduced recognition memory in the forebrain-specific $Ca_V2.1$ knockout mice (Mallmann et al. 2013). Interestingly these channels are also believed to be involved in AD pathology as Aβ downregulates $Ca_V2.1$ channels, thus blocking Ca^{2+} current (Nimmrich et al. 2008; Nimmrich and Ebert 2009). Furthermore, Gangarossa suggested the potential functional involvement of these specific Ca^{2+} channels in many brain disorders as the physiological activity of these specific Ca^{2+} channels is required for affective and cognitive behaviors (Gangarossa et al. 2014). H_2S can modulate the activity of voltage-gated Ca^{2+} channels. A recent study conducted by Sekiguchi et al. demonstrated that the function of these channels is tonically enhanced by endogenous H_2S synthesized by CSE in HEK293 cells transfected with Cav3.2, and that exogenous H_2S is capable of enhancing Cav3.2 function when endogenous H_2S production by CSE is inhibited (Sekiguchi et al. 2014).

5.4 Neuroinflammation and Other Factors

Neuroinflammation involves elevated production of myriad of pro-inflammatory cytokines, ROS, and activation of apoptotic pathways. These events lead to neurodegeneration as they negatively impact the functions of neurons, which affect the processes of learning and memory. In various behavior studies, it has been seen that the H_2S reverses neuroinflammatory changes, ultimately repairing the cognitive and memory impairments.

In the initial studies done in rat models of neuroinflammation, exogenous administration of H_2S in the form of NaHS (Gong et al. 2010) or *S*-propargyl-cysteine (SPRC) (Gong et al. 2011) ameliorated LPS-induced cognitive impairment as evaluated in the Morris water maze test. Shortly afterwards, Tang et al. demonstrated that the disturbances in endogenous H_2S generation played a pivotal role in formaldehyde-induced deficits in memory and cognition in animals (Tang et al. 2013). These results were confirmation of the findings of other study in which the neurotoxic effects of Hcy were investigated. Elevated plasma level of Hcy is a known risk factor for AD. The spatial memory acquisition and spatial learning as evaluated by probe trial and hidden-platform acquisition tests, respectively, showed that Hcy decreases spatial learning ability and memory of rats. They also employed novel object recognition test to study short-term, declarative memory and attention. Hcy, by decreasing endogenous production of H_2S, significantly decreased the discrimination index of rats in novel object recognition test highlighting the unsuccessful retention of memory of familiar objects (Tang et al. 2011). Furthermore, the intraperitoneal administration of H_2S attenuated the spatial memory impairment in Aβ rat model of AD (Xuan et al. 2012). In continuation of their previous study, Xuan et al. demonstrated that the application of exogenous H_2S resulted in improved spatial learning and memory acquisition in double-transgenic APP/PS1 mice, an established animal model of AD (He et al. 2014)

6 Toxicity of H_2S

Although H_2S is produced endogenously in the brain and plays important biological functions, one should bear in mind that long-term treatment of neurodegenerative diseases may potentially cause brain H_2S accumulation and therefore neurotoxicity. Previous studies have shown that the exposure-response curve of H_2S is steep, and thus concentration of inhaled gas is more important compared to the duration of exposure (Prior et al. 1988; Guidotti 1996). The approximate concentrations (exposure levels) of inhaled H_2S for the major toxicological effects are given in Fig. 3. The toxidrome (i.e., a set of symptoms and signs associated with a particular poison) of H_2S is often considered as one of the most unusual and reliable toxidromes (Wang 1989; Milby and Baselt 1999). It is characterized by the "knockdown" (acute central neurotoxicity), pulmonary edema, conjunctivitis, and odor perception followed by respiratory paralysis (Guidotti 2010). Acute toxicity leading to reversible unconsciousness caused by H_2S inhalation is called as "knockdown" (Guidotti 1996). Although knockdowns can be fatal in the cases of prolonged high-concentration exposure (about 500–1000 ppm), the transient exposure is often reversible and apparently complete functionally (Burnett et al. 1977). Pulmonary edema is a well-recognized effect of acute H_2S toxicity. As H_2S has relatively low solubility, it penetrates deeply into respiratory track, causing alveolar injury culminating in acute pulmonary edema (Guidotti 2010). The conjunctivitis caused by prolonged low-concentration exposure (about 20 ppm) (Lambert et al. 2006) is peculiarly associated with reversible chromatic distortion and visual changes. These symptoms are sometimes accompanied by blepharospasm and photophobia (Tansy et al. 1981; Milby and Baselt 1999). H_2S is an odorous gas at low concentration (0.01–0.3 ppm). As the concentration increases, however, the victims start to experience olfactory fatigue. It is a sensory adaptation where the victims get accustomed to strong odor. At around 100 ppm concentration, H_2S paralyzes the olfactory mechanism, preventing perception of any smell. This phenomenon removes the primary warning sign of H_2S exposure (Ronk and White 1985; Turner et al. 1990).

The primary mechanism underlying H_2S toxicity is the inhibition of mitochondrial respiratory chain by HS^- (a hydrogen sulfide ion). HS^- binds to ferric iron (Fe^{3+}) of cytochrome oxidase C. It culminates into sulfmethemoglobinemia, lactic acidosis, and hypoxia (Sastre et al. 2013). H_2S is not completely oxidized by brain, making it most vulnerable for the toxic effects. Initially, this pattern seemed identical to that observed in case of cyanide (HCN) poisoning (Dorman et al. 2002).

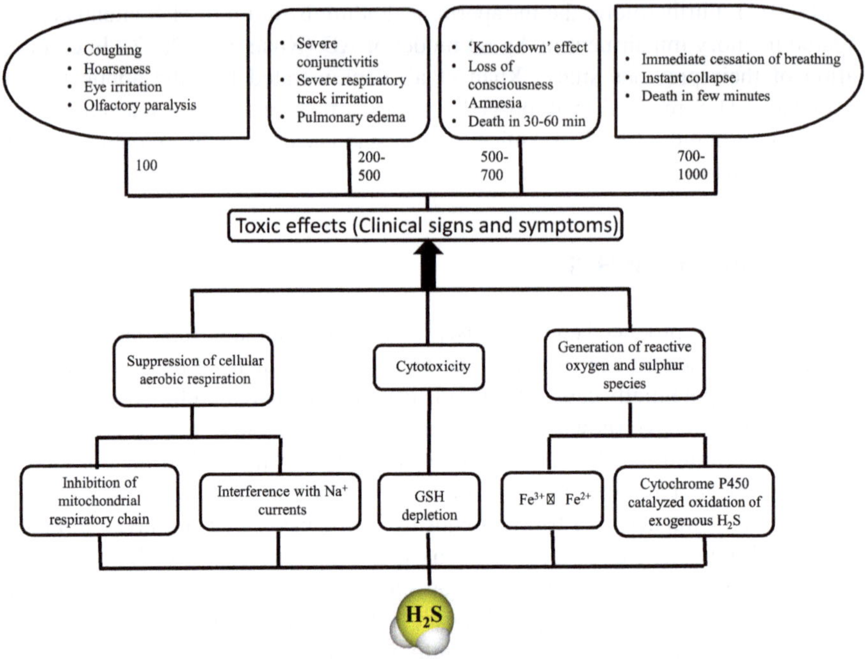

Fig. 3 Toxicity of H_2S. The spectrum of toxic effects of H_2S spans from mild irritation of eyes and respiratory tract to immediate deaths. The severity of symptoms depends on the exposure levels as depicted in the figure. The "toxidrome" of H_2S is often considered as fairly constant and reliable. The underlying molecular mechanisms for toxicity are varied. Along with suppression of cellular aerobic respiration, excessive generation of reactive molecules is proposed as primary mechanisms. The depletion of cytoprotective GSH also makes cells vulnerable for cellular injury

However, the anomalies began to appear when methemoglobin treatment did not yield effective results in cases of H_2S poisonings, despite its successful use in HCN poisoning (Truong et al. 2006). Hence, many other underlying mechanisms were proposed. Warenycia et al. demonstrated that H_2S interferes with sodium currents by abolishing sodium channel function, which might be responsible for depletion of cellular respiration in H_2S poisoning (Warenycia et al. 1989). H_2S was also found responsible for reduction of bound form of iron (ferric, Fe^{3+}) to free ferrous (Fe^{2+}) form. This along with cytochrome P450-catalyzed oxidation of the exogenous H_2S compound resulted into reactive sulfur and oxygen species generation. The concurrent depletion of GSH made neuronal cells more susceptible to neurotoxicity (Truong et al. 2006).

7 Concluding Remarks

The prevalence and morbidity of neurodegenerative diseases are increasing at rapid pace across the globe. Despite the investigations studying the therapeutic effects of H_2S are still in their infancy, plentiful evidence has proven a protective role for this gasotransmitter in the pathology of neurodegenerative diseases. However, our present knowledge on neuroprotection offered by H_2S mainly comes from cell and animal studies with the use of H_2S donors and inhibitors of endogenous H_2S. Whether the therapeutic effects of these donors and inhibitors in "bench" studies can be transferred to "bedside" clinical studies needs to be explored. Additionally, more information about drug safety and toxicity due to long-term H_2S-based therapeutic approaches is desirable. The development of H_2S-releasing drug with a sustained and controlled release is necessary due to non-physiological rapid generation of H_2S by most H_2S donors. Thorough understanding of these problems will enable us to study underlying mechanisms still deeper, and formulate H_2S-based therapeutic interventions to treat neurodegenerative diseases.

Acknowledgment This work was supported by research grants from National University Health System (NUHS B2B research grant-NUHSRO/2011/012/STB/B2B-08) and National Kidney Foundation (NKFRC/2011/01/04).

References

Abe K, Kimura H (1996) The possible role of hydrogen sulfide as an endogenous neuromodulator. J Neurosci 16(3):1066–1071

Bauer EP, Schafe GE, LeDoux JE (2002) NMDA receptors and L-type voltage-gated calcium channels contribute to long-term potentiation and different components of fear memory formation in the lateral amygdala. J Neurosci 22(12):5239–5249

Berzofsky JA, Peisach J, Blumberg WE (1971) Sulfheme proteins. I. Optical and magnetic properties of sulfmyoglobin and its derivatives. J Biol Chem 246(10):3367–3377

Brown WR, Blair RM, Moody DM, Thore CR, Ahmed S, Robbins ME, Wheeler KT (2007) Capillary loss precedes the cognitive impairment induced by fractionated whole-brain irradiation: a potential rat model of vascular dementia. J Neurol Sci 257(1–2):67–71

Burnett WW, King EG, Grace M, Hall WF (1977) Hydrogen sulfide poisoning: review of 5 years' experience. Can Med Assoc J 117(11):1277–1280

Butterfield DA, Castegna A, Lauderback CM, Drake J (2002) Evidence that amyloid beta-peptide-induced lipid peroxidation and its sequelae in Alzheimer's disease brain contribute to neuronal death. Neurobiol Aging 23(5):655–664

Calvert JW, Jha S, Gundewar S, Elrod JW, Ramachandran A, Pattillo CB, Kevil CG, Lefer DJ (2009) Hydrogen sulfide mediates cardioprotection through Nrf2 signaling. Circ Res 105(4): 365–374

Chance B, Sies H, Boveris A (1979) Hydroperoxide metabolism in mammalian organs. Physiol Rev 59(3):527–605

Chen CC, Shen JW, Chung NC, Min MY, Cheng SJ, Liu IY (2012) Retrieval of context-associated memory is dependent on the Ca(v)3.2 T-type calcium channel. PLoS One 7(1), e29384

Clarke R, Smith AD, Jobst KA, Refsum H, Sutton L, Ueland PM (1998) Folate, vitamin B12, and serum total homocysteine levels in confirmed Alzheimer disease. Arch Neurol 55(11): 1449–1455 d'Emmanuele di Villa Bianca R, Sorrentino R, Coletta C, Mitidieri E, Rossi A, Vellecco V, Pinto A, Cirino G, Sorrentino R (2011) Hydrogen sulfide-induced dual vascular effect involves arachidonic acid cascade in rat mesenteric arterial bed. J Pharmacol Exp Ther 337(1):59–64

Diwakar L, Ravindranath V (2007) Inhibition of cystathionine-γ-lyase leads to loss of glutathione and aggravation of mitochondrial dysfunction mediated by excitatory amino acid in the CNS. Neurochem Int 50(2):418–426

Dorman DC, Moulin FJ-M, McManus BE, Mahle KC, James RA, Struve MF (2002) Cytochrome oxidase inhibition induced by acute hydrogen sulfide inhalation: correlation with tissue sulfide concentrations in the rat brain, liver, lung, and nasal epithelium. Toxicol Sci 65(1):18–25

Dwyer BE, Raina AK, Perry G, Smith MA (2004) Homocysteine and Alzheimer's disease: a modifiable risk? Free Radic Biol Med 36(11):1471–1475

Enokido Y, Suzuki E, Iwasawa K, Namekata K, Okazawa H, Kimura H (2005) Cystathionine β-synthase, a key enzyme for homocysteine metabolism, is preferentially expressed in the radial glia/astrocyte lineage of developing mouse CNS. FASEB J 19(13):1854–1856

Eto K, Asada T, Arima K, Makifuchi T, Kimura H (2002) Brain hydrogen sulfide is severely decreased in Alzheimer's disease. Biochem Biophys Res Commun 293(5):1485–1488

Ferri CP, Prince M, Brayne C, Brodaty H, Fratiglioni L, Ganguli M, Hall K, Hasegawa K, Hendrie H, Huang Y, Jorm A, Mathers C, Menezes PR, Rimmer E, Scazufca M (2005) Global prevalence of dementia: a Delphi consensus study. Lancet 366(9503):2112–2117

Floyd RA, Carney JM (1992) Free radical damage to protein and DNA: mechanisms involved and relevant observations on brain undergoing oxidative stress. Ann Neurol 32(Suppl):S22–S27

Gangarossa G, Laffray S, Bourinet E, Valjent E (2014) T-type calcium channel Cav3.2 deficient mice show elevated anxiety, impaired memory and reduced sensitivity to psychostimulants. Front Behav Neurosci 8:92

Garzon J, Rodriguez-Munoz M, Sanchez-Blazquez P (2012) Direct association of Mu-opioid and NMDA glutamate receptors supports their cross-regulation: molecular implications for opioid tolerance. Curr Drug Abuse Rev 5(3):199–226

Giuliani D, Ottani A, Zaffe D, Galantucci M, Strinati F, Lodi R, Guarini S (2013) Hydrogen sulfide slows down progression of experimental Alzheimer's disease by targeting multiple pathophysiological mechanisms. Neurobiol Learn Mem 104:82–91

Goedert M, Spillantini MG (2006) A century of Alzheimer's disease. Science 314(5800):777–781

Gong Q-H, Wang Q, Pan L-L, Liu X-H, Huang H, Zhu Y-Z (2010) Hydrogen sulfide attenuates lipopolysaccharide-induced cognitive impairment: a pro-inflammatory pathway in rats. Pharmacol Biochem Behav 96(1):52–58

Gong QH, Wang Q, Pan LL, Liu XH, Xin H, Zhu YZ (2011) S-propargyl-cysteine, a novel hydrogen sulfide-modulated agent, attenuates lipopolysaccharide-induced spatial learning and memory impairment: involvement of TNF signaling and NF-kappaB pathway in rats. Brain Behav Immun 25(1):110–119

Gorelick PB, Scuteri A, Black SE, Decarli C, Greenberg SM, Iadecola C, Launer LJ, Laurent S, Lopez OL, Nyenhuis D, Petersen RC, Schneider JA, Tzourio C, Arnett DK, Bennett DA, Chui HC, Higashida RT, Lindquist R, Nilsson PM, Roman GC, Sellke FW, Seshadri S (2011) Vascular contributions to cognitive impairment and dementia: a statement for healthcare professionals from the american heart association/american stroke association. Stroke 42(9): 2672–2713

Guidotti TL (1996) Hydrogen sulphide. Occup Med (Lond) 46(5):367–371

Guidotti TL (2010) Hydrogen sulfide: advances in understanding human toxicity. Int J Toxicol 29(6):569–581

Guo W, Kan JT, Cheng ZY, Chen JF, Shen YQ, Xu J, Wu D, Zhu YZ (2012) Hydrogen sulfide as an endogenous modulator in mitochondria and mitochondria dysfunction. Oxid Med Cell Longev 2012:878052

He XL, Yan N, Zhang H, Qi YW, Zhu LJ, Liu MJ, Yan Y (2014) Hydrogen sulfide improves spatial memory impairment and decreases production of Abeta in APP/PS1 transgenic mice. Neurochem Int 67:1–8

Hegde A, Bhatia M (2011) Hydrogen sulfide in inflammation: friend or foe? Inflamm Allergy Drug Targets 10(2):118–122
Hildebrandt TM, Grieshaber MK (2008) Three enzymatic activities catalyze the oxidation of sulfide to thiosulfate in mammalian and invertebrate mitochondria. FEBS J 275(13): 3352–3361
Hu LF, Wong PT, Moore PK, Bian JS (2007) Hydrogen sulfide attenuates lipopolysaccharide-induced inflammation by inhibition of p38 mitogen-activated protein kinase in microglia. J Neurochem 100(4):1121–1128
Hu LF, Pan TT, Neo KL, Yong QC, Bian JS (2008) Cyclooxygenase-2 mediates the delayed cardioprotection induced by hydrogen sulfide preconditioning in isolated rat cardiomyocytes. Pflugers Arch 455(6):971–978
Hu LF, Lu M, Wu ZY, Wong PT, Bian JS (2009) Hydrogen sulfide inhibits rotenone-induced apoptosis via preservation of mitochondrial function. Mol Pharmacol 75(1):27–34
Hu LF, Lu M, Tiong CX, Dawe GS, Hu G, Bian JS (2010) Neuroprotective effects of hydrogen sulfide on Parkinson's disease rat models. Aging Cell 9(2):135–146
Hu LF, Lu M, Hon Wong PT, Bian JS (2011) Hydrogen sulfide: neurophysiology and neuropathology. Antioxid Redox Signal 15(2):405–419
Iadecola C (2013) The pathobiology of vascular dementia. Neuron 80(4):844–866
Ishigami M, Hiraki K, Umemura K, Ogasawara Y, Ishii K, Kimura H (2009) A source of hydrogen sulfide and a mechanism of its release in the brain. Antioxid Redox Signal 11(2):205–214
Jiang LH, Wang J, Wei XL, Liang QY, Cheng TT (2012) Exogenous sodium hydrosulfide can attenuate naloxone-precipitated withdrawal syndromes and affect cAMP signaling pathway in heroin-dependent rat's nucleus accumbens. Eur Rev Med Pharmacol Sci 16(14):1974–1982
Kabil O, Banerjee R (2010) Redox biochemistry of hydrogen sulfide. J Biol Chem 285(29): 21903–21907
Kabil O, Banerjee R (2014) Enzymology of H2S biogenesis, decay and signaling. Antioxid Redox Signal 20(5):770–782
Kandel ER (2001) The molecular biology of memory storage: a dialogue between genes and synapses. Science 294(5544):1030–1038
Kida K, Yamada M, Tokuda K, Marutani E, Kakinohana M, Kaneki M, Ichinose F (2011) Inhaled hydrogen sulfide prevents neurodegeneration and movement disorder in a mouse model of Parkinson's disease. Antioxid Redox Signal 15(2):343–352
Kimura H (2011) Hydrogen sulfide: its production, release and functions. Amino Acids 41(1): 113–121
Kimura H (2014) Production and physiological effects of hydrogen sulfide. Antioxid Redox Signal 20(5):783–793
Kimura Y, Kimura H (2004) Hydrogen sulfide protects neurons from oxidative stress. FASEB J 18(10):1165–1167
Kimura Y, Dargusch R, Schubert D, Kimura H (2006) Hydrogen sulfide protects HT22 neuronal cells from oxidative stress. Antioxid Redox Signal 8(3-4):661–670
Kimura Y, Goto Y, Kimura H (2010) Hydrogen sulfide increases glutathione production and suppresses oxidative stress in mitochondria. Antioxid Redox Signal 12(1):1–13
Kohn C, Dubrovska G, Huang Y, Gollasch M (2012) Hydrogen sulfide: potent regulator of vascular tone and stimulator of angiogenesis. Int J Biomed Sci 8(2):81–86
Lambert TW, Goodwin VM, Stefani D, Strosher L (2006) Hydrogen sulfide (H_2S) and sour gas effects on the eye. A historical perspective. Sci Total Environ 367(1):1–22
Lee SW, Hu YS, Hu LF, Lu Q, Dawe GS, Moore PK, Wong PT, Bian JS (2006) Hydrogen sulphide regulates calcium homeostasis in microglial cells. Glia 54(2):116–124
Lee SW, Cheng Y, Moore PK, Bian J-S (2007) Hydrogen sulphide regulates intracellular pH in vascular smooth muscle cells. Biochem Biophys Res Commun 358(4):1142–1147
Lee M, Schwab C, Yu S, McGeer E, McGeer PL (2009) Astrocytes produce the antiinflammatory and neuroprotective agent hydrogen sulfide. Neurobiol Aging 30(10):1523–1534

Lee M, Tazzari V, Giustarini D, Rossi R, Sparatore A, Del Soldato P, McGeer E, McGeer PL (2010) Effects of hydrogen sulfide-releasing L-DOPA derivatives on glial activation: potential for treating Parkinson disease. J Biol Chem 285(23):17318–17328

Lim JJ, Liu YH, Khin ES, Bian JS (2008) Vasoconstrictive effect of hydrogen sulfide involves downregulation of cAMP in vascular smooth muscle cells. Am J Physiol Cell Physiol 295(5): C1261–C1270

Liu C, Wu J, Gu J, Xiong Z, Wang F, Wang J, Wang W, Chen J (2007) Baicalein improves cognitive deficits induced by chronic cerebral hypoperfusion in rats. Pharmacol Biochem Behav 86(3):423–430

Liu XQ, Liu XQ, Jiang P, Huang H, Yan Y (2008) Plasma levels of endogenous hydrogen sulfide and homocysteine in patients with Alzheimer's disease and vascular dementia and the significance thereof. Zhonghua Yi Xue Za Zhi 88(32):2246–2249

Liu YH, Lu M, Hu LF, Wong PT, Webb GD, Bian JS (2012) Hydrogen sulfide in the mammalian cardiovascular system. Antioxid Redox Signal 17(1):141–185

Lu M, Hu LF, Hu G, Bian JS (2008) Hydrogen sulfide protects astrocytes against H(2)O(2)-induced neural injury via enhancing glutamate uptake. Free Radic Biol Med 45(12):1705–1713

Malaguarnera L, Motta M, Di Rosa M, Anzaldi M, Malaguarnera M (2006) Interleukin-18 and transforming growth factor-beta 1 plasma levels in Alzheimer's disease and vascular dementia. Neuropathology 26(4):307–312

Mallmann RT, Elgueta C, Sleman F, Castonguay J, Wilmes T, van den Maagdenberg A, Klugbauer N (2013) Ablation of Ca(V)2.1 voltage-gated Ca(2)(+) channels in mouse forebrain generates multiple cognitive impairments. PLoS One 8(10), e78598

Mark G, Naumov S, von Sonntag C (2011) The reaction of ozone with bisulfide (HS−) in aqueous solution – mechanistic aspects. Ozone Sci Eng 33(1):37–41

Mikami Y, Shibuya N, Kimura Y, Nagahara N, Yamada M, Kimura H (2011) Hydrogen sulfide protects the retina from light-induced degeneration by the modulation of Ca^{2+} influx. J Biol Chem 286(45):39379–39386

Milby TH, Baselt RC (1999) Hydrogen sulfide poisoning: clarification of some controversial issues. Am J Ind Med 35(2):192–195

Milner B, Squire LR, Kandel ER (1998) Cognitive neuroscience and the study of memory. Neuron 20(3):445–468

Morrison LD, Smith DD, Kish SJ (1996) Brain S-adenosylmethionine levels are severely decreased in Alzheimer's disease. J Neurochem 67(3):1328–1331

Nagai Y, Tsugane M, Oka J, Kimura H (2004) Hydrogen sulfide induces calcium waves in astrocytes. FASEB J 18(3):557–559

Nagpure BV, Bian JS (2014) Hydrogen sulfide inhibits A2A adenosine receptor agonist induced beta-amyloid production in SH-SY5Y neuroblastoma cells via a cAMP dependent pathway. PLoS One 9(2), e88508

Nimmrich V, Ebert U (2009) Is Alzheimer's disease a result of presynaptic failure? Synaptic dysfunctions induced by oligomeric beta-amyloid. Rev Neurosci 20(1):1–12

Nimmrich V, Grimm C, Draguhn A, Barghorn S, Lehmann A, Schoemaker H, Hillen H, Gross G, Ebert U, Bruehl C (2008) Amyloid beta oligomers (A beta(1-42) globulomer) suppress spontaneous synaptic activity by inhibition of P/Q-type calcium currents. J Neurosci 28(4): 788–797

O'Suilleabhain PE, Sung V, Hernandez C, Lacritz L, Dewey RB Jr, Bottiglieri T, Diaz-Arrastia R (2004) Elevated plasma homocysteine level in patients with Parkinson disease: motor, affective, and cognitive associations. Arch Neurol 61(6):865–868

Pan TT, Feng ZN, Lee SW, Moore PK, Bian JS (2006) Endogenous hydrogen sulfide contributes to the cardioprotection by metabolic inhibition preconditioning in the rat ventricular myocytes. J Mol Cell Cardiol 40(1):119–130

Polhemus DJ, Calvert JW, Butler J, Lefer DJ (2014) The cardioprotective actions of hydrogen sulfide in acute myocardial infarction and heart failure. Scientifica (Cairo) 2014:768607

Prior MG, Sharma AK, Yong S, Lopez A (1988) Concentration-time interactions in hydrogen sulphide toxicity in rats. Can J Vet Res 52(3):375

Reitz C, Brayne C, Mayeux R (2011) Epidemiology of Alzheimer disease. Nat Rev Neurol 7(3): 137–152

Robert K, Vialard F, Thiery E, Toyama K, Sinet PM, Janel N, London J (2003) Expression of the cystathionine beta synthase (CBS) gene during mouse development and immunolocalization in adult brain. J Histochem Cytochem 51(3):363–371

Ronk R, White MK (1985) Hydrogen sulfide and the probabilities of 'inhalation' through a tympanic membrane defect. J Occup Environ Med 27(5):337–340

Sastre C, Baillif-Couniou V, Kintz P, Cirimele V, Bartoli C, Christia-Lotter M-A, Piercecchi-Marti M-D, Leonetti G, Pelissier-Alicot A-L (2013) Fatal accidental hydrogen sulfide poisoning: a domestic case. J Forensic Sci 58:S280–S284

Sekiguchi F, Miyamoto Y, Kanaoka D, Ide H, Yoshida S, Ohkubo T, Kawabata A (2014) Endogenous and exogenous hydrogen sulfide facilitates T-type calcium channel currents in Cav3.2-expressing HEK293 cells. Biochem Biophys Res Commun 445(1):225–229

Seoane A, Massey PV, Keen H, Bashir ZI, Brown MW (2009) L-type voltage-dependent calcium channel antagonists impair perirhinal long-term recognition memory and plasticity processes. J Neurosci 29(30):9534–9544

Shibuya N, Kimura H (2013) Production of hydrogen sulfide from d-cysteine and its therapeutic potential. Front Endocrinol (Lausanne) 4:87

Shibuya N, Tanaka M, Yoshida M, Ogasawara Y, Togawa T, Ishii K, Kimura H (2009) 3-Mercaptopyruvate sulfurtransferase produces hydrogen sulfide and bound sulfane sulfur in the brain. Antioxid Redox Signal 11(4):703–714

Singh S, Padovani D, Leslie RA, Chiku T, Banerjee R (2009) Relative contributions of cystathionine beta-synthase and gamma-cystathionase to H2S biogenesis via alternative trans-sulfuration reactions. J Biol Chem 284(33):22457–22466

Sparatore A, Santus G, Giustarini D, Rossi R, Del Soldato P (2011) Therapeutic potential of new hydrogen sulfide-releasing hybrids. Expert Rev Clin Pharmacol 4(1):109–121

Stein A, Bailey SM (2013) Redox biology of hydrogen sulfide: implications for physiology, pathophysiology, and pharmacology. Redox Biol 1(1):32–39

Stephan BC, Brayne C (2008) Vascular factors and prevention of dementia. Int Rev Psychiatry 20(4):344–356

Tang XQ, Yang CT, Chen J, Yin WL, Tian SW, Hu B, Feng JQ, Li YJ (2008) Effect of hydrogen sulphide on beta-amyloid-induced damage in PC12 cells. Clin Exp Pharmacol Physiol 35(2): 180–186

Tang XQ, Shen XT, Huang YE, Chen RQ, Ren YK, Fang HR, Zhuang YY, Wang CY (2011) Inhibition of endogenous hydrogen sulfide generation is associated with homocysteine-induced neurotoxicity: role of ERK1/2 activation. J Mol Neurosci 45(1):60–67

Tang X-Q, Zhuang Y-Y, Zhang P, Fang H-R, Zhou C-F, Gu H-F, Zhang H, Wang C-Y (2013) Formaldehyde impairs learning and memory involving the disturbance of hydrogen sulfide generation in the hippocampus of rats. J Mol Neurosci 49(1):140–149

Tansy MF, Kendall FM, Fantasia J, Landin WE, Oberly R, Sherman W (1981) Acute and subchronic toxicity studies of rats exposed to vapors of methyl mercaptan and other reduced-sulfur compounds. J Toxicol Environ Health 8(1-2):71–88

Thompson AD, Scaglione KM, Prensner J, Gillies AT, Chinnaiyan A, Paulson HL, Jinwal UK, Dickey CA, Gestwicki JE (2012) Analysis of the tau-associated proteome reveals that exchange of Hsp70 for Hsp90 is involved in tau degradation. ACS Chem Biol 7(10): 1677–1686

Tiong CX, Lu M, Bian JS (2010) Protective effect of hydrogen sulphide against 6-OHDA-induced cell injury in SH-SY5Y cells involves PKC/PI3K/Akt pathway. Br J Pharmacol 161(2): 467–480

Truong DH, Eghbal MA, Hindmarsh W, Roth SH, O'Brien PJ (2006) Molecular mechanisms of hydrogen sulfide toxicity. Drug Metab Rev 38(4):733–744

Turner RM, Fairhurst S, Britain G (1990) Toxicology of substances in relation to major hazards: hydrogen sulphide. HM Stationery Office, London
Uttara B, Singh AV, Zamboni P, Mahajan RT (2009) Oxidative stress and neurodegenerative diseases: a review of upstream and downstream antioxidant therapeutic options. Curr Neuropharmacol 7(1):65–74
van Kampen EJ, Zijlstra WG (1983) Spectrophotometry of hemoglobin and hemoglobin derivatives. Adv Clin Chem 23:199–257
Vorobets VS, Kovach SK, Kolbasov GY (2002) Distribution of Ion species and formation of ion pairs in concentrated polysulfide solutions in photoelectrochemical transducers. Rus J Appl Chem 75(2):229–234
Voss K, Combs B, Patterson KR, Binder LI, Gamblin TC (2012) Hsp70 alters tau function and aggregation in an isoform specific manner. Biochemistry 51(4):888–898
Wang D (1989) A review of 152 cases of acute poisoning of hydrogen sulfide. Zhonghua Yi Xue Za Zhi 23(6):330–332
Warenycia MW, Steele JA, Karpinski E, Reiffenstein RJ (1989) Hydrogen sulfide in combination with taurine or cysteic acid reversibly abolishes sodium currents in neuroblastoma cells. Neurotoxicology 10(2):191–199
Whiteman M, Winyard PG (2011) Hydrogen sulfide and inflammation: the good, the bad, the ugly and the promising. Expert Rev Clin Pharmacol 4(1):13–32
Whiteman M, Armstrong JS, Chu SH, Jia-Ling S, Wong BS, Cheung NS, Halliwell B, Moore PK (2004) The novel neuromodulator hydrogen sulfide: an endogenous peroxynitrite 'scavenger'? J Neurochem 90(3):765–768
Xie L, Tiong CX, Bian J-S (2012) Hydrogen sulfide protects SH-SY5Y cells against 6-hydroxydopamine-induced endoplasmic reticulum stress. Am J Physiol Cell Physiol 303(1):C81–C91
Xie L, Hu LF, Teo XQ, Tiong CX, Tazzari V, Sparatore A, Del Soldato P, Dawe GS, Bian JS (2013) Therapeutic effect of hydrogen sulfide-releasing L-Dopa derivative ACS84 on 6-OHDA-induced Parkinson's disease rat model. PLoS One 8(4), e60200
Xuan A, Long D, Li J, Ji W, Zhang M, Hong L, Liu J (2012) Hydrogen sulfide attenuates spatial memory impairment and hippocampal neuroinflammation in beta-amyloid rat model of Alzheimer's disease. J Neuroinflammation 9:202
Yamanishi T, Tuboi S (1981) The mechanism of the L-cystine cleavage reaction catalyzed by rat liver gamma-cystathionase. J Biochem 89(6):1913–1921
Yang G, Wu L, Jiang B, Yang W, Qi J, Cao K, Meng Q, Mustafa AK, Mu W, Zhang S, Snyder SH, Wang R (2008) H2S as a physiologic vasorelaxant: hypertension in mice with deletion of cystathionine gamma-lyase. Science 322(5901):587–590
Yang HY, Wu ZY, Wood M, Whiteman M, Bian JS (2013) Hydrogen sulfide attenuates opioid dependence by suppression of adenylate cyclase/cAMP pathway. Antioxid Redox Signal 20(1):31–41
Yin WL, He JQ, Hu B, Jiang ZS, Tang XQ (2009) Hydrogen sulfide inhibits MPP(+)-induced apoptosis in PC12 cells. Life Sci 85(7-8):269–275
Yong QC, Choo CH, Tan BH, Low CM, Bian JS (2010) Effect of hydrogen sulfide on intracellular calcium homeostasis in neuronal cells. Neurochem Int 56(3):508–515
Zhang LM, Jiang CX, Liu DW (2009) Hydrogen sulfide attenuates neuronal injury induced by vascular dementia via inhibiting apoptosis in rats. Neurochem Res 34(11):1984–1992
Zhang H, Gao Y, Zhao F, Dai Z, Meng T, Tu S, Yan Y (2011) Hydrogen sulfide reduces mRNA and protein levels of beta-site amyloid precursor protein cleaving enzyme 1 in PC12 cells. Neurochem Int 58(2):169–175
Zhao Y, Biggs TD, Xian M (2014) Hydrogen sulfide (HS) releasing agents: chemistry and biological applications. Chem Commun (Camb)
Zhou X, Cao Y, Ao G, Hu L, Liu H, Wu J, Wang X, Jin M, Zheng S, Zhen X, Alkayed NJ, Jia J, Cheng J (2014) CaMKKbeta-dependent activation of AMP-activated protein kinase is critical

to suppressive effects of hydrogen sulfide on neuroinflammation. Antioxid Redox Signal 21(12):1741–1758

Zhu L, Chen X, He X, Qi Y, Yan Y (2014) Effect of exogenous hydrogen sulfide on BACE-1 enzyme expression and beta-amyloid peptide metabolism in high-glucose primary neuronal culture. Nan Fang Yi Ke Da Xue Xue Bao 34(4):504–506, 510

Zoccolella S, dell'Aquila C, Specchio LM, Logroscino G, Lamberti P (2010) Elevated homocysteine levels in Parkinson's disease: is there anything besides L-dopa treatment? Curr Med Chem 17(3):213–221

H_2S and Pain: A Novel Aspect for Processing of Somatic, Visceral and Neuropathic Pain Signals

Yuka Terada and Atsufumi Kawabata

Contents

Abstract

Hydrogen sulfide (H_2S) formed by multiple enzymes including cystathionine-γ-lyase (CSE) targets $Ca_v3.2$ T-type Ca^{2+} channels (T-channels) and transient receptor potential ankyrin-1 (TRPA1). Intraplantar and intracolonic administration of H_2S donors promotes somatic and visceral pain, respectively, via activation of $Ca_v3.2$ and TRPA1 in rats and/or mice. Injection of H_2S donors into the plantar tissues, pancreatic duct, colonic lumen, or bladder causes T-channel-dependent excitation of nociceptors, determined as phosphorylation of ERK or expression of Fos in the spinal dorsal horn. Electrophysiological studies demonstrate that exogenous and/or endogenous H_2S facilitates membrane currents through T-channels in NG108-15 cells and isolated mouse dorsal root ganglion

Y. Terada • A. Kawabata (✉)
Division of Pharmacology and Pathophysiology, Kinki University School of Pharmacy, Higashi-Osaka 577-8502, Japan
e-mail: kawabata@phar.kindai.ac.jp

P.K. Moore, M. Whiteman (eds.), *Chemistry, Biochemistry and Pharmacology of Hydrogen Sulfide*, Handbook of Experimental Pharmacology 230,
DOI 10.1007/978-3-319-18144-8_11

(DRG) neurons that abundantly express $Ca_v3.2$ and also in $Ca_v3.2$-transfected HEK293 cells. In mice with cerulein-induced pancreatitis and cyclophosphamide-induced cystitis, visceral pain and/or referred hyperalgesia are inhibited by CSE inhibitors and by pharmacological blockade or genetic silencing of $Ca_v3.2$, and CSE protein is upregulated in the pancreas and bladder. In rats with neuropathy induced by L5 spinal nerve cutting or by repeated administration of paclitaxel, an anticancer drug, the neuropathic hyperalgesia is reversed by inhibitors of CSE or T-channels and by silencing of $Ca_v3.2$. Upregulation of $Ca_v3.2$ protein in DRG is detectable in the former, but not in the latter, neuropathic pain models. Thus, H_2S appears to function as a nociceptive messenger by facilitating functions of $Ca_v3.2$ and TRPA1, and the enhanced function of the CSE/H_2S/$Ca_v3.2$ pathway is considered to be involved in the pancreatitis- and cystitis-related pain and in neuropathic pain.

Keywords
H_2S • $Ca_v3.2$ T-type calcium channel • TRPA1 • Pain

1 Introduction

Hydrogen sulfide (H_2S), an endogenous gasotransmitter, is generated from L-cysteine mainly by cystathionine-β-synthase (CBS), cystathionine-γ-lyase (CSE), or 3-mercaptopyruvate sulfurtransferase (3-MST) along with cysteine aminotransferase (CAT) throughout the mammalian body (Kimura 2010) including the heart (Geng et al. 2004), blood vessels (Zhao et al. 2001), and central nervous system (CNS) (Warenycia et al. 1989). H_2S appears to play dual or complex roles in many tissues/organs. In the nervous system, there is evidence not only for the neurotoxicity of H_2S through activation of glutamate receptors including NMDA receptors (Cheung et al. 2007) and activation of the MEK/ERK pathway (Kurokawa et al. 2011) but also for the neuroprotection by H_2S via multiple mechanisms such as activation of ATP-sensitive K^+ (K_{ATP}) and Cl^- channels and elevation of intracellular glutathione levels (Kimura and Kimura 2004; Kimura et al. 2006). Endogenous H_2S is also involved in neuronal excitation via the Ca^{2+}/calmodulin-mediated pathway (Eto et al. 2002) and modification of long-term potentiation (Abe and Kimura 1996) via enhancement of the cAMP-induced NMDA receptor response (Kimura 2000).

H_2S appears to cause excitation of sensory nerves and play important roles in neurogenic inflammation, since NaHS, an H_2S donor, increases neuropeptide release in guinea-pig airways, an effect attenuated by capsaicin desensitization of C-fiber neurons and by the transient receptor potential vanilloid-1 (TRPV1) antagonist capsazepine. In addition, NaHS induces contraction of isolated airways, which is abolished by TRPV1 antagonism and pretreatment with the combination of tachykinin NK_1 and NK_2 receptor antagonists (Trevisani et al. 2005). It is consistent with the evidence that H_2S stimulates capsaicin-sensitive primary afferent nerve terminals, from which tachykinins are released to produce concentration-dependent contractile responses by activating NK_1 and NK_2 receptors in the rat urinary bladder (Patacchini et al. 2004). Nonetheless, TRPV1 is not considered a direct target for H_2S, and $Ca_v3.2$ T-type Ca^{2+} channels and transient receptor potential ankyrin-1(TRPA1) rather appear responsible for excitation of sensory neurons (Miyamoto et al. 2011; Andersson et al. 2012; Pozsgai et al. 2012; Sekiguchi and Kawabata 2013; Donatti et al. 2014). Sensory plasticity in nociception, known as sensitization, is thought to contribute significantly to the production of persistent, often pathological pain. Here, we review the role of H_2S in somatic, visceral, and neuropathic pain processing.

2 Targets of H_2S in Pain Processing

2.1 $Ca_v3.2$ T-Type Ca^{2+} Channels

$Ca_v3.2$ T-type Ca^{2+} channels (T-channels) are activated at near-resting membrane potential and play a crucial role in excitability of both central and peripheral neurons. T-channels are expressed in dorsal root ganglion (DRG) neurons including the peripheral and central axons/endings, suggesting the role of T-channels in sensory transmission (Fig. 1) (Nelson et al. 2005; Todorovic and Jevtovic-Todorovic 2011; Rose et al. 2013). Three isoforms of T-channels have been

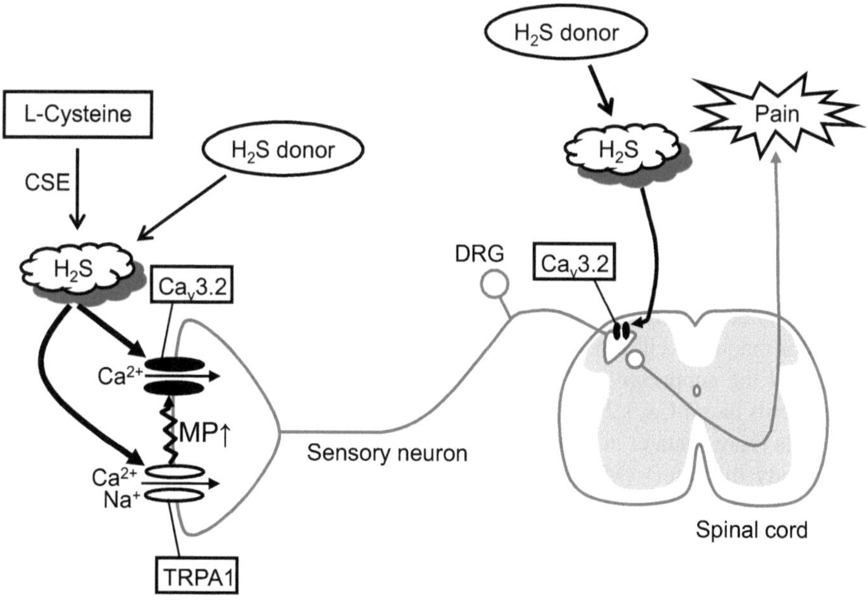

Fig. 1 Roles of H_2S in nociceptive processing at the peripheral and central terminals of the primary afferent neurons. *CSE* cystathionine-γ-lyase, *MP* membrane potential, *DRG* dorsal root ganglion

identified, i.e., $Ca_v3.1$, $Ca_v3.2$, and $Ca_v3.3$ (Perez-Reyes 2003). Studies using *in situ* hybridization have demonstrated that mRNA for $Ca_v3.2$ is the most abundant isoform of T-channels in peripheral sensory neurons (Todorovic and Jevtovic-Todorovic 2011). Bourinet and his coworkers have demonstrated thermal and mechanical analgesia in rats with molecular knockdown of $Ca_v3.2$ T-channels in DRG neurons (Bourinet et al. 2005). $Ca_v3.2$ T-channels are much more sensitive to inhibition by metals, such as zinc, copper, and nickel, than $Ca_v3.1$ or $Ca_v3.3$ (Todorovic and Jevtovic-Todorovic 2011; Sekiguchi and Kawabata 2013). It is now clear that a histidine residue at position 191 (His191) in the second extracellular loop of domain I of $Ca_v3.2$ is a critical determinant for the trace metal inhibition and is not conserved in $Ca_v3.1$ or $Ca_v3.3$ (Iftinca and Zamponi 2009). $Ca_v3.2$ T-channels appear to be tonically exposed to inhibition by Zn^{2+} in cultured cells and possibly in mammalian tissues under physiological conditions, and compounds that chemically interact with Zn^{2+}, such as L-cysteine and H_2S, selectively enhance the channel functions of $Ca_v3.2$ among the three T-channel isoforms (Nelson et al. 2007a; Matsunami et al. 2011; Sekiguchi and Kawabata 2013). Moreover, L-ascorbic acid, a reducing compound, at physiologically relevant concentrations selectively inhibits native T-currents in the DRG and thalamus and diminishes T-channel-dependent burst firing in reticular thalamic cells. This highly subtype-specific effect was achieved via metal-catalyzed oxidation of critical metal-binding and $Ca_v3.2$-unique His191 residue (Nelson et al. 2007b). NaHS, an H_2S donor, facilitates membrane currents through T-channels in NG108-15 cells that naturally express $Ca_v3.2$ (Kawabata et al. 2007; Nagasawa et al. 2009) and in isolated mouse DRG neurons that abundantly express $Ca_v3.2$ (Matsunami et al. 2009). In $Ca_v3.2$-transfected HEK293 cells, T-currents are not altered by H_2S donors, NaHS at 1.5 mM and Na_2S at 0.1 mM, but strongly suppressed by DL-propargylglycine (PPG), an inhibitor of CSE, an H_2S-forming enzyme. Further, in the presence of the CSE inhibitor, the H_2S donors, NaHS at 1.5 mM and Na_2S at 0.1–0.3 mM, enhance the decreased $Ca_v3.2$ T-currents, and CSE is abundantly expressed in both $Ca_v3.2$-transfected and background HEK293 cells (Sekiguchi et al. 2014). Collectively, endogenous and exogenous H_2S appears to positively regulate the functions of $Ca_v3.2$ T-channels (Fig. 1). An independent group has reported that NaHS at low (0.1–1 mM) and high concentrations (3 mM) suppresses and enhances T-currents by 25–30 %, respectively, in $Ca_v3.2$-transfected HEK293 cells or randomly chosen rat DRG cells (Elies et al. 2014), whereas we have never detected the suppressive effect of NaHS or Na_2S even at low concentrations on T-currents in our $Ca_v3.2$-transfected HEK293 cells (Sekiguchi et al. 2014), NG108-15 cells (Kawabata et al. 2007), or small mouse DRG neurons (less than 30 μm in cell body diameter) (Matsunami et al. 2009), known to be rich in T-type Ca^{2+} channels (Todorovic and Jevtovic-Todorovic 2011). Particularly, Na_2S at 0.1–0.3 mM rather augments T-currents in $Ca_v3.2$-transfected HEK293 cells pretreated with the CSE inhibitor (Sekiguchi et al. 2014), as mentioned above, being inconsistent to the suppressive effect of NaHS at equivalent concentrations in their report (Elies et al. 2014). The discrepancy has yet to be interpreted, although the effect of H_2S on $Ca_v3.2$ channels in vitro and in vivo might be dependent on the abundance of H_2S-forming enzymes including CSE, the formation or contamination of polysulfides, and the concentrations of Zn^{2+} and ascorbic acid that negatively regulate $Ca_v3.2$ functions (Todorovic and Jevtovic-Todorovic 2011; Kimura et al. 2013; Sekiguchi and Kawabata 2013; Sekiguchi et al. 2014). Particularly, levels of Zn^{2+} would have great impact on the $Ca_v3.2$ channel activity and its regulation by H_2S, since Zn^{2+} binds to not only His191 of $Ca_v3.2$ (Todorovic and Jevtovic-Todorovic 2011; Sekiguchi and Kawabata 2013) but also H_2S itself (Munchberg et al. 2007). Given that $Ca_v3.2$ is involved in somatic and visceral nociceptive processing and in neuropathic pain (Todorovic and Jevtovic-Todorovic 2011; Francois et al. 2014), regulation of $Ca_v3.2$ by endogenous and exogenous H_2S is considered to have great impact on the pathophysiology of inflammatory and neuropathic pain (Sekiguchi et al. 2014). The T-channel suppression by NaHS at low doses, if any, might be overcome by TRPA1 activation by NaHS in vivo, as described below.

2.2 Transient Receptor Potential Ankyrin-1

Transient receptor potential ankyrin-1 (TRPA1), a member of the TRP channel family, is abundantly expressed in a subpopulation of unmyelinated nociceptors that also express TRPV1. It is activated by a diverse assortment of pungent or irritating reactive chemical compounds including those found in mustard oil (allylisothiocyanate), cinnamon oil (cinnamaldehyde), exhaust (acrolein), raw garlic, and onions (allicin). Moreover, TRPA1 has been put forth as a putative transducer of natural physical stimuli including both cold (<17 °C) and mechanical force (Stucky et al. 2009). It has been reported that TRPA1 is sensitized by proteinase-activated receptor-2 (PAR2) activation, contributing to inflammatory pain (Dai et al. 2007). We have provided evidence that pharmacological inhibition of TRPA1 prevents the spinal Fos expression following infusion of SLIGRL-NH_2, a PAR2-activating peptide, into the pancreatic duct, suggesting the pronociceptive role of TRPA1 as a downstream signal of PAR2 activation in the pancreatic nociceptors (Terada et al. 2013). Most interestingly, it has been reported that NaHS evokes a time- and concentration-dependent increase in cytosolic calcium concentration in TRPA1-transfected cells (Streng et al. 2008) and that NaHS induces calcium influx and inward currents in rat DRG cells, an effect inhibited by a TRPA1 antagonist (Miyamoto et al. 2011; Ogawa et al. 2012). The H_2S-induced TRPA1 activation appears to involve H_2S modulation of two cysteine residues located in the N-terminal intracellular domain of the TRPA1 (Ogawa et al. 2012). These findings suggest that TRPA1 in collaboration with $Ca_v3.2$ mediates H_2S-induced excitation of nociceptors, participating in pain processing (Fig. 1).

3 Somatic Pain

We first reported that intraplantar (i.pl.) administration of NaHS produces prompt hyperalgesia in rats, accompanied by expression of Fos protein, a marker for neuronal excitation, in the spinal dorsal horn, and that the H_2S-evoked hyperalgesia is blocked by ethosuximide and mibefradil, T-channel blockers (Kawabata et al. 2007). Another H_2S donor, Na_2S, also facilitates hyperalgesia by i.pl. administration, an effect completely blocked by pretreatment with NNC 55-0396 (NNC), a T-channel blocker (Sekiguchi et al. 2014). L-Cysteine, an endogenous source for H_2S, given i.pl., mimics the H_2S-induced hyperalgesia, an effect being abolished by PPG and β-cyanoalanine (BCA), inhibitors of CSE, in rats (Kawabata et al. 2007). The NaHS-induced hyperalgesia is also suppressed by pretreatment with i.pl. zinc chloride and/or ascorbic acid, known to selectively inhibit $Ca_v3.2$ among three T-channel isoforms, and by silencing of $Ca_v3.2$ T-channels in the sensory neurons by repeated intrathecal (i.t.) administration of antisense oligodeoxynucleotides (ODNs) targeting $Ca_v3.2$ T-channels in rats or mice (Maeda et al. 2009; Okubo et al. 2012). Interestingly, i.t. administration of NaHS also rapidly decreases nociceptive threshold in rats, as determined by the paw pressure method. The hyperalgesia caused by i.t. NaHS is also abolished by co-administration of mibefradil and pretreatment with i.t. zinc chloride and by silencing of $Ca_v3.2$ protein in the DRG (Maeda et al. 2009). These findings are consistent with the evidence that presynaptic $Ca_v3.2$ T-channel regulates spontaneous excitatory synaptic neurotransmitter release from the central terminal of nociceptors in the spinal dorsal horn, in addition to neuronal excitability in the peripheral endings of nociceptors (Jacus et al. 2012). Thus, positive regulation of $Ca_v3.2$ T-channel by H_2S is considered to have a great impact on nociceptive processing at both peripheral and central endings of the primary afferent neurons (Fig. 1). We have also demonstrated that the NaHS-induced hyperalgesia/allodynia in mice is significantly suppressed by AP18, a TRPA1 blocker, and by silencing of TRPA1 in the sensory neurons (Okubo et al. 2012), suggesting that TRPA1 and $Ca_v3.2$ mediate the pronociceptive effect of H_2S (Fig. 1). The involvement of TRPA1 in H_2S-induced hyperalgesia has been confirmed by independent studies (Andersson et al. 2012; Ogawa et al. 2012). It is likely that H_2S causes activation of TRPA1 followed by elevation of membrane potentials that activates $Ca_v3.2$ and then enhances the evoked $Ca_v3.2$ T-channel currents, leading to increased nociceptor excitability and hyperalgesia. Thus, both TRPA1 and $Ca_v3.2$ are considered to participate in the H_2S-induced hyperalgesia (Fig. 1).

We have shown that PPG and BCA, CSE inhibitors, partially block the hyperalgesia induced by i.pl. lipopolysaccharide (LPS), an effect being reversed by i.pl. NaHS (Kawabata et al. 2007). The suppression of LPS-induced hyperalgesia by CSE inhibitors might be associated with the prevention of neutrophil recruitment to the plantar tissue (Cunha et al. 2008). H_2S is also pronociceptive in the formalin model of persistent inflammatory pain in rats. Injection of formalin into the hindpaw evokes a dose-dependent increase in the concentration of H_2S in the paw tissue. Both nociceptive flinching and paw edema induced by formalin are attenuated by pretreatment with PPG. Furthermore, PPG pretreatment attenuates the induction of c-Fos in spinal laminae I–II following injection of formalin (Lee et al. 2008). It has also been reported that pretreatment of mice with PPG reduces zymosan-induced articular hyperalgesia (Cunha et al. 2008). Recent evidence has shown that the inflammatory pain in temporomandibular joint by injection of complete Freund's adjuvant in rats is attenuated by subcutaneous administration of aminooxyacetic acid (AOAA), known as a CBS inhibitor (Miao et al. 2014), although the selectivity of AOAA for CBS is now questioned (Asimakopoulou et al. 2013). Thus, there is plenty of evidence that H_2S participates in inflammatory pain processing. Nevertheless, it is noteworthy that exogenously applied H_2S reveals antinociceptive activity in certain experimental conditions through activation of K_{ATP} channels or unknown mechanisms (Cunha et al. 2008).

4 Visceral Pain

4.1 Pancreatic Pain

Clinically, pancreatic pain is a serious problem for patients with pancreatitis or pancreatic cancer. In humans, pain elicited from the pancreas is often referred to the upper abdominal area and radiates to the back, and these skin areas are usually tender to touch. In pancreatic acinar cells isolated from mice, treatment with cerulein increases the levels of H_2S and CSE mRNA expression (Tamizhselvi et al. 2007). Furthermore, the activity and protein expression of CSE in the pancreatic tissues increase after the development of cerulein-induced pancreatitis in mice (Nishimura et al. 2009). Given the prevention by PPG, a CSE inhibitor, of cerulein-induced pancreatitis and referred hyperalgesia/allodynia (Bhatia et al. 2005; Nishimura et al. 2009), H_2S is considered to play a critical role in the development of acute pancreatitis and related pain. NaHS infused into the pancreatic duct causes the expression of Fos protein (Nishimura et al. 2009) and ERK phosphorylation (Fukushima et al. 2010), delayed and rapid markers for neuronal activation, respectively, in the superficial layers of spinal cord in anesthetized mice or rats, which are attenuated by i.p. administration of mibefradil, a T-channel blocker. Mibefradil also suppresses the referred hyperalgesia/allodynia in mice with cerulein-induced pancreatitis (Nishimura et al. 2009). Together, H_2S produced by upregulated CSE during the development of pancreatitis augments the function of T-channels, contributing to pancreatitis-related pain. As described above for somatic pain processing, TRPA1 in addition to T-channels appears to mediate H_2S-induced pancreatic nociception in naïve animals, because the TRPA1 inhibitor AP18 or the T-channel inhibitor NNC 55-0396 prevents spinal Fos expression following injection of NaHS into the pancreatic duct in mice without pancreatitis. In contrast, in mice with cerulein-induced pancreatitis, NNC 55-0396 or genetic silencing of $Ca_v3.2$ abolishes the referred hyperalgesia accompanying cerulein-induced pancreatitis, while AP18 or genetic TRPA1 silencing just facilitates the effect of NNC 55-0396, but had no significant effect on the pancreatitis-related pain by itself (Terada et al. 2014). Given the downregulation of TRPA1, but not $Ca_v3.2$, in the pancreatic tissue following cerulein treatment (Terada et al. 2014), $Ca_v3.2$ is considered to primarily participate in the pancreatitis-related pain, while TRPA1 appears to play a secondary role in nociceptive signaling during pancreatitis.

4.2 Colonic Pain

In the colonic lumen, H_2S is also produced by sulfate-reducing bacteria and may be associated with the pathogenesis of inflammatory bowel disease (IBD) and colorectal cancer (Roediger et al. 1997; Ohge et al. 2003; Taniguchi et al. 2009). Nonetheless, there is evidence for a protective role of H_2S against IBD (Wallace et al. 2009). H_2S regulates the secretion of HCO_3^- in the duodenum (Ise et al. 2011) and Cl^- in the colon (Schicho et al. 2006), predicting possible involvement of H_2S in visceral nociception. Distrutti et al. (Distrutti et al. 2006) have described that i.p. injection of H_2S donors suppressed responses to colorectal distention by activating K_{ATP} channels. However, it has been reported that CBS in colonic sensory neurons is upregulated and contributes to visceral hypersensitivity in a rat model of irritable bowel syndrome (IBS) (Xu et al. 2009) and that upregulation of CBS expression by NF-κB contributes to visceral hypersensitivity in rats (Li et al. 2012). We have demonstrated that intracolonic (i.col.) administration of NaHS induces visceral nociceptive behavior accompanied by referred hyperalgesia/allodynia and phosphorylation of ERK in the spinal cord in mice (Matsunami et al. 2009). The behavioral effects of i.col. NaHS are abolished by mibefradil and NNC 55-0396; T-channel blockers; and zinc chloride, known to selectively inhibit $Ca_v3.2$ among three isoforms of T-channels by genetic knockdown of $Ca_v3.2$ and also to be attenuated by AP18, a TRPA1 blocker, but not verapamil, an L-type Ca^{2+} channel blocker, or glibenclamide, a K_{ATP} channel blocker (Matsunami et al. 2009; Matsunami et al. 2011). Interestingly, $Ca_v3.2$ is constantly exposed to Zn^{2+} inhibition under physiological conditions, and zinc chelators applied exogenously enhance $Ca_v3.2$ functions by cancelling the Zn^{2+} inhibition (Nelson et al. 2007a). We have shown that two distinct Zn^{2+} chelators, *N*,*N*,*N*′, *N*′-tetrakis (2-pyridylmethyl)-ethylenediamine (TPEN) and dipicolinic acid, when administered intracolonically, mimics the NaHS-evoked visceral nociceptive behavior, referred abdominal hyperalgesia/allodynia, and phosphorylation of ERK in the spinal dorsal horn, which are inhibited by T-channel blockers (Matsunami et al. 2011). The contribution of $Ca_v3.2$ T-channels to colonic hypersensitivity has been confirmed by an independent group (Marger et al. 2011). Collectively, H_2S plays critical roles in colonic pain signaling through facilitation of $Ca_v3.2$ T-channel functions by cancelling the Zn^{2+} inhibition and also activation of TRPA1 channels. Our recent study has also shown that $Ca_v3.2$-dependent excitation of capsaicin-sensitive sensory neurons by H_2S contributes to colonic mucosal cytoprotection in rats with 2,4,6-trinitrobenzenesulfonic acid (TNBS)-induced colitis (Matsunami et al. 2012a).

4.3 Bladder Pain

The most common visceral pain in the field of urology is bladder pain, a symptom of lower urinary tract diseases such as cystitis, cystolithiasis, and bladder cancer, which impairs the quality of life in patients. Interstitial cystitis or painful bladder syndrome is an idiopathic disease, presenting with bladder pain and urinary frequency or urgency (Kelada and Jones 2007). The bladder pain accompanying interstitial cystitis is regional and chronic and spreads over the lower pelvic and suprapubic area and mimics neuropathic pain (Theoharides et al. 2008). In isolated rat urinary bladder, H_2S stimulates capsaicin-sensitive primary afferent neurons and produces contractile responses (Patacchini et al. 2004). There is also evidence that intravesical NaHS produces changes in urodynamic parameters after chemical disruption of the urothelial barrier in rats (Streng et al. 2008). Intraperitoneal administration of cyclophosphamide induces cystitis in mice, accompanied by bladder pain-like nociceptive behavior and referred hyperalgesia in the region between the anus and urethral opening, which is now recognized as an animal model of human interstitial cystitis (Wantuch et al. 2007).

We have shown that inhibition of CSE by pre-administration of PPG prevents cyclophosphamide-induced bladder nociception, i.e., bladder pain-like nociceptive behavior and referred hyperalgesia, and that expression of CSE protein is markedly increased in the bladder tissue of mice with cyclophosphamide-induced cystitis (Matsunami et al. 2012b), as observed in the pancreatic tissue of mice with cerulein-induced pancreatitis (Nishimura et al. 2009). Post-administration of T-channel blockers, mibefradil and NNC 55-0396, and the knockdown of $Ca_v3.2$ T-channels by the antisense method suppress the cyclophosphamide-induced bladder nociception, while neither an L-type Ca^{2+} channel blocker, verapamil, nor TRPA1 blockers, AP18 and HC-030031, exert such effect (Matsunami et al. 2012b). The T-channel-dependent pronociceptive role of H_2S in bladder nociceptive processing is also supported by our evidence that intravesical administration of NaHS causes phosphorylation of ERK in the spinal cord, an effect prevented by pretreatment with the T-channel blocker NNC 55-0396 (Matsunami et al. 2012b).

5 Neuropathic Pain

Neuropathic pain is the consequence of damage to peripheral nerves or to the CNS and common in clinical practice and greatly impairs the quality of life of patients. The neuropathic syndromes include painful diabetic neuropathy, postherpetic neuralgia, phantom limb pain, complex regional pain syndrome (CRPS), etc. The molecular mechanisms underlying neuropathic pain involve abnormal functions and/or altered expression of voltage-gated ion channels such as sodium (McGowan et al. 2009), calcium (Boroujerdi et al. 2011), and potassium channels (Cao et al. 2010) in sensory neurons. There is evidence for upregulation of T-current density in small DRG cells isolated from rats with neuropathic pain due to chronic constrictive injury (CCI) of the sciatic nerve (Jagodic et al. 2008) and that genetic silencing of $Ca_v3.2$ T-channels reverses CCI-induced neuropathic pain (Bourinet et al. 2005). We have shown that neuropathic hyperalgesia/allodynia caused by L5 spinal nerve cutting (L5SNC) is strongly suppressed by i.p. administration of CSE inhibitors and by i.pl. administration of T-channel blockers or genetic silencing of $Ca_v3.2$ T-channels (Takahashi et al. 2010). Given that the protein levels of $Ca_v3.2$ T-channels, but not CSE, are dramatically upregulated in the DRG of rats with L5SNC (Takahashi et al. 2010), our data suggest that the upregulated $Ca_v3.2$ T-channels in sensory neurons after spinal nerve injury are exposed to positive regulation by CSE-derived endogenous H_2S, contributing to the maintenance of neuropathic pain. H_2S is also considered to be involved in diabetes-associated peripheral neuropathy, since systemic administration of inhibitors of CSE or CBS and silencing of $Ca_v3.2$ T-channels in DRG reverse mechanical allodynia/ hyperalgesia in streptozotocin (STZ)-induced diabetic neuropathy rats (Todorovic et al. 2001; Velasco-Xolalpa et al. 2013). It is also to be noted that T-channels, particularly of the $Ca_v3.2$ isoform, are upregulated and involved in the hyperalgesia/allodynia in the laboratory animals with the diabetic neuropathy (Messinger et al. 2009). Upregulation of T-channel current density has also been shown in DRG cells isolated from diabetic Bio-Bred/Worchester (BB/W) rats and from rats with STZ-induced diabetic neuropathy (Todorovic and Jevtovic-Todorovic 2011). Thus, the upregulation of $Ca_v3.2$ T-channels in the sensory neurons appears to cause diabetes-induced hyperexcitability of the nociceptors, contributing to the pathogenesis of diabetic neuropathy.

Many anticancer agents, such as vincristine, paclitaxel, and oxaliplatin, often cause painful peripheral neuropathies that limit their use in cancer therapy. It has been demonstrated that ethosuximide, a T-channel blocker, reverses paclitaxel- and vincristine-induced painful peripheral neuropathy in laboratory animals (Flatters and Bennett 2004). We have reported that paclitaxel-induced neuropathic hyperalgesia is reversed by i.pl. administration of NNC 55-0396 or mibefradil, systemic administration of PPG or BCA, CSE inhibitors, and genetic silencing of $Ca_v3.2$. Unlike L5SNC rats, neither $Ca_v3.2$ nor CSE are upregulated at protein levels in the DRG, spinal cord, or peripheral tissues including the hindpaws, whereas H_2S content in hindpaw tissues is elevated (Okubo et al. 2011). Together, $Ca_v3.2$, a target for H_2S formed by CSE, is a key molecule in the maintenance of distinct types of neuropathic pain.

6 Conclusion

H_2S plays important roles in processing of somatic, visceral, and neuropathic pain signals by targeting $Ca_v3.2$ T-channels and TRPA1. In this context, H_2S-generating enzymes, $Ca_v3.2$, and TRPA1 could be molecular targets for treatment of inflammatory and neuropathic pain including visceral pain. However, endogenous H_2S plays important protective roles in the cardiovascular and gastrointestinal systems, CNS, and so on. Therefore, we must consider the risk of side effects of CSE and

CBS inhibitors, especially in the case of their long-term use. Thus, blockade of $Ca_v3.2$ T-channels and/or TRPA1 would have greater advantage than inhibition of H_2S formation. In particular, the development of selective $Ca_v3.2$ inhibitors should impact on improvement of therapeutic strategy for neuropathic pain, although no isoform selective T-channel inhibitors are available at present.

References

Abe K, Kimura H (1996) The possible role of hydrogen sulfide as an endogenous neuromodulator. J Neurosci 16:1066–1071

Andersson DA, Gentry C, Bevan S (2012) TRPA1 has a key role in the somatic pro-nociceptive actions of hydrogen sulfide. PLoS One 7, e46917

Asimakopoulou A, Panopoulos P, Chasapis CT, Coletta C, Zhou Z, Cirino G et al (2013) Selectivity of commonly used pharmacological inhibitors for cystathionine beta synthase (CBS) and cystathionine gamma lyase (CSE). Br J Pharmacol 169:922–932

Bhatia M, Wong FL, Fu D, Lau HY, Moochhala SM, Moore PK (2005) Role of hydrogen sulfide in acute pancreatitis and associated lung injury. FASEB J 19:623–625

Boroujerdi A, Zeng J, Sharp K, Kim D, Steward O, Luo ZD (2011) Calcium channel alpha-2-delta-1 protein upregulation in dorsal spinal cord mediates spinal cord injury-induced neuropathic pain states. Pain 152:649–655

Bourinet E, Alloui A, Monteil A, Barrere C, Couette B, Poirot O et al (2005) Silencing of the Cav3.2 T-type calcium channel gene in sensory neurons demonstrates its major role in nociception. EMBO J 24:315–324

Cao XH, Byun HS, Chen SR, Cai YQ, Pan HL (2010) Reduction in voltage-gated K^+ channel activity in primary sensory neurons in painful diabetic neuropathy: role of brain-derived neurotrophic factor. J Neurochem 114:1460–1475

Cheung NS, Peng ZF, Chen MJ, Moore PK, Whiteman M (2007) Hydrogen sulfide induced neuronal death occurs via glutamate receptor and is associated with calpain activation and lysosomal rupture in mouse primary cortical neurons. Neuropharmacology 53:505–514

Cunha TM, Dal-Secco D, Verri WA Jr, Guerrero AT, Souza GR, Vieira SM et al (2008) Dual role of hydrogen sulfide in mechanical inflammatory hypernociception. Eur J Pharmacol 590:127–135

Dai Y, Wang S, Tominaga M, Yamamoto S, Fukuoka T, Higashi T et al (2007) Sensitization of TRPA1 by PAR2 contributes to the sensation of inflammatory pain. J Clin Invest 117:1979–1987

Distrutti E, Sediari L, Mencarelli A, Renga B, Orlandi S, Antonelli E et al (2006) Evidence that hydrogen sulfide exerts antinociceptive effects in the gastrointestinal tract by activating K_{ATP} channels. J Pharmacol Exp Ther 316:325–335

Donatti AF, Araujo RM, Soriano RN, Azevedo LU, Leite-Panissi CA, Branco LG (2014) Role of hydrogen sulfide in the formalin-induced orofacial pain in rats. Eur J Pharmacol 738:49–56

Elies J, Scragg JL, Huang S, Dallas ML, Huang D, MacDougall D et al (2014) Hydrogen sulfide inhibits $Ca_v3.2$ T-type Ca^{2+} channels. FASEB J 28(12):5376–5387

Eto K, Ogasawara M, Umemura K, Nagai Y, Kimura H (2002) Hydrogen sulfide is produced in response to neuronal excitation. J Neurosci 22:3386–3391

Flatters SJ, Bennett GJ (2004) Ethosuximide reverses paclitaxel- and vincristine-induced painful peripheral neuropathy. Pain 109:150–161

Francois A, Laffray S, Pizzoccaro A, Eschalier A, Bourinet E (2014) T-type calcium channels in chronic pain: mouse models and specific blockers. Pflugers Arch 466:707–717

Fukushima O, Nishimura S, Matsunami M, Aoki Y, Nishikawa H, Ishikura H et al (2010) Phosphorylation of ERK in the spinal dorsal horn following pancreatic pronociceptive stimuli

with proteinase-activated receptor-2 agonists and hydrogen sulfide in rats: evidence for involvement of distinct mechanisms. J Neurosci Res 88:3198–3205
Geng B, Yang J, Qi Y, Zhao J, Pang Y, Du J et al (2004) H_2S generated by heart in rat and its effects on cardiac function. Biochem Biophys Res Commun 313:362–368
Iftinca MC, Zamponi GW (2009) Regulation of neuronal T-type calcium channels. Trends Pharmacol Sci 30:32–40
Ise F, Takasuka H, Hayashi S, Takahashi K, Koyama M, Aihara E et al (2011) Stimulation of duodenal HCO_3^- secretion by hydrogen sulphide in rats: relation to prostaglandins, nitric oxide and sensory neurones. Acta Physiol (Oxf) 201:117–126
Jacus MO, Uebele VN, Renger JJ, Todorovic SM (2012) Presynaptic $Ca_v3.2$ channels regulate excitatory neurotransmission in nociceptive dorsal horn neurons. J Neurosci 32:9374–9382
Jagodic MM, Pathirathna S, Joksovic PM, Lee W, Nelson MT, Naik AK et al (2008) Upregulation of the T-type calcium current in small rat sensory neurons after chronic constrictive injury of the sciatic nerve. J Neurophysiol 99:3151–3156
Kawabata A, Ishiki T, Nagasawa K, Yoshida S, Maeda Y, Takahashi T et al (2007) Hydrogen sulfide as a novel nociceptive messenger. Pain 132:74–81
Kelada E, Jones A (2007) Interstitial cystitis. Arch Gynecol Obstet 275:223–229
Kimura H (2000) Hydrogen sulfide induces cyclic AMP and modulates the NMDA receptor. Biochem Biophys Res Commun 267:129–133
Kimura H (2010) Hydrogen sulfide: from brain to gut. Antioxid Redox Signal 12:1111–1123
Kimura Y, Kimura H (2004) Hydrogen sulfide protects neurons from oxidative stress. FASEB J 18:1165–1167
Kimura Y, Dargusch R, Schubert D, Kimura H (2006) Hydrogen sulfide protects HT22 neuronal cells from oxidative stress. Antioxid Redox Signal 8:661–670
Kimura Y, Mikami Y, Osumi K, Tsugane M, Oka J, Kimura H (2013) Polysulfides are possible H_2S-derived signaling molecules in rat brain. FASEB J 27:2451–2457
Kurokawa Y, Sekiguchi F, Kubo S, Yamasaki Y, Matsuda S, Okamoto Y et al (2011) Involvement of ERK in NMDA receptor-independent cortical neurotoxicity of hydrogen sulfide. Biochem Biophys Res Commun 414:727–732
Lee AT, Shah JJ, Li L, Cheng Y, Moore PK, Khanna S (2008) A nociceptive-intensity-dependent role for hydrogen sulphide in the formalin model of persistent inflammatory pain. Neuroscience 152:89–96
Li L, Xie R, Hu S, Wang Y, Yu T, Xiao Y et al (2012) Upregulation of cystathionine beta-synthetase expression by nuclear factor-kappa B activation contributes to visceral hypersensitivity in adult rats with neonatal maternal deprivation. Mol Pain 8:89
Maeda Y, Aoki Y, Sekiguchi F, Matsunami M, Takahashi T, Nishikawa H et al (2009) Hyperalgesia induced by spinal and peripheral hydrogen sulfide: evidence for involvement of $Ca_v3.2$ T-type calcium channels. Pain 142:127–132
Marger F, Gelot A, Alloui A, Matricon J, Ferrer JF, Barrere C et al (2011) T-type calcium channels contribute to colonic hypersensitivity in a rat model of irritable bowel syndrome. Proc Natl Acad Sci U S A 108:11268–11273
Matsunami M, Tarui T, Mitani K, Nagasawa K, Fukushima O, Okubo K et al (2009) Luminal hydrogen sulfide plays a pronociceptive role in mouse colon. Gut 58:751–761
Matsunami M, Kirishi S, Okui T, Kawabata A (2011) Chelating luminal zinc mimics hydrogen sulfide-evoked colonic pain in mice: possible involvement of T-type calcium channels. Neuroscience 181:257–264
Matsunami M, Kirishi S, Okui T, Kawabata A (2012a) Hydrogen sulfide-induced colonic mucosal cytoprotection involves T-type calcium channel-dependent neuronal excitation in rats. J Physiol Pharmacol 63:61–68
Matsunami M, Miki T, Nishiura K, Hayashi Y, Okawa Y, Nishikawa H et al (2012b) Involvement of the endogenous hydrogen sulfide/$Ca_v3.2$ T-type Ca^{2+} channel pathway in cystitis-related bladder pain in mice. Br J Pharmacol 167:917–928

McGowan E, Hoyt SB, Li X, Lyons KA, Abbadie C (2009) A peripherally acting $Na_v1.7$ sodium channel blocker reverses hyperalgesia and allodynia on rat models of inflammatory and neuropathic pain. Anesth Analg 109:951–958

Messinger RB, Naik AK, Jagodic MM, Nelson MT, Lee WY, Choe WJ et al (2009) In vivo silencing of the Ca(V)3.2 T-type calcium channels in sensory neurons alleviates hyperalgesia in rats with streptozocin-induced diabetic neuropathy. Pain 145:184–195

Miao X, Meng X, Wu G, Ju Z, Zhang HH, Hu S et al (2014) Upregulation of cystathionine-beta-synthetase expression contributes to inflammatory pain in rat temporomandibular joint. Mol Pain 10:9

Miyamoto R, Otsuguro K, Ito S (2011) Time- and concentration-dependent activation of TRPA1 by hydrogen sulfide in rat DRG neurons. Neurosci Lett 499:137–142

Munchberg U, Anwar A, Mecklenburg S, Jacob C (2007) Polysulfides as biologically active ingredients of garlic. Org Biomol Chem 5:1505–1518

Nagasawa K, Tarui T, Yoshida S, Sekiguchi F, Matsunami M, Ohi A et al (2009) Hydrogen sulfide evokes neurite outgrowth and expression of high-voltage-activated Ca^{2+} currents in NG108-15 cells: involvement of T-type Ca^{2+} channels. J Neurochem 108:676–684

Nelson MT, Joksovic PM, Perez-Reyes E, Todorovic SM (2005) The endogenous redox agent L-cysteine induces T-type Ca^{2+} channel-dependent sensitization of a novel subpopulation of rat peripheral nociceptors. J Neurosci 25:8766–8775

Nelson MT, Woo J, Kang HW, Vitko I, Barrett PQ, Perez-Reyes E et al (2007a) Reducing agents sensitize C-type nociceptors by relieving high-affinity zinc inhibition of T-type calcium channels. J Neurosci 27:8250–8260

Nelson MT, Joksovic PM, Su P, Kang HW, Van Deusen A, Baumgart JP et al (2007b) Molecular mechanisms of subtype-specific inhibition of neuronal T-type calcium channels by ascorbate. J Neurosci 27:12577–12583

Nishimura S, Fukushima O, Ishikura H, Takahashi T, Matsunami M, Tsujiuchi T et al (2009) Hydrogen sulfide as a novel mediator for pancreatic pain in rodents. Gut 58:762–770

Ogawa H, Takahashi K, Miura S, Imagawa T, Saito S, Tominaga M et al (2012) H_2S functions as a nociceptive messenger through transient receptor potential ankyrin 1 (TRPA1) activation. Neuroscience 218:335–343

Ohge H, Furne JK, Springfield J, Sueda T, Madoff RD, Levitt MD (2003) The effect of antibiotics and bismuth on fecal hydrogen sulfide and sulfate-reducing bacteria in the rat. FEMS Microbiol Lett 228:137–142

Okubo K, Takahashi T, Sekiguchi F, Kanaoka D, Matsunami M, Ohkubo T et al (2011) Inhibition of T-type calcium channels and hydrogen sulfide-forming enzyme reverses paclitaxel-evoked neuropathic hyperalgesia in rats. Neuroscience 188:148–156

Okubo K, Matsumura M, Kawaishi Y, Aoki Y, Matsunami M, Okawa Y et al (2012) Hydrogen sulfide-induced mechanical hyperalgesia and allodynia require activation of both $Ca_v3.2$ and TRPA1 channels in mice. Br J Pharmacol 166:1738–1743

Patacchini R, Santicioli P, Giuliani S, Maggi CA (2004) Hydrogen sulfide (H_2S) stimulates capsaicin-sensitive primary afferent neurons in the rat urinary bladder. Br J Pharmacol 142:31–34

Perez-Reyes E (2003) Molecular physiology of low-voltage-activated t-type calcium channels. Physiol Rev 83:117–161

Pozsgai G, Hajna Z, Bagoly T, Boros M, Kemeny A, Materazzi S et al (2012) The role of transient receptor potential ankyrin 1 (TRPA1) receptor activation in hydrogen-sulphide-induced CGRP-release and vasodilation. Eur J Pharmacol 689:56–64

Roediger WE, Moore J, Babidge W (1997) Colonic sulfide in pathogenesis and treatment of ulcerative colitis. Dig Dis Sci 42:1571–1579

Rose KE, Lunardi N, Boscolo A, Dong X, Erisir A, Jevtovic-Todorovic V et al (2013) Immunohistological demonstration of $Ca_v3.2$ T-type voltage-gated calcium channel expression in soma of dorsal root ganglion neurons and peripheral axons of rat and mouse. Neuroscience 250:263–274

Schicho R, Krueger D, Zeller F, Von Weyhern CW, Frieling T, Kimura H et al (2006) Hydrogen sulfide is a novel prosecretory neuromodulator in the Guinea-pig and human colon. Gastroenterology 131:1542–1552

Sekiguchi F, Kawabata A (2013) T-type calcium channels: functional regulation and implication in pain signaling. J Pharmacol Sci 122:244–250

Sekiguchi F, Miyamoto Y, Kanaoka D, Ide H, Yoshida S, Ohkubo T et al (2014) Endogenous and exogenous hydrogen sulfide facilitates T-type calcium channel currents in Cav3.2-expressing HEK293 cells. Biochem Biophys Res Commun 445:225–229

Streng T, Axelsson HE, Hedlund P, Andersson DA, Jordt SE, Bevan S et al (2008) Distribution and function of the hydrogen sulfide-sensitive TRPA1 ion channel in rat urinary bladder. Eur Urol 53:391–399

Stucky CL, Dubin AE, Jeske NA, Malin SA, McKemy DD, Story GM (2009) Roles of transient receptor potential channels in pain. Brain Res Rev 60:2–23

Takahashi T, Aoki Y, Okubo K, Maeda Y, Sekiguchi F, Mitani K et al (2010) Upregulation of $Ca_v3.2$ T-type calcium channels targeted by endogenous hydrogen sulfide contributes to maintenance of neuropathic pain. Pain 150:183–191

Tamizhselvi R, Moore PK, Bhatia M (2007) Hydrogen sulfide acts as a mediator of inflammation in acute pancreatitis: in vitro studies using isolated mouse pancreatic acinar cells. J Cell Mol Med 11:315–326

Taniguchi E, Matsunami M, Kimura T, Yonezawa D, Ishiki T, Sekiguchi F et al (2009) Rhodanese, but not cystathionine-gamma-lyase, is associated with dextran sulfate sodium-evoked colitis in mice: a sign of impaired colonic sulfide detoxification? Toxicology 264:96–103

Terada Y, Fujimura M, Nishimura S, Tsubota M, Sekiguchi F, Nishikawa H et al (2013) Contribution of TRPA1 as a downstream signal of proteinase-activated receptor-2 to pancreatic pain. J Pharmacol Sci 123:284–287

Terada Y, Fujimura M, Nishimura S, Tsubota M, Sekiguchi F, Kawabata A (2014) Roles of $Ca_v3.2$ and TRPA1 channels targeted by hydrogen sulfide in pancreatic nociceptive processing in mice with or without acute pancreatitis. J Neurosci Res 93(2):361–369

Theoharides TC, Whitmore K, Stanford E, Moldwin R, O'Leary MP (2008) Interstitial cystitis: bladder pain and beyond. Expert Opin Pharmacother 9:2979–2994

Todorovic SM, Jevtovic-Todorovic V (2011) T-type voltage-gated calcium channels as targets for the development of novel pain therapies. Br J Pharmacol 163:484–495

Todorovic SM, Jevtovic-Todorovic V, Meyenburg A, Mennerick S, Perez-Reyes E, Romano C et al (2001) Redox modulation of T-type calcium channels in rat peripheral nociceptors. Neuron 31:75–85

Trevisani M, Patacchini R, Nicoletti P, Gatti R, Gazzieri D, Lissi N et al (2005) Hydrogen sulfide causes vanilloid receptor 1-mediated neurogenic inflammation in the airways. Br J Pharmacol 145:1123–1131

Velasco-Xolalpa ME, Barragan-Iglesias P, Roa-Coria JE, Godinez-Chaparro B, Flores-Murrieta FJ, Torres-Lopez JE et al (2013) Role of hydrogen sulfide in the pain processing of non-diabetic and diabetic rats. Neuroscience 250:786–797

Wallace JL, Vong L, McKnight W, Dicay M, Martin GR (2009) Endogenous and exogenous hydrogen sulfide promotes resolution of colitis in rats. Gastroenterology 137:569–578

Wantuch C, Piesla M, Leventhal L (2007) Pharmacological validation of a model of cystitis pain in the mouse. Neurosci Lett 421:250–252

Warenycia MW, Goodwin LR, Benishin CG, Reiffenstein RJ, Francom DM, Taylor JD et al (1989) Acute hydrogen sulfide poisoning. Demonstration of selective uptake of sulfide by the brainstem by measurement of brain sulfide levels. Biochem Pharmacol 38:973–981

Xu GY, Winston JH, Shenoy M, Zhou S, Chen JD, Pasricha PJ (2009) The endogenous hydrogen sulfide producing enzyme cystathionine-beta synthase contributes to visceral hypersensitivity in a rat model of irritable bowel syndrome. Mol Pain 5:44

Zhao W, Zhang J, Lu Y, Wang R (2001) The vasorelaxant effect of H_2S as a novel endogenous gaseous K_{ATP} channel opener. EMBO J 20:6008–6016

Part VI

H_2S, Angiogenesis and Cancer

Hydrogen Sulfide and Cancer

Mark R. Hellmich and Csaba Szabo

Contents

Abstract

Recent studies revealed increased expression of various hydrogen sulfide (H_2S)-producing enzymes in cancer cells of various tissue types, and new roles of H_2S in the pathophysiology of cancer have emerged. This is particularly evident in cancers of the colon and ovaries, where the malignant cells both overexpress cystathionine-β-synthase (CBS) and produce increased amounts of H_2S, which enhances tumor growth and spread by (a) stimulating cellular bioenergetics, (b) activating proliferative, migratory, and invasive signaling pathways, and (c) enhancing tumor angiogenesis. Importantly, in preclinical models of these cancers, either pharmacological inhibition or genetic silencing of CBS was shown to be sufficient to suppress cancer cell bioenergetics in vitro, inhibit tumor growth and metastasis in vivo, and enhance the antitumor efficacy of frontline chemotherapeutic agents, providing a strong rationale for the

M.R. Hellmich • C. Szabo (✉)
Departments of Surgery and Anesthesiology, University of Texas Medical Branch, Galveston, TX, USA
e-mail: mhellmic@utmb.edu; csszabo@utmb.edu

P.K. Moore, M. Whiteman (eds.), *Chemistry, Biochemistry and Pharmacology of Hydrogen Sulfide*, Handbook of Experimental Pharmacology 230,
DOI 10.1007/978-3-319-18144-8_12

development of CBS-targeted inhibitors as anticancer therapies. However, the observation that inhibition of H_2S biosynthesis exerts anticancer effects is contradicted by other studies showing that increasing H_2S with exogenous donors also exerts antitumor actions. Herein, we present a brief review of the scientific literature documenting the function of H_2S, H_2S donors, and transsulfuration enzymes in cancers from various tissue types, and propose that the paradoxical actions of H_2S can be resolved by considering the bell-shaped pharmacology of H_2S, whereby lower (endogenous) H_2S production tends to promote, while higher (generated from exogenously added H_2S donors) tends to inhibit cancer cell proliferation. Finally, we suggest areas for future investigations to expand our knowledge of this nascent field.

Keywords
Cancer • Hydrogen sulfide • Mitochondria • Bioenergetics • Angiogenesis • Proliferation • Signaling

1 Biological Roles of H_2S: A Brief Overview

Work over the last decade recognized the importance of endogenously produced H_2S in a variety of biological functions in the nervous, cardiovascular, and immune system, in health and disease (reviewed in: Szabo 2007; Kimura 2011; Whiteman et al. 2011; Wang 2012). The three principal enzymes involved in the physiological production of H_2S are CBS, CSE, and 3-mercaptopyruvate sulfurtransferase (3-MST). For the purposes of the current article, we focus on the biological effects of H_2S that are relevant for cancer biology; these include the regulation of vascular function (physiologically: vasorelaxation and stimulation of angiogenesis) (reviewed in: Wang et al. 2010; Szabo and Papapetropoulos 2011), regulation of cellular bioenergetics (physiologically: stimulation of mitochondrial electron transport and maintenance of cellular energetics) (reviewed in: Szabo et al. 2014; Módis et al. 2014a), and the regulation of intracellular signaling and cell death (physiologically, acting as a direct and indirect antioxidant and inhibiting oxidative damage and cell death in response to diverse stimuli) (reviewed in: Wang 2012; Kolluru et al. 2013). Appreciating the complexities of H_2S biology requires familiarity with its pharmacology, including the fact that it is a labile, diffusible gas, and that it has a bell-shaped (or biphasic) dose–response curve, whereby lower concentrations (or lower rates of production) can exert markedly different (often, opposing) effects compared to the effects of H_2S seen at higher concentrations (or higher rates of production) (reviewed in: Szabo et al. 2014).

2 Role of Endogenous H_2S Production in Colon Cancer

In 2013 we have compared human colon cancer specimens with patient-matched normal mucosa tissue and discovered that there is a selective upregulation of CBS in the cancer tissue, while the non-cancerous peritumor tissue has low CBS expression levels. The expression of the other two H_2S-producing enzymes, CSE and 3-MST, did not show an upregulation in the tumor tissue. When subsequently checking several colon adenocarcinoma-derived cell lines (HCT-116, HT-29, LoVo), we have also observed a selective upregulation of CBS, as compared to the non-malignant colonic epithelial cell line NCM356. We have also conducted cell fractionation studies to test the localization of CBS in colon cancer cells. While CBS is traditionally viewed as a cytosolic enzyme, it can also be translocated to the mitochondria (Teng et al. 2013; Szabo et al. 2014). Our results revealed that CBS in the HCT116 cancer cells is present both in the cytosol and in the mitochondria. As expected, homogenates of the patient-derived colon tumor specimens as well as homogenates of the colon cancer-derived cell lines showed increased rates of H_2S production, and this response was inhibited by the prototypical CBS inhibitor compound aminooxyacetic acid (AOAA) (Szabo et al. 2013).

Next, we have studied the functional role of CBS-derived H_2S in the control of colon cancer cell proliferation, migration, and invasion in vitro, by a combination of genetic (shRNA-mediated stable silencing of CBS in HCT116 cells or adenoviral overexpression of CBS in NCM356 cells) and pharmacologic (CBS inhibition by AOAA) approaches. Genetic silencing or pharmacological inhibition of CBS suppressed HCT116 cell proliferation, migration, and invasion (Szabo et al. 2013). We have subsequently also used *S*-adenosyl-L-methionine (SAM), an allosteric activator of CBS; this compound, at low concentrations, increased HCT116 cell proliferation (Módis et al. 2014b).

Part of the proliferative and pro-migratory effects of CBS-derived H_2S are likely to be due to stimulation of Akt/PI3K signaling, as prior studies have demonstrated that exogenous H_2S donors stimulate HCT116 cell migration via activation of these pathways (Szabo and Hellmich 2013). In addition, part of the effect of CBS-derived H_2S is due to mitochondrial, bioenergetic stimulatory effects. We have observed that both the silencing of CBS and the inhibition of CBS with AOAA suppressed HCT116 bioenergetic functions (including basal electron transport and a bioenergetic parameter known as "respiratory reserve capacity," which is quantified as the increase in mitochondrial oxygen consumption in response to a mitochondrial uncoupling agent) (Szabo et al. 2013). CBS inhibition not only suppressed mitochondrial function but also glycolytic function in HCT116 cells (Szabo et al. 2013), an effect that may be attributable to the known stimulatory role of H_2S on the activity of GAPDH (Mustafa et al. 2009), an essential enzyme in the glycolytic pathway. Similar to the effects of the allosteric CBS activator SAM on proliferation, SAM, at low concentrations, increased HCT116 cell bioenergetic functions (Módis et al. 2014b).

Subsequent studies in nude mice bearing xenografts of either HCT116 cells or patient-derived tumor tissue (PDTX) extended the findings into in vivo models.

Silencing of CBS expression and/or pharmacological inhibition of CBS with AOAA significantly reduced the growth rate of the tumor xenografts. While we conclude that part of the effects of CBS inhibition seen in vivo are likely related to intratumoral mechanisms (i.e. inhibition of cancer cell metabolism and signaling), part of the effect may also involve paracrine mechanisms in the tumor microenvironment, because CBS silencing or CBS inhibition suppressed the density and complexity of CD31-positive blood vessels within the tumor tissue (indicative of reduced tumor angiogenesis). Furthermore, in line with the role of H_2S as a local vasodilator, direct injection of AOAA into the tumor parenchyma reduced peritumor blood flow (Szabo et al. 2013). In addition to reducing primary tumor growth, inhibition of CBS with AOAA decreased the metastatic spread of HCT116 cell from the cecum to the liver (i.e., decrease the number of metastatic lesions per area) in an orthotopic xenograft model in nude mice, and AOAA synergizes with the anti-metastasis effects of oxaliplatin in the same model (Bohanon et al. 2014).

In HCT116 cells, silencing or pharmacological inhibition of CSE did not exert any effects on HCT116 proliferation, migration, or tumor growth in vitro or in vivo (Szabo et al. 2013). In contrast, in another human colon cancer cell line (SW480), high expression levels of CSE were observed; these levels were further increased by activation of the Wnt pathway in these cells. Moreover, pharmacological inhibition of CSE (with propargylglycine [PAG]) or genetic silencing of CSE attenuated cell proliferation in vitro. In addition, SW480 cells with CSE silencing tended to reduce tumor growth (significant reduction in tumor volume, but not in tumor weight) when injected into tumor-bearing nude mice (Fan et al. 2014).

The above findings are consistent with the conclusion that increased H_2S production (from CBS, but in other cell lines also from CSE) plays an essential role in colon cancer cell proliferation.

3 Role of Endogenous H_2S Production in Ovarian Cancer

Similar to colon cancer cells, CBS has been found to be overexpressed in primary epithelial ovarian cancer tissues, as well as in multiple ovarian cancer cell lines. When examining a collection of more than 200 patients' tissue microarrays constructed from primary epithelial ovarian cancers, Bhattacharyya and colleagues we found high expression of CBS in primary ovarian tumors, particularly in serous carcinoma, the most common histologic variant. Tumors that had serous histology and higher-grade cancers tended to contain higher levels of CBS. CBS expression was already strong in most of the early-stage (FIGO stages I and II) ovarian cancers studied (Bhattacharyya et al. 2013). In additional studies, quantitative RT-PCR and immunoblotting were used to compare the expression of CBS mRNA and protein levels in a variety of ovarian cancer cell lines, when compared to a control, non-malignant ovarian surface epithelial cell line (OSE). Most ovarian cancer cell lines studied showed high CBS expression (both at protein and mRNA level). CSE was not found to be overexpressed in ovarian cancers, but it was found in the normal ovarian epithelial cell line. Similar to the colon cancer study (see above),

CBS exhibited significant mitochondrial localization in A2780 cells (Bhattacharyya et al. 2013).

Next, Bhattacharyya and colleagues studied the functional role of CBS-derived H_2S in the control of ovarian cancer cell proliferation, migration, and invasion in vitro, by a combination of genetic (siRNA-mediated stable silencing of CBS in A2780, A2780/CP-70, OV202, and SKOV3 cells) and pharmacologic (CBS inhibition by AOAA) approaches. Downregulation or inhibition of CBS was found to inhibit cell proliferation in vitro, and AOAA treatment (especially at higher concentrations) also reduced cell viability. When studying the intracellular mechanisms responsible for these actions, Bhattacharyya and colleagues found that downregulation or inhibition of CBS reduces the intracellular content of the key antioxidant glutathione (GSH + GSSG) and triggers apoptotic cascades. This latter effect may well be the consequence of the intracellular antioxidant depletion after CBS inhibition/silencing. Another important consequence of the CBS silencing or CBS inhibition was an increase in cellular reactive oxygen species levels; this effect may well be secondary to antioxidant depletion (see above) or it may also be related to changes in mitochondrial function (see below). Finally, silencing CBS in A2780 cells also affected intracellular signaling pathways: Bhattacharyya and colleagues found that CBS silencing increases the expression of p53, while the expression of the RelA/p65 subunit of NF-κB was decreased (Bhattacharyya et al. 2013).

Similar to the study in colon cancer cells, H_2S in ovarian cancer cells supports mitochondrial function and cellular bioenergetics. Bhattacharyya and colleagues found that silencing of CBS reduced mitochondrial oxygen consumption, and similar effects were seen when ovarian cancer cell lines were treated with the CBS inhibitor AOAA. Additional consequences of CBS silencing and/or CBS inhibition were (a) an increase in mitochondrial ROS production, (b) a decrease in intracellular NAD/NADH ratio, (c) a reduction in ATP synthesis, and an increase in ADP/ATP ratio (Bhattacharyya et al. 2013).

Subsequent studies in nude mice transplanted with A2780/CP-20 xenografts extended the findings into in vivo models. Silencing of CBS resulted in a significant approximately 40 % reduction in tumor weight and an even more marked (approximately 70 %) decrease in the number of tumor nodules. The reduction by CBS silencing of the proliferative capacity of the cancer cells was confirmed with Ki-67 staining. In addition (and similar to the colon cancer study discussed above), CBS silencing resulted in an inhibition of peritumor angiogenesis, as evidenced by a reduction of CD31 staining (Bhattacharyya et al. 2013).

Last, but not least, Bhattacharyya and colleagues also demonstrated that inhibition of CBS sensitizes the cancer cells to concomitant chemotherapy in vitro and in vivo. In vitro, CBS silencing of A2780 cells shifted the IC_{50} of cisplatin from 13.1 to 7.9 mM. In vivo, the combination therapy of CBS siRNA and cisplatin produced a dramatic reduction in tumor weight and number of tumor nodules (both effects approximately 80–90 %), when compared to the group that received cisplatin alone (Bhattacharyya et al. 2013).

4 Lack of Functional Role of Endogenous H_2S Production in Melanoma

In contrast to colon cancer cells and ovarian cancer cells, Panza and colleagues found that CBS is not overexpressed in human nevi (compound, junctional, or dysplastic forms). Likewise, in samples of human primitive melanoma, CBS and 3-MST displayed a variable, but always very low level of expression. Furthermore, four distinct human melanoma cell lines (A375, Sk-Mel-5, Sk-Mel-28, and PES 43), when compared in normal human epidermal melanocytes (NHEM), showed no increase in CBS expression. In contrast, another H_2S-producing enzyme, CSE, exhibited significantly higher expression in all of the nevi, melanoma samples, and melanoma cell lines studied (Panza et al. 2015).

In order to evaluate whether the various H_2S-producing enzymes had a functional effect on the proliferation of human melanoma cell line, A375 cells were transiently transfected with siRNAs for either CBS, CSE, or 3-MST. Silencing of these enzymes did not affect melanoma cell proliferation (Panza et al. 2015). These data indicate that in melanoma (as opposed to colon cancer and ovarian cancer), the expression of H_2S-producing enzymes does not have an endogenous stimulatory role on cell proliferation.

5 CBS Silencing Accelerates the Development of Glioma

To assess the role of CBS in glioma tumorigenesis, Takano and colleagues established a subclone of U87-MG glioma cells with stable silencing of CBS. When subcutaneously or orthotopically injected into the flank of SCID mice, the subclone with CB silencing exhibited a shorter latency period for tumor growth, as well as a more pronounced overall tumor growth rate. In addition to greater tumor volume, the CBS-silenced tumors exhibited increased depth of invasion, vascular density, cell proliferation, and more apoptosis (Takano et al. 2014). These findings unveil a role of CBS in glioma that is markedly different from its role in colon cancer or ovarian cancer. As far as the mechanisms responsible for the observed effects, in vitro studies by Takano and colleagues showed that glioma cells respond to CBS silencing with increased VEGF and ANGPTL4 levels, higher HIF2a expression, and increased anchorage-independent cell growth (Takano et al. 2014).

6 Changes in H_2S-Producing Enzymes in Other Forms of Cancer

A limited number of studies are available that detect significant changes in various H_2S-producing enzymes in other forms of cancer. Perhaps the most detailed study was conducted by Guo and colleagues in human prostatic tissues and prostate cancer tissues. In normal prostatic tissue, CBS and CSE were both detected in the prostatic epithelium, while the periacinar stroma cells contained CSE, but not CBS.

LNCaP (an androgen-dependent prostate cancer cell line) exhibited marked CBS and CSE expression; this is in marked contrast with the low expression of CBS and CSE in RWPE-1 (a normal prostatic peripheral zone epithelial cell line). A lesser degree of CBS and CSE expression was seen in several other prostatic cancer lines. Both CBS and CSE were identified primarily cytoplasmic. Guo and colleagues did not investigate the effect of silencing or pharmacological modulation of CBS or CSE on the proliferation rate of LNCaP cells. However, they have tested dihydrotestosterone (DHT) on the expression of CBS and CSE and found that it increased the expression of both CBS and CSE in LNCaP cells (Guo et al. 2012). Other cancer cell types where the expression of CBS has been demonstrated include myeloma (De Vos et al. 2002), biliary track carcinoma (Hansel et al. 2003), and a significant portion of the cells contained in the NCI60 collection, with most consistent/highest increases seen in most breast cancers and most renal cancers present in the collection (Zhang et al. 2005). None of these studies have investigated the functional effect of modulation of H_2S synthesis in these cancer cell lines.

7 H_2S Donors and Cancer

There is an apparent paradox in the literature, since many reports (discussed in the previous sections) show that inhibition of H_2S biosynthesis exerts anticancer effects, while many studies show that H_2S donors of various types exert anticancer actions in vitro and in vivo (overviewed in: Hellmich et al. 2015). In addition, treatment of the recipients with high doses of the CBS/CSE substrate L-cysteine has been shown to increase the growth of melanoma (Panza et al. 2015). This paradox can be resolved by considering the bell-shaped pharmacology of H_2S, whereby lower (endogenous) H_2S production tends to promote, while higher (generated from exogenously added H_2S donors) tends to inhibit cancer cell proliferation (see for review: Hellmich et al. 2015).

8 Conclusions and Future Directions

Although limited in terms of quantity and mechanistic detail, there is a reasonable body of evidence suggesting that endogenous H_2S production is important for the growth and proliferation of at least two types of cancer: colon cancer and ovarian cancer. The data show that melanoma does not rely on endogenous H_2S production for its growth and proliferation, while in glioma, CBS silencing, in fact, accelerates tumor proliferation. It is reasonable to hypothesize that cancers from different organs and tissues will utilize different H_2S-associated pathways to regulate cell proliferation and survival mechanisms in a tumor cell-type dependent manner. The field of H_2S in cancer is a young one, where (as overviewed in Hellmich et al. 2015) additional work is necessary in many areas, including (a) studies to further delineate the mechanism of upregulation of H_2S-producing enzymes in cancer cells; (b) studies aimed to "place" the H_2S-related energetic mechanisms into the overall scheme of tumor cell bioenergetics; (c) additional work on clinical samples, including biomarkers of H_2S production in patients with cancer; (d) studies aimed at "connecting" the H_2S pathway in cancer cells to various known pathways of cell proliferation and death/survival signaling; (e) further studies focusing on the interaction of tumor cell-derived H_2S with its microenvironment including cancer stem cells; (f) investigations not only into changes in the expression/activity of H_2S-producing enzymes, but also enzymes involved in its metabolism and degradation; (g) studies exploring potential interactions of the H_2S system with other gaseous mediators in cancer, such as nitric oxide and carbon monoxide; and (h) pharmacological and drug discovery studies into improved CBS inhibitors for the therapy of cancer, in forms of cancer where the preclinical data warrant such expansion.

Acknowledgments The work is supported by the National Institutes of Health (R01 CA175803) and the Institute for Translational Sciences at the University of Texas Medical Branch, and Clinical and Translational Science Award (8UL1TR000071).

Conflict of Interest Disclosure Statement M.R.H. and C.S. have stock ownership in CBS Therapeutic Inc., a for-profit organization involved in the development of CBS inhibitor therapies for cancer.

References

Bhattacharyya S, Saha S, Giri K, Lanza IR, Nair KS, Jennings NB, Rodriguez-Aguayo C, Lopez-Berestein G, Basal E, Weaver AL, Visscher DW, Cliby W, Sood AK, Bhattacharya R, Mukherjee P (2013) Cystathionine beta-synthase (CBS) contributes to advanced ovarian cancer progression and drug resistance. PLoS One 8:e79167

Bohanon FJ, Mrazek AA, Porro JL, Spratt H, Hellmich MR, Chao C (2014) Aminooxyacetic acid in combination with oxaliplatin significantly decreases colorectal liver metastasis in vivo. J Am Coll Surg 219(Suppl):S132

De Vos J, Thykjaer T, Tarte K, Ensslen M, Raynaud P, Requirand G, Pellet F, Pantesco V, Rème T, Jourdan M, Rossi JF, Ørntoft T, Klein B (2002) Comparison of gene expression profiling between malignant and normal plasma cells with oligonucleotide arrays. Oncogene 21:6848–6857

Fan K, Lin N, Qi J, Yin P, Zhao C, Wang L, Li Z, Zha X (2014) Cell Signal 26(12):2801–2808

Guo H, Gai JW, Wang Y, Jin HF, Du JB, Jin J (2012) Characterization of hydrogen sulfide and its synthases, cystathionine β-synthase and cystathionine γ-lyase, in human prostatic tissue and cells. Urology 79:483.e1–483.e5

Hansel DE, Rahman A, Hidalgo M, Thuluvath PJ, Lillemoe KD, Shulick R, Ku JL, Park JG, Miyazaki K, Ashfaq R, Wistuba II, Varma R, Hawthorne L, Geradts J, Argani P, Maitra A (2003) Identification of novel cellular targets in biliary tract cancers using global gene expression technology. Am J Pathol 163:217–229

Hellmich MR, Coletta C, Chao C, Szabo C (2015) The therapeutic potential of cystathionine β-synthetase/hydrogen sulfide inhibition in cancer. Antioxid Redox Signal 87(2):142–149

Kimura H (2011) Hydrogen sulfide: its production, release and functions. Amino Acids 41:113–121

Kolluru GK, Shen X, Bir SC, Kevil CG (2013) Hydrogen sulfide chemical biology: pathophysiological roles and detection. Nitric Oxide 35:5–20

Módis K, Bos EM, Calzia E, van Goor H, Coletta C, Papapetropoulos A, Hellmich MR, Radermacher P, Bouillaud F, Szabo C (2014a) Regulation of mitochondrial bioenergetic function by hydrogen sulfide. Part II. Pathophysiological and therapeutic aspects. Br J Pharmacol 171:2123–2146

Módis K, Coletta C, Asimakopoulou A, Szczesny B, Chao C, Papapetropoulos A, Hellmich MR, Szabo C (2014b) Effect of S-adenosyl-L-methionine (SAM), an allosteric activator of cystathionine-β-synthase (CBS) on colorectal cancer cell proliferation and bioenergetics in vitro. Nitric Oxide 41:146–156

Mustafa AK, Gadalla MM, Sen N, Kim S, Mu W, Gazi SK, Barrow RK, Yang G, Wang R, Snyder SH (2009) H_2S signals through protein S-sulfhydration. Sci Signal 2:ra72

Panza E, De Cicco P, Armogida C, Scognamiglio G, Gigantino V, Botti G, Germano D, Napolitano M, Papapetropoulos A, Bucci M, Cirino G, Ianaro A (2015) Role of the cystathionine γ lyase/hydrogen sulfide pathway in human melanoma progression. Pigment Cell Melanoma Res 28(1):61–72

Szabo C (2007) Hydrogen sulphide and its therapeutic potential. Nat Rev Drug Discov 6:917–935

Szabo C, Hellmich MR (2013) Endogenously produced hydrogen sulfide supports tumor cell growth and proliferation. Cell Cycle 12:2915–2916

Szabo C, Papapetropoulos A (2011) Hydrogen sulphide and angiogenesis: mechanisms and applications. Br J Pharmacol 164:853–865

Szabo C, Coletta C, Chao C, Módis K, Szczesny B, Papapetropoulos A, Hellmich MR (2013) Tumor-derived hydrogen sulfide, produced by cystathionine-β-synthase, stimulates bioenergetics, cell proliferation, and angiogenesis in colon cancer. Proc Natl Acad Sci U S A 110:12474–12479

Szabo C, Ransy C, Módis K, Andriamihaja M, Murghes B, Coletta C, Olah G, Yanagi K, Bouillaud F (2014) Regulation of mitochondrial bioenergetic function by hydrogen sulfide. Part I. Biochemical and physiological mechanisms. Br J Pharmacol 171:2099–2122

Takano N, Sarfraz Y, Gilkes DM, Chaturvedi P, Xiang L, Suematsu M, Zagzag D, Semenza GL (2014) Decreased expression of cystathionine β-synthase promotes glioma tumorigenesis. Mol Cancer Res 12:1398–1406

Teng H, Wu B, Zhao K, Yang G, Wu L, Wang R (2013) Oxygen-sensitive mitochondrial accumulation of cystathionine β-synthase mediated by Lon protease. Proc Natl Acad Sci U S A 110:12679–12684

Wang R (2012) Physiological implications of hydrogen sulfide: a whiff exploration that blossomed. Physiol Rev 92:791–896

Wang MJ, Cai WJ, Zhu YC (2010) Mechanisms of angiogenesis: role of hydrogen sulphide. Clin Exp Pharmacol Physiol 37:764–771

Whiteman M, Le Trionnaire S, Chopra M, Fox B, Whatmore J (2011) Emerging role of hydrogen sulfide in health and disease: critical appraisal of biomarkers and pharmacological tools. Clin Sci (Lond) 121:459–488

Zhang W, Braun A, Bauman Z, Olteanu H, Madzelan P, Banerjee R (2005) Expression profiling of homocysteine junction enzymes in the NCI60 panel of human cancer cell lines. Cancer Res 65:1554–1560

Role of H_2S Donors in Cancer Biology

Zheng-Wei Lee and Lih-Wen Deng

Contents

Abstract

Hydrogen sulfide (H_2S) donors including organosulfur compounds (OSC), inorganic sulfide salts, and synthetic compounds are useful tools in studies to elucidate the effects of H_2S in cancer biology. Studies using such donors have shown the ability of H_2S to suppress tumor growth both in vitro and in vivo, with some of them suggesting the selectivity of its cytotoxic effects to cancer cells. In addition to promoting cancer cell death, H_2S donors were also found to inhibit cancer angiogenesis and metastasis. The underlying mechanisms for the anticancer activities of H_2S involve (1) cell signaling pathways, such as MAPK and STAT; (2) cell cycle regulation; (3) microRNAs regulation; and (4) cancer metabolism and pH regulation. Altogether, compiling evidences have demonstrated the great potential of using H_2S donors as anticancer agents. Nevertheless, the application and development of H_2S for therapy are still facing

Z.-W. Lee • L.-W. Deng (✉)
Department of Biochemistry, Yong Loo Lin School of Medicine, National University of Singapore, Singapore 117596, Singapore
e-mail: bchdlw@nus.edu.sg

P.K. Moore, M. Whiteman (eds.), *Chemistry, Biochemistry and Pharmacology of Hydrogen Sulfide*, Handbook of Experimental Pharmacology 230,
DOI 10.1007/978-3-319-18144-8_13

challenges as identification of molecular targets of H_2S awaits further investigation.

Keywords
Hydrogen sulfide donor • Organosulfur • GYY4137 • HS-NSAID • Cancer • Cell viability • Metastasis

Regarded as a hazardous gas with a pungent rotten egg smell, hydrogen sulfide (H_2S) has long been known to possess inhibitory effects on cellular mitochondrial activity and is toxic to central nervous and respiratory systems. Nevertheless, this gas molecule is found to be synthesized endogenously at least via three enzymatic pathways (cystathionine β-synthase, CBS; cystathionine γ-lyase, CSE; and 3-mercaptopyruvate sulfurtransferase, 3-MPST) in various mammalian tissues. Since then, there has been an upsurge of interest to study the role of H_2S in normal physiology and the pathophysiology of disease. H_2S has been reported to regulate various physiological processes including cardiovascular and neuronal functions as well as inflammatory responses. Yet, the roles of H_2S in cancer are less explored. This chapter gives an overview of the roles of H_2S in cancer biology, aiming to provide insights into utilizing H_2S donors as an anticancer agent.

1 Classification of H_2S Donors

The biological functions of H_2S have been mostly discovered through studies that made use of a variety of H_2S donors, either naturally existing or synthetically made. Classification of numerous H_2S donors and their respective bioactivities have been summarized in several reviews (Gu and Zhu 2011; Kashfi and Olson 2013; Song et al. 2014; Zhao et al. 2014b). Briefly, naturally occurring organosulfur compounds (OSCs) have been widely studied for their anticancer activities. Similarly, synthetic compounds such as anethole trithione (ADT-OH) derivatives and those that contain active P–S bonds, including Lawesson's reagent and a slow-release H_2S donor, GYY4137, have also been used in cancer studies (Table 1). This chapter will describe in depth the bioactivities of these H_2S donors, particularly their effects on cancer cell viability as well as cancer angiogenesis and metastasis (Table 2). The underlying mechanisms and the associated signaling pathways responsible for the actions of H_2S in cancer will be summarized (Fig. 1). Lastly, some perspectives and future challenges pertaining to employment of H_2S as an anticancer agent will be discussed.

Table 1 Major classes of H_2S donors

Category	Example of donor(s)
Organosulfur compounds (OSCs)	Allicin
	Ajoene
	Allyl sulfides
	Diallyl sulfide (DAS), diallyl disulfide (DADS), diallyl trisulfide (DATS)
	Isothiocyanate
	Erucin, iberin, sulforaphane
Inorganic	Sulfide salts
	Calcium sulfide (CaS)
	Sodium hydrosulfide (NaHS)
	Sodium sulfide (Na_2S)
Synthetic	Anethole trithione (ADT-OH) derivatives
	ACS 5, ACS 48, ACS50, etc.
	HS-NSAIDs: HS-sulindac, HS-ibuprofen, HS-naproxen, HS-aspirin
	Cysteine-activated
	N-(Benzoylthio)benzamide
	Cysteine analogs
	S-propyl cysteine, *S*-allyl cysteine, *N*-acetyl cysteine, *S*-propargyl cysteine
	Lawesson's reagent
	(*p*-Methoxyphenyl) morpholino-phosphinodithioic acid (GYY4137)
	AP39 (mitochondria targeted)

Table 2 Effects of H_2S donors on cancer cell viability

Model	Results	Signaling/mechanism(s)	References
Organosulfurs			
Aged garlic extracts (AGE)			
Human patients with colorectal adenomas	Suppressed size and number of colon adenomas after 12 months of 2.4 mL/day treatment		Tanaka et al. (2006)
S-allylmercapto-cysteine			
Prostate cancer (PC3) orthotropic tumor mice	Inhibited primary tumor growth, reduced lung and adrenal metastasis	Upregulated E-cadherin adhesion molecule	Howard et al. (2007)

(continued)

Table 2 (continued)

Model	Results	Signaling/mechanism(s)	References
Garlic oil			
Pancreatic cancer (AsPC-1, MiaPaCa-2, PANC-1)	Inhibited cell proliferation, induced cell cycle arrest, and apoptosis		Lan et al. (2013)
Fresh garlic extracts			
Breast cancer (MCF7)	Altered cellular morphology, decreased cell growth	Deregulated E-cadherin, cytokeratin 8/18, and beta-catenin; downregulated cyclin D1, decreased ERK1 phosphorylation, and increased eIF2-alpha phosphorylation	Modem et al. (2012)
Ajoene derivatives (p-methoxybenzyl-substituted end group)			
Esophageal cancer (WHCO1)	Inhibited cell growth, induced G2/M cell cycle arrest	Induced apoptosis by caspase 3 activation	Kaschula et al. (2012)
Allicin			
Gastric cancer (SGC-7901)	Inhibited cell proliferation, induced G2/M cell cycle arrest, and apoptosis		Tao et al. (2014)
Colon cancer (HCT116, LS174T, HT29, Caco2)	Decreased cell growth, induced hypodiploid DNA, and apoptosis	Decreased Bcl-2 and increased Bax expression, increased cytochrome c release from mitochondria, induced translocation of Nrf2 into nucleus	Bat-Chen et al. (2010)
L5178Y lymphoma-bearing mice	Inhibited cell proliferation and induced apoptosis, increased survival rate of tumor-bearing mice	Increased caspase-3 activity	Padilla-Camberos et al. (2010)
Tetra-arsenic tetra-sulfide (As_4S_4)			
Retinoid acid (RA)-resistant human acute promyelocytic leukemic NB4-R1	Induced apoptosis	Decreased SET gene expression led to increased protein phosphatase 2 (PP2A) and decreased PML-RARa level	Liu et al. (2014)

(continued)

Table 2 (continued)

Model	Results	Signaling/mechanism(s)	References
Isothiocyanate erucin (ER)			
Prostate cancer (PC3)	ER decreased cell growth and proliferation	ER increased p21 expression and ERK1/2 phosphorylation	Melchini et al. (2013)
Organosulfur compounds (OSCs)			
Breast carcinoma (MCF7, Vcr-R)	Induced G2/M cell cycle arrest	Diallyltetrasulfide and dipropyltetrasulfide emerged as irreversible inhibitors of Cdc25 isoforms A and C	Viry et al. (2011)
Garlic-derived organosulfurs			
Colon cancer (HCT116)	Diallyltrisulfide and diallyltetrasulfide treatment reduced cell viability, induced cell cycle arrest, and apoptosis	Induction of apoptosis dependent on redox state of the cell, dispensable of p53; growth arrest and apoptosis associated with reduction of the level of cdc25c	Busch et al. (2010)
Diallyl sulfide (DAS)			
Colon cancer (Colo 320)	Induced G2/M cell cycle arrest and apoptosis	Increased ROS, upregulated NF-κB expression, promoted caspase-3 expression, suppressed ERK-2 activity	Sriram et al. (2008)
Human cervical cancer (HeLa)	Induced G0/G1 cell cycle arrest and sub-G1 apoptosis phase; DNA damage and fragmentation	Induced cytochrome c release and increased expression of pro-caspase-3 and pro-caspase-9	Wu et al. (2011)
Diallyl disulfide (DADS)			
Colon cancer (HCT116)	Induced G2/M cell cycle arrest and apoptosis	Increased ROS, increased expression level of cyclin B1	Song et al. (2009)
Neuroblastoma (SH-SY5Y)	Induced morphological changes and depolymerization of microtubules	Induced cytoskeleton oxidation, microfilaments reduction, and Tau dephosphorylation	Aquilano et al. (2010)

(continued)

Table 2 (continued)

Model	Results	Signaling/mechanism(s)	References
Colon cancer (HCT116)	Induced cell growth inhibition and G2/M arrest	Increased cyclin B1 expression but not affected by p53 silencing	Jo et al. (2008)
Prostate cancer (LNCaP)	Induced apoptosis and nuclear fragmentation		Gunadharini et al. (2006)
Lung carcinoma (A549)	DADS-induced G2/M cell cycle arrest and apoptosis	Increased ROS	Wu et al. (2005)
Colon cancer (HCT116, HCT15) and prostate cancer (PC3)	Inhibited cell proliferation	Induced p53 and nonsteroidal anti-inflammatory drug (NSAID)-activated gene (NAG1)	Bottone et al. (2002)
Gastric cancer (MGC-803)	Antitumor effect in vitro and in vivo	Upregulated miR-200b and miR-22 that target Wnt-1	Tang et al. (2013)
Gastric cancer (BGC823)	Inhibited cell growth, induced G2/M cell cycle arrest	Decreased Cdc25c and cyclin B1, increased p21, GADD45-alpha and p53; decreased Chk1 phosphorylation	Bo et al. (2014)
Leukemia (HL60)	Inhibited cell growth, induced apoptosis	Activated p38, reduced ERK activity	Tan et al. (2008)
Gastric cancer (MGC-803)	Inhibited proliferation, induced G2/M cell cycle arrest	Induced differentiation and decreased ERK1/2 phosphorylation	Ling et al. (2006)
Gastric cancer (MGC-803)	Inhibited cell growth, induced G2/M cell cycle arrest	Increased p38 phosphorylation, decreased Cdc25c expression	Yuan et al. (2004)
Colon cancer (HCT15)	Induced G2/M cell cycle arrest	Increased ERK phosphorylation	Knowles and Milner (2003)
Diallyl trisulfide (DATS)			
Osteosarcoma (U2OS, SaOS-2, MG-63)	Inhibited proliferation by triggering cell cycle arrest and apoptosis; suppressed wound-healing capacity, invasion and cancer media-induced HUVEC tube formation	Decreased Notch-1 expression; increased tumor-suppressive miR-34a, miR-143, miR-145, and miR-200b/c	Li et al. (2013)

(continued)

Table 2 (continued)

Model	Results	Signaling/mechanism(s)	References
Benzo(a)pyrene (BaP)-induced precancerous MCF10A	Inhibited cell proliferation, decreased G2/M cell cycle transition	Reduced BaP-induced peroxide formation and DNA strand breaks, thus preventing BaP-induced carcinogenesis	Nkrumah-Elie et al. (2012)
Basal cell carcinoma (BCC) and melanoma (A375)	Induced G2/M cell cycle arrest and apoptosis	Increased ROS, induced cytosolic Ca^{2+} mobilization, decreased mitochondrial membrane potential	Wang et al. (2010)
Prostate cancer (LNCaP, C4-2, TRAMP-C1)	Inhibited cell proliferation	Decreased androgen receptor (AR) protein level and AR promoter activity	Stan and Singh (2009)
Transgenic adenocarcinoma of mouse prostate (TRAMP)	Inhibited progression to invasive carcinoma and pulmonary metastasis	Induced cyclin B1 and securin protein expression, suppressed protein expression of neuroendocrine marker synaptophysin	Singh et al. (2008)
Prostate cancer (PC3)	Oral gavage DATS retarded growth of PC3 xenografts	Induction of proapoptotic Bax and Bak, decreased VEGFR2 protein level	Xiao et al. (2006)
Gastric cancer (MGC803, SGC7901)	Decreased G0/G1 and increased G2/M phase cells	Upregulated p21 gene expression level	Ha et al. (2005)
Colon cancer (HCT15 and DLD-1)	Decreased G0/G1 and increased G2/M phase cells	Induced disruption of tubulin polymerization	Hosono et al. (2005)
Prostate cancer (PC3 and DU145)	Induced G2/M cell cycle arrest	Inhibited cyclin-dependent kinase 1 and hyperphosphorylation of Cdc25c	Herman-Antosiewicz and Singh (2005)
Inorganic and synthetic donors			
NaHS			
Prostate cancer (LNCaP)	Decreased proliferation of LNCaP and antiandrogen-resistant LNCaP-B cells	H_2S sulfhydrated androgen receptor, inhibiting AR binding to DNA and transactivation of its downstream genes	Zhao et al. (2014a)

(continued)

Table 2 (continued)

Model	Results	Signaling/mechanism(s)	References
TGF-β1-treated MCF7	Decreased proliferation, G0/G1 phase cell cycle arrest, and apoptosis	Decreased TGF-β1-induced p38 phosphorylation	Lv et al. (2014)
Colon epithelial (YAMC) and colon cancer (HT-29, SW1116, HCT116)	Inhibited proliferation, migration, and G1/S transition; increased LC3B autophagic vacuoles and acidic vesicular organelles	Increased AMPK phosphorylation and decreased mTOR and S6 kinase phosphorylation; induced p21 expression	Wu et al. (2012)
Systemic lupus erythematosus lymphocytes	Inhibited proliferation and S phase distribution of cell cycle	Decreased phytohemagglutinin (PHA)-induced CDK2 expression, AKT (S473) and GSK3b (S9) phosphorylation, increased p27 and p21 expression	Han et al. (2013)
Oral squamous carcinoma (Cal27, GNM, WSU-HN6)	Induced proliferation and accelerated cell cycle progression of synchronized cells	Increased Akt and ERK1/2 phosphorylation	Ma et al. (2014)
Colon cancer (HCT116, SW480)	Promoted cell proliferation, decreased G0/G1 phase, and increased S phase cells	Increased Akt and ERK phosphorylation; inhibited expression of p21, and reduced NO metabolite	Cai et al. (2010)
GYY4137			
Hepatocellular carcinoma (HepG2, Bel7402)	Suppressed cell proliferation and tumor growth, inhibited G1/S cell cycle transition	Decreased Stat3 (Y705) activation; altered Bcl-2, cyclin D1, Mcl-1, survivin, VEGF, and HIF-1a expression; increased cleaved caspase-9, caspase-3, and PARP	Lu et al. (2014)
HeLa, HCT116, HepG2, HL60, MCF7, MV4-11, U2OS	GYY4137 but not NaHS decreased cancer cell survival; noncancer cells were less affected; reduced xenograft tumor growth	Induced apoptosis and partial G2/M cell cycle arrest	Lee et al. (2011)

(continued)

Table 2 (continued)

Model	Results	Signaling/mechanism(s)	References
Breast cancer (MCF7), liver cancer (HepG2)	Continual exposure but not single exposure to low concentration of NaHS decreased MCF7 and HepG2 viability; noncancer MCF10A and WI38 were less affected	Induced increased lactate production and impaired pH regulation; caused intracellular acidification	Lee et al. (2014)
S-propargyl-cysteine (SPRC)			
Gastric cancer (SGC-7901)	Decreased cell viability and suppressed proliferation and migration caused G1/S cell cycle arrest; reduced tumor weight and volumes	Increased p53 and Bax expression; increased CSE expression and tumoral CSE activity	Ma et al. (2011)
HS-NSAIDs			
Human colon, breast, pancreatic, prostate, and lung cancer and leukemia	Inhibited cell proliferation, caused G0/G1 cell cycle block	Induction of apoptosis	Chattopadhyay et al. (2012b)
HS-aspirin (HS-ASA)			
Leukemia (Jurkat)	Inhibited proliferation, induced apoptosis, G0/G1 cell cycle arrest	Downregulated B-catenin, cyclin D1, and c-Myc	Chattopadhyay et al. (2013)
Triple negative breast cancer (MDA-MB-231)	Suppressed cell growth, induced G0/G1 cell cycle arrest and apoptosis; reduced xenograft tumor volume and mass	Downregulated NF-κB, reduced tioredoxin reductase activity and increased ROS	Chattopadhyay et al. (2012a)
NOSH-aspirin (NBS-1120)			
Colon cancer (HT29)	Inhibited cell proliferation, induced apoptosis and G0/G1 cell cycle block; xenograft tumor volume reduction of 85 %		Chattopadhyay et al. (2012c)
NOSH-naproxen (AVT-219) and NOSH-sulindac (AVT-18A)			
Colon (HT-29), pancreas adenomatous (BxPC-3), and breast cancer (MCF7) and lymphocytic leukemia (Jurkat)	Inhibited proliferation, G0/G1 or G2/M block in cell cycle		Kodela et al. (2013)

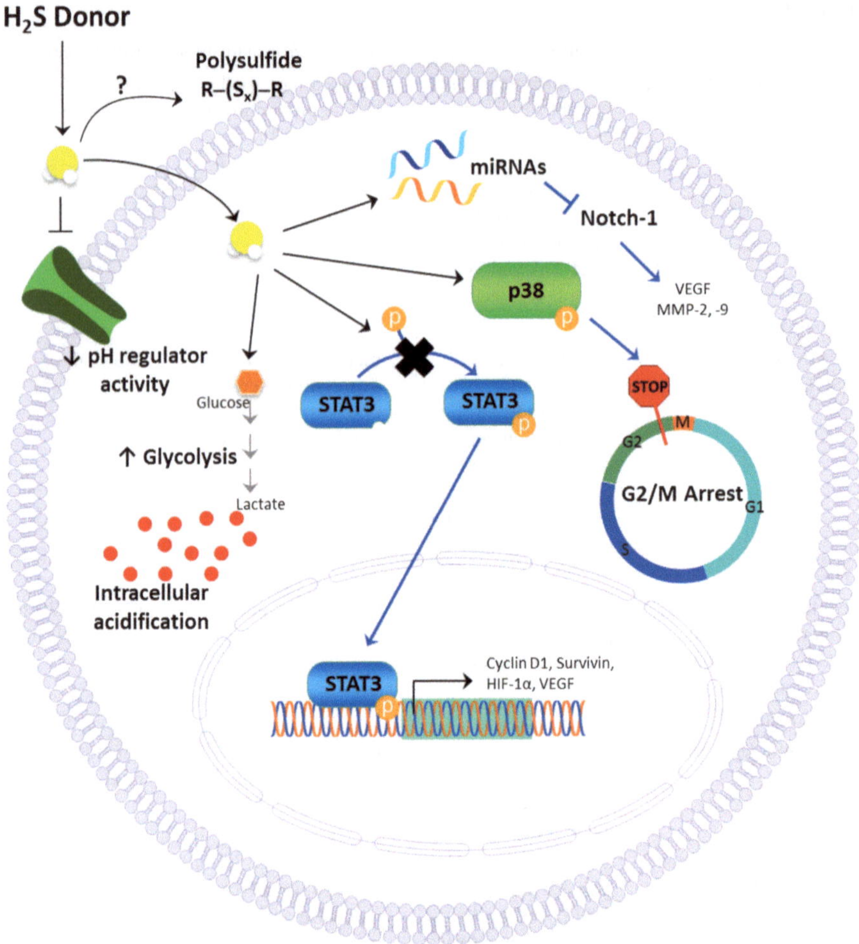

Fig. 1 H_2S-mediated pathways in cancer cells. H_2S potentially decreases STAT activation, increases p38 phosphorylation, induces tumor-suppressive miRNAs, impairs pH regulator activity, and increases glycolysis resulting in intracellular acidification in cancer cells

2 Effects of H_2S Donors on Cancer Cell Viability

2.1 Organosulfur Compounds (OSCs)

OSCs including allicin and ajoenes that are abundantly found in *Allium* vegetables, especially garlic, have been recognized to confer health benefits throughout recorded human history, dating back since ancient Egypt, Chinese, and Hindu civilizations (Block 1985). Consumption of garlic and other *Allium* vegetables is believe to reduce serum cholesterol and blood glucose levels (Sher et al. 2012) and to stimulate immune responses against microbial, viral, and pathogen infection

(Feng et al. 2012; Lau et al. 1991). In the mid-twentieth century, garlic extracts were shown to effectively inhibit growth of sarcoma 180 ascites tumor in CFW Swiss mice (Weisberger and Pensky 1957). Since then, increasing evidences for anticancer effects of OSCs from population-based case–control studies have suggested that greater intake of *Allium* vegetables reduces risks of developing cancer. For instance, a study carried out in Chinese population revealed that those who have the highest garlic intake experienced nearly 60 % reduction in gastric cancer occurrence as compared to those who consumed the least (You et al. 1989). In another double-blind, randomized clinical trial, Tanaka et al. reported that high dose treatment of aged garlic extracts significantly suppressed the progression of colorectal adenomas in patients (Tanaka et al. 2004, 2006). The effects of garlic or garlic oil in inhibiting cancer growth are also evidenced in several other cancer types including androgen-independent prostate cancer, pancreatic cancer, liver cancer, and breast cancer (Howard et al. 2007; Lan et al. 2013; Modem et al. 2012; Zhang et al. 2013).

Garlic is a complex mixture of different OSCs. The major forms of OSCs present in intact garlic are γ-glutamyl-*S*-alk(en)yl-L-cysteines and *S*-alk(en)yl-L-cysteine sulfoxides. Upon exposure to atmospheric water and oxygen, the cysteine sulfoxides rapidly get hydrolyzed and oxidized into several hundreds of thiosulfinates, of which 70–80 % is allicin. However, allicin is highly unstable and susceptible to further oxidation. In a rat liver perfusion model, 90 % of the allicin disappears from the tissue within the first 3 min while less than 1 % remains after 6 min. Decomposition of allicin produces the oil-soluble volatiles such as diallyl sulfides (DAS), diallyl disulfides (DADS), diallyl trisulfides (DATS), methyl allyl sulfides, and ajoene (Block 1985; Fenwick and Hanley 1985). Allicin and these volatile OSCs are the compounds that are responsible for the unpleasant odor of garlic.

Given the complexity of garlic chemistry, the observed biological activities mentioned above might be contributed by either one or collective effects of the OSC mixture. Therefore, to identify and characterize the OSC that is responsible for the anticancer activity, researchers have purified and isolated these individual compounds. Intriguingly, compiling evidences suggest that most of these OSCs possess anticancer activities even when they are singly administered. Allicin significantly inhibits cell proliferation and induces apoptosis in gastric cancer cell line SGC-7901 (Tao et al. 2014). Likewise, allicin effectively decreases cell proliferation of several colon cancer cell lines, including HCT-116, LS174T, HT-29, and Caco-2 (Bat-Chen et al. 2010). The inhibitory effect of allicin on cancer growth has also been demonstrated in an in vivo mouse model. Using L5178Y lymphoma-bearing mice, Padilla-Camberos and colleagues showed that administration of allicin significantly inhibited tumor growth. Additionally, normal splenocytes were able to tolerate with at least two times higher than the doses used, suggesting the low toxicity of allicin (Padilla-Camberos et al. 2010).

As mentioned previously, the half-life of allicin is very short. Therefore, the observed anticancer growth effects are unlikely due to the activity of allicin alone. Allyl sulfides produced from oxidative decomposition of allicin are comparably stable. In a study to evaluate the reactivity of different allyl sulfides in altering erythrocytes oxidative balance, tetrasulfides were found to be the most reactive species. The reactivity decreased according to the order of tetra- > tri- > disulfides, while monosulfides exhibited no significant effect on the examining parameters (Munday et al. 2003). This suggests that the reactivity of the OSC is directly related to the number of sulfur atom present in the molecule. In fact, most of the studies used DADS or DATS rather than DAS.

DADS treatment alone was reported to be effective in inhibiting cancer cell growth, inducing apoptosis and causing partial G2/M cell cycle arrest in cancers of colon, pancreas, prostate, lung, brain, and stomach (Aquilano et al. 2010; Bo et al. 2014; Bottone et al. 2002; Gunadharini et al. 2006; Song et al. 2009; Wu et al. 2005). In particular, the anticancer proliferation effects were reported to be associated with an increase in reactive oxygen species (ROS) production. For instance, DADS was shown to induce cytoskeleton oxidation and microtubule depolymerization, thence morphological changes in neuroblastoma SH-SY5Y cells (Aquilano et al. 2010). However, these effects were reversed by administration of exogenous precursors of cellular antioxidants, *N*-acetylcysteine (NAC) or glutathione ester.

With an extra sulfur atom, DATS is thought to be more reactive than DADS. Nkrumah-Elie and colleagues reported that DATS effectively inhibited benzo(a) pyrene (BaP)-induced carcinogenesis in noncancer breast epithelial MCF10A (Nkrumah-Elie et al. 2012). BaP is a potent mutagen and carcinogen found in coal tar and cigarettes (Kaiserman and Rickert 1992). The author reported that either co-treatment or pretreatment of DATS on BaP-treated MCF10A significantly reduced cell proliferation. Furthermore, DATS decreased BaP-induced peroxide formation and DNA strand breaks, suggesting that DATS might be effective in preventing BaP-induced genomic instability and carcinogenesis. Concomitantly, anticancer activity of DATS has also been shown in other cancers, including colon, gastric, and prostate cancers (Ha et al. 2005; Herman-Antosiewicz and Singh 2005; Hosono et al. 2005). Not surprisingly, the involvement of ROS in DATS-mediated anticancer effects was evidenced by Busch's study that the apoptosis induced by DATS was able to be reversed by ascorbate or NAC (Busch et al. 2010). Moreover, DATS might selectively target cancer cells, as shown by Wang and colleagues that 25–100 μM of DATS significantly decreased viability of basal cell carcinoma BCC and melanoma A375 cells but had minimal effects on human keratinocytes HaCaT (Wang et al. 2010). Collectively, OSC (at least for DADS and DATS) exerts anticancer activity in cancers of various tissue origins.

Despite mounting studies reporting the anticancer activity of OSC or garlic derivatives, the contribution of H_2S remains elusive. Many other active side groups or moieties in the OSC molecules might play a part in the anticancer effects. By comparing the effects of DAS, DADS, DATS, diproylsulfide, dipropyl disulfide, and allyl mercaptan on the reduction of androgen receptor expression in prostate cancer cells, Stan et al. revealed that a subtle change in the structure and number of sulfur atom of OSC greatly affects their anticancer effects (Stan and Singh 2009). DATS with the three sulfur atoms flanking with allyl chains appeared to be the most effective OSC. The importance of these allyl side chains in determining OSC activities was evident as the allyl side chain produced from the breakage of S–S bond in DATS was able to oxidize the cysteine residue of tubulin protein, resulting in the disruption of tubulin polymerization thence causing cell cycle arrest in colon cancer cell HCT15 (Hosono et al. 2005). Nevertheless, diallyl polysulfides present in garlic can also react with cysteine or glutathione to produce H_2S in human body, suggesting that H_2S might also be involved in the abovementioned bioactivities (Benavides et al. 2007; Jacob 2006).

2.2 Inorganic and Synthetic H_2S Donors

Inorganic salts of sulfide such as sodium hydrosulfide (NaHS) and sodium sulfide (Na_2S) are the simplest H_2S donors being utilized. In contrast to OSC or garlic derivatives, NaHS or Na_2S dissolves instantaneously in aqueous solution and produces only sulfide species. This simple chemistry allows one to easily correlate the observed bioactivities with H_2S. NaHS was shown to decrease breast cancer MCF7 and prostate cancer LNCaP cell proliferation, coupled with an increase in apoptosis (Lv et al. 2014; Zhao et al. 2014a). In contrast, NaHS stimulated proliferation and accelerated cell cycle progression of synchronized oral squamous carcinoma Cal27, GNM, and WSU-HN6 cell lines (Ma et al. 2014). This ambiguous role of NaHS has even been observed in the same colon cancer cell line, HCT116. Cai et al. reported that 200 μM of NaHS treatment induced HCT116 cell proliferation and promoted S phase progression, while Wu et al. observed that 1 mM NaHS inhibited cell proliferation and migration (Cai et al. 2010; Wu et al. 2012). It appears that cancer cells could tolerate H_2S derived from low micromolar concentrations of NaHS and get stimulated to proliferate faster. However, at a high concentration of NaHS exposure, pro-survival signaling was suppressed and cell death was triggered. In line with this hypothesis, HCT116 exposed to low micromolar concentration of NaHS increases Akt and ERK phosphorylation stimulating cell survival, while millimolar range of NaHS decreases mTOR and S6 kinase activity, leading to autophagy and apoptosis (Cai et al. 2010; Wu et al. 2012). These disparate results suggest that the anti- or pro-cancer effect of NaHS appears to depend on the concentration and duration of exposure.

One concern of utilizing sulfide salts is that it releases massive amounts of the gas almost instantaneously, likely leading the effect of the gas on cells to be acute and toxic. In addition, the amount released by NaHS or Na_2S does not mimic physiological levels. In view of this, researchers have been working on chemically synthesizing H_2S donor drugs with the ability to release H_2S at different rates. Synthetic H_2S donors such as (*p*-methoxyphenyl)morpholino-phosphinodithioic acid (GYY4137) were synthesized in order to achieve a prolonged and consistent H_2S release in culture as compared to the sulfide salts (Li et al. 2008). In a study done by Lee and colleagues, a wide range of cell lines were treated with either GYY4137 or NaHS (400 μM), and their viability was examined after 5 days of incubation (Lee et al. 2011). GYY4137 was able to reduce cell viability in various cancer cell lines including HeLa, HCT116, HepG2, HL60, MCF7, MV4-11, and U2OS but spare noncancer cell lines, IMR90 and WI38. In contrast, NaHS marginally affected the cancer cell viability. The differential effects of GYY4137 and NaHS are speculated to be resulted from their differential H_2S-release manner. As opposed to NaHS which releases H_2S instantaneously, GYY4137 slowly releases H_2S and sustains over days in culture. This observation proposed that a prolonged and low exposure to H_2S might be the key to selective anticancer activity of H_2S. In a follow-up study, the authors demonstrated that the viability of cancer cells, exposed to low concentration (5–20 μM) of fresh NaHS solution every 2 h over 5 days, was significantly decreased (Lee et al. 2014). On the other hand, noncancer cells tolerated the same H_2S treatment regime. It is also worth mentioning that a high and continual H_2S exposure (>80 μM) killed both cancer and noncancer cells, likely due to the long known inhibitory effects of H_2S on mitochondrial activity (Savolainen 2010).

Besides GYY4137, a group of synthetic H_2S-releasing hybrid compounds was also shown to exhibit anticancer effects. Among these, nonsteroidal anti-inflammatory drugs (NSAIDs) appear to be good candidates. NSAIDs have been recorded to exhibit a certain extent of anticancer activity. Patients undergoing aspirin medication had significantly lower colorectal cancer incidence, corrected for age and gender (Kune et al. 1988). Analysis of data generated from 51 randomized trials also revealed that daily aspirin intake reduces cancer-associated death and incidence (Rothwell et al. 2012). However, long-term medication on NSAIDs may cause significant side effects, mainly gastrointestinal and renal damages (Harirforoosh and Jamali 2009; Sostres et al. 2013). Therefore, owing to the potential cytoprotective effects of H_2S (reviewed in other chapters), new hybrids of HS-NSAIDs were designed (Kashfi 2014).

Four HS-NSAIDs including HS-sulindac, HS-ibuprofen, HS-naproxen, and HS-aspirin (HS-ASA) were used to study their activities in 11 cancer cell lines derived from six tissue origins, including colon (HT29, HCT15, SW480), breast (MCF7, MDA-MB-231, SKBR3), leukemia (Jurkat), pancreas (BxPC3, MIAPaCa-2), prostate (LNCaP), and lung (A549) (Chattopadhyay et al. 2012b). All four HS-NSAIDs are effective in inhibiting the growth of the 11 cancer cell lines, whereas their respective NSAID treatment alone requires 38–1300 times higher concentrations to achieve the same extent of inhibitory effect. Among the four HS-NSAIDs, HS-ASA is the most potent hybrid even though ASA is the least potent NSAID in anticancer cell growth. The author speculates that the conjugation of ASA with HS moiety forms a hybrid which releases H_2S over a long period of time. Taken together, the authors conclude that HS-NSAID, as a H_2S donor, is a potential anticancer agent independent of cancer tissue types.

3 Underlying Mechanisms of H_2S Donors in Suppressing Cancer Cell Growth

Mitogen-Activated Protein Kinase (MAPK) Signaling and Cell Cycle Regulation

MAPK signaling plays a central role in regulating cell proliferation. The specific activation and deactivation of MAPK in response to H_2S donors have been documented. For instance, p38 was found to be activated but ERK activity was reduced in DADS-induced apoptotic HL60 cells (Tan et al. 2008). Antagonists against p38 or ERK are able to decrease or enhance toxicity and apoptosis induced by DADS, respectively, suggesting direct involvement of these two MAPKs in mediating DADS anticancer activity. Consistently, DADS induces differentiation or G_2/M arrest in gastric cancer MGC803 cell line, and such activity is mediated by an increase in p38 activation and a decrease in phosphorylation of ERK (Ling et al. 2006; Yuan et al. 2004). However, in colon cancer HCT15 model, an elevated level of ERK phosphorylation has been shown to be an early signaling event that led to G_2/M arrest induced by DADS (Knowles and Milner 2003). All these studies converge to the fact that anticancer activity of DADS is p38 and more importantly ERK mediated.

STAT Pathway

Signal transducer and activator of transcription (STAT) pathway is implicated in many pathophysiologies, including cancer. Specifically, constitutive activation of STAT3 was reported to be involved in maintaining progression of many malignancies and solid tumors (Siveen et al. 2014). Therefore, inhibition of STAT3 pathway in cancer potentially serves as a therapeutic strategy (Santos and Costa-Pereira 2011). It has been reported that GYY4137 suppressed hepatocellular carcinoma tumor growth via inhibition of STAT3 pathway both in vitro and in vivo (Lu et al. 2014). GYY4137 significantly decreased phospho-STAT3 (Y705) level in HepG2 and Bel7402 cells, leading to decreased expression of its downstream genes, including Bcl-2, cyclin D1, Mcl-1, survivin, vascular endothelial growth factor (VEGF), and hypoxia-inducible factor-1α (HIF-1α). Altered levels of these molecules thence contributed to the inhibitory effect of GYY4137 on hepatocellular carcinoma progression. Having known that STAT3 is widely activated in various types of cancer, the potential of using GYY4137 to deactivate STAT3 in other cancers is worth exploring.

miRNA Regulation

Several studies have reported that miRNA is involved in the bioactivity of H_2S donors. miR-200b and miR-22 were found to be upregulated in DADS-treated MGC803 gastric cancer cells (Tang et al. 2013). The authors further identified Wnt1 to be a target of these two miRNAs and reported a synergistic effect between miR-200b and miR-22 in the inhibition of gastric cancer growth in vitro and in vivo. Another study showed that DATS treatment on osteosarcoma U2OS, SaOS-2, and MG-63 cell lines increased expression of several tumor-suppressive miRNA such as miR-34a, miR-143, miR-145, and miR-200b/c (Li et al. 2013). Re-expression of miR-34a into the osteosarcoma cells resulted in the attenuation of Notch-1 activity and downregulation of its downstream genes such as VEGF and matrix metalloproteinase (MMP)-2 and matrix metalloproteinase (MMP)-9. Moreover, miR-200b/c was found to be downregulated in osteosarcoma. miR-200b/c is known to suppress cancer epithelial to mesenchymal transition. DATS treatment successfully re-expressed these miRNAs that in turn inhibited osteosarcoma growth. The role of miRNA in modulating anticancer activity of H_2S donors requires more future studies.

Cancer Metabolism and pH Regulation

A recent report by Lee and colleagues showed that a continual and low exposure to H_2S, either by one-time administration of GYY4137 or NaHS replaced every 2 h for over 5 days, significantly increased glycolysis causing overproduction of metabolic acid lactate (Lee et al. 2014). Simultaneously, such low and prolonged exposure of H_2S impaired the ability of cancer cells to regulate their own intracellular pH by inhibiting anion exchanger and sodium/proton exchanger activity. Combinatorial effects of increased glycolysis and defective pH regulation cause accumulation of metabolic acid, resulting in an uncontrolled intracellular acidification leading to cancer cell death (Lee et al. 2014). Neither increased metabolic acid production nor inhibition of pH regulator alone is sufficient to induce cancer cell death. Importantly, these mechanisms do not operate in normal cells, and hence, H_2S has the capability to kill cancer cells but spare normal cells. This study suggests a possibility of developing novel selective anticancer drugs based on the principle of H_2S donation and their effect on a fundamental difference in the energy-producing mechanisms of cancer, vis-a-vis, normal cells.

4 Effects of H_2S Donors on Cancer Angiogenesis and Metastasis

The vasodilatory role of H_2S has been well documented and so has the modulatory effects of different H_2S donors on endothelial cells (reviewed in other chapters). However, the effects of H_2S donors specifically on cancer angiogenesis and metastasis have been poorly studied. Nonetheless, a study by Thejass and Kuttan (2007) provided evidences that DAS exhibited anticancer angiogenesis property. B16F-10 melanoma cells were injected intradermally on the ventral skin of mouse and were treated with DAS for 5 consecutive days. After 9 days of tumor induction, the mice were sacrificed and ventral skin was dissected for quantification of tumor-directed capillaries. The DAS treatment reduced almost 67 % of the number of tumor-directed capillaries as compared to the control. The anticancer angiogenic effect of DAS was also evidenced using ex vivo rat aortic ring assay. Conditioned medium collected from DAS-pre-treated B16F-10 melanoma cells resulted in a more profound reduction of microvessel outgrowth as compared to the non-treated conditioned medium. This may imply that DAS treatment alters the release of cancer-associated angiogenic factors. One important factor identified in the study was VEGF as DAS treatment significantly decreased mRNA expression of VEGF in the melanoma cells.

With respect to cancer metastasis, NaHS treatment significantly reduced migration of transforming growth factor-β1 (TGF-β1)-stimulated MCF7 in Boyden chamber assay (Lv et al. 2014). TGF-β1 potentiates epithelial-mesenchymal transition (EMT), a cellular process that allows cancer cells to acquire higher motility and invasiveness. The ability of NaHS to suppress TGF-β1-induced cancer migration suggests that H_2S may be an anti-EMT agent. Inhibition of cancer invasion by NaHS was then identified partly due to the downregulation of Snail protein and reduction of p38 phosphorylation. Besides, OSCs including DAS, DADS, and DATS were shown to inhibit migration and invasion of colon cancer Colo 205 (Lai et al. 2013). The number of OSC-treated cells which migrated into the lower part of the Boyden chamber was nearly threefold lesser than that of non-treated cells. The inhibitory effect was not only concentration dependent, but also enhanced with the number of sulfur atom of OSC. In the same study, the author reported that these OSCs significantly decreased expression level of proteins in MAPK pathway leading to inhibition of MMP-2, MMP-7, and MMP-9 activities. MMPs are proteinases expressed by cancer cells in order to degrade extracellular matrix, thereby facilitating their extravasion and migration to distant sites (Deryugina and Quigley 2006; Itoh and Nagase 2002). Downregulation of MMP by OSC underlies the antimetastatic property of the H_2S donor. In an in vivo transgenic adenocarcinoma of mouse prostate (TRAMP) mice model, DATS administration significantly suppressed incidence and pulmonary metastatic event of poorly differentiated prostate cancer (Singh et al. 2008). Dorsolateral prostate of TRAMP mice given DATS were harvested, and the percentage of the area occupied by poorly differentiated prostate cancer cells was quantitated. 1 or 2 mg treatment of DATS, thrice weekly, lowered the area of poorly differentiated prostate cancer by 34 and 41 %, respectively, as compared to non-treated control. Poorly differentiated cancer resembles high-grade cancer with high metastatic event. The inhibitory activity of DATS on prostate cancer development and metastasis is encouraging and more work is needed to further explore the anti-angiogenic and antimetastatic activities of other H_2S donors using various in vivo and in vitro cancer models.

5 Therapeutic Potential and Challenges of Utilizing H_2S Donors as Anticancer Agents

Organosulfur compounds (OSCs), GYY4137 and HS-NSAID have been shown by many independent groups to suppress the proliferation of various cancer cell lines but exhibit minimal effects on the noncancer cells. The lower toxicity of H_2S donors, as compared to other cytotoxic agents on noncancer cells, is of a great value. Wang's group showed that DATS was tolerated by normal keratinocytes, but it inhibited basal cell carcinoma and melanoma cell growth (Wang et al. 2010). Similarly, allicin spared normal splenocytes at more than double the dosages used to kill lymphoma cells (Padilla-Camberos et al. 2010). The slow-releasing H_2S donor, GYY4137, was also shown to be selective in inhibiting proliferation of various cancer cells lines including colon, liver, breast, and cervical cancers and leukemia (Lee et al. 2011). In vivo studies have demonstrated that these compounds were effective in reducing tumor volume and mass. At whole animal level, although not measured objectively, the authors reported that no detrimental effects on animal physiology were observed after H_2S donor treatment. These evidences clearly show that low toxicity chemotherapy is possibly achieved using H_2S donors. Furthermore, several studies observed increased ROS production upon OSC treatment, and some clearly showed that ROS was required for the anticancer activity. Therefore, further examinations are needed to evaluate if the increase in ROS can exert detrimental effects to normal physiology.

One of the major challenges in developing H_2S donors as an anticancer agent is the identification of its molecular targets. Being a lipophilic gas, H_2S is cell permeable; therefore, it is difficult to estimate its distribution and to identify the exact site of its actions. Nevertheless, using octanol and hexane organic solvents as membrane bilayer mimics, Cuevasanta et al. demonstrated that H_2S is highly soluble at hydrophobic core and therefore likely to exhibit greater activity within lipid membrane layers (Cuevasanta et al. 2012). Given the dissociation constant of H_2S, about two-third of H_2S exists as HS^- at physiological pH. One possible modulation of protein structure and activity by H_2S is via a posttranslational modification, sulfhydration, on the cysteine residue (Mustafa et al. 2009). HS^- can covalently link with sulfur atom of cysteine amino acid, forming S–SH bond. Another mechanism in which H_2S may modulate protein functions is via breaking of disulfide linkage between two cysteine residues. Tao et al. reported that Cys1045–1024 disulfide bond of VEGFR2 was labile and susceptible to H_2S-mediated reduction to form S–SH modification (Tao et al. 2013). Sulfhydration has also been detected in glyceraldehyde 3-phosphate dehydrogenase (GAPDH) and actin and was shown to alter the protein activity (Paul and Snyder 2012). Based on these findings, it is reasonable to speculate that the effects of H_2S on cancer cell glycolysis and pH can arise from its sulfhydration on membrane-bound proteins, such as the glucose transporters and pH regulators (Lee et al. 2014).

As mentioned previously, OSC in polysulfide form could react with reduced thiols present in the cells to produce H_2S (Benavides et al. 2007). However, given the high reactivity of these OSCs, and competition by other reducing agents in the cells, H_2S production from OSC is subtle. More importantly, sustainability of H_2S-release from these compounds is a major concern as the key to selective anticancer activity of H_2S appears to result from a low and continual exposure to H_2S (Lee et al. 2014). In this context, synthetic H_2S donors such as GYY4137 or HS-NSAID, which can slowly release H_2S over long period of time, would be useful.

Whether H_2S is the sulfur species that gives rise to the observed physiological responses remains debatable. Recently, Greiner et al. reported that H_2S being a sulfur-containing molecule with the lowest oxidation state is unlikely to be the active species that can lead to oxidative thiol modification on protein cysteine residue (Greiner et al. 2013). They analyzed the effect of protein thiol modification by several different H_2S donors and found the formation of polysulfide in aqueous solution, regardless of the donors. The formed polysulfide then readily modified protein inside the cells. Kimura also reported that H_2S-derived polysulfide was able to activate transient receptor potential ankyrin-1 (TRPA1) channels in rat astrocytes (Kimura et al. 2013). In this study, polysulfide was found to be more potent than its parental H_2S. Taken together, these studies suggest that polysulfide may be the molecule that mediates sulfide signaling. Hence, identification of the sulfide-signaling molecule may greatly enhance our understanding and knowledge in using H_2S donors in cancer therapy.

References

Aquilano K, Vigilanza P, Filomeni G, Rotilio G, Ciriolo MR (2010) Tau dephosphorylation and microfilaments disruption are upstream events of the anti-proliferative effects of DADS in SH-SY5Y cells. J Cell Mol Med 14:564–577

Bat-Chen W, Golan T, Peri I, Ludmer Z, Schwartz B (2010) Allicin purified from fresh garlic cloves induces apoptosis in colon cancer cells via Nrf2. Nutr Cancer 62:947–957

Benavides GA, Squadrito GL, Mills RW, Patel HD, Isbell TS, Patel RP, Darley-Usmar VM, Doeller JE, Kraus DW (2007) Hydrogen sulfide mediates the vasoactivity of garlic. Proc Natl Acad Sci 104:17977–17982

Block E (1985) The chemistry of garlic and onions. Sci Am 252:114–119

Bo S, Hui H, Li W, Hui L, Hong X, Lin D, Dai WX, Wu YH, Ai XH, Hao J, Qi S (2014) Chk1, but not Chk2, is responsible for G2/M phase arrest induced by diallyl disulfide in human gastric cancer BGC823 cells. Food Chem Toxicol 68:61–70

Bottone FG Jr, Baek SJ, Nixon JB, Eling TE (2002) Diallyl disulfide (DADS) induces the antitumorigenic NSAID-activated gene (NAG-1) by a p53-dependent mechanism in human colorectal HCT 116 cells. J Nutr 132:773–778

Busch C, Jacob C, Anwar A, Burkholz T, Aicha Ba L, Cerella C, Diederich M, Brandt W, Wessjohann L, Montenarh M (2010) Diallylpolysulfides induce growth arrest and apoptosis. Int J Oncol 36:743–749

Cai WJ, Wang MJ, Ju LH, Wang C, Zhu YC (2010) Hydrogen sulfide induces human colon cancer cell proliferation: role of Akt, ERK and p21. Cell Biol Int 34:565–572

Chattopadhyay M, Kodela R, Nath N, Barsegian A, Boring D, Kashfi K (2012a) Hydrogen sulfide-releasing aspirin suppresses NF-kappaB signaling in estrogen receptor negative breast cancer cells in vitro and in vivo. Biochem Pharmacol 83:723–732

Chattopadhyay M, Kodela R, Nath N, Dastagirzada YM, Velazquez-Martinez CA, Boring D, Kashfi K (2012b) Hydrogen sulfide-releasing NSAIDs inhibit the growth of human cancer cells: a general property and evidence of a tissue type-independent effect. Biochem Pharmacol 83:715–722

Chattopadhyay M, Kodela R, Olson KR, Kashfi K (2012c) NOSH-aspirin (NBS-1120), a novel nitric oxide- and hydrogen sulfide-releasing hybrid is a potent inhibitor of colon cancer cell growth in vitro and in a xenograft mouse model. Biochem Biophys Res Commun 419:523–528

Chattopadhyay M, Nath N, Kodela R, Sobocki T, Metkar S, Gan ZY, Kashfi K (2013) Hydrogen sulfide-releasing aspirin inhibits the growth of leukemic Jurkat cells and modulates beta-catenin expression. Leuk Res 37:1302–1308

Cuevasanta E, Denicola A, Alvarez B, Moller MN (2012) Solubility and permeation of hydrogen sulfide in lipid membranes. PLoS One 7, e34562

Deryugina EI, Quigley JP (2006) Matrix metalloproteinases and tumor metastasis. Cancer Metastasis Rev 25:9–34

Feng Y, Zhu X, Wang Q, Jiang Y, Shang H, Cui L, Cao Y (2012) Allicin enhances host pro-inflammatory immune responses and protects against acute murine malaria infection. Malar J 11:268

Fenwick GR, Hanley AB (1985) The genus Allium. Part 2. Crit Rev Food Sci Nutr 22:273–377

Greiner R, Palinkas Z, Basell K, Becher D, Antelmann H, Nagy P, Dick TP (2013) Polysulfides link H2S to protein thiol oxidation. Antioxid Redox Signal 19:1749–1765

Gu X, Zhu YZ (2011) Therapeutic applications of organosulfur compounds as novel hydrogen sulfide donors and/or mediators. Expert Rev Clin Pharmacol 4:123–133

Gunadharini DN, Arunkumar A, Krishnamoorthy G, Muthuvel R, Vijayababu MR, Kanagaraj P, Srinivasan N, Aruldhas MM, Arunakaran J (2006) Antiproliferative effect of diallyl disulfide (DADS) on prostate cancer cell line LNCaP. Cell Biochem Funct 24:407–412

Ha MW, Ma R, Shun LP, Gong YH, Yuan Y (2005) Effects of allitridi on cell cycle arrest of human gastric cancer cells. World J Gastroenterol 11:5433–5437

Han Y, Zeng F, Tan G, Yang C, Tang H, Luo Y, Feng J, Xiong H, Guo Q (2013) Hydrogen sulfide inhibits abnormal proliferation of lymphocytes via AKT/GSK3beta signal pathway in systemic lupus erythematosus patients. Cell Physiol Biochem 31:795–804

Harirforoosh S, Jamali F (2009) Renal adverse effects of nonsteroidal anti-inflammatory drugs. Expert Opin Drug Saf 8:669–681

Herman-Antosiewicz A, Singh SV (2005) Checkpoint kinase 1 regulates diallyl trisulfide-induced mitotic arrest in human prostate cancer cells. J Biol Chem 280:28519–28528

Hosono T, Fukao T, Ogihara J, Ito Y, Shiba H, Seki T, Ariga T (2005) Diallyl trisulfide suppresses the proliferation and induces apoptosis of human colon cancer cells through oxidative modification of beta-tubulin. J Biol Chem 280:41487–41493

Howard EW, Ling M-T, Chua CW, Cheung HW, Wang X, Wong YC (2007) Garlic-derived S-allylmercaptocysteine is a novel in vivo antimetastatic agent for androgen-independent prostate cancer. Clin Cancer Res 13:1847–1856

Itoh Y, Nagase H (2002) Matrix metalloproteinases in cancer. Essays Biochem 38:21–36

Jacob C (2006) A scent of therapy: pharmacological implications of natural products containing redox-active sulfur atoms. Nat Prod Rep 23:851–863

Jo HJ, Song JD, Kim KM, Cho YH, Kim KH, Park YC (2008) Diallyl disulfide induces reversible G2/M phase arrest on a p53-independent mechanism in human colon cancer HCT-116 cells. Oncol Rep 19:275–280

Kaiserman MJ, Rickert WS (1992) Carcinogens in tobacco smoke: benzo[a]pyrene from Canadian cigarettes and cigarette tobacco. Am J Public Health 82:1023–1026

Kaschula CH, Hunter R, Stellenboom N, Caira MR, Winks S, Ogunleye T, Richards P, Cotton J, Zilbeyaz K, Wang Y, Siyo V, Ngarande E, Parker MI (2012) Structure-activity studies on the anti-proliferation activity of ajoene analogues in WHCO1 oesophageal cancer cells. Eur J Med Chem 50:236–254

Kashfi K (2014) Anti-cancer activity of new designer hydrogen sulfide-donating hybrids. Antioxid Redox Signal 20:831–846

Kashfi K, Olson KR (2013) Biology and therapeutic potential of hydrogen sulfide and hydrogen sulfide-releasing chimeras. Biochem Pharmacol 85:689–703

Kimura Y, Mikami Y, Osumi K, Tsugane M, Oka J, Kimura H (2013) Polysulfides are possible H2S-derived signaling molecules in rat brain. FASEB J 27:2451–2457

Knowles LM, Milner JA (2003) Diallyl disulfide induces ERK phosphorylation and alters gene expression profiles in human colon tumor cells. J Nutr 133:2901–2906

Kodela R, Chattopadhyay M, Kashfi K (2013) Synthesis and biological activity of NOSH-naproxen (AVT-219) and NOSH-sulindac (AVT-18A) as potent anti-inflammatory agents with chemotherapeutic potential. MedChemComm 4:1472–1481

Kune GA, Kune S, Watson LF (1988) Colorectal cancer risk, chronic illnesses, operations, and medications: case control results from the Melbourne Colorectal Cancer Study. Cancer Res 48: 4399–4404

Lai KC, Hsu SC, Kuo CL, Yang JS, Ma CY, Lu HF, Tang NY, Hsia TC, Ho HC, Chung JG (2013) Diallyl sulfide, diallyl disulfide, and diallyl trisulfide inhibit migration and invasion in human colon cancer colo 205 cells through the inhibition of matrix metalloproteinase-2, -7, and -9 expressions. Environ Toxicol 28:479–488

Lan X, Sun H, Liu J, Lin Y, Zhu Z, Han X, Sun X, Li X, Zhang H, Tang Z (2013) Effects of garlic oil on pancreatic cancer cells. Asian Pac J Cancer Prev 14:5905–5910

Lau BH, Yamasaki T, Gridley DS (1991) Garlic compounds modulate macrophage and T-lymphocyte functions. Mol Biother 3:103–107

Lee ZW, Zhou J, Chen CS, Zhao Y, Tan CH, Li L, Moore PK, Deng LW (2011) The slow-releasing hydrogen sulfide donor, GYY4137, exhibits novel anti-cancer effects in vitro and in vivo. PLoS One 6:e21077

Lee ZW, Teo XY, Tay EY, Tan CH, Hagen T, Moore PK, Deng LW (2014) Utilizing hydrogen sulfide as a novel anti-cancer agent by targeting cancer glycolysis and pH imbalance. Br J Pharmacol 171:4322–4336

Li L, Whiteman M, Guan YY, Neo KL, Cheng Y, Lee SW, Zhao Y, Baskar R, Tan CH, Moore PK (2008) Characterization of a novel, water-soluble hydrogen sulfide-releasing molecule (GYY4137): new insights into the biology of hydrogen sulfide. Circulation 117:2351–2360

Li Y, Zhang J, Zhang L, Si M, Yin H, Li J (2013) Diallyl trisulfide inhibits proliferation, invasion and angiogenesis of osteosarcoma cells by switching on suppressor microRNAs and inactivating of Notch-1 signaling. Carcinogenesis 34:1601–1610

Ling H, Zhang LY, Su Q, Song Y, Luo ZY, Zhou XT, Zeng X, He J, Tan H, Yuan JP (2006) Erk is involved in the differentiation induced by diallyl disulfide in the human gastric cancer cell line MGC803. Cell Mol Biol Lett 11:408–423

Liu Y, He P, Liu F, Zhou N, Cheng X, Shi L, Zhu H, Zhao J, Wang Y, Zhang M (2014) Tetra-arsenic tetra-sulfide (As4S 4) promotes apoptosis in retinoid acid -resistant human acute promyelocytic leukemic NB4-R1 cells through downregulation of SET protein. Tumour Biol 35:3421–3430

Lu S, Gao Y, Huang X, Wang X (2014) GYY4137, a hydrogen sulfide (H(2)S) donor, shows potent anti-hepatocellular carcinoma activity through blocking the STAT3 pathway. Int J Oncol 44:1259–1267

Lv M, Li Y, Ji MH, Zhuang M, Tang JH (2014) Inhibition of invasion and epithelial-mesenchymal transition of human breast cancer cells by hydrogen sulfide through decreased phospho-p38 expression. Mol Med Rep 10:341–346

Ma K, Liu Y, Zhu Q, Liu CH, Duan JL, Tan BK, Zhu YZ (2011) H2S donor, S-propargyl-cysteine, increases CSE in SGC-7901 and cancer-induced mice: evidence for a novel anti-cancer effect of endogenous H2S? PLoS One 6:e20525

Ma Z, Bi Q, Wang Y (2014) Hydrogen sulfide accelerates cell cycle progression in oral squamous cell carcinoma cell lines. Oral Dis 21:156–162

Melchini A, Traka MH, Catania S, Miceli N, Taviano MF, Maimone P, Francisco M, Mithen RF, Costa C (2013) Antiproliferative activity of the dietary isothiocyanate erucin, a bioactive compound from cruciferous vegetables, on human prostate cancer cells. Nutr Cancer 65: 132–138

Modem S, Dicarlo SE, Reddy TR (2012) Fresh garlic extract induces growth arrest and morphological differentiation of MCF7 breast cancer cells. Genes Cancer 3:177–186

Munday R, Munday JS, Munday CM (2003) Comparative effects of mono-, di-, tri-, and tetrasulfides derived from plants of the Allium family: redox cycling in vitro and hemolytic activity and Phase 2 enzyme induction in vivo. Free Radic Biol Med 34:1200–1211

Mustafa AK, Gadalla MM, Sen N, Kim S, Mu W, Gazi SK, Barrow RK, Yang G, Wang R, Snyder SH (2009) H2S signals through protein S-sulfhydration. Sci Signal 2:ra72

Nkrumah-Elie YM, Reuben JS, Hudson A, Taka E, Badisa R, Ardley T, Israel B, Sadrud-Din SY, Oriaku E, Darling-Reed SF (2012) Diallyl trisulfide as an inhibitor of benzo(a)pyrene-induced precancerous carcinogenesis in MCF-10A cells. Food Chem Toxicol 50:2524–2530

Padilla-Camberos E, Zaitseva G, Padilla C, Puebla AM (2010) Antitumoral activity of allicin in murine lymphoma L5178Y. Asian Pac J Cancer Prev 11:1241–1244

Paul BD, Snyder SH (2012) H(2)S signalling through protein sulfhydration and beyond. Nat Rev Mol Cell Biol 13:499–507

Rothwell PM, Price JF, Fowkes FG, Zanchetti A, Roncaglioni MC, Tognoni G, Lee R, Belch JF, Wilson M, Mehta Z, Meade TW (2012) Short-term effects of daily aspirin on cancer incidence, mortality, and non-vascular death: analysis of the time course of risks and benefits in 51 randomised controlled trials. Lancet 379:1602–1612

Santos CI, Costa-Pereira AP (2011) Signal transducers and activators of transcription-from cytokine signalling to cancer biology. Biochim Biophys Acta 1816:38–49

Savolainen H (2010) High toxicity of hydrogen sulfide by the inhibition of mitochondrial respiration. J Biol Chem 285:le9

Sher A, Fakhar-Ul-Mahmood M, Shah SN, Bukhsh S, Murtaza G (2012) Effect of garlic extract on blood glucose level and lipid profile in normal and alloxan diabetic rabbits. Adv Clin Exp Med 21:705–711

Singh SV, Powolny AA, Stan SD, Xiao D, Arlotti JA, Warin R, Hahm E-R, Marynowski SW, Bommareddy A, Potter DM, Dhir R (2008) Garlic constituent diallyl trisulfide prevents development of poorly differentiated prostate cancer and pulmonary metastasis multiplicity in TRAMP mice. Cancer Res 68:9503–9511

Siveen KS, Sikka S, Surana R, Dai X, Zhang J, Kumar AP, Tan BK, Sethi G, Bishayee A (2014) Targeting the STAT3 signaling pathway in cancer: role of synthetic and natural inhibitors. Biochim Biophys Acta 1845:136–154

Song JD, Lee SK, Kim KM, Park SE, Park SJ, Kim KH, Ahn SC, Park YC (2009) Molecular mechanism of diallyl disulfide in cell cycle arrest and apoptosis in HCT-116 colon cancer cells. J Biochem Mol Toxicol 23:71–79

Song ZJ, Ng MY, Lee ZW, Dai W, Hagen T, Moore PK, Huang D, Deng LW, Tan C-H (2014) Hydrogen sulfide donors in research and drug development. MedChemComm 5:557–570

Sostres C, Gargallo CJ, Lanas A (2013) Nonsteroidal anti-inflammatory drugs and upper and lower gastrointestinal mucosal damage. Arthritis Res Ther 15(Suppl 3):S3

Sriram N, Kalayarasan S, Ashokkumar P, Sureshkumar A, Sudhandiran G (2008) Diallyl sulfide induces apoptosis in Colo 320 DM human colon cancer cells: involvement of caspase-3, NF-kappaB, and ERK-2. Mol Cell Biochem 311:157–165

Stan SD, Singh SV (2009) Transcriptional repression and inhibition of nuclear translocation of androgen receptor by diallyl trisulfide in human prostate cancer cells. Clin Cancer Res 15: 4895–4903

Tan H, Ling H, He J, Yi L, Zhou J, Lin M, Su Q (2008) Inhibition of ERK and activation of p38 are involved in diallyl disulfide induced apoptosis of leukemia HL-60 cells. Arch Pharm Res 31: 786–793

Tanaka S, Haruma K, Kunihiro M, Nagata S, Kitadai Y, Manabe N, Sumii M, Yoshihara M, Kajiyama G, Chayama K (2004) Effects of aged garlic extract (AGE) on colorectal adenomas: a double-blinded study. Hiroshima J Med Sci 53:39–45

Tanaka S, Haruma K, Yoshihara M, Kajiyama G, Kira K, Amagase H, Chayama K (2006) Aged garlic extract has potential suppressive effect on colorectal adenomas in humans. J Nutr 136: 821S–826S

Tang H, Kong Y, Guo J, Tang Y, Xie X, Yang L, Su Q (2013) Diallyl disulfide suppresses proliferation and induces apoptosis in human gastric cancer through Wnt-1 signaling pathway by up-regulation of miR-200b and miR-22. Cancer Lett 340:72–81

Tao BB, Liu SY, Zhang CC, Fu W, Cai WJ, Wang Y, Shen Q, Wang MJ, Chen Y, Zhang LJ, Zhu YZ, Zhu YC (2013) VEGFR2 functions as an H2S-targeting receptor protein kinase with its novel Cys1045-Cys1024 disulfide bond serving as a specific molecular switch for hydrogen sulfide actions in vascular endothelial cells. Antioxid Redox Signal 19:448–464

Tao M, Gao L, Pan J, Wang X (2014) Study on the inhibitory effect of allicin on human gastric cancer cell line SGC-7901 and its mechanism. Afr J Tradit Complement Altern Med 11: 176–179

Thejass P, Kuttan G (2007) Antiangiogenic activity of diallyl sulfide (DAS). Int Immunopharmacol 7(3):295–305

Viry E, Anwar A, Kirsch G, Jacob C, Diederich M, Bagrel D (2011) Antiproliferative effect of natural tetrasulfides in human breast cancer cells is mediated through the inhibition of the cell division cycle 25 phosphatases. Int J Oncol 38:1103–1111

Wang HC, Yang J-H, Hsieh S-C, Sheen L-Y (2010) Allyl sulfides inhibit cell growth of skin cancer cells through induction of DNA damage mediated G2/M arrest and apoptosis. J Agric Food Chem 58:7096–7103

Weisberger AS, Pensky J (1957) Tumor-inhibiting effects derived from an active principle of Garlic (*Allium sativum*). Science 126:1112–1114

Wu XJ, Kassie F, Mersch-Sundermann V (2005) The role of reactive oxygen species (ROS) production on diallyl disulfide (DADS) induced apoptosis and cell cycle arrest in human A549 lung carcinoma cells. Mutat Res 579:115–124

Wu PP, Chung HW, Liu KC, Wu RS, Yang JS, Tang NY, Lo C, Hsia TC, Yu CC, Chueh FS, Lin SS, Chung JG (2011) Diallyl sulfide induces cell cycle arrest and apoptosis in HeLa human cervical cancer cells through the p53, caspase- and mitochondria-dependent pathways. Int J Oncol 38:1605–1613

Wu YC, Wang XJ, Yu L, Chan FK, Cheng AS, Yu J, Sung JJ, Wu WK, Cho CH (2012) Hydrogen sulfide lowers proliferation and induces protective autophagy in colon epithelial cells. PLoS One 7:e37572

Xiao D, Lew KL, Kim YA, Zeng Y, Hahm ER, Dhir R, Singh SV (2006) Diallyl trisulfide suppresses growth of PC-3 human prostate cancer xenograft in vivo in association with Bax and Bak induction. Clin Cancer Res 12:6836–6843

You W-C, Blot WJ, Chang Y-S, Ershow A, Yang ZT, An Q, Henderson BE, Fraumeni JF, Wang T-G (1989) Allium vegetables and reduced risk of stomach cancer. J Natl Cancer Inst 81:162–164

Yuan JP, Wang GH, Ling H, Su Q, Yang YH, Song Y, Tang RJ, Liu Y, Huang C (2004) Diallyl disulfide-induced G2/M arrest of human gastric cancer MGC803 cells involves activation of p38 MAP kinase pathways. World J Gastroenterol 10:2731–2734

Zhang CL, Zeng T, Zhao XL, Xie KQ (2013) Garlic oil attenuated nitrosodiethylamine-induced hepatocarcinogenesis by modulating the metabolic activation and detoxification enzymes. Int J Biol Sci 9:237–245

Zhao K, Li S, Wu L, Lai C, Yang G (2014a) Hydrogen sulfide represses androgen receptor transactivation by targeting at the second zinc finger module. J Biol Chem 289:20824–20835

Zhao Y, Biggs TD, Xian M (2014b) Hydrogen sulfide (H2S) releasing agents: chemistry and biological applications. Chem Commun (Camb) 50:11788–11805

Part VII

H_2S and Ageing

H_2S: A New Approach to Lifespan Enhancement and Healthy Ageing?

Bedoor Qabazard and Stephen R. Stürzenbaum

Contents

Abstract

Ageing, a progressive structural and functional decline, is considered to be a major risk factor for virtually all ageing-associated pathologies and disabilities, including Alzheimer's disease, Parkinson's disease, stroke, diabetes, athero-sclerosis and certain cancers. Biogerontology research has now been largely

B. Qabazard
MRC-HPA Centre for Environment and Health, Analytical and Environmental Sciences Division, Faculty of Life Sciences and Medicine, King's College London, 150 Stamford Street, London SE1 9NH, UK

Department of Pharmacology and Therapeutics, Faculty of Pharmacy, Kuwait University, Kuwait City, Kuwait

S.R. Stürzenbaum (✉)
MRC-HPA Centre for Environment and Health, Analytical and Environmental Sciences Division, Faculty of Life Sciences and Medicine, King's College London, 150 Stamford Street, London SE1 9NH, UK
e-mail: stephen.sturzenbaum@kcl.ac.uk

P.K. Moore, M. Whiteman (eds.), *Chemistry, Biochemistry and Pharmacology of Hydrogen Sulfide*, Handbook of Experimental Pharmacology 230,
DOI 10.1007/978-3-319-18144-8_14

directed towards finding novel drug targets to decelerate the ageing process and attain healthy ageing in order to delay the onset of all ageing-related diseases. H_2S has been reported to exert vasodilatory, antioxidant, antiapoptotic and anti-inflammatory actions and has been shown to act as a signalling molecule, neuromodulator and cytoprotectant. Intriguingly, H_2S has been reported to regulate cell cycle and survival in healthy cells which suggests that it may regulate cell fate and hence the ageing process. This chapter sets out to provide an overview of the current knowledge regarding the involvement of H_2S in ageing, with a specific focus on the invertebrate model nematode *C. elegans*.

Keywords
H_2S • Ageing • Lifespan • *C. elegans* • Nematode

1 The Ageing Process: An Overview

Ageing, or senescence, is the progressive structural and functional organismal decline which emerges as an organism grows older (Golden and Melov 2007). Ageing is characterised by homeostatic imbalances and an impaired ability to respond to stress, thereby increasing the incidence of pathology and probability of death, the ultimate inevitable consequence of ageing. Importantly, ageing is the major risk factor for the development of age-associated diseases, such as cancer, neurodegenerative illnesses, cardiovascular pathologies, dementias and other structural and functional derangements.

1.1 Theories of Ageing: Why Do We Age?

Our understanding of the ageing process has expanded considerably over the last few decades, and it is now apparent that ageing is not driven by a single causative gene or a key defective body system but rather a complex, multidisciplinary event controlled by several interacting cellular, molecular and environmental factors (Weinert and Timiras 2003). Numerous evolutionary-, molecular-, cellular- and systemic-ageing theories are available. For example, ageing has been extensively studied in light of the evolutionary theory, with the most influential theories being the disposal soma theory (Kirkwood 1977) and antagonistic pleiotropy (Williams 1957). Other theories postulate that life expectancy can be influenced by neuroendocrine control (Hammerman 1987), metabolic rate (Pearl 1928) or telomere length (Campisi et al. 2001).

1.2 Ageing Pathways: The Genetic Control

The fundamental discovery that mutations in single genes (e.g. in *age-1* or *daf-2*) can drastically increase the lifespan of the nematode *Caenorhabditis elegans* (Klass

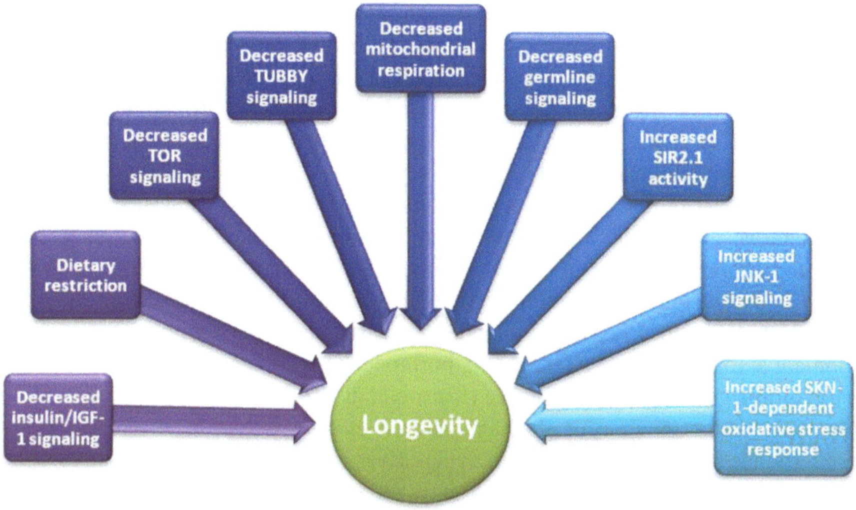

Fig. 1 Multiple evolutionarily conserved pathways are involved in the complex regulation of lifespan. Decreased or increased activity of the following signalling pathways and processes has been shown to modulate ageing (reviewed in Kenyon 2010)

1977; Friedman and Johnson 1988, 2002; Kenyon et al. 1993) supports the notion that ageing is part of a programmed biological route and subject to regulation by specific genetic pathways which confer plasticity to the rate of ageing. The process of lifespan regulation relies on evolutionarily conserved genetic pathways (Fig. 1), including the insulin/insulin growth factor-1 (IGF-1)-like signalling pathway (IIS), DAF-16/FOXO transcription factor, c-Jun N-terminal kinase (JNK) signalling, stress-response pathways, mitochondrial control, dietary restriction, SIR2 deacetylase activity, germ line signalling, sensory perception, TOR (target of rapamycin), TUBBY signalling as well as telomere length. Mutations in genes affecting these signalling pathways protect against stress, reduce disease susceptibility and dramatically extend the lifespan of organisms (Kenyon 2010).

Mutations in *daf-2* (an insulin receptor/IGF-1 receptor (Kimura et al. 1997)) or *age-1* (a phosphatidylinositol 3-kinase (PI3K) subunit (Morris et al. 1996)), key genes within the insulin/IGF-1 signalling (IIS) pathway, are known to modulate hormonal signalling and extend the lifespan of animals by more than 100 % (Kenyon 2010). The IIS signalling cascade is initiated after DAF-2 receptor activates the PI3K subunit AGE-1, which results in activation of Akt/PKB homolog PDK-1 (Paradis and Ruvkun 1998) and subsequently the phosphorylation and activation of AKT-1, AKT-2 and SGK-1 (Paradis and Ruvkun 1998). These kinases inactivate DAF-16, a Forkhead FOXO transcription factor homolog, by retaining it in the cytoplasm and preventing its nuclear translocation (Ogg et al. 1997). DAF-16 is the major downstream target of IIS that, when present in the nucleus, promotes life extension and stress resistance (Lin et al. 1997; Henderson and Johnson 2001; Landis and Murphy 2010). Thus, interventions that reduce the activity and signalling of the upstream components of the IIS pathway will reduce the repressive signal on DAF-16 and hence enhance the protective pro-survival downstream of DAF-16 targets. In contrast, a null mutation of *daf-16* marginally shortens lifespan and antagonises lifespan extension caused by mutations in the upstream IIS components (e.g. *daf-2*, *age-1*, *akt1/2*, *aap-1*, *pdk-1* or *sgk-1*) (Ogg et al. 1997; Henderson and Johnson 2001). Microarray studies have identified several genes downstream of DAF-16, namely, stress-response genes (e.g. catalases, superoxide dismutases, glutathione *S*-transferases and metallothioneins), antimicrobial genes and chaperones (Lee et al. 2003; McElwee et al. 2003; Murphy et al. 2003). Another hormonal contribution to ageing regulation is believed to originate from the novel age-related gene *Klotho*, which encodes a cell-surface protein with an extracellular domain acting similar to a circulating hormone (Shiraki-Iida et al. 1998). Suppression of this gene accelerates ageing (Kuro-o et al. 1997), whilst upregulation slows ageing in mice (Kurosu et al. 2005) and *C. elegans* (Château et al. 2010). The underlying mechanism of the anti-ageing effect of *Klotho* is poorly understood but may involve IIS inhibition/DAF-16 activation and reduction of oxidative stress (Yamamoto et al. 2005; Château et al. 2010) and angiotensin II (AngII) suppression (Ishizaka et al. 2002).

Another evolutionarily conserved intervention influencing longevity is the caloric or dietary restriction (DR) in the absence of malnutrition (Mair and Dillin 2008). DR is the most efficient environmental intervention proven to delay ageing. An important nutrient-sensing component in this process is the Sirtuins family of NAD^+-dependent protein deacetylases, which respond to metabolic changes in the cell (i.e. nutrient and energy availability and stress) and mediate the beneficial effects of DR on lifespan in yeast, worms and flies (Rogina and Helfand 2004; Wang et al. 2006). Overexpression of *sir-2.1*, the *C. elegans* ortholog of the yeast SIR2, increases lifespan by 15–50 % via a *daf-16*-dependent action (Tissenbaum and Guarente 2001), but more recent findings have challenged this effect (Burnett et al. 2011). The Sirtuins pathway has been shown to interact with the IIS pathway and directly deacetylate and activate DAF-16 (Berdichevsky et al. 2006). Pharmacological activation of Sirtuins has therefore been widely accepted as a potential DR mimetic approach to slow ageing, an example is the polyphenol drug resveratrol (Wood et al. 2004). Despite the evidence that resveratrol longevity and *sir-2.1* are interlinked in *C. elegans* (Viswanathan et al. 2005), Bass et al. (2007) did not observe lifespan extension with resveratrol in *Drosophila* in seven independent trials and reported only a slight increase in lifespan in wild-type and *sir-2.1* mutant *C. elegans* in some but not all trials (Bass et al. 2007).

Another nutrient- and amino acid-sensing pathway linked to DR is the TOR (target of rapamycin) pathway. TOR is a protein kinase that phosphorylates ribosomal S6 kinase and translation initiation factor 4E-binding protein 1 (4E BP1) in response to nutrients, thus stimulating growth and inhibiting autophagy when food is abundant (Kapahi and Zid 2004; Kenyon 2010). Mutations that reduce TOR signalling extend the lifespan of flies (Kapahi et al. 2004), yeast (Kaeberlein et al. 2005), mice (Harrison et al. 2009) and worms (Vellai et al. 2003) and increase resistance to environmental stress (Hansen et al. 2007). Since the lifespan extension produced by TOR inhibition is not further increased by DR (Kapahi et al. 2004; Kaeberlein et al. 2005; Bjedov et al. 2010), TOR inhibition is thought to mimic the effects of DR (Kenyon 2010). Recently, the TOR inhibitor rapamycin was shown to return promising anti-ageing effects in mice (Wilkinson et al. 2012). Interestingly, TOR can be inhibited by other longevity-promoting energy sensors such as AMPK (AMP-activated protein kinase) (Inoki et al. 2003) or resveratrol (Baur et al. 2006; Dasgupta and Milbrandt 2007). Likewise, the antidiabetic drug metformin can extend mouse lifespan via AMPK activation (Anisimov et al. 2008). The role of AMPK in longevity is not clear, but there is evidence that its activation mimics DR and has been found to extend the lifespan of *C. elegans* (Apfeld et al. 2004; Greer et al. 2007a, b) and mice (Ingram et al. 2004; Anisimov et al. 2005) partly via FOXO/DAF-16 phosphorylation (Greer et al. 2007a, b).

The stress-induced MAP kinase, JNK, prolongs survival in worms (Oh et al. 2005) and flies (Wang et al. 2003). In *C. elegans*, *jnk-1* overexpression leads to a DAF-16-dependent increase in survival and improved resistance to heat and oxidative stress (Oh et al. 2005). JNK-1 interacts with IIS by directly phosphorylating/activating DAF-16, thereby inducing the expression of DAF-16 target genes which are crucial for longevity and stress tolerance. Another important stress-related pathway is SKN-1/Nrf2, which protects against oxidative stress by activating the conserved phase II detoxification system through constitutive and stress-inducible mechanisms (An and Blackwell 2003). Phase II enzymes mediate the synthesis of glutathione, scavenge free radicals and help to detoxify the reactive intermediates from the phase I reactions. Hence, SKN-1 induces the expression of genes involved in detoxification, cell repair and pathogen resistance and suppresses genes that decrease survival and stress tolerance (Oliveira et al. 2009). SKN-1 is expressed constitutively in ASI chemosensory neurons under normal conditions and induced in the intestinal nuclei (by reduced IIS signalling) in response to stress (An and Blackwell 2003; An et al. 2005; Tullet et al. 2008). Moreover, SKN-1 contributes to the longevity and stress resistance associated with IIS inhibition, and, likewise, DAF-16 is essential for the induction of some SKN-1 targets under reduced IIS signalling (Tullet et al. 2008), suggesting a possible crosstalk between the two pathways in regulating stress response and lifespan. In addition, an interaction between SKN-1 and p38 MAPK has also been suggested as p38 MAPK represses the inhibitory signal from glycogen synthase kinase-3 (GSK-3) on SKN-1 leading to intestinal SKN-1 expression and activation under oxidative stress (Inoue et al. 2005).

2 The Role of H_2S in Ageing and Age-Associated Diseases

2.1 H_2S: A Prehistoric Entity

H_2S is thought to have caused a mass extinction in the Permian-Triassic periods, some 250 million years ago (Kump et al. 2005; Ward 2006). Massive volcanic eruptions in Siberia and a cascade of environmental destructions resulted in a severe depletion of oxygen levels in the oceans and atmosphere, which caused the majority of organisms to become extinct (Kump et al. 2005). However, certain species managed to survive these harsh habitats, thanks to their ability to utilise H_2S in the process of chemosynthesis (Lavu et al. 2011). For example, the anaerobic green sulfur bacteria (*Chlorobium*) respirated, instead of oxygen, the sulfate dissolved in water and in doing so continued to generate more lethal H_2S gas (Wang 2012). Upon the subsequent environmental depletion of H_2S, primitive species gradually switched to use water instead of H_2S by means of photosynthesis (Lavu et al. 2011). Blackstone et al. (2005) found that very low levels of the toxic H_2S gas can reduce the oxygen demands and prolong survival in mice by inducing a hypometabolic hibernation-like or 'suspended animation' state that extend their lives.

2.2 Induced Hypothermia and Suspended Animation: From Animals to Humans

According to Blackstone et al. (2005), house mice (*Mus musculus*) exposed to low (non-toxic) doses of H_2S enter a reversible state of suspended animation-like hypothermia. In detail, H_2S exposure dose-dependently decreases the metabolic rate by 50 % with a 90 % reduction in oxygen demand. The core body temperature declines from 37 °C to about 15 °C, and the respiration falls from 120 to 10 breaths per minute. The mice survive this procedure for 6 h and show no negative health consequences afterwards (Blackstone et al. 2005). In this context it should be noted that mice do not naturally hibernate but can undergo a clinical torpor state during food deprivation. Further experiments uncovered that a pretreatment with H_2S primes the mice to tolerate and survive hypoxia. Mice exposed to 150 ppm H_2S for 20 min are more resistant to an oxygen challenge, possibly due to the reduction in metabolic rate and whole-body oxygen demand (Blackstone and Roth 2007). Furthermore, rats display a reduced degree of brain damage when exposed to H_2S for 48 h following an experimentally induced stroke (Florian et al. 2008). In these animals the expression of pro-inflammatory and pro-apoptotic genes is suppressed in the areas surrounding the infarct region, and the mice are characterised by superior memory and learning profiles. Likewise, H_2S inhalation (80 ppm) in mice (Faller et al. 2010) and NaSH infusion in rats (Aslami et al. 2010) protect against ventilator-induced lung injury.

Further studies with larger animals yielded conflicting results. For example, studies failed to induce hibernation or hypometabolism in pigs (Simon et al. 2008) and sheep (Haouzi et al. 2008), leading the authors to conclude that the effects seen in mice may not be achievable in larger mammals. One reason could be due to the difference in the ratio between surface area and body mass which may render large animals less responsive to the same elative dose of H_2S (Wang 2012). However, if H_2S-induced hibernation would be feasible in humans, it would promise potential therapeutic benefits regarding the emergency management of severely injured patients and in the conservation of donated organs during transplant surgery. Indeed, in 2010 it was announced that H_2S-induced hypothermia had completed phase I clinical trials, but the trials were withdrawn and terminated for unknown reasons.

2.3 H_2S: Protective Effects That Promote Healthy Ageing

Researchers have recently explored the physiological importance of endogenously produced H_2S in cellular senescence and organism longevity and, hence, its contribution to ageing and age-related diseases. In addition to the effects of H_2S as an anti-inflammatory, antiapoptotic and antioxidant mediator, several lines of evidence support the notion that H_2S acts as a cytoprotective mediator in the ageing process. Plasma levels of H_2S in humans decrease with age (Chen et al. 2005). Similarly, the activities of CBS and CSE enzymes decline progressively in the livers of ageing mice with a corresponding decrease in cysteine and glutathione levels (Dröge et al. 2006). CSE gene and protein expression and enzyme activity are attenuated in the lenses of old rats, leading to cataract formation in vitro (Sastre et al. 2005). H_2S replacement promotes endothelial protection in hyperglycaemia due to the preservation of mitochondrial function (Suzuki et al. 2011). Likewise, the protective effect of calorie restriction in blood vessels of ageing rats has been attributed to the maintenance of normal H_2S levels (Predmore et al. 2010). Moreover, the suggested antioxidant role of H_2S seems to align well with the free radical theory of ageing and may contribute to its therapeutic potential in ageing-associated diseases, such as AD, PD, atherosclerosis and others. H_2S acts as a reducing agent that can scavenge other oxidative and nitrosative species (Whiteman et al. 2004, 2005), but additional mechanisms are believed to be responsible for the antioxidant role of H_2S. Importantly, H_2S can increase the intracellular GSH pool, the major cellular antioxidant, by stimulating γ-glutamylcysteine synthase and cysteine transport in neurons (Kimura and Kimura 2004) and glutamate uptake in astrocytes (Lu et al. 2008). NaSH confers protection to a murine hippocampal HT22 neuronal cell line from oxidative glutamate toxicity by opening K_{ATP} and Cl^- channels in addition to increasing GSH levels (Kimura et al. 2006). In agreement, the H_2S-releasing derivative of sildenafil, ACS6, inhibits superoxide formation and NADH oxidation in endothelial cells (Muzaffar et al. 2008). Moreover, the slow-releasing H_2S donor GYY4137 protects cultured chondrocytes and mesenchymal progenitor cells from oxidative and nitrative stress induced by SIN-1 and 4-HNE (Fox et al. 2012).

Molecular studies revealed that certain age-related genes are involved in the H_2S response, notably *Sirtuins* (Miller and Roth 2007) and *Klotho* (Zhang et al. 2013). H_2S also affects transcription factors, namely, Nrf-2 (nuclear factor erythroid 2-related factor 2), STAT3 (signal transducer and activator of transcription 3) and Hif-1 (hypoxia inducible factor 1). Nrf-2 regulates the expression of several cytoprotective and antioxidant enzymes (e.g. thioredoxin and heme oxygenase), and activation of this transcription factor contributes to the cardioprotective effect of H_2S (Calvert et al. 2009) and mediates H_2S responses in the nematode *C. elegans* (Miller et al. 2011). STAT3, the cell survival and proliferation regulator, is activated by GYY4137 treatment in LPS-injected rats (Li et al. 2009). Lastly, enhanced Hif-1 activity and a genetic requirement for this factor were reported in *C. elegans* after H_2S exposure (Budde and Roth 2010; Miller et al. 2011). Recently, H_2S has been shown to prevent H_2O_2-induced senescence in human umbilical vein endothelial cells through SIRT1 activation (Suo et al. 2013) and protect against cellular ageing via *S*-sulfhydration of Keap1 and Nrf-2 activation in mouse embryonic fibroblasts (Yang et al. 2013). In SH-SY5Y neuronal cells, H_2S confers protection against D-galactose-induced cell injury by suppression of advanced glycation end-product formation and oxidative stress (Liu et al. 2013). H_2S has been associated with mitogen-activated protein kinase (MAPK); however, whilst most papers suggest an activation, others claim that H_2S is involved in the inhibition of the MAPK pathway. For instance, H_2S induces apoptosis of human aorta smooth muscle cells (Yang et al. 2004) and stimulates angiogenesis in cultured human umbilical vein endothelial cells (Papapetropoulos et al. 2009) via p38MAPK activation, whereas the antiapoptotic effect of H_2S in human polymorphonuclear leukocytes is likely due to the inhibition of p38MAPK (Rinaldi et al. 2006). In addition, H_2S activates and promotes the phosphorylation of ERK1/2 (extracellular signal-regulated kinase 1/2) in HEK-239 cells (Yang et al. 2004), human colon cancer cells (Cai et al. 2010) and rat vascular smooth muscle cells (Jeong et al. 2006). Moreover, akt activation/phosphorylation has also been suggested as a mechanism for H_2S-induced angiogenesis (Cai et al. 2007) and cardioprotection (Yong et al. 2008). JNK (c-Jun NH_2 terminal kinase) upregulation by NaSH increases rat intestinal epithelial cell proliferation (Deplancke and Gaskins 2003), whilst JNK downregulation by NaSH prevents apoptosis in a human neuroblastoma cell line (Hu et al. 2009).

S-Sulfhydration is a signalling mechanism for H_2S involving posttranslational modification of cellular proteins and enzymes (Mustafa et al. 2009). The mechanism entails covalent modification of reactive cysteine residues via the addition of one sulphur (from H_2S) to the thiol group of cysteine (-SH) to yield a hydropersulfide molecule (-SSH). This process results in an increased activity of the modified protein bearing the -SSH moiety (Mustafa et al. 2009). H_2S can alter, via this process, the function of a wide range of cellular proteins and enzymes in multiple settings, most notably in cell survival and stress pathways. For example, the sulfhydration of cysteine residues can shield them from oxidation under oxidative stress conditions, thereby preserving protein function and preventing permanent damage (Paul and Snyder 2012). H_2S-mediated *S*-sulfhydration of p65 at cysteine-38 (Sen et al. 2012) and PTP1B phosphatase (Krishnan et al. 2011) inhibit the apoptotic effect of NF-κB and modulate ER stress responses, respectively. More recently, H_2S has been shown to regulate cellular redox signalling via electrophile *S*-sulfhydration (Nishida et al. 2012). Further studies are required to confirm and validate this mechanism and to what extent the various H_2S-mediated responses correlate with health and disease.

2.4 H_2S: A Novel Therapeutic Agent in Age-Associated Pathologies?

A considerable body of evidence suggests that H_2S acts as a protective mediator in the heart, blood vessels, brain and other body tissues and compartments by virtue of its possible antioxidant, antiapoptotic and anti-inflammatory properties. Although the role of H_2S in inflammation is controversial (Whiteman and Winyard 2011) with some studies advocating a pro-inflammatory effect of H_2S by activating leukocyte recruitment (Zhang and Bhatia 2008), numerous studies describe H_2S as an anti-inflammatory molecule. For example, H_2S was shown to counteract endothelial inflammation by inhibiting leukocyte adherence (Zanardo et al. 2006). Moreover, H_2S accelerates the resolution of inflammation by enhancing neutrophil survival through p38MAPK inhibition (Rinaldi et al. 2006). NaSH treatment reverses LPS (lipopolysaccharide)-induced iNOS expression and NO release (Oh et al. 2006; Muzaffar et al. 2008) and prevents homocysteine-induced toxicity and subsequent ROS formation in VSMCs (Yan et al. 2006). Likewise, exogenous H_2S attenuates oxidised low-density lipoprotein (oxLDL)-induced cytotoxicity in cultured HUVECs and was shown to protect against atherosclerosis (Laggner et al. 2007a, b; Jeney et al. 2009). The anti-atherosclerotic effect of H_2S is further supported by the reported inhibition of platelet aggregation (Zagli et al. 2007), inhibition of VSMCs proliferation (Du et al. 2004) and reduction of neointima formation of balloon-injured carotid arteries (Meng et al. 2007). More compellingly, H_2S appears to activate antiapoptotic signalling to protect against ischaemia/reperfusion (I/R) injury (Sivarajah et al. 2006; Jha et al. 2008). Activation of Nrf-2 (Calvert et al. 2009), Akt (Cai et al. 2007; Yong et al. 2008) and ERK1/2 and PI3K (Hu et al. 2008) signalling pathways and preservation of mitochondrial structure and function (Elrod et al. 2007) have all been suggested to contribute to H_2S-induced cardioprotection. H_2S may also mediate cardioprotection by influencing intracellular organelles such as the endoplasmic reticulum (ER), as evidenced by the suppression of cardiac myocyte death in hyperhomocysteinaemic rats (Wei et al. 2010).

Exogenous H_2S has been shown to protect against detrimental pro-inflammatory oxidants, such as peroxynitrite ($ONOO^-$) (Whiteman et al. 2004), hydrogen peroxide (H_2O_2) (Muzaffar et al. 2008), hypochlorous acid (HOCl) (Whiteman et al. 2005) and β-amyloid (Tang et al. 2008). Given that H_2S is not very abundant suggests that the potential for scavenging activity is limited and/or that other mechanisms participate in the cytoprotective action of this gas. In agreement, H_2S protects the mouse HT22 neuronal cell line from peroxynitrite-induced oxidative stress by activating K_{ATP} and Cl^- channels and increasing GSH levels (Kimura and Kimura 2004; Kimura et al. 2006). In addition, H_2S-inhibited HOCl provokes cytotoxicity, protein oxidation and lipid peroxidation in the brain (Whiteman et al. 2005) and protects brain endothelial cells against methionine-induced oxidative stress (Tyagi et al. 2009). The interaction with cystic fibrosis transmembrane conductance regulator (CFTR) Cl^- channels may also be a mechanism for H_2S protection against oxidative stress (Lahousse et al. 2003). It has been suggested that Cl^- channels play a critical role in maintaining normal cell survival as the reduction of Cl^- current causes cell death in hippocampal neurons, and indeed a downregulation of CFTR gene expression is observed in the brains of Alzheimer's disease (AD) patients (Lahousse et al. 2003). In fact, CBS has been stipulated to be a potential risk factor for the pathogenesis of AD (Beyer et al. 2004). H_2S levels are also reduced in rat models of Parkinson's disease (PD) and administering H_2S alleviates the condition (Hu et al. 2010). In animal models of PD, H_2S reduces the levels of TNFα and IL-6 (Lee et al. 2010); however, in Down syndrome more H_2S is produced (Kamoun et al. 2003). Interestingly, H_2S has been found to influence Cl^- conductance in the CNS leading to neuroprotection (Kimura et al. 2005). In support of this, H_2S activates CFTR Cl^- channels in HT22 neuronal cell lines leading to neuroprotection during oxytosis, which is reversed by the CFTR blockers NPPB and IAA-94 (Kimura et al. 2006). Moreover, H_2S activates Cl^-/HCO_3^- transporters in smooth muscle cells (Lee et al. 2007). In addition to the interaction with ion channels in the CNS, H_2S can affect neurotransmitters, namely, GABA (the major inhibitory neurotransmitter) and glutamate (the major excitatory neurotransmitter). As such, H_2S ameliorates hippocampal damage associated with recurrent febrile seizures by upregulating the expression of $GABA_B$ receptors (in Ca^{2+}-dependent manner) which in turn enhances GABAergic inhibition (Han et al. 2005a). It can be therapeutically important during seizures/epilepsy to restore the disrupted balance between CNS excitation and inhibition (Han et al. 2005b). In contrast, H_2S seems to potentiate the effect of glutamate on the NMDA receptor with the subsequent toxicity and neuronal death due to Ca^{2+} overload (Gagliardi 2000; Garcia-Bereguiain et al. 2008).

H_2S demonstrates a notable growth-suppressive effect on prostate cancer (Pei et al. 2011), breast cancer (Chattopadhyay et al. 2012a) and colon cancer (Chattopadhyay et al. 2012b) cell lines. Lee and colleagues (2011) suggest that the slow-releasing H_2S donor GYY4137 is a potential anticancer agent owing to its ability to kill seven different human cancer cell lines (HeLa, HCT-116, Hep G2, HL-60, MCF-7, MV4-11 and U2OS) in a concentration-dependent manner. Moreover, H_2S-releasing nonsteroidal anti-inflammatory drugs (NSAID) have been shown to constrain the growth of various human cancer cells by inhibiting cell proliferation, inducing apoptosis, and cause G_0/G_1 cell cycle block (Chattopadhyay et al. 2012c).

However, the precise role of H_2S in ageing and its contribution to age-related pathologies awaits further elucidation but promises to return new therapeutic leads to be added to the growing list of the beneficial effects of this gas.

2.5 H_2S-Mediated Lifespan Extension: Insights from *Caenorhabditis elegans*

Given that the monitoring of complete survival profiles in humans is challenging, model organisms such as the nematode *C. elegans* are valuable surrogates and arguably the premier animal model for lifespan studies and ageing research.

Miller and Roth (2007) observed that exposure to a small amount of H_2S (50 ppm, in a controlled atmosphere) extends the lifespan of *C. elegans* and increases thermotolerance without any major change in metabolic function. H_2S-treated worms exhibit a 70 % increase in survival rate compared to the untreated controls, and the authors suggest that the mechanism of lifespan extension is dependent on the SIR2 (the silent information regulator 2) pathway. However, it is noteworthy that the response to H_2S is not identical among different species. For example, whilst an exposure to H_2S does not affect the metabolic function in *C. elegans*, hypothermia and hypometabolism are observed in rodents; however, this may also be due to differences in the relative dose exposures and uptake dynamics. Nevertheless, most ageing pathways are conserved between man and worm, and thus *C. elegans* represents a great advantage in exploring the role of H_2S in the ageing process, which can subsequently be extrapolated to higher animals in order to unravel the riddles of longevity and allow a more targeted approach to combat ageing-associated diseases.

A recent study, which explored exogenous H_2S effects and endogenous H_2S metabolism in the nematode *C. elegans*, demonstrated that a slow-releasing H_2S donor (GYY4137) is able to extend the median survival by up to 23 % and increases tolerance towards oxidative and endoplasmic reticulum (ER) stress (Qabazard et al. 2013). These authors also found that *cysl-2*, a sulfhydrylase/cysteine synthase in *C. elegans*, is transcriptionally upregulated by GYY4137 treatment and that the deletion of *cysl-2* results in a significant reduction in lifespan which is partially recovered by the supplementation of GYY4137. Likewise, in a mammalian cell culture system, GYY4137 is able to protect bovine aortic endothelial cells (BAECs) from oxidative stress and (H_2O_2)-induced cell death (Qabazard et al. 2013). Taken together, this provides further support that H_2S exerts a protective function which is consistent with the longevity dividend theory (Olshansky et al. 2006).

Moreover, a related study revealed that the *C. elegans* genome encodes orthologues of the three H_2S synthesising enzymes, namely, CBS, CSE and 3-MST, and is capable of the endogenous synthesis of H_2S (Qabazard et al. 2014). Genetic deficiency of *mpst-1* (3-MST orthologue 1), but not *cth-2* (CSE orthologue), reduces the lifespan of *C. elegans*, an effect which is reversed in worms supplemented with GYY4137 (Qabazard et al. 2014). Treatment of *C. elegans* with GYY4137 increases the expression of several age-related, stress-response and antioxidant genes, increases the lifespan in short-lived *mev-1* mutants (which are characterised by an elevated level of oxidative stress) and protects wild-type *C. elegans* against paraquat poisoning (Qabazard et al. 2014). The authors conclude that the lifespan-prolonging and health-promoting effects of H_2S in *C. elegans* are likely due to the antioxidant action of this highly cell-permeable gas (Qabazard et al. 2014). Whether H_2S prolongs directly the lifespan and promotes healthy ageing in mammals is a key question which has yet to be fully explored.

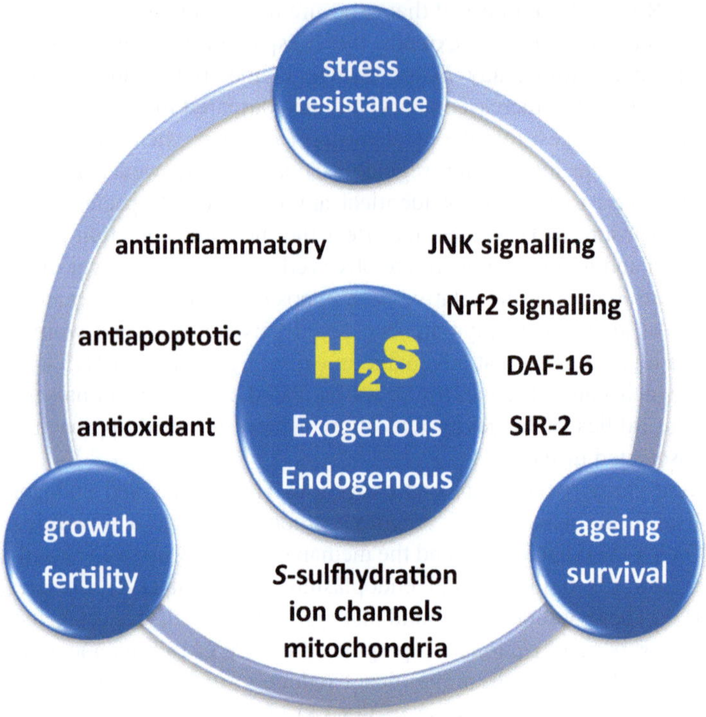

Fig. 2 Summary of H_2S contribution to ageing

3 Conclusions

Numerous findings suggest that endogenous H_2S plays a critical regulatory role in ageing and the cellular response to stress (Fig. 2), which in concert highlight that exogenous H_2S donors may prove valuable in the pharmacological intervention to delay the ageing process, the protection against age-associated diseases and ultimately the extension of a healthy lifespan.

References

An JH, Blackwell TK (2003) SKN-1 links *C. elegans* mesendodermal specification to a conserved oxidative stress response. Genes Dev 17(15):1882–1893

An JH, Vranas K, Lucke M, Inoue H, Hisamoto N, Matsumoto K, Blackwell TK (2005) Regulation of the *Caenorhabditis elegans* oxidative stress defense protein SKN-1 by glycogen synthase kinase-3. Proc Natl Acad Sci U S A 102(45):16275–16280

Anisimov VN, Berstein LM, Egormin PA, Piskunova TS, Popovich IG, Zabezhinski MA, Kovalenko IG, Poroshina TE, Semenchenko AV, Provinciali M (2005) Effect of metformin on life span and on the development of spontaneous mammary tumors in HER-2/neu transgenic mice. Exp Gerontol 40(8):685–693

Anisimov VN, Berstein LM, Egormin PA, Piskunova TS, Popovich IG, Zabezhinski MA, Tyndyk ML, Yurova MV, Kovalenko IG, Poroshina TE (2008) Metformin slows down aging and extends life span of female SHR mice. Cell Cycle 7(17):2769–2773

Apfeld J, O'Connor G, McDonagh T, DiStefano PS, Curtis R (2004) The AMP-activated protein kinase AAK-2 links energy levels and insulin-like signals to lifespan in *C. elegans*. Genes Dev 18(24):3004–3009

Aslami H, Heinen A, Roelofs JJ, Zuurbier CJ, Schultz MJ, Juffermans NP (2010) Suspended animation inducer hydrogen sulfide is protective in an in vivo model of ventilator-induced lung injury. Intensive Care Med 36(11):1946–1952

Bass TM, Weinkove D, Houthoofd K, Gems D, Partridge L (2007) Effects of resveratrol on lifespan in *Drosophila* melanogaster and *Caenorhabditis elegans*. Mech Ageing Dev 128(10): 546–552

Baur JA, Pearson KJ, Price NL, Jamieson HA, Lerin C, Kalra A, Prabhu VV, Allard JS, Lopez-Lluch G, Lewis K (2006) Resveratrol improves health and survival of mice on a high-calorie diet. Nature 444(7117):337–342

Berdichevsky A, Viswanathan M, Horvitz HR, Guarente L (2006) *C. elegans* SIR-2.1 interacts with 14-3-3 proteins to activate DAF-16 and extend life span. Cell 125(6):1165–1177

Beyer K, Lao JI, Carrato C, Rodriguez-Vila A, Latorre P, Mataro M, Llopis MA, Mate JL, Ariza A (2004) Cystathionine beta synthase as a risk factor for Alzheimer disease. Curr Alzheimer Res 1(2):127–133

Bjedov I, Toivonen JM, Kerr F, Slack C, Jacobson J, Foley A, Partridge L (2010) Mechanisms of life span extension by rapamycin in the fruit fly *Drosophila melanogaster*. Cell Metab 11(1): 35–46

Blackstone E, Roth MB (2007) Suspended animation-like state protects mice from lethal hypoxia. Shock 27(4):370–372

Blackstone E, Morrison M, Roth MB (2005) H_2S induces a suspended animation-like state in mice. Science 308(5721):518

Budde MW, Roth MB (2010) Hydrogen sulfide increases hypoxia-inducible factor-1 activity independently of von Hippel-Lindau tumor suppressor-1 in *C. elegans*. Sci Signal 21(1): 212–217

Burnett C, Valentini S, Cabreiro F, Goss M, Somogyvári M, Piper MD, Hoddinott M, Sutphin GL, Leko V, McElwee JJ, Vazquez-Manrique RP, Orfila AM, Ackerman D, Au C, Vinti G, Riesen M, Howard K, Neri C, Bedalov A, Kaeberlein M, Soti C, Partridge L, Gems D (2011) Absence of effects of Sir2 overexpression on lifespan in *C. elegans* and *Drosophila*. Nature 477(7365):482–485

Cai WJ, Wang MJ, Moore PK, Jin HM, Yao T, Zhu YC (2007) The novel proangiogenic effect of hydrogen sulfide is dependent on Akt phosphorylation. Cardiovasc Res 76(1):29–40

Cai W, Wang M, Ju L, Wang C, Zhu Y (2010) Hydrogen sulfide induces human colon cancer cell proliferation: role of Akt, ERK and p21. Cell Biol Int 34:565–572

Calvert JW, Jha S, Gundewar S, Elrod JW, Ramachandran A, Pattillo CB, Kevil CG, Lefer DJ (2009) Hydrogen sulfide mediates cardioprotection through Nrf2 signaling. Circ Res 105(4): 365–374

Campisi J, Kim S-H, Lim C-S, Rubio M (2001) Cellular senescence, cancer and aging: the telomere connection. Exp Gerontol 36(10):1619–1638

Château M-T, Araiz C, Descamps S, Galas S (2010) Klotho interferes with a novel FGF-signalling pathway and insulin/Igf-like signalling to improve longevity and stress resistance in *Caenorhabditis elegans*. Aging 2(9):567–581

Chattopadhyay M, Kodela R, Nath N, Barsegian A, Boring D, Kashfi K (2012a) Hydrogen sulfide-releasing aspirin suppresses NF-kappaB signaling in estrogen receptor negative breast cancer cells in vitro and in vivo. Biochem Pharmacol 83:723–732

Chattopadhyay M, Kodela R, Nath N, Dastagirzada YM, Velazquez-Martinez CA, Boring D, Kashfi K (2012b) Hydrogen sulfide-releasing NSAIDs inhibit the growth of human cancer cells: a general property and evidence of a tissue type-independent effect. Biochem Pharmacol 83:715–722
Chattopadhyay M, Kodela R, Olson KR, Kashfi K (2012c) NOSH-aspirin (NBS-1120), a novel nitric oxide- and hydrogen sulfide-releasing hybrid is a potent inhibitor of colon cancer cell growth in vitro and in a xenograft mouse model. Biochem Biophys Res Commun 419:523–528
Chen Y-H, Yao W-Z, Geng B, Ding Y-L, Lu M, Zhao M-W, Tang C-S (2005) Endogenous hydrogen sulfide in patients with COPD. Chest J 128(5):3205–3211
Dasgupta B, Milbrandt J (2007) Resveratrol stimulates AMP kinase activity in neurons. Proc Natl Acad Sci U S A 104(17):7217–7222
Deplancke B, Gaskins HR (2003) Hydrogen sulfide induces serum-independent cell cycle entry in nontransformed rat intestinal epithelial cells. FASEB J 17(10):1310–1312
Dröge W, Kinscherf R, Hildebrandt W, Schmitt T (2006) The deficit in low molecular weight thiols as a target for antiageing therapy. Curr Drug Targets 7(11):1505–1512
Du J, Hui Y, Cheung Y, Bin G, Jiang H, Chen X, Tang C (2004) The possible role of hydrogen sulfide as a smooth muscle cell proliferation inhibitor in rat cultured cells. Heart Vessels 19(2): 75–80
Elrod JW, Calvert JW, Morrison J, Doeller JE, Kraus DW, Tao L, Jiao X, Scalia R, Kiss L, Szabo C, Kimura H, Chow CW, Lefer DJ (2007) Hydrogen sulfide attenuates myocardial ischemia-reperfusion injury by preservation of mitochondrial function. Proc Natl Acad Sci U S A 104(39):15560–15565
Faller S, Ryter SW, Choi AM, Loop T, Schmidt R, Hoetzel A (2010) Inhaled hydrogen sulfide protects against ventilator-induced lung injury. Anesthesiology 113(1):104–115
Florian B, Vintilescu R, Balseanu AT, Buga A-M, Grisk O, Walker LC, Kessler C, Popa-Wagner A (2008) Long-term hypothermia reduces infarct volume in aged rats after focal ischemia. Neurosci Lett 438(2):180–185
Fox B, Schantz JT, Haigh R, Wood ME, Moore PK, Viner N, Spencer JP, Winyard PG, Whiteman M (2012) Inducible hydrogen sulfide synthesis in chondrocytes and mesenchymal progenitor cells: is H_2S a novel cytoprotective mediator in the inflamed joint? J Cell Mol Med 16(4):896–910
Friedman DB, Johnson TE (1988) Three mutants that extend both mean and maximum life span of the nematode, *Caenorhabditis elegans*, define the *age-1* gene. J Gerontol 43(4):B102–B109
Friedman DB, Johnson TE (2002) A mutation in the *age-1* gene in *Caenorhabditis elegans* lengthens life and reduces hermaphrodite fertility. Sci Aging Knowledge Environ 38:22–30
Gagliardi RJ (2000) Neuroprotection, excitotoxicicity and NMDA antagonists. Arq Neuro-psiquiatr 58(2B):583–588
Garcia-Bereguiain MA, Samhan-Arias AK, Martin-Romero FJ, Gutierrez-Merino C (2008) Hydrogen sulfide raises cytosolic calcium in neurons through activation of L-type Ca^{2+} channels. Antioxid Redox Signal 10(1):31–42
Golden TR, Melov S (2007) Gene expression changes associated with aging in *C. elegans*. WormBook 12:1–12
Greer EL, Dowlatshahi D, Banko MR, Villen J, Hoang K, Blanchard D, Gygi SP, Brunet A (2007a) An AMPK-FOXO pathway mediates longevity induced by a novel method of dietary restriction in *C. elegans*. Curr Biol 17(19):1646–1656
Greer EL, Oskoui PR, Banko MR, Maniar JM, Gygi MP, Gygi SP, Brunet A (2007b) The energy sensor AMP-activated protein kinase directly regulates the mammalian FOXO3 transcription factor. J Biol Chem 282(41):30107–30119
Hammerman M (1987) Insulin-like growth factors and aging. Endocrinol Metab Clin North Am 16(4):995–1011
Han Y, Qin J, Chang X, Yang Z, Bu D, Du J (2005a) Modulating effect of hydrogen sulfide on gamma-aminobutyric acid B receptor in recurrent febrile seizures in rats. Neurosci Res 53(2): 216–219

Han Y, Qin J, Chang X, Yang Z, Tang X, Du J (2005b) Hydrogen sulfide may improve the hippocampal damage induced by recurrent febrile seizures in rats. Biochem Biophys Res Commun 327(2):431–436

Hansen M, Taubert S, Crawford D, Libina N, Lee SJ, Kenyon C (2007) Lifespan extension by conditions that inhibit translation in *Caenorhabditis elegans*. Aging Cell 6(1):95–110

Haouzi P, Notet V, Chenuel B, Chalon B, Sponne I, Ogier V, Bihain B (2008) H_2S induced hypometabolism in mice is missing in sedated sheep. Respir Physiol Neurobiol 160(1): 109–115

Harrison DE, Strong R, Sharp ZD, Nelson JF, Astle CM, Flurkey K, Nadon NL, Wilkinson JE, Frenkel K, Carter CS (2009) Rapamycin fed late in life extends lifespan in genetically heterogeneous mice. Nature 460(7253):392–395

Henderson ST, Johnson TE (2001) Daf-16 integrates developmental and environmental inputs to mediate aging in the nematode *Caenorhabditis elegans*. Curr Biol 11(24):1975–1980

Hu Y, Chen X, Pan TT, Neo KL, Lee SW, Khin ESW, Moore PK, Bian JS (2008) Cardioprotection induced by hydrogen sulfide preconditioning involves activation of ERK and PI3K/Akt pathways. Pflügers Arch Eur J Physiol 455(4):607–616

Hu L-F, Lu M, Wu Z-Y, Wong PT-H, Bian J-S (2009) Hydrogen sulfide inhibits rotenone-induced apoptosis via preservation of mitochondrial function. Mol Pharmacol 75(1):27–34

Hu LF, Lu M, Tiong CX, Dawe GS, Hu G, Bian JS (2010) Neuroprotective effects of hydrogen sulfide on Parkinson's disease rat models. Aging Cell 9(2):135–146

Ingram DK, Anson RM, Cabo R, Mamczarz J, Zhu M, Mattison J, Lane MA, Roth GS (2004) Development of calorie restriction mimetics as a prolongevity strategy. Ann N Y Acad Sci 1019(1):412–423

Inoki K, Zhu T, Guan K-L (2003) TSC2 mediates cellular energy response to control cell growth and survival. Cell 115(5):577–590

Inoue H, Hisamoto N, An JH, Oliveira RP, Nishida E, Blackwell TK, Matsumoto K (2005) The *C. elegans* p38 MAPK pathway regulates nuclear localization of the transcription factor SKN-1 in oxidative stress response. Genes Dev 19(19):2278–2283

Ishizaka N, Mitani H, Nagai R (2002) Angiotensin II regulates Klotho gene expression. Jpn J Clin Med 60(10):1935–1939

Jeney V, Komódi E, Nagy E, Zarjou A, Vercellotti GM, Eaton JW, Balla G, Balla J (2009) Suppression of hemin-mediated oxidation of low-density lipoprotein and subsequent endothelial reactions by hydrogen sulfide (H_2S). Free Radic Biol Med 46(5):616–623

Jeong SO, Pae HO, Oh GS, Jeong GS, Lee BS, Lee S, Kim DY, Rhew HY, Lee KM, Chung HT (2006) Hydrogen sulfide potentiates interleukin-1β-induced nitric oxide production via enhancement of extracellular signal-regulated kinase activation in rat vascular smooth muscle cells. Biochem Biophys Res Commun 345(3):938–944

Jha S, Calvert JW, Duranski MR, Ramachandran A, Lefer DJ (2008) Hydrogen sulfide attenuates hepatic ischemia-reperfusion injury: role of antioxidant and antiapoptotic signaling. Am J Physiol Heart Circ Physiol 295(2):H801–H806

Kaeberlein M, Powers RW III, Steffen KK, Westman EA, Hu D, Dang N, Kerr EO, Kirkland KT, Fields S, Kennedy BK (2005) Regulation of yeast replicative life span by TOR and Sch9 in response to nutrients. Sci Signal 310(5751):1193–1196

Kamoun P, Belardinelli MC, Chabli A, Lallouchi K, Chadefaux-Vekemans B (2003) Endogenous hydrogen sulfide overproduction in Down syndrome. Am J Med Genet 116(3):310–311

Kapahi P, Zid B (2004) TOR pathway: linking nutrient sensing to life span. Sci Aging Knowledge Environ 36:34

Kapahi P, Zid BM, Harper T, Koslover D, Sapin V, Benzer S (2004) Regulation of lifespan in *Drosophila* by modulation of genes in the TOR signaling pathway. Curr Biol 14(10):885–890

Kenyon CJ (2010) The genetics of ageing. Nature 464(7288):504–512

Kenyon C, Chang J, Gensch E, Rudner A, Tabtiang R (1993) A *C. elegans* mutant that lives twice as long as wild type. Nature 366(6454):461–464

Kimura Y, Kimura H (2004) Hydrogen sulfide protects neurons from oxidative stress. FASEB J 18 (10):1165–1167

Kimura KD, Tissenbaum HA, Liu Y, Ruvkun G (1997) Daf-2, an insulin receptor-like gene that regulates longevity and diapause in *Caenorhabditis elegans*. Science 277(5328):942–946

Kimura H, Nagai Y, Umemura K, Kimura Y (2005) Physiological roles of hydrogen sulfide: synaptic modulation, neuroprotection, and smooth muscle relaxation. Antioxid Redox Signal 7(5–6):795–803

Kimura Y, Dargusch R, Schubert D, Kimura H (2006) Hydrogen sulfide protects HT22 neuronal cells from oxidative stress. Antioxid Redox Signal 8(3–4):661–670

Kirkwood TB (1977) Evolution of ageing. Nature 270(5635):301–304

Klass MR (1977) Aging in the nematode *Caenorhabditis elegans*: major biological and environmental factors influencing life span. Mech Ageing Dev 6:413–429

Krishnan N, Fu C, Pappin DJ, Tonks NK (2011) H_2S-induced sulfhydration of the phosphatase PTP1B and its role in the endoplasmic reticulum stress response. Sci Signal 4(203):86

Kump LR, Pavlov A, Arthur MA (2005) Massive release of hydrogen sulfide to the surface ocean and atmosphere during intervals of oceanic anoxia. Geology 33(5):397–400

Kuro-o M, Matsumura Y, Aizawa H, Kawaguchi H, Suga T, Utsugi T, Ohyama Y, Kurabayashi M, Kaname T, Kume E (1997) Mutation of the mouse Klotho gene leads to a syndrome resembling ageing. Nature 390(6655):45–51

Kurosu H, Yamamoto M, Clark JD, Pastor JV, Nandi A, Gurnani P, McGuinness OP, Chikuda H, Yamaguchi M, Kawaguchi H (2005) Suppression of aging in mice by the hormone Klotho. Sci Signal 309(5742):1829–1833

Laggner H, Hermann M, Esterbauer H, Muellner MK, Exner M, Gmeiner BM, Kapiotis S (2007a) The novel gaseous vasorelaxant hydrogen sulfide inhibits angiotensin-converting enzyme activity of endothelial cells. J Hypertens 25(10):2100–2104

Laggner H, Muellner MK, Schreier S, Sturm B, Hermann M, Exner M (2007b) Hydrogen sulphide: a novel physiological inhibitor of LDL atherogenic modification by HOCl. Free Radic Res 41 (7):741–747

Lahousse SA, Stopa EG, Mulberg AE, de la Monte SM (2003) Reduced expression of the cystic fibrosis transmembrane conductance regulator gene in the hypothalamus of patients with Alzheimer's disease. J Alzheimers Dis 5(6):455–462

Landis JN, Murphy CT (2010) Integration of diverse inputs in the regulation of *Caenorhabditis elegans* DAF-16/FOXO. Dev Dyn 239(5):1405–1412

Lavu M, Bhushan S, Lefer D (2011) Hydrogen sulfide-mediated cardioprotection: mechanisms and therapeutic potential. Clin Sci 120:219–229

Lee SS, Kennedy S, Tolonen AC, Ruvkun G (2003) DAF-16 target genes that control *C. elegans* life-span and metabolism. Sci Signal 300(5619):644–647

Lee SW, Cheng Y, Moore PK, Bian J-S (2007) Hydrogen sulphide regulates intracellular pH in vascular smooth muscle cells. Biochem Biophys Res Commun 358(4):1142–1147

Lee M, Tazzari V, Giustarini D, Rossi R, Sparatore A, Del Soldato P, McGeer E, McGeer PL (2010) Effects of hydrogen sulfide-releasing l-DOPA derivatives on glial activation potential for treating Parkinson disease. J Biol Chem 285(23):17318–17328

Lee ZW, Zhou J, Chen CS, Zhao Y, Tan CH, Li L, Moore PK, Deng LW (2011) The slow-releasing hydrogen sulfide donor, GYY4137, exhibits novel anti-cancer effects in vitro and in vivo. PLoS One 6:e21077

Li L, Salto-Tellez M, Tan C-H, Whiteman M, Moore PK (2009) GYY4137, a novel hydrogen sulfide-releasing molecule, protects against endotoxic shock in the rat. Free Radic Biol Med 47(1):103–113

Lin K, Dorman JB, Rodan A, Kenyon C (1997) Daf-16: an HNF-3/forkhead family member that can function to double the life-span of *Caenorhabditis elegans*. Science 278(5341):1319–1322

Liu YY, Nagpure BV, Wong PT, Bian JS (2013) Hydrogen sulfide protects SH-SY5Y neuronal cells against d-galactose induced cell injury by suppression of advanced glycation end products formation and oxidative stress. Neurochem Int 62(5):603–609

Lu M, Hu L-F, Hu G, Bian J-S (2008) Hydrogen sulfide protects astrocytes against H_2O_2-induced neural injury via enhancing glutamate uptake. Free Radic Biol Med 45(12):1705–1713

Mair W, Dillin A (2008) Aging and survival: the genetics of life span extension by dietary restriction. Annu Rev Biochem 77:727–754

McElwee J, Bubb K, Thomas JH (2003) Transcriptional outputs of the *Caenorhabditis elegans* forkhead protein DAF-16. Aging Cell 2(2):111–121

Meng QH, Yang G, Yang W, Jiang B, Wu L, Wang R (2007) Protective effect of hydrogen sulfide on balloon injury-induced neointima hyperplasia in rat carotid arteries. Am J Pathol 170(4): 1406–1414

Miller DL, Roth MB (2007) Hydrogen sulfide increases thermotolerance and lifespan in *Caenorhabditis elegans*. Proc Natl Acad Sci U S A 104(51):20618–20622

Miller DL, Budde MW, Roth MB (2011) HIF-1 and SKN-1 coordinate the transcriptional response to hydrogen sulfide in *Caenorhabditis elegans*. PLoS One 6(9):e25476

Morris JZ, Tissenbaum HA, Ruvkun G (1996) A phosphatidylinositol-3-OH kinase family member regulating longevity and diapause in *Caenorhabditis elegans*. Nature 382:536–539

Murphy CT, McCarroll SA, Bargmann CI, Fraser A, Kamath RS, Ahringer J, Li H, Kenyon C (2003) Genes that act downstream of DAF-16 to influence the lifespan of *Caenorhabditis elegans*. Nature 424(6946):277–283

Mustafa AK, Gadalla MM, Sen N, Kim S, Mu W, Gazi SK, Barrow RK, Yang G, Wang R, Snyder SH (2009) H_2S signals through protein S-sulfhydration. Sci Signal 2(96):72

Muzaffar S, Shukla N, Bond M, Newby AC, Angelini GD, Sparatore A, Del Soldato P, Jeremy JY (2008) Exogenous hydrogen sulfide inhibits superoxide formation, NOX-1 expression and Rac1 activity in human vascular smooth muscle cells. J Vasc Res 45(6):521–528

Nishida M, Sawa T, Kitajima N, Ono K, Inoue H, Ihara H, Motohashi H, Yamamoto M, Suematsu M, Kurose H (2012) Hydrogen sulfide anion regulates redox signaling via electrophile sulfhydration. Nat Chem Biol 8(8):714–724

Ogg S, Paradis S, Gottlieb S, Patterson GI, Lee L, Tissenbaum HA, Ruvkun G (1997) The Fork head transcription factor DAF-16 transduces insulin-like metabolic and longevity signals in *C. elegans*. Nature 389(6654):994–999

Oh SW, Mukhopadhyay A, Svrzikapa N, Jiang F, Davis RJ, Tissenbaum HA (2005) JNK regulates lifespan in *Caenorhabditis elegans* by modulating nuclear translocation of forkhead transcription factor/DAF-16. Proc Natl Acad Sci U S A 102(12):4494–4499

Oh G-S, Pae H-O, Lee B-S, Kim B-N, Kim J-M, Kim H-R, Jeon SB, Jeon WK, Chae H-J, Chung H-T (2006) Hydrogen sulfide inhibits nitric oxide production and nuclear factor-kappaB via heme oxygenase-1 expression in RAW2647 macrophages stimulated with lipopolysaccharide. Free Radic Biol Med 41(1):106–119

Oliveira RP, Porter Abate J, Dilks K, Landis J, Ashraf J, Murphy CT, Blackwell TK (2009) Condition-adapted stress and longevity gene regulation by *Caenorhabditis elegans* SKN-1/Nrf. Aging Cell 8(5):524–541

Olshansky SJ, Perry D, Miller RA, Butler RN (2006) In pursuit of the longevity dividend. The Scientist 20:28–36

Papapetropoulos A, Pyriochou A, Altaany Z, Yang G, Marazioti A, Zhou Z, Jeschke MG, Branski LK, Herndon DN, Wang R (2009) Hydrogen sulfide is an endogenous stimulator of angiogenesis. Proc Natl Acad Sci U S A 106(51):21972–21977

Paradis S, Ruvkun G (1998) *Caenorhabditis elegans* Akt/PKB transduces insulin receptor-like signals from AGE-1 PI3 kinase to the DAF-16 transcription factor. Genes Dev 12(16): 2488–2498

Paul BD, Snyder SH (2012) H_2S signalling through protein sulfhydration and beyond. Nat Rev Mol Cell Biol 13:499–507

Pearl R (1928) The rate of living. University Press, New York

Pei Y, Wu B, Cao Q, Wu L, Yang G (2011) Hydrogen sulfide mediates the anti-survival effect of sulforaphane on human prostate cancer cells. Toxicol Appl Pharmacol 257:420–428

Predmore BL, Alendy MJ, Ahmed KI, Leeuwenburgh C, Julian D (2010) The hydrogen sulfide signaling system: changes during aging and the benefits of caloric restriction. Age 32(4): 467–481

Qabazard B, Ahmed S, Li L, Arlt VM, Moore PK, Sturzenbaum SR (2013) *C. elegans* aging is modulated by hydrogen sulfide and the sulfhydrylase/cysteine synthase *cysl-2*. PLoS One 8(11):e80135

Qabazard B, Li L, Gruber J, Peh MT, Ng LF, Kumar SD, Rose P, Tan CH, Dymock BW, Wei F, Swain SC, Halliwell B, Sturzenbaum SR, Moore PK (2014) Hydrogen sulfide is an endogenous regulator of aging in *Caenorhabditis elegans*. Antioxid Redox Signal 20(16):2621–2630

Rinaldi L, Gobbi G, Pambianco M, Micheloni C, Mirandola P, Vitale M (2006) Hydrogen sulfide prevents apoptosis of human PMN via inhibition of p38 and caspase 3. Lab Invest 86(4): 391–397

Rogina B, Helfand SL (2004) Sir2 mediates longevity in the fly through a pathway related to calorie restriction. Proc Natl Acad Sci U S A 101(45):15998–16003

Sastre J, Martín JA, Gómez-Cabrera M-C, Pereda J, Borrás C, Pallardó FV, Viña J (2005) Age-associated oxidative damage leads to absence of γ-cystathionase in over 50 % of rat lenses: relevance in cataractogenesis. Free Radic Biol Med 38(5):575–582

Sen N, Paul BD, Gadalla MM, Mustafa AK, Sen T, Xu R, Kim S, Snyder SH (2012) Hydrogen sulfide-linked sulfhydration of NF-κB mediates its anti-apoptotic actions. Mol Cell 45(1): 13–24

Shiraki-Iida T, Aizawa H, Matsumura Y, Sekine S, Iida A, Anazawa H, Nagai R, Kuro-o M, Nabeshima Y-I (1998) Structure of the mouse Klotho gene and its two transcripts encoding membrane and secreted protein. FEBS Lett 424(1):6–10

Simon F, Giudici R, Duy CN, Schelzig H, Öter S, Gröger M, Wachter U, Vogt J, Speit G, Szabo C (2008) Hemodynamic and metabolic effects of hydrogen sulfide during porcine ischemia/reperfusion injury. Shock 30(4):359–364

Sivarajah A, McDonald M, Thiemermann C (2006) The production of hydrogen sulfide limits myocardial ischemia and reperfusion injury and contributes to the cardioprotective effects of preconditioning with endotoxin, but not ischemia in the rat. Shock 26(2):154–161

Suo R, Zhao ZZ, Tang ZH, Ren Z, Liu X, Liu LS, Wang Z, Tang CK, Wei DH, Jiang ZS (2013) Hydrogen sulfide prevents H_2O_2-induced senescence in human umbilical vein endothelial cells through SIRT1 activation. Mol Med Rep 7(6):1865–1870

Suzuki K, Olah G, Modis K, Coletta C, Kulp G, Gerö D, Szoleczky P, Chang T, Zhou Z, Wu L (2011) Hydrogen sulfide replacement therapy protects the vascular endothelium in hyperglycemia by preserving mitochondrial function. Proc Natl Acad Sci U S A 108(33):13829–13834

Tang XQ, Yang CT, Chen J, Yin WL, Tian SW, Hu B, Feng J, Li YJ (2008) Effect of hydrogen sulfide on β-amyloid-induced damage in PC12 cells. Clin Exp Pharmacol Physiol 35(2): 180–186

Tissenbaum HA, Guarente L (2001) Increased dosage of a *sir-2* gene extends lifespan in *Caenorhabditis elegans*. Nature 410:227–230

Tullet JM, Hertweck M, An JH, Baker J, Hwang JY, Liu S, Oliveira RP, Baumeister R, Blackwell TK (2008) Direct inhibition of the longevity-promoting factor SKN-1 by insulin-like signaling in *C. elegans*. Cell 132(6):1025–1038

Tyagi N, Moshal KS, Sen U, Vacek TP, Kumar M, Hughes WM Jr, Kundu S, Tyagi SC (2009) H_2S protects against methionine-induced oxidative stress in brain endothelial cells. Antioxid Redox Signal 11(1):25–33

Vellai T, Takacs-Vellai K, Zhang Y, Kovacs AL, Orosz L, Müller F (2003) Influence of TOR kinase on lifespan in *C. elegans*. Nature 426(6967):620

Viswanathan M, Kim SK, Berdichevsky A, Guarente L (2005) A role for SIR-2.1 regulation of ER stress response genes in determining *C. elegans* life span. Dev Cell 9(5):605–615

Wang R (2012) Physiological implications of hydrogen sulfide: a whiff exploration that blossomed. Physiol Rev 92(2):791–896

Wang MC, Bohmann D, Jasper H (2003) JNK signaling confers tolerance to oxidative stress and extends lifespan in *Drosophila*. Dev Cell 5(5):811–816

Wang Y, Oh SW, Deplancke B, Luo J, Walhout A, Tissenbaum HA (2006) *C. elegans* 14-3-3 proteins regulate life span and interact with SIR-2.1 and DAF-16/FOXO. Mech Ageing Dev 127(9):741–747

Ward PD (2006) Impact from the deep. Sci Am 295(4):64–71

Wei H, Zhang R, Jin H, Liu D, Tang X, Tang C, Du J (2010) Hydrogen sulfide attenuates hyperhomocysteinemia-induced cardiomyocytic endoplasmic reticulum stress in rats. Antioxid Redox Signal 12(9):1079–1091

Weinert BT, Timiras PS (2003) Invited review: theories of aging. J Appl Physiol 95(4):1706–1716

Whiteman M, Winyard PG (2011) Hydrogen sulfide and inflammation: the good, the bad, the ugly and the promising. Expert Rev Clin Pharmacol 4(1):13–32

Whiteman M, Armstrong JS, Chu SH, Jia-Ling S, Wong BS, Cheung NS, Halliwell B, Moore PK (2004) The novel neuromodulator hydrogen sulfide: an endogenous peroxynitrite 'scavenger'? J Neurochem 90(3):765–768

Whiteman M, Cheung NS, Zhu YZ, Chu SH, Siau JL, Wong BS, Armstrong JS, Moore PK (2005) Hydrogen sulphide: a novel inhibitor of hypochlorous acid-mediated oxidative damage in the brain? Biochem Biophys Res Commun 326(4):794–798

Wilkinson JE, Burmeister L, Brooks SV, Chan CC, Friedline S, Harrison DE, Hejtmancik JF, Nadon N, Strong R, Wood LK (2012) Rapamycin slows aging in mice. Aging Cell 11(4): 675–682

Williams GC (1957) Pleiotropy, natural selection, and the evolution of senescence. Evolution 11(4):398–411

Wood JG, Rogina B, Lavu S, Howitz K, Helfand SL, Tatar M, Sinclair D (2004) Sirtuin activators mimic caloric restriction and delay ageing in metazoans. Nature 430(7000):686–689

Yamamoto M, Clark JD, Pastor JV, Gurnani P, Nandi A, Kurosu H, Miyoshi M, Ogawa Y, Castrillon DH, Rosenblatt KP (2005) Regulation of oxidative stress by the anti-aging hormone Klotho. J Biol Chem 280(45):38029–38034

Yan SK, Chang T, Wang H, Wu L, Wang R, Meng QH (2006) Effects of hydrogen sulfide on homocysteine-induced oxidative stress in vascular smooth muscle cells. Biochem Biophys Res Commun 351(2):485–491

Yang G, Sun X, Wang R (2004) Hydrogen sulfide-induced apoptosis of human aorta smooth muscle cells via the activation of mitogen-activated protein kinases and caspase-3. FASEB J 18(14):1782–1784

Yang G, Zhao K, Ju Y, Mani S, Cao Q, Puukila S, Khaper N, Wu L, Wang R (2013) Hydrogen sulfide protects against cellular senescence via S-sulfhydration of Keap1 and activation of Nrf2. Antioxid Redox Signal 18(15):1906–1919

Yong QC, Lee SW, Foo CS, Neo KL, Chen X, Bian J-S (2008) Endogenous hydrogen sulphide mediates the cardioprotection induced by ischemic postconditioning. Am J Physiol Heart Circ Physiol 295(3):H1330–H1340

Zagli G, Patacchini R, Trevisani M, Abbate R, Cinotti S, Gensini GF, Masotti G, Geppetti P (2007) Hydrogen sulfide inhibits human platelet aggregation. Eur J Pharmacol 559(1):65–68

Zanardo RC, Brancaleone V, Distrutti E, Fiorucci S, Cirino G, Wallace JL (2006) Hydrogen sulfide is an endogenous modulator of leukocyte-mediated inflammation. FASEB J 20(12): 2118–2120

Zhang H, Bhatia M (2008) Hydrogen sulfide: a novel mediator of leukocyte activation. Immunopharmacol Immunotoxicol 30(4):631–645

Zhang Y, Tang Z-H, Ren Z, Qu S-L, Liu M-H, Liu L-S, Jiang Z-S (2013) Hydrogen sulfide, the next potent preventive and therapeutic agent in aging and age-associated diseases. Mol Cell Biol 33(6):1104–1113

Wang MC, Bohmann D, Jasper H (2003) JNK signaling confers tolerance to oxidative stress and extends lifespan in Drosophila. Dev Cell 5(5):811–816
Wang Y, Oh SW, Deplancke B, Luo J, Walhout AJ, Tissenbaum HA (2006) C. elegans 14-3-3 proteins regulate life span and interact with SIR-2.1 and DAF-16/FOXO. Mech Ageing Dev 127(9):741–747
[illegible]

Part VIII

H_2S, Measurement and Possible Therapeutics?

Fluorescent Probes for H_2S Detection and Quantification

Wei Feng and Brian W. Dymock

Contents

Abstract
Many diverse, sensitive and structurally novel fluorescent probes have recently been reported for H_2S detection. Quantification of H_2S requires a selective chemosensor which will react only with H_2S against a background of high concentrations of other thiols or reducing agents. Most published probes are able to quantify H_2S selectively in a simple in vitro system with the most

W. Feng
Department of Pharmacy, National University of Singapore, Block S4A #03,
18 Science Drive 4, Singapore 117543, Singapore

Centre for Life Sciences, Life Sciences Institute, National University of Singapore, #05-02,
28 Medical Drive, Singapore 117456, Singapore

B.W. Dymock (✉)
Department of Pharmacy, National University of Singapore, Block S4A #03,
18 Science Drive 4, Singapore 117543, Singapore
e-mail: phadbw@nus.edu.sg

P.K. Moore, M. Whiteman (eds.), *Chemistry, Biochemistry and Pharmacology of Hydrogen Sulfide*, Handbook of Experimental Pharmacology 230,
DOI 10.1007/978-3-319-18144-8_15

sensitive probes able to detect H_2S at below 100 nM concentrations. A subset of probes also have utility in sensing H_2S in living cells, and there are now several with specific sub-cellular localization and a few cases of in vivo applications. Biologists studying H_2S now have a wide range of tools to assist them to aid further understanding of the role of H_2S in biology.

Keywords
Hydrogen sulfide • Fluorescent probe • Chemosensor

1 Introduction

Fluorescent probe molecules which can chemically detect or 'sense' H_2S are often called 'chemosensors' or 'chemodosimeters'. Quantification of H_2S requires a very selective chemosensor which will react only with H_2S against a background of high concentrations of other thiols or reducing agents, as is found in a cell. Most published chemosensors are able to quantify H_2S in a simple in vitro system, but application in the cell is still very challenging in most cases. A wide range of diverse chemosensing probes have been reported with most being simple probe molecules that selectively react with H_2S to produce a fluorescent product that can be quantified using a standard laboratory fluorimeter. For application in a wide range of biological situations, a good probe molecule must fulfil a range of criteria:

- Rapidly react with H_2S under physiological conditions
- Selectively react with H_2S over high concentrations of biological thiols and anions
- Be sensitive to low concentrations of H_2S (high quantum yield)
- Linear response within physiological H_2S concentration range
- Be biocompatible with good permeability and intracellular stability
- Is not toxic and does not produce toxic products
- Be chemically stable for storage and longer-term studies
- Be straightforward to synthesise or be commercially available

With those objectives in mind, most authors have addressed several but rarely adequately solved all of these challenges. This chapter discusses published probes operating by a fluorescence readout and will critically appraise their performance against the aforementioned key criteria. The probes are organised into three categories: (1) probes which react with H_2S through a nucleophilic reaction, (2) probes which undergo selective reduction by H_2S and (3) probes which trap H_2S using a coordinated metal.

The probes discussed herein are especially selected from the more recent literature from the past 2–3 years, for earlier material readers are directed to other reviews (Lin and Chang 2012; Peng and Xian 2014; Yu et al. 2014).

2 Probes Which React with H_2S Through a Nucleophilic Reaction

Under physiological conditions, H_2S (pK_a c.6.9) dissociates to form HS^- as the major form (estimated at 72 % (Li and Lancaster 2013)), which is a very good nucleophile. A number of reports on H_2S fluorescent probes which are based on this nucleophilic feature have been published recently. When the H_2S amount is as small as endogenous levels, the side fluorescence turn-on can be comparable to that resulting from H_2S. As a non-substituted thiol, H_2S is different from mono-substituted biothiols, such as cysteine and glutathione (GSH), in that H_2S can undergo nucleophilic reaction twice instead of once for normal biothiols. It seems that probes that will turn on fluorescence only after two nucleophilic reactions can distinguish H_2S from biothiols. The general design of nucleophilic substitution–cyclisation-based fluorescent probes has been reviewed in 2014 by Xian as shown in Fig. 1 (Peng et al. 2014). H_2S reacts with the electrophilic component in probe **A** to deliver an intermediate **B1** which contains a free thiol. In the presence of another electrophilic centre, such as an ester group, the free thiol will undergo intramolecular cyclisation spontaneously to form product **P** and release the fluorophore which can be detected by fluorescence measurement. Probe **A** may also react with biothiols to form intermediate **B2**. However, the cyclisation will not happen due to the lack of a free SH in **B2**. Biothiols are also not reactive enough to undergo intermolecular attack on the ester group. As a result, the fluorophore cannot be released.

In 2011, He and co-workers reported the synthesis and application of bis-electrophilic centred probes SFP 1 (**1**) and SFP 2 (**3**) (Fig. 2) (Qian et al. 2011). 1,3,5-Triaryl-2-pyrazoline and 4,4-difluoro-1,3,5,7-tetramethyl-4-bora-3a,4a-diaza-*s*-indacene (BODIPY) were employed as fluorophores for each probe, respectively. Aldehyde groups of **1** and **3** react with HS^- to form an intermediate hemithioacetal group which then undergoes Michael addition to a

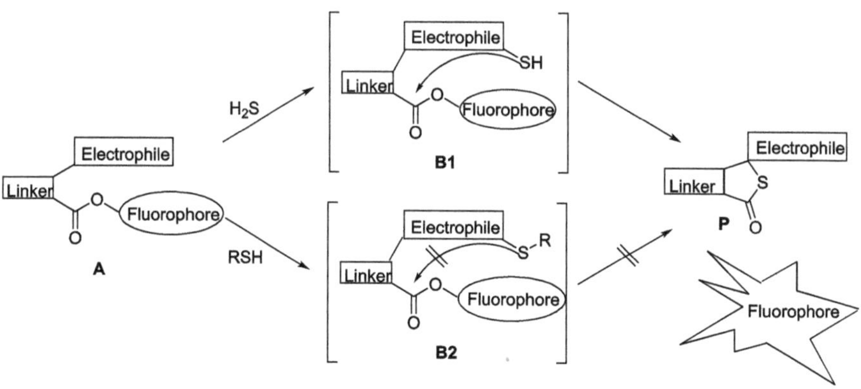

Fig. 1 General design of the nucleophilic substitution–cyclisation-based fluorescent probes for H_2S

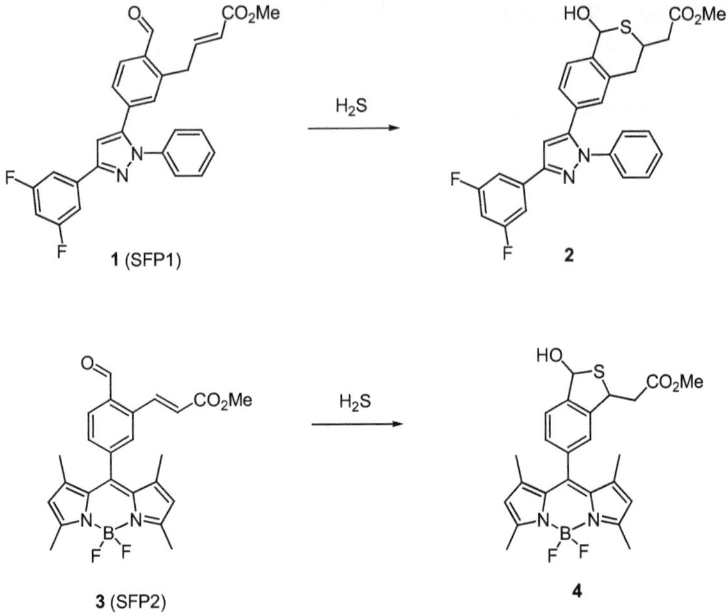

Fig. 2 Structures of probe SFP 1 and SFP 2 and their mechanisms

nearby acrylate ester give cyclised products **2** and **4**, respectively. Fluorescence response of these probes to sulphide (50 μM) is about fourfold than that of biothiols (1 mM). However, biothiols do turn on fluorescence, which is much higher than control experiments. Imaging of sulphide with these two probes in HeLa cells was successful. Images were obtained by incubating 10 μM probes with Na_2S of different concentrations ranging from 10 to 200 μM. Furthermore, as well as addition of exogenous Na_2S, addition of GSH and cysteine can also lead to fluorescence turn-on in HeLa cells following increased production of H_2S. However, better sensitivity of H_2S detection in cells is required.

Also based on an aldehyde-Michael acceptor system, the recently reported probe P3 (**5**) was shown to form a five-membered carbocyclic product following a three-step mechanism that the authors propose as a revision to the generally conceived mechanism of initial Michael addition (Fig. 3) (Singha et al. 2015). Probe **5** was optimised using a computational approach which added two methoxy groups to the aromatic ring increasing the electrophilicity of the enone system as well as creating a steric hindrance to attack at the enone, hence favouring the smaller HS^- anion attack. **5** (10 μM) reacts with Na_2S (100 μM) in only 5 min (pH 7.4 in Hepes buffer). Very good selectivity over other biological thiols is achieved with cysteine (200 μM) giving a 7 % response. However, with 10 % FBS added to the media, the selectivity over cysteine was improved indicating promise for cellular application. **5** was reported to have a LOD of only 50 nM with S/N ratio of 3 making it one of the most sensitive probes published to date. In HeLa cells, **5** (10 μM) was sensitive enough to detect endogenous levels of H_2S following a 30 min incubation,

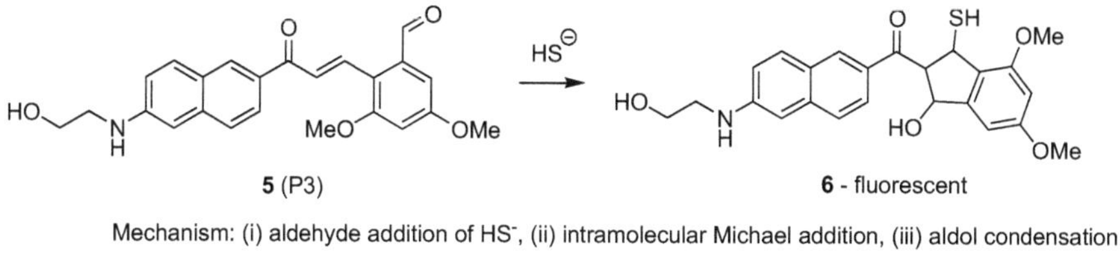

Fig. 3 Structure of probe P3 and the conversion to its fluorescent carbocyclic product via a three-step mechanism

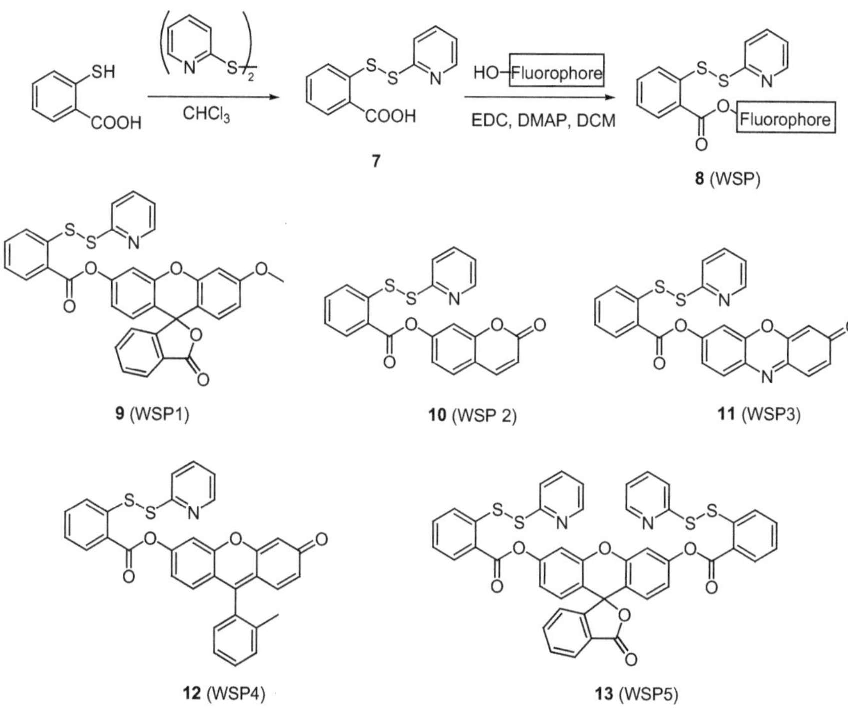

Fig. 4 Synthesis of WSP probes

further supported by a reduction in signal when H_2S synthesis inhibitors phorbol myristate acetate or propargyl glycine were added. With the addition of exogenous Na_2S (300 μM), a very strong fluorescence signal was obtained. Being a sensitive and selective probe, **5** demonstrates that other less selective probes have overestimated the levels of endogenously produced H_2S.

A series of WSP (**8**) probes were synthesised by Xian (Liu et al. 2011; Peng et al. 2014). Intermediate 2-(2-pyridinyldithio)-benzoic acid (**7**) was easily prepared in good yield. It was coupled to different fluorophores, such as methoxy fluorescein, 7-hydroxycoumarin, resorufin and 2-methyl Tokyo Green, through an ester formation reaction (Fig. 4). WSP 1 (**9**), WSP 2 (**10**), WSP 3 (**11**) and WSP

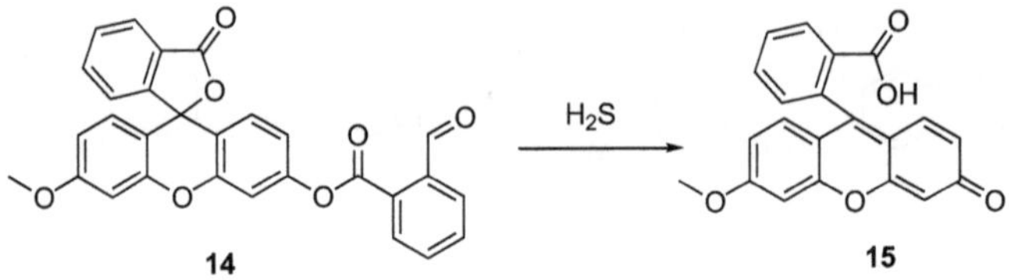

Fig. 5 Structure of probe **15** and its mechanism

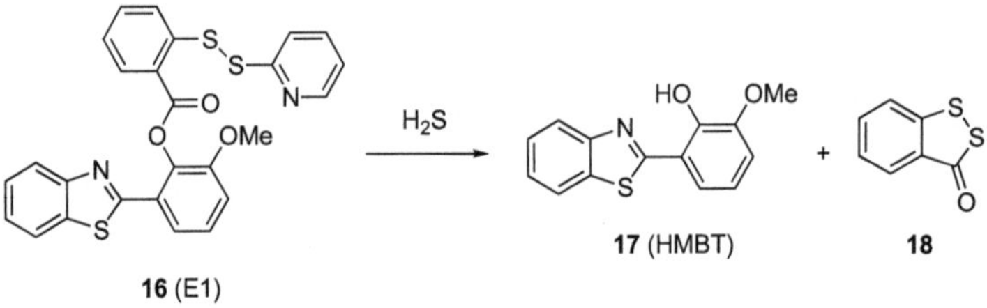

Fig. 6 Structure of probe E1 and its mechanism

5 (**13**) have low detection limits ranging from 47 to 79 nM, and WSP 4 (**12**) has a detection limit of 266 nM. A time period of 30 min is required for the fluorescence intensity to reach the maximum. However, it can be shortened to 5 min when a small amount of surfactant is present. All of these probes showed good selectivity of H_2S over other reactive sulphur species (200 μM) including cysteine, GSH, homocysteine, SO_3^{2-} and $S_2O_3^{2-}$. Good fluorescence images were obtained when **12** (30 μM) and **13** (50 μM) were incubated together with NaSH in HeLa cells, and no fluorescence was observed when no NaSH was added, indicating a good selectivity of H_2S over biothiols.

In 2014, Zhang and co-workers reported the synthesis and application of probe **14** (Fig. 5) (Liu et al. 2014b). It has a 2-formylbenzoate group which can react with HS^- to form a hemithioacetal. This probe is applicable for H_2S detection with H_2S concentrations ranging from 2.5 μM to 1000 μM. Confocal fluorescence imaging was achieved when probe **14** (5 μM) was incubated together with Na_2S (1 mM) in HeLa cells.

Probe E1 (**16**) was developed by Qian in 2012 (Xu et al. 2012). It is based on excited-state intramolecular proton transfer (ESIPT). 2-(20-Hydroxyphenyl) benzothiazole (HMBT) was chosen as the fluorophore, and the bis-electrophilic centres are disulphide and ester groups (Fig. 6). Following reaction of HS^- with the disulphide, a hydrogenpersulphide is formed which then attacks the ester forming the fluorescent HMBT (**17**) and by-product **18**. Probe **16** responds to H_2S within 2 min with a 30-fold fluorescence enhancement. The limit of detection can be as low as 120 nM. It is also applicable for H_2S imaging in HeLa cells, and the fluorescence can be significantly enhanced by addition of cetyltrimethylammonium bromide (CTAB). However, there is a disadvantage of these probes. Although biothiols, such as cysteine and GSH, cannot react to elicit fluorescence, they do

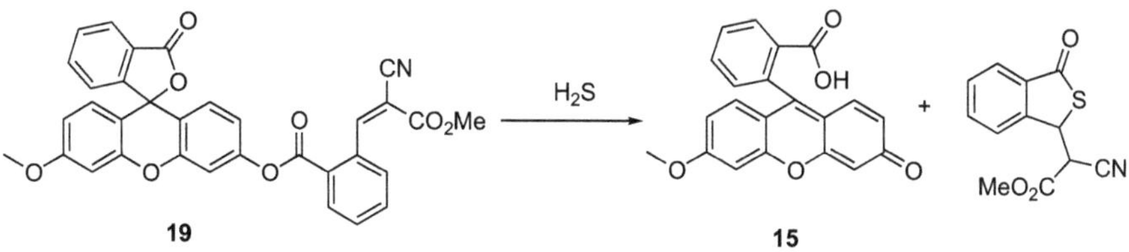

Fig. 7 Structure of probe **19** and its mechanism

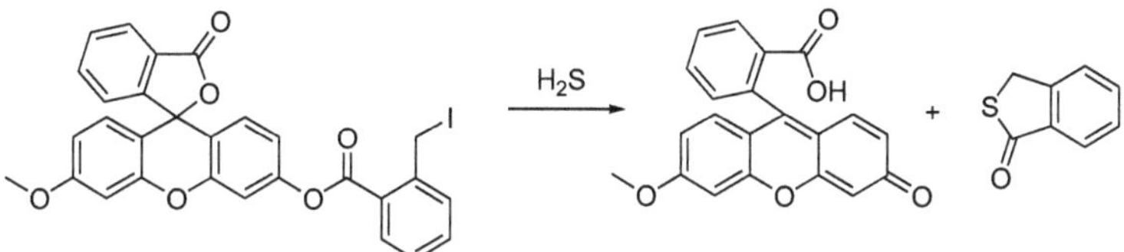

Fig. 8 Structure of probe **21** and its mechanism

react with the disulphide or aldehyde moieties, thus leading to the requirement of high probe loadings. Since H_2S and biothiols have different pKa values in aqueous solution (approximately 6.9 for H_2S and 8.5 for thiols), a probe that has bis-electrophilic centres which are both reactive to H_2S but not to biothiols is possible.

In 2012, Xian developed probe **19** which has bis-electrophilic centres: a Michael acceptor and an ester group (Fig. 7) (Liu et al. 2012). Experiments showed that the Michael acceptor does not react with cysteine or GSH. Upon treatment with NaSH, probe **19** responds with an 11-fold fluorescence increase within 30 min, generating fluorescent product **15**. However, when **19** was used in COS-7 cells for H_2S imaging, only a weak fluorescence image was observed.

In 2013, Guo and co-workers reported the design and synthesis of probe **20**, which employs an alkyl iodide as the first electrophilic centre (Fig. 8) (Zhang et al. 2013). Experiments showed that cysteine or GSH (200 μM) does not react with **20** (20 μM) in PBS:MeCN (1:1, v/v), while reaction of **20** with H_2S is complete within 1 h. The capability of **20** (5 μM) to image H_2S (100 μM NaSH) in a biological system was demonstrated in COS-7 cells. Fluorescence intensity can be further enhanced by adding CTAB without any interference from biothiols.

Probes DSP 1 (**21**), DSP 2 (**22**) and DSP 3 (**23**) were also developed by Xian (Fig. 9) (Liu et al. 2014a). **23** gave a much stronger fluorescence response than **21** or **22**. This probe was designed for the detection of hydrogen polysulphide H_2S_n through aromatic substitution and cyclisation reactions. With **23**, H_2S_n can be detected with a very low LOD of 71 nM. When H_2S was mixed together with reactive oxygen species (ROS), fluorescence enhancement was also observed. However, **23** does not respond to H_2S itself.

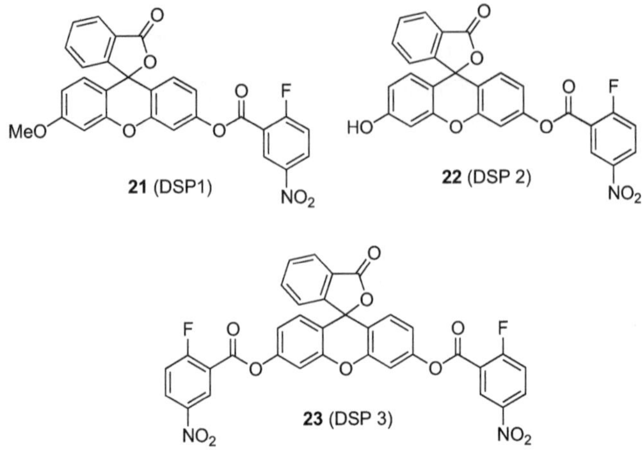

Fig. 9 Structures of probes DSP 1, DSP 2, DSP 3

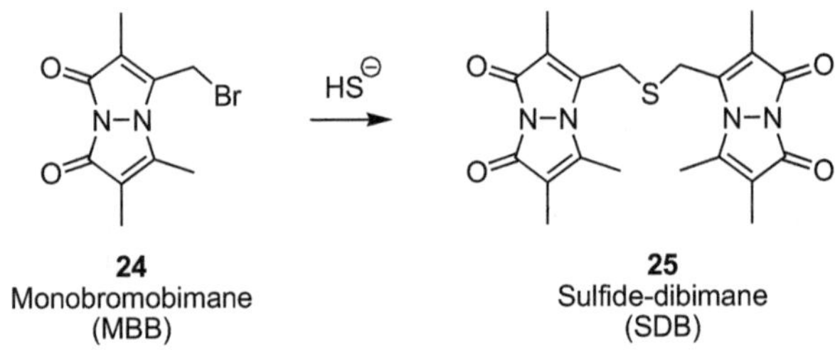

Fig. 10 Reaction of MBB with sulphide to produce fluorescent SDB which is accurately quantified by reversed-phase HPLC using a fluorescence detector

Probably the most sensitive detection system yet published is the protocol developed by Kevil for reaction of sulphide with monobromobimane (MBB, **24**) (Shen et al. 2011, 2012). Following alkylation with 2 MBB molecules, the resulting product, sulfide-bimane (SDB, **25**), is detected by reversed-phase HPLC with fluorescence detection (excitation 390 nm, emission 475 nm) (Fig. 10). This complex series of experiments requires many precautions to be taken, such as rigorous exclusion of oxygen, to achieve the clamed LOD of 5 nM for sulphide. Using this method, plasma levels of total H_2S were quantified at around 1.5 μM where free H_2S levels were in the range of 200 nM (Shen et al. 2012).

3 Probes Which Undergo Selective Reduction by H_2S

Certain functional groups can be reduced by H_2S to reveal a new functionality in a very rapid and selective manner. When the product function of this reductive process is part of a fluorescent molecule, the overall reaction sequence is an

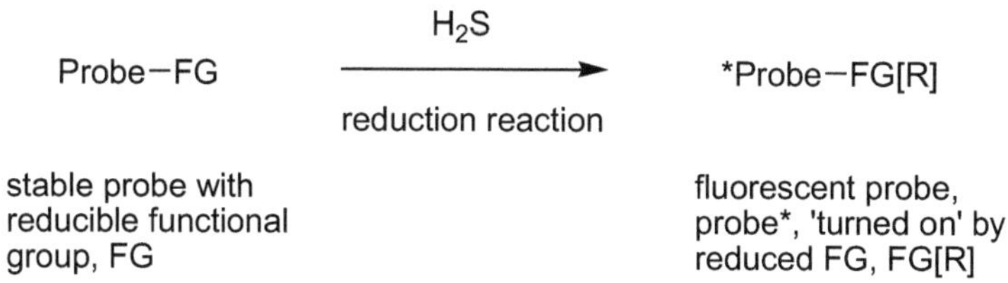

Fig. 11 Scheme showing reduction of a functional group (FG) with H_2S to give a fluorescent product

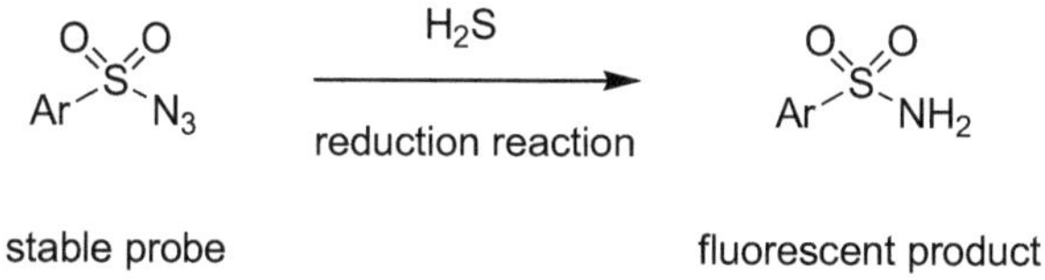

Fig. 12 General reduction of aromatic sulphonyl azides by H_2S gives fluorescent primary sulphonamide products

effective fluorescent turn-on signal upon reaction between the probe and H_2S (Fig. 11). These probe molecules are in most cases simpler structures than the other categories discussed herein, increasing the likelihood of easier access to the probe and better probe stability. The reduction of one functional group by H_2S is a well-known chemical reaction but rarely used in modern synthesis mostly due to the hazards of handling H_2S gas. However, for the sensitive detection of H_2S, this process is well suited. It can be viewed as a mild reductive process so only certain more susceptible functional groups would react rapidly and completely. One of the most popular groups in probe molecules is the azide (N_3). Usually azide is attached to a carbonyl, sulphonyl and/or aromatic system so that the resulting NH_2 product function is conjugated with the rest of the system, the overall product being a new fluorophore. For H_2S detection, azide directly attached to an aromatic system is by far the most studied. However, aromatic sulphonyl azide ($ArSO_2N_3$) undergoes rapid and complete reduction when exposed to H_2S in a wide range of media (Fig. 12) and will be discussed first.

3.1 Dansyl Sulfonyl Azides

Wang and co-workers have reported two sulphonyl azide probes based on the dansyl template. In this series, the first reported probe was 1,5-dansyl azide (**26**) (Fig. 13) (Peng et al. 2011). This probe has a very simple structure with a central naphthalene ring bearing only two substituents: a dimethylamino group and the reactive sulphonyl azide. **26** can be prepared easily in a single step from the sulphonyl chloride with sodium azide. The authors selected the sulphonyl azide group on the assumption that

Fig. 13 Dansylazide probes reported for specific use in H_2S quantification

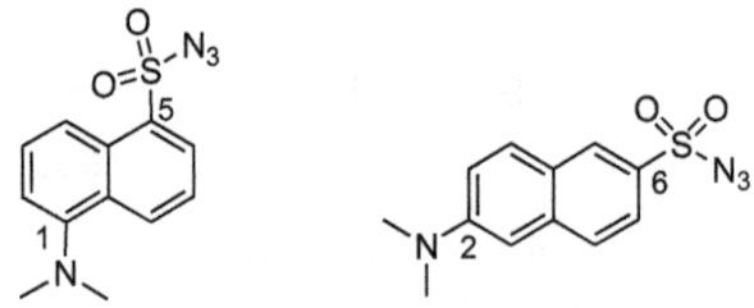

1,5-dansyl azide (**26**) 2,6-dansyl azide (**27**)

the azide moiety would react rapidly with H_2S since it is activated by the electron withdrawing naphthylsulphonyl. This reaction does indeed take place with a lower detection limit of 1 μM H_2S with signal:noise ratio of 3. Experiments were conducted in 20 mM sodium phosphate buffer with 0.5 % tween-20. A 40-fold increase in fluorescence was recorded with 25 μM sulphide (in the form of sodium sulphide, Na_2S). The response was seen to be linear at up to 100 μM sulphide. Good selectivity was observed with over 18 anions studied at or above their physiological concentrations, and only HSO_3^-, $S_2O_4^{2-}$ and $S_2O_5^{2-}$ led to small responses. Thiols such as benzyl mercaptan did give stronger responses but still only one fifth of that of H_2S. **26** was also unreactive towards very high concentrations of glycine or lysine. In bovine serum, rapid and clear responses were obtained leading these authors to study the quantification of H_2S levels in mouse blood. This was found to be 31.9 ± 9.4 μM; however, this level does not agree with other authors such as the work from the Kevil group who show in their sensitive method that plasma levels of free H_2S are well below 1 μM (Shen et al. 2012). Although probe **26** appears to fulfil all the desirable criteria, there may be other reactions in blood which lead to an increase in signal. The same authors then published a variant of **26** with a different arrangement of the same two substituents. Probe **27**, 2,6-dansyl azide, is an isomer of **26**, but has a much higher quantum yield, 40-fold higher than **26** (Wang et al. 2014). This desirable property increases the sensitivity of **27** towards reduction by H_2S. Furthermore, the authors have shown that **27** does not require any added surfactant to boost the signal. Another benefit of **27** is its short and straightforward synthesis, well within the capabilities of most chemistry labs. When studied in only phosphate buffer (no added surfactant), **27** (20 μM) detected H_2S from a sample of Na_2S (10 μM) with a 100-fold increase in fluorescence within 1 h. The corresponding increase with **26** is reported as eightfold in phosphate buffer. Probe **27** shows a good overall anion selectivity as for **26**, including glutathione and cysteine at 100 μM. A linear response is seen in the reaction of **27** with H_2S at up to 10 μM with a limit of detection of 1 μM and S/N ratio of 3:1, in aqueous media. **27** gives a similar response in fetal bovine serum (FBS) indicating its promise in cellular applications.

3.2 Aromatic Azides

Direct attachment of azide to an aromatic group produces a fluorescent aniline product upon reduction by H_2S. Lippert et al. have reported two probes based on azide-caged rhodamines with either a carbamate (sulphide fluor-1, SF1, **28**) or urea

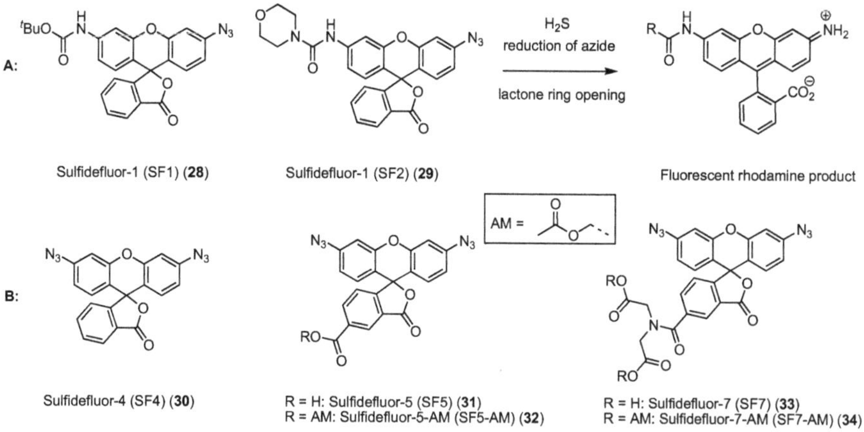

Fig. 14 Azide-caged rhodamine probes undergo lactone ring opening upon reduction by H_2S (**A**). Second-generation bis-azide rhodamines with intracellular hydrolysable 'cell-trappable' AM pro-drug (**B**)

(sulphide fluor-2, SF2, **29**) side chain (Fig. 14) (Lippert et al. 2011). **28** and **29** were synthesised in two steps from rhodamine 110, a commercially available starting material. These probes are described as 'caged' since the fluorescent rhodamine product is only revealed upon lactone ring opening triggered following azide reduction by H_2S. Following reaction of either probe (10 μM) with NaSH (100 μM) for 1 h in aqueous Hepes buffer at pH 7.4, excitation at 488 nm and emission at 525 nm produced a seven- to ninefold turn-on response. Under these conditions, the LOD was 5–10 μM. **29** was found to be preferred due to its higher selectivity for other reactive species such as glutathione (fivefold selective). In a live-cell imaging study in HEK293T cells, **28** (5 μM) was found to give a slightly stronger response to H_2S possibly due to its higher cLogP (1.98) compared to **29** (1.21). The Chang group have further developed their 'SF' probes with the report of bis-azides SF4 (**30**), SF5 (**31**), SF5-AM (**32**), SF7 (**33**) and SF7-AM (**34**) with enhanced sensitivity (Fig. 14) (Lin et al. 2013). In aqueous buffer (20 mM Hepes, pH 7.4), probes **30**, **31** and **33** (10 μM) produced a 40-, 4- and 20-fold increase in fluorescence intensity when incubated with NaSH (100 μM) for 1 h. From these probes, **30** displays the lowest LOD of 125 nM. These probes showed good selectivity against 100 μM of a range of reactive sulphur, nitrogen and oxygen species, 500 μM cysteine and 5 mM glutathione. In live HUVEC cells, **30**, **32** and **34** were compared with the first generation probe SF2 before and after replacing the cell media. Compounds with the 'cell-trappable' group, an acetoxymethyl (AM) ester, retained their brightness and penetrated the nucleus, whereas **29** and **30** did not. In addition **29** and **30** were only present in the cytosol. Bis-AM esters SF5-AM (**32**) and SF7-AM (**34**) underwent intracellular saponification to reveal a bis-carboxylic acid, with two negative charges, which was unable to exit the cell. In this way, the 'AM' probes are 'trapped' in the cell; hence, brightness is retained for

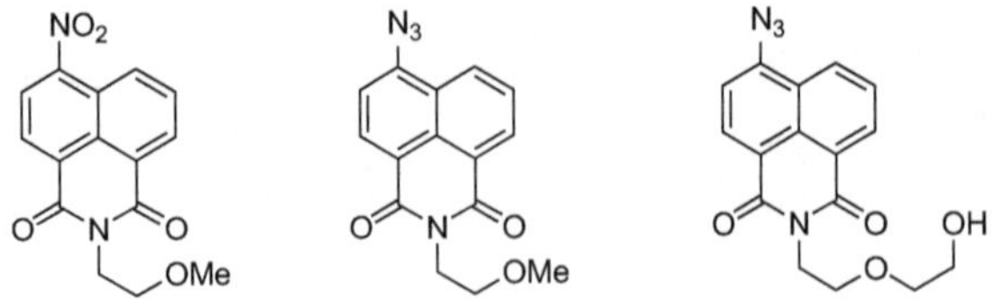

Fig. 15 Naphthalimide probes with methoxyethyl (HSN2) and hydroxyethoxyethyl ('probe 1') and side chains. Nitro derivative HSN1 also reduces to a fluorescent aniline in the presence of H_2S

a much longer period of time. SF7-AM was the most sensitive cell-trappable probe with signal still evident with only a 1 μM dose of NaSH.

Naphthalimide probes based on the Lucifer yellow family are attractive due to their synthetic accessibility and large Stokes shift. Three probes have been described in this family, two of them azides, with a methoxyethyl HSN2 (**36**) (Montoya and Pluth 2012) and hydroxyethoxyethyl (**37**) side chain (Guo et al. 2014; Montoya and Pluth 2012) (Fig. 15). A nitro derivative, HSN1 (**35**), capable of being reduced by H_2S to the fluorescent aniline, was also described, a rare example in probe structures. However, nitro **35** (5 μM) only experienced a 15-fold increase in fluorescence when exposed to 500 μM H_2S (50 mM PIPES, 100 mM KCl, pH 7.4) over 90 min, whereas with azide **36**, a 60-fold increase was observed in 45 min. Under these conditions, with the more sensitive **36**, the LOD is 1–5 μM. Furthermore, **36** is more selective: with 100 equiv. (500 μM) of reactive oxygen species and biological thiols, both probes are selective with S/N of about 7:1; however, with 10 mM of cysteine or glutathione, only **36** retains selectivity over a 60 min reaction time. In live HeLa cells incubated with both probes (5 μM) for 30 min, convincing increases in intracellular fluorescence were observed when H_2S (250 μM) was added. Interestingly, **36** gave a signal without any addition of H_2S, whereas nitro **35** did not despite being less selective towards biological thiols. Hydroxyethoxyethyl derivative ('probe 1', **37**), prepared in a two-step synthesis, when tested at a concentration of 10 μM (in 1:1 DMSO:20 mM Hepes buffer, pH 7.4) with 10 equiv. of Na_2S (100 μM), gave a 25-fold increase in fluorescence intensity following a 60 min reaction time (Guo et al. 2014). With **37** the LOD was 0.37 μM. Interestingly, **37** was very selective against cysteine (1 mM) and a wide range of other anions but less selective against sulphate (100 μM), iodide (1 mM) and chlorate (1 mM) with a fluorescence enhancement of about 9–12-fold in those cases. **37** was used to assess a series of water samples, including mineral, lake and tap water, for their H_2S content. Spiked samples with 200, 300 and 400 ppb of added Na_2S were detected in all the samples with good recoveries. Lake water had the highest H_2S content with 17 ppb. No cellular experiments were reported with **37**.

Tang has reported two diphenylethene-based probes, DPE-Az (**38**) and TPE-Az (**39**) (Fig. 16) which operate by a photostable process known as aggregate-induced emission (AIE) (Kwok et al. 2014), unlike other reported probes (Cai et al. 2014). Both probes become aggregates at 10 μM concentrations in a 2:8 mixture of DMSO:Hepes buffer. When 5 equiv. of NaSH are added to **39** (10 μM), after

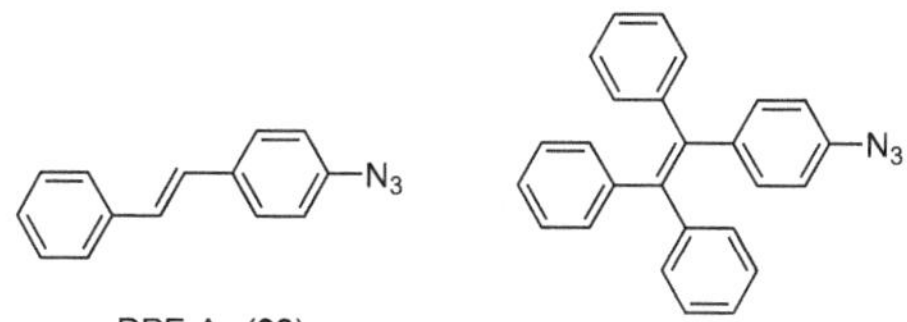

Fig. 16 Diphenylethene (DPE) and tetraphenylethene (TPE) probes based on aggregation-induced emission (AIE)

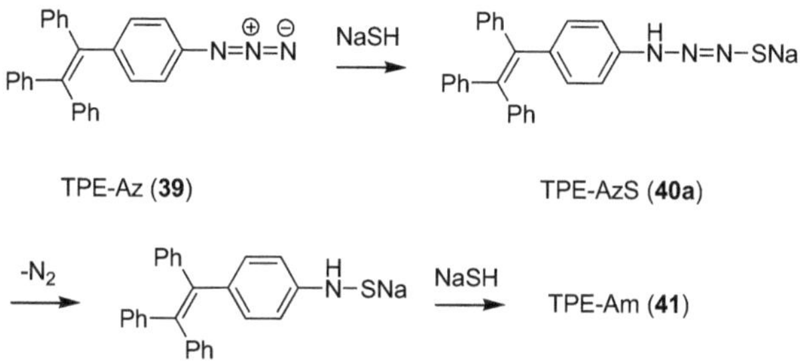

Fig. 17 Proposed mechanism of reaction of TPE-Az with NaSH explaining why more than 1 equiv. of H_2S is required before a signal can be detected

only a 5 min incubation, a 60-fold increase in emission at 493 nm was recorded (340 nm excitation). This reaction is complete after only 2 min, much faster than many other probes. Furthermore, the preferred probe, **39**, is selective for H_2S over other species such as cysteine (1 mM) and glutathione (5 mM). Importantly, **39** does not give a signal until *more than one equivalent* of H_2S has been added to the system. To explain this unusual phenomenon, the authors propose that one molecule of HS^- adds to the azide, to give TPE-AzS (**40a**), which then undergoes loss of nitrogen to form a sulfurated aniline (**40b**), which is then reduced to fluorescent TPE-Am (**41**) by a second equivalent of HS^- (Fig. 17). By varying the amount of the probe, the level where the signal starts will vary. In other words just by using different concentrations of **39**, the point at which a signal is obtained means that more than 1 equiv. of H_2S has been detected. In this way, **39** can directly quantify H_2S *without a separate calibration being required.* However, the authors only studied a lower concentration of 10 μM of **39** and did not report any studies in cells.

Han and co-workers reported a cyanine-based near-infrared fluorescence probe, Cy-N_3 (**42**), based on a strategy of internal charge transfer (ICT) (Fig. 18) (Yu et al. 2012). They reasoned that a change of substitution on a cyanine dye scaffold, upon reduction of azide to amine, would shift the absorption/emission wavelength. Indeed when **42** (10 μM) was treated with NaSH (100 μM) in Hepes (40 mM, pH 7.4) within 20 min, the NIR fluorescence maximum changed from 710 to 750 nm, following excitation at 625 nm. The increase in emission intensity

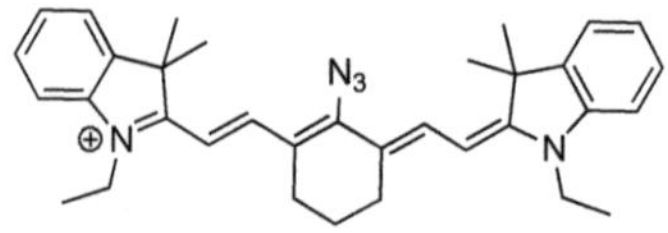

Cy-N_3 (**42**)

Fig. 18 Structure of near-infrared cyanine fluorescence probe, Cy-N_3, selective for H_2S

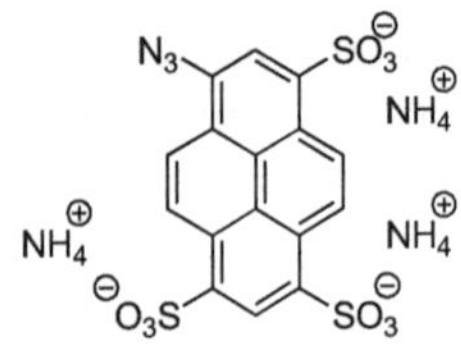

Azidopyrene trisulfonic acid (N_3-PTS) (**43**)

Fig. 19 Structure of an azidopyrene-based fluorescence probe selective for H_2S

was twofold with an excellent LOD of 80 nM. This represents one of the most sensitive fluorescence probes reported to date. In addition **42** was selective for 24 different anions and reactive oxygen, nitrogen and sulphur species, including cysteine and glutathione at 1 mM concentrations. Intensities from the most challenging species were still only 10 % of the sulphide response. This probe (10 μM) was able to quantify H_2S from 0.08 to 100 μM, with a linear response in both water and fetal bovine serum. Detection of H_2S released from a small molecule donor, ADT-OH (Li et al. 2007) (500 μM), with **42** (10 μM) was achieved in fetal bovine serum (40 mM Hepes, pH 7.4) at 37 °C. Release from ADT-OH at this concentration was equivalent to NaSH at 100 μM. Moreover **42** was sensitive and selective enough to detect H_2S in cells. RAW264.7 cells (mouse macrophages) were incubated with 100 μM H_2S then 10 μM **42** for 20 min at 37 °C. Quantifiable increases in intracellular H_2S were clearly visualised. In summary **42** has been extensively validated as a sensitive, selective and cell-applicable fluorescence probe for the quantification of H_2S.

Highly water-soluble probe N_3-PTS (**43**), based on a pyrene-1,3,6-trisulfonic acid scaffold, was prepared in one step from known fluorescent dye APTS (Fig. 19) (Hartman and Dcona 2012). As a tris-ammonium salt, **43** has water solubility of >100 mM. Following addition of NaSH (50 μM) to **43** (100 μM), a sevenfold increase in fluorescence intensity (emission at 505 nm, excitation at 435 nm) was detected after 90 min of reaction in PBS buffer. When compared with NaSH (50 μM), **43** gave at least a fivefold greater fluorescence intensity compared with a range of other reactive species, including cysteine and glutathione at 1 mM. A linear response is obtained when **43** (100 μM) is reacted with NaSH from 2 to 100 μM, suggesting the LOD is around 2 μM or less. In fetal bovine serum, very little fluorescence quenching was observed and the concentration of NaSH required to increase the signal by twofold was 9 μM. Generally the H_2S sensitivity in FBS was about 20 % that of PBS, and the authors suggested that new derivatives of **43**

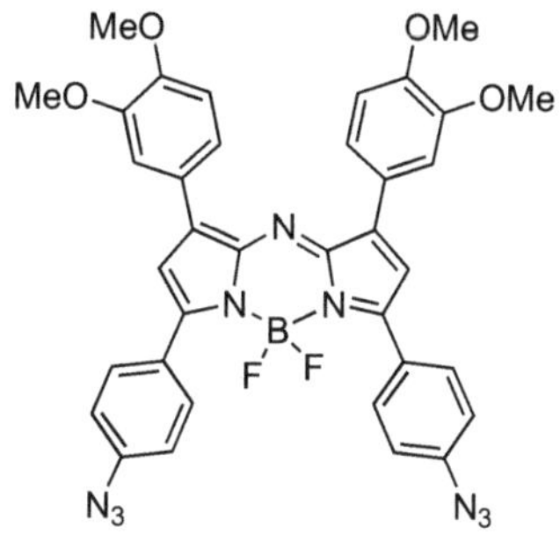

44

Fig. 20 Aza-BODIPY bis-aromatic azide dye

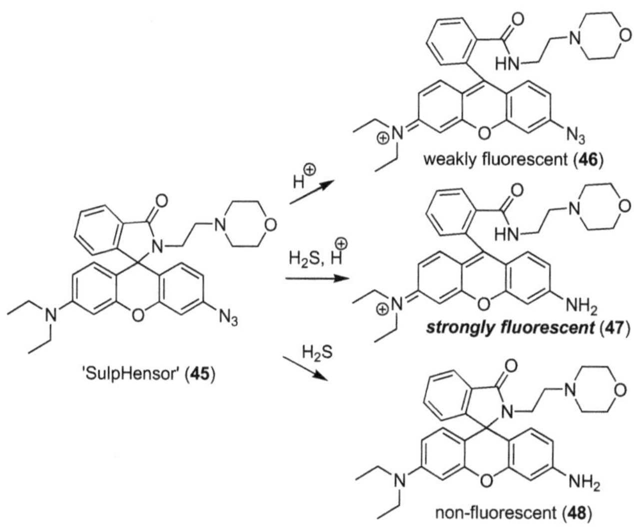

Fig. 21 Rhodamine probes modified with a spirolactam (acid sensitive) and an aromatic azide (H_2S senstitive) and a lysosome-targeting morpholine group

could increase the serum sensitivity. No cellular studies were carried out with this probe.

An aza-BODIPY dye (**44**) has been reported by Ramaiah and co-workers (Fig. 20) (Adarsh et al. 2014). Reaction of **44** (12 μM) in THF with 2 equiv. of Na_2S in water was complete in only 30 s and was selective for other ions; however, data on cysteine or glutathione were not reported. No data in cells was reported.

Yang and co-workers have devised a dual-mechanism acid-sensitive probe, 'SulpHensor' (**45**) targeted to acidic lysosomes (pH 4–6) in cells (Fig. 21) (Yang et al. 2014). **45** can react separately, and reversibly, with either acid or H_2S leading to separate products with little fluorescence (**46/48**). However, when both events

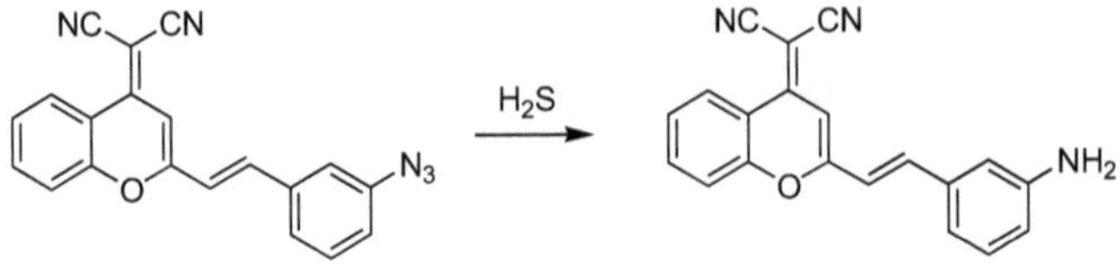

Fig. 22 Conjugated benzopyranazide probe **1** with two-photon absorbing properties undergoes reduction with H_2S to give fluorescent product 1-NH_2

happen, a strongly fluorescent product is formed (**47**). Hence in cells, **45** only visualises H_2S in acidic organelles. The rapid acid/base equilibration to give a nonfluorescent product at neutral pH was designed to lead to a clear selectivity within cells for the lysosomal compartment. **45** (5 μM) gave a 15-fold increase in fluorescence when reacted with NaSH (50 μM) at pH 4.5. With other reactive species, **45** was quite selective with the least being glutathione (5 mM) with an approximate 7:1 S/N ratio. In HeLa cells, **45** (5 μM), with a 30 min pre-incubation, was confirmed to detect H_2S (50 μM) selectively in lysosomes. Boosting endogenous production of H_2S with cysteine pretreatment in HeLa cells was confirmed with **45** demonstrating the specific useful applications of this probe.

Two-photon microscopy, as opposed to one-photon of high energy, utilises two photons of low energy to obtain the excited state of a pharmacophore, potentially allowing deeper tissue penetration of the probe (Kim and Cho 2009). Peng and co-workers developed two-photon probe **49**, which reacts with H_2S to give fluorescent 1-NH_2 (**50**) in the NIR wavelength range (Fig. 22) (Sun et al. 2013). Unlike other NIR probes, **49** was shown to be photostable when subjected to a 500 W lamp for 4 h. When a solution of **49** (5 μM) in 100 mM pH 7.4 PBS buffer, with 50 % DMSO, was reacted at 37 °C with NaSH (50 equiv.), a 65-fold fluorescence turn-on response was observed. **49** was generally selective for other anions with homocysteine being the least selective at about 20:1 S/N ratio. A pH range from 2.5 to 10 was found to be tolerable by this probe, and it could be used in serum (FBS) as well. Incubation of MCF-7 cells with **49** (5 μM) at 37 °C in PBS with NaSH (50 equiv.) for 60 min gave a strong signal with emission 575–630 nm. **49** was also tested in live animals, the first reported probe to have been studied in vivo. Using the skin-popping methodology, **49** (40 μM) and NaSH (25 equiv.) were administered to ICR mice. At the 4 h timepoint, a clear increase in signal was seen compared to treated groups indicating the in vivo applicability of the two-photon strategy.

3.3 Aliphatic Azides

In contrast to the aromatic azides, aliphatic azides are not directly conjugated to an extended chromophore system. However, they are still reduced efficiently to aliphatic amines which are then good nucleophiles for intramolecular processes

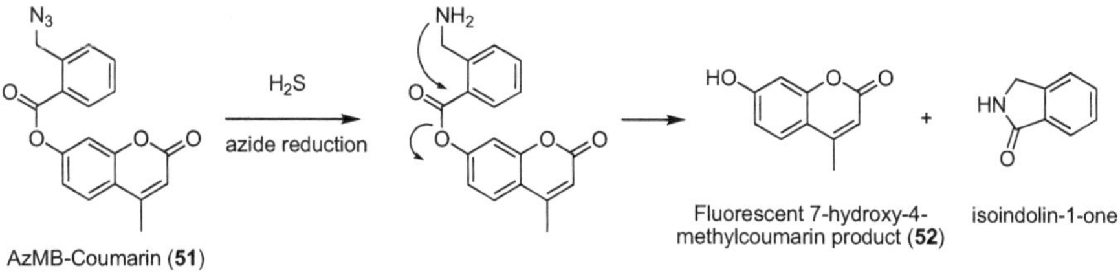

Fig. 23 Aliphatic azide in AzMB-coumarin is reduced by H_2S triggering an intramolecular cyclisation and ejection of fluorescent coumarin

which ultimately result in a fluorescent product. An elegant example is described by Wu et al. where probe molecule AzMB-coumarin (**51**) undergoes reduction followed by intramolecular attack on the nearby ester ejecting fluorescent coumarin (**52**) and the innocuous by-product isoindolin-1-one (Fig. 23) (Wu et al. 2012). The reaction is complete within 20–60 min either in acetonitrile or 20 % v/v sodium phosphate (pH 7.4, 100 mM) buffered acetonitrile. Control of pH is important since the hydrosulphide ion, HS^- (pka of H_2S is c.6.9), is present as the major species at pH 7.4. Using **51** (20 μM) and NaSH, following a 1 h incubation with excitation at 365 nm and emission at 450 nm, the LOD was 100 μM in aqueous acetonitrile and 10 μM in acetonitrile. **51** (100 μM) is selective over biological thiols such as glutathione and cysteine (1 mM) with an S/N ratio of about 6:1. In HeLa cells treated with **51** (1 mM) for 10 min then treated with or without NaSH (1 mM), a clear signal was seen with the treated cells visualised by fluorescence microscopy.

3.4 Probes Based on Reduction of a Nitro Group

Hydrogen sulphide can also reduce nitro groups to anilines; however, when the nitro is part of an aromatic phenol, a thiolytic process occurs, where the resulting aniline product is cleaved to reveal a fluorescent phenol. Using this principle, a three-channel-based fluorescent probe **53** (Fig. 24) was developed by Lin and co-workers in 2013 (Wang et al. 2013). This white light-emitting probe was constructed by connecting a blue fluorescent dye, 1*H*-phenanthro[9,10-*d*]imidazole, and an ESIPT (excited-state intramolecular proton transfer) component, 3-hydroxychromone. The reactive site was modified with a 2,4-dinitrophenyl group. When exposed to NaSH, **53** reacts through the dinitrophenol carbon ejecting fluorophore **54** responding with 3-, 6- and 16-fold emission enhancements at three channels around 440, 510 and 570 nm. **53** is stable over a pH range of 5.5–8.5 with a LOD of 10 μM at pH 7.4 in PBS:DMSO (v/v, 4/1) with 0.5 % tween-20. Selectivity of **53** (5 μM) for H_2S is good for a range of anions (S/N up to 20:1), wherein cysteine (1 mM) and glutathione (10 mM) responses were two- to threefold higher than for other anions (S/N 5:1). An MG63 cell line study indicated that **53** is not toxic at concentrations below 50 μM. Application of **53** (5 μM) for endogenous H_2S

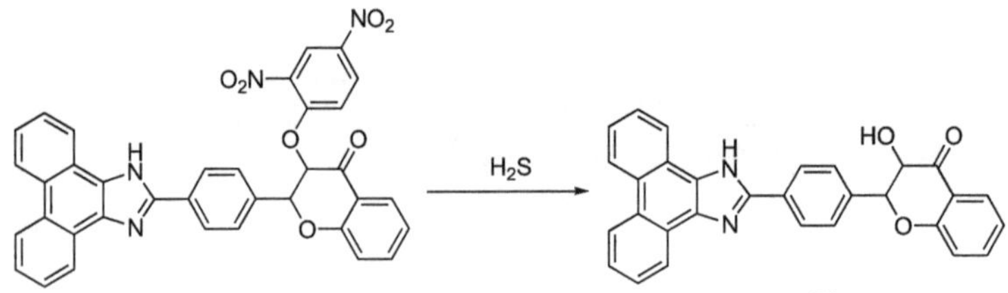

53 **54**

Fig. 24 Structure of probe **1** and its mechanism

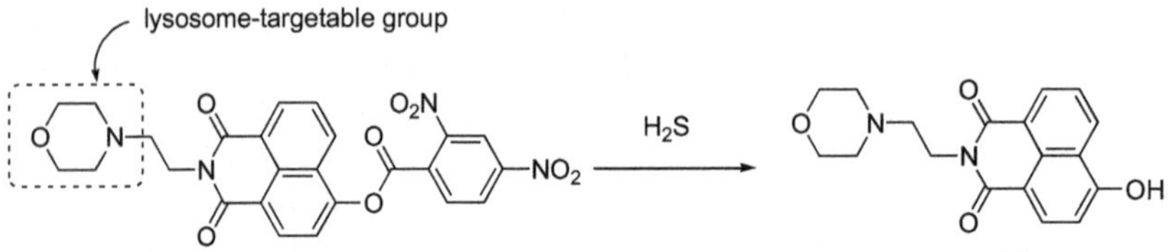

55 **56**

Fig. 25 Structure of probe **3** and its mechanism

(50 μM) detection was carried out in MG63 cells. Bright fluorescence images of blue, green and red colours were obtained, featuring the first three-channel sensor for hydrogen sulphide. However, incubation of **53** with cysteine and GSH also led to bright fluorescence images at three channels.

A 1,8-naphthalimide-derived fluorescent probe **55** (Fig. 25) for H_2S based on the thiolysis of a dinitrophenyl benzoate ester, to produce fluorescent **56**, was developed by Cui and co-workers in 2013 (Liu et al. 2013). **55** is applicable to H_2S detection over a pH range of 4–8 in bovine serum with a LOD of 0.48 μM. The response of 10 μM **55** to 1 mM NaHS was 12-fold of control and sixfold of that to 1 mM cysteine and GSH in less than 20 min. **55** was also shown to be suitable for H_2S detection in live cells with minimal toxicity. Because of the presence of a lysosome-targetable group, 4-(2-aminoethyl)morpholine, **55** is able to specifically localise in lysosomes at 5 μM and detect H_2S in lysosomes in live cells.

Lin and co-workers have created dinitrophenol ether attached to a BODIPY core (Fig. 26) (Cao et al. 2012). The resulting probe, NIR-H_2S (**57**) (5 μM), has an 18-fold turn-on response when reacted with NaSH (40 μM) in 50 mM PBS buffer with 3 mM cetyltrimethylammonium bromide and 10 % ethanol. **57** was selective with a range of other reactive species, the most challenging being cysteine (1 mM) and glutathione (10 mM), where the S/N was approximately ten- and sevenfold, respectively. In MCF-7 cells treated with **57** (5 μM) for 20 min then treated with NaSH (0–100 μM) for 10 min clear images were obtained showing H_2S in the cells, primarily in the mitochondria.

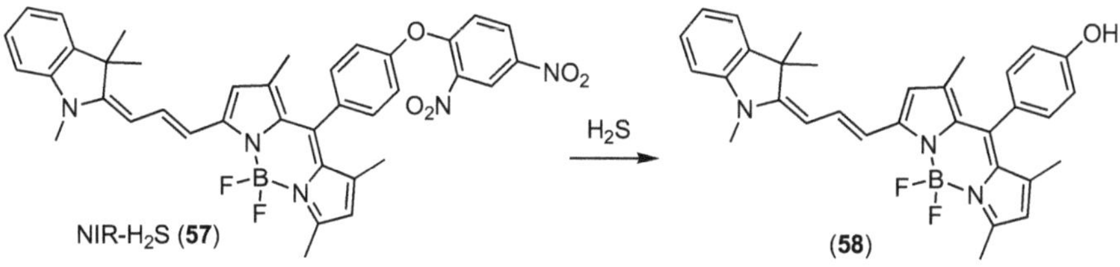

Fig. 26 Dinitrophenylether NIR-H_2S is reduced by H_2S triggering an intramolecular cyclisation and ejection of fluorescent coumarin

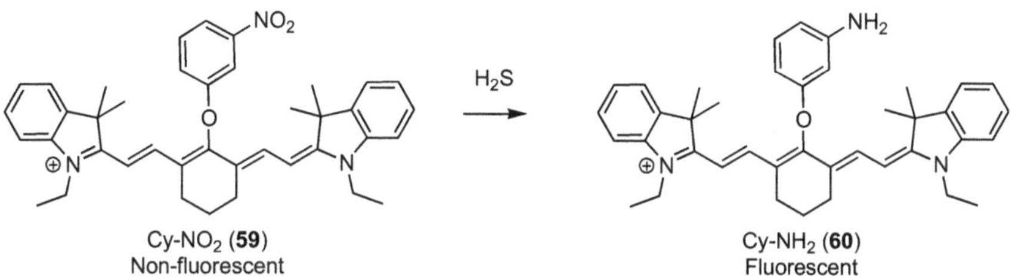

Fig. 27 Nitrophenylether-substituted cyanine dyeCy-$NO_2$is reduced by H_2S to a fluorescent aniline product, Cy-NH_2

Earlier, a cyanine dye-based probe, **42** (Cy-N_3), was discussed (Fig. 18) where an aromatic azide was reduced by H_2S to give a fluorescent product. A nitro reduction design of a similar cyanine probe, Cy-NO_2 (**59**), was described by Chen and co-workers (Fig. 27) (Wang et al. 2012). Probe **59** was not affected by pH in the range 4.2–8.2 and operated well under physiological conditions in the NIR range (ex/em 755/789). Hence, **59** (10 μM) in 40 mM Hepes buffer at 37 °C gave a 12.7-fold fluorescence increase when treated with Na_2S (350 μM). Selectivity was also good with cysteine and glutathione being the most reactive, but still with a very good window. **59** was also effective in FBS which led on to studies in RAW264.7 cells. Cells were treated with **59** (10 μM) for 10 min, washed then treated with Na_2S (50–350 μM) then imaged by confocal fluorescence microscopy. A clear dose response was obtained with brighter images at higher concentrations of Na_2S. Further studies revealed that **59** is predominately located in the cytoplasm.

4 Probes Which Trap H_2S Using a Coordinated Metal

A fluorescein-based probe conjugated with an azamacrocyclic copper complex has been reported to specifically react with H_2S through the copper centre resulting in an increase in fluorescence (Fig. 28). Cyclen (1,4,7,10-tetraazacyclododecane,

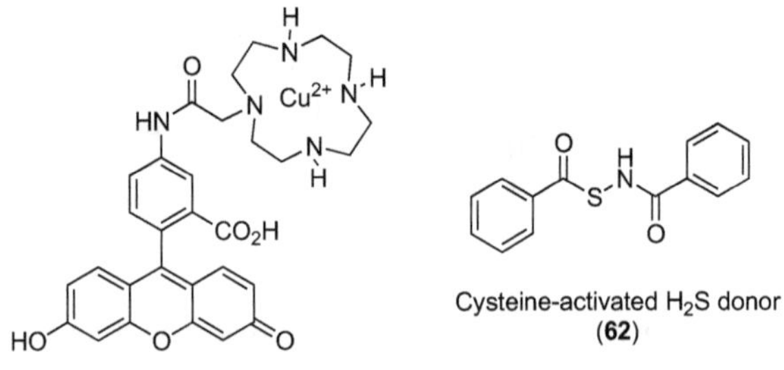

Fig. 28 Fluorescein conjugated with an azamacrocyclic 'cyclen' copper complex. When H_2S binds the Cu^{2+}, the copper is released resulting in an increase in fluorescence

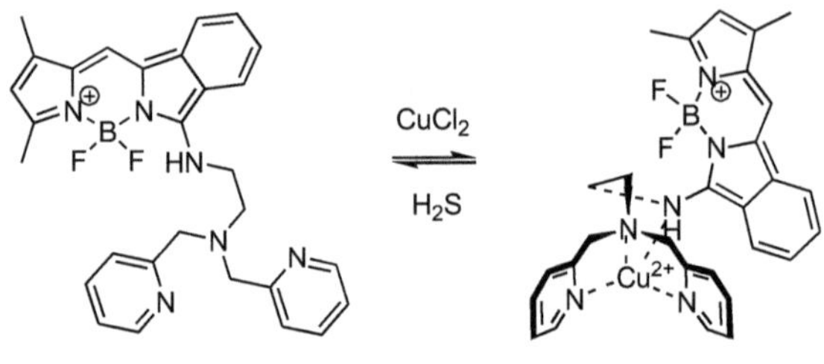

Fig. 29 *N*-Bis [2-pyridylmethyl] aminoethene attached to BODIPY (ZS2) forms a copper complex 'quenched ZS2-Cu'. When H_2S binds the Cu^{2+}, the copper is released resulting in an increase in fluorescence

HSip-1, **61**) was found to be the most sensitive and selective of four different macrocyclic ring sizes studied (Sasakura et al. 2011). At 10 μM concentration, **61** gave a 50-fold turn-on response (ex/em 491/516 nm) when treated with NaSH (100 μM) in 30 mM Hepes buffer, whereas with high concentration of GSH (10 mM), almost no increase in fluorescence was detected. When present together, response to NaSH was not affected by 10 mM glutathione. With added cysteine (1 mM), a >10-fold S/N ratio was achieved; all other tested anions were even more selective. **61** (1 μM) was also tested in vitro with a cysteine-activated H_2S donor drug (50 μM) (Zhao et al. 2011). In HeLa cells **61** was impermeable, so it had to be diacetylated, then it was able to image exogenously added H_2S (200 μM).

Probes which undergo nucleophilic reaction with H_2S are often too slow to be useful in cells. For example, the reactivity of the aldehyde–acrylate reaction (see Fig. 2) takes 50 min to reach maximum fluorescence (Li et al. 2013). Another disadvantage is their poor water solubility requiring non-biocompatible solvents. Taking advantage of the strong affinity between sulphur and copper has produced probes such as **61** (Sasakura et al. 2011) and Cu-1 (Qu et al. 2013). A new probe, ZS2-Cu (**64**), has good water solubility (1.28 mg/ml) and membrane permeability and is highly sensitive (Fig. 29) (Li et al. 2014). BODIPY-based **63** is complexed with Cu^{2+} through a bis-pyridylethylenendiamine group. BODIPY is membrane permeable and the bis-pyridyl feature provides good water solubility. In fact

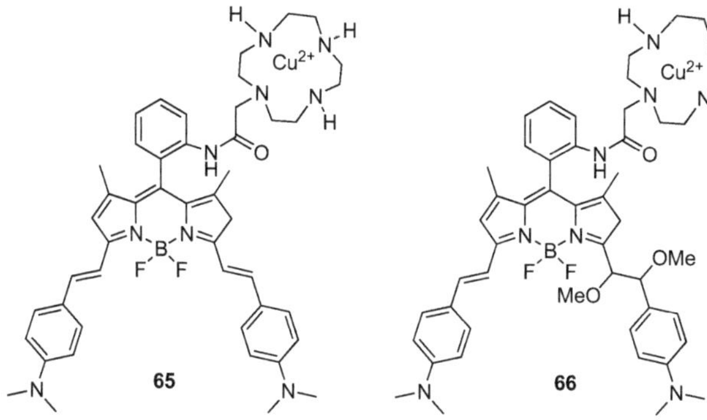

Fig. 30 Cyclen copper complexes, probe **7** and **8**, attached to BODIPY modified with substitutions to create an NIR-fluorescent probe upon complexation of Cu with S

experiments are carried out purely in aqueous solution without the need for an organic co-solvent. Fluorescence turn-off response of this probe was specific to copper; a range of 10 other cations did not quench the fluorescence of **63** even when mixed in the presence of copper. **64** exhibited over 30-fold selectivity towards H_2S over other anions and biothiols such as cysteine and glutathione and a 15–18-fold increase in fluorescence when NaSH was added in the presence of other anions or biothiols. When 2 equiv. of NaSH was added to **64** (10 μM), a 19-fold increase in fluorescence intensity was measured at 546 nm. The reaction occurs very rapidly with a lower limit of detection of 250 nM. In HeLa cells incubated with **64** (5 μM) for 15 min, weak fluorescence was observed, but when cells were washed and NaSH (100 μM) added, a strong increase in intracellular signal was observed confirming the good membrane permeability of **64**.

Among another series of BODIPY-based probes published recently appears to be a rare example of in vivo application of H_2S sensing using a fluorescence probe. Huang and co-workers have designed near IR (NIR) fluorescent probes **65** and **66** (Fig. 30) with excitation at 680/600 nm and emission at 680/765 nm, respectively (Wu et al. 2014). **65** was selective for reactive species with the lowest selectivity for 1 mM cysteine (about 10:1 S/N) and 10 mM glutathione (about 20:1 S/N). The LOD for **65** was 80 nM with a limit of quantitation (LOQ) determined at 0.27 μM. In RAW264.7 cells **65** (10 μM) was not cell penetrant and hence required liposomes as a carrier to achieve cell penetration. However, this was effective and led to an eightfold fluorescence enhancement upon exogenous application of Na_2S (200 μM). Probe **66**, also using the liposome formulation, was shown to be effective in MCF-7 cells and was able to detect H_2S evolved from diallyltrisulphide. In mice, **65** (40 μM) was studied using a skin-popping technique to administer the compound into the tissue near the skin. A dose of Na_2S (25 equiv.) was injected and detected with a sevenfold increase in fluorescence within 5 h.

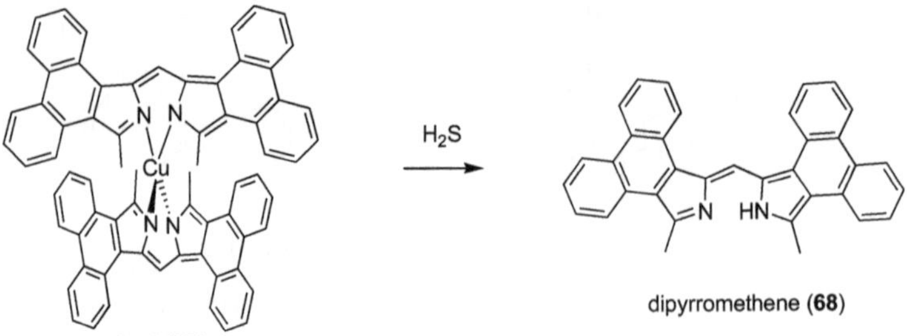

Fig. 31 Bis-phenanthrene-fused dipyrromethene complex, Cu-1, reacts with H_2S to give fluorescent product **1** with fluorescence in the NIR range

Another NIR probe based on the favourable copper-sulphide affinity, with potential for deep tissue application in vivo, was reported by Shen and co-workers (Fig. 31) (Qu et al. 2013). Dipyrromethene-based copper complex, Cu-1 (**67**), reacts instantly with H_2S (1.5 equiv.) in DMSO:water (1:1) solution to liberate fluorescent **68** with a 14-fold increase in fluorescence intensity. There is some interference when cations such as Zn^{2+} are added where the reaction rate slows down to about 5 h. However, good selectivity is seen with other reactive sulphur species with glutathione (10 equiv.) still giving an 8:1 S:N ratio. In HeLa cells, **67** (50 μM) treatment for 30 min with addition of NaSH (50 μM) gave bright images and indicated that the fluorescence product was located in the cytoplasm. In addition, **67** was found to be nontoxic up to 100 μM concentrations assessed by MTT assay.

5 Conclusions

There are now many diverse, sensitive fluorescent probes reported for H_2S detection and quantification with a particular acceleration in the past few years highlighting the upsurge in interest in H_2S biology in general. Table 1 summarises selected probes showing the range of detection limits (some down to low nanomolar), detection wavelengths, reaction times and conditions and application in a biological system. There are now many probes which have some utility in living cells and several with specific subcellular localisation. Biologists studying H_2S now have a wide range of tools to assist them which should lead to yet further progress in the understanding of the role of H_2S in biology and how to apply this knowledge therapeutically.

Table 1 Selected H_2S-sensitive fluorescent probes listed according to mechanism: probes which react nucleophilically with H_2S, probes which are reduced by H_2S and probes which involve metal coordination

Probe	Probe name	Probe structure	Fluorescent product	LOD (μM)	Wavelength (EX/EM, nm)	Reaction time (mins)	pH	Medium	Application in biosystem	References
Probes which react nucleophilically with H_2S										
1	SFP1	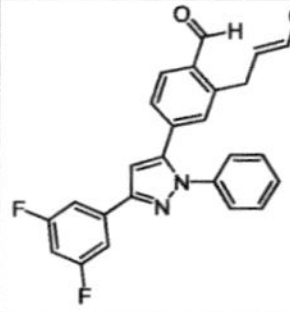	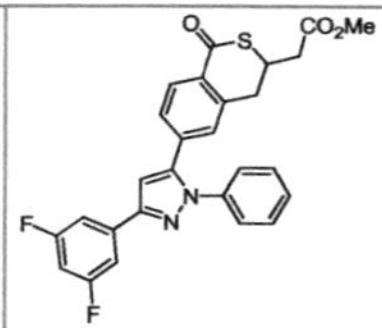	<5	310/550	20	7.4	PBS	HeLa cells	Qian et al. (2011)
3	SFP2	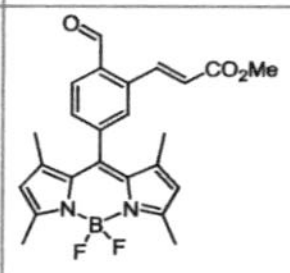	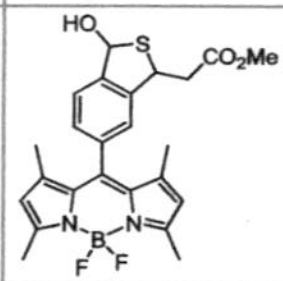	<5	465/480–580	20	7.4	PBS	HeLa cells	Qian et al. (2011)
5	P3	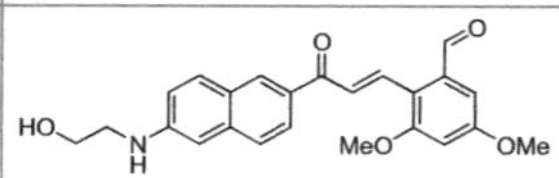	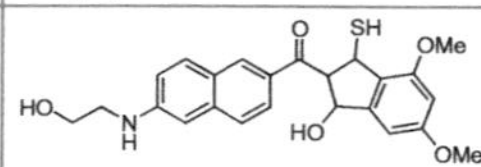	0.05	375/505	5	7.4	Hepes, 10 % FBS	HeLa cells	Singha et al. (2015)
9	WSP1	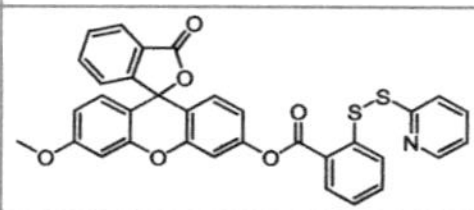	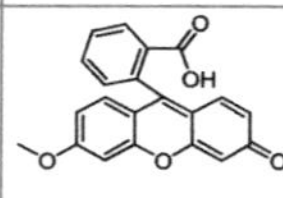	0.06	476/516	60	7.4	10 mM PBS/MeCN (1:1) + Hexadecyltrimethylammonium bromide (CTAB)	COS7 cells	Liu et al. (2011), Peng et al. (2014)
10	WSP2	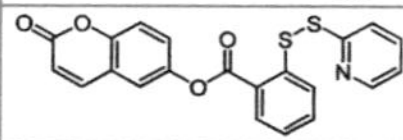	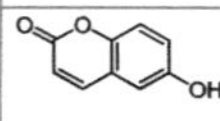	0.079	385/456	5	7.4	10 mM PBS/MeCN (1:1) + Hexadecyl trimethyl ammonium bromide (CTAB)	–	Peng et al. (2014)
11	WSP3	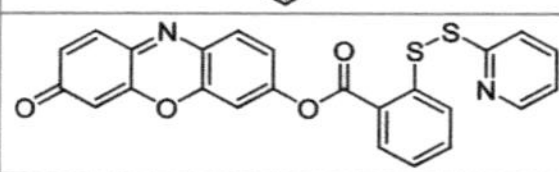	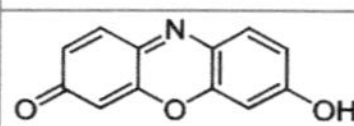	0.047	550/586	5	7.4	10 mM PBS/MeCN (1:1) + Hexadecyl trimethyl ammonium bromide (CTAB)	–	Peng et al. (2014)

(continued)

Table 1 (continued)

Probe	Probe name	Probe structure	Fluorescent product	LOD (μM)	Wavelength (EX/EM, nm)	Reaction time (mins)	pH	Medium	Application in biosystem	References
12	WSP4	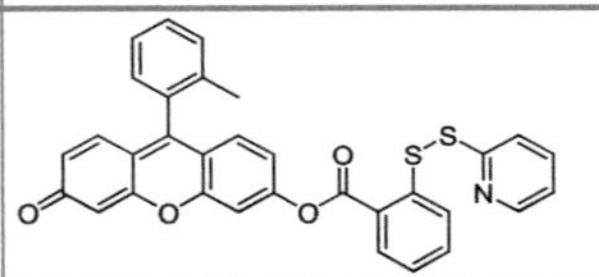	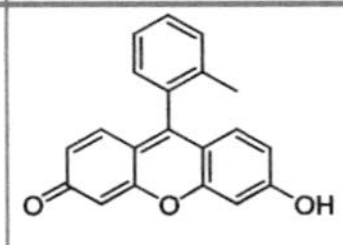	0.266	512/531	5	7.4	10 mM PBS/MeCN (1:1) + Hexadecyl trimethyl ammonium bromide (CTAB)	HeLa cells	Peng et al. (2014)
13	WSP5	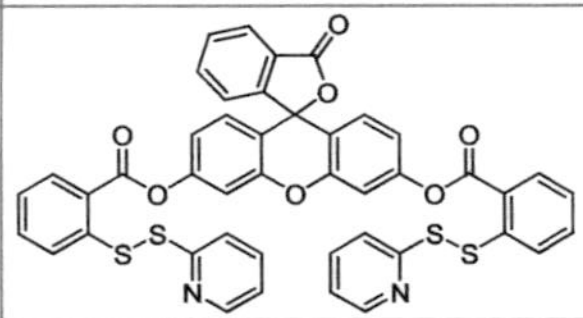	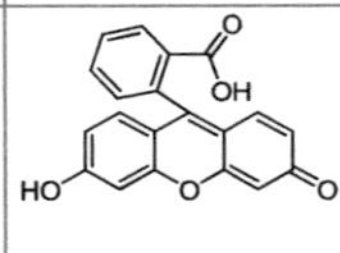	0.047	502/525	5	7.4	10 mM PBS/MeCN (1:1) + Hexadecyl trimethyl ammonium bromide (CTAB)	HeLa cells	Peng et al. (2014)
14	–	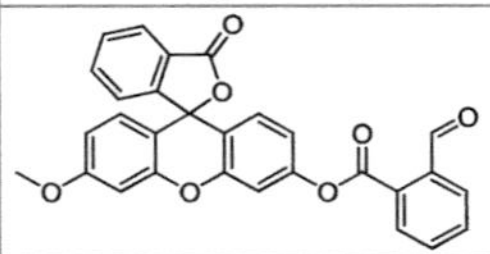	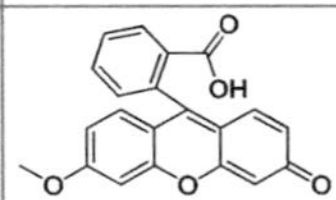	2.5	517	30	7,4	EtOH/PBS 1/1 (V/V)	HeLa cell	Liu et al. (2014b)
16	E1	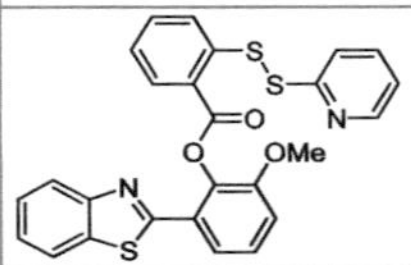	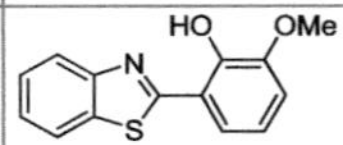	0.12	295/487	5	7.4	Tris–HCl buffer	HeLa cells	Xu et al. (2012)
19		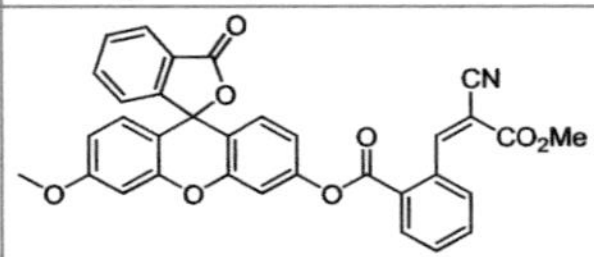	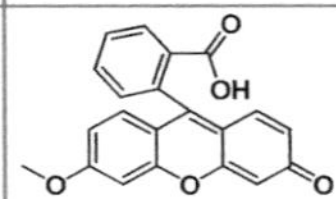	1	517	30	7.4	PBS	COS-7 cells	Liu et al. (2012)
20	–	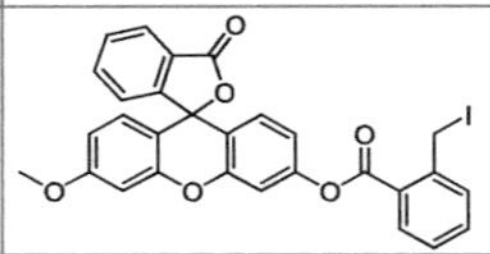	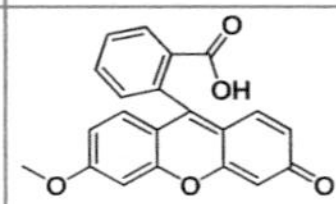	0.1	517	40	7.4	PBS/MeCN (1:1, v/v)	COS7 cells	Zhang et al. (2013)

23	DSP3	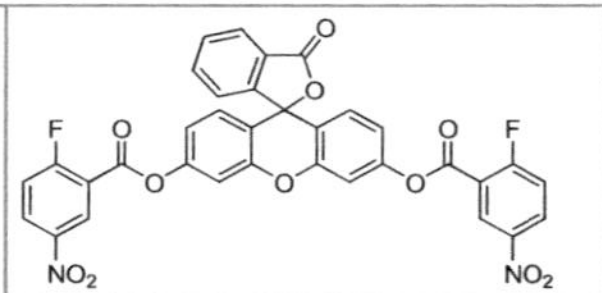	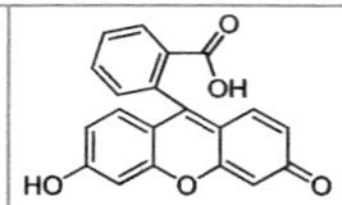	0.07	517	10	7.4	PBS	HeLa cell	Liu et al. (2014a)
24	MBB	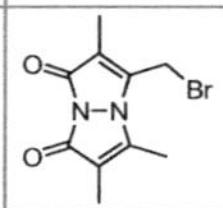	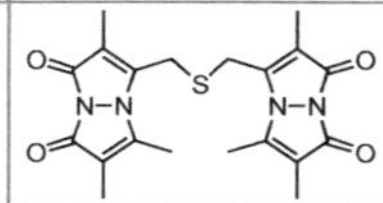	0.005	390/475	30	9.5	100 mM Tris–HCl buffer	Can measure plasma H_2S levels. Clinical trial on-going	Shen et al. (2011, 2012)
Aromatic azide probes which are reduced by H_2S to produce fluorescent aniline products										
26	1,5-Dns-Az	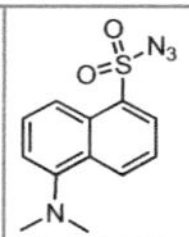	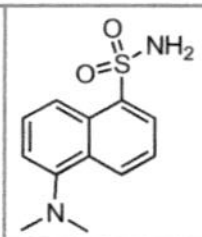	1	340/535	60	7.5	20 mM sodium phosphate buffer with 0.5 % tween-20	–	Peng et al. (2011)
27	2,6-Dns-Az	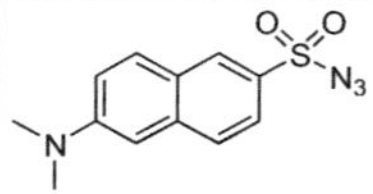	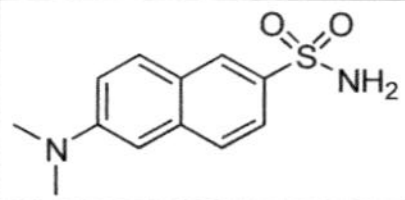	1 (6 μM in FBS)	325/450	60	7.4	PBS	–	Wang et al. (2014)
30	SF4	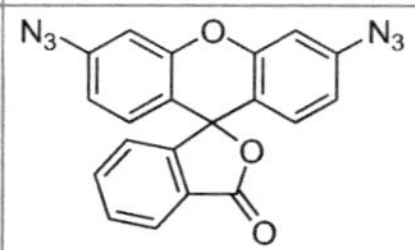	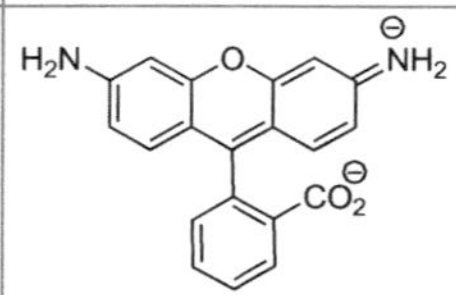	0.125	496/517	60	7.4	20 mM Hepes	–	Lin et al. (2013)
31	SF5	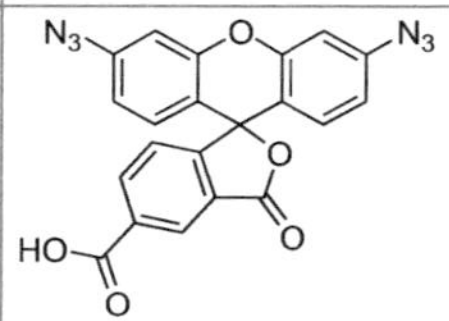	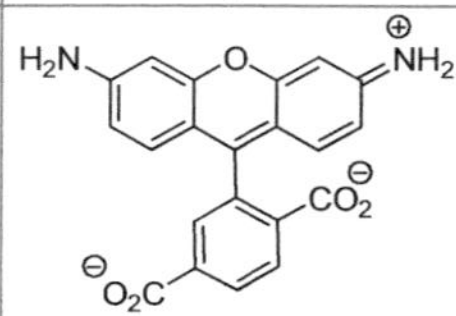	–	498/521	60	7.4	20 mM Hepes	–	Lin et al. (2013)

(continued)

Table 1 (continued)

Probe	Probe name	Probe structure	Fluorescent product	LOD (μM)	Wavelength (EX/EM, nm)	Reaction time (mins)	pH	Medium	Application in biosystem	References
32	SF5-AM	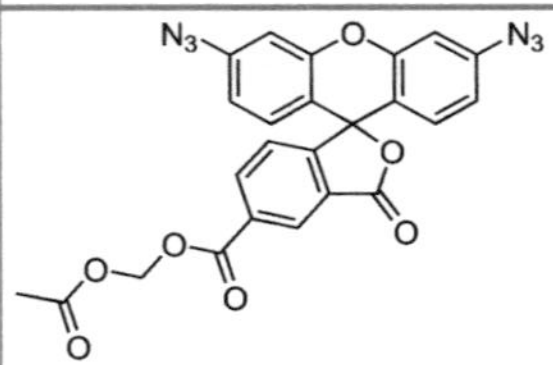	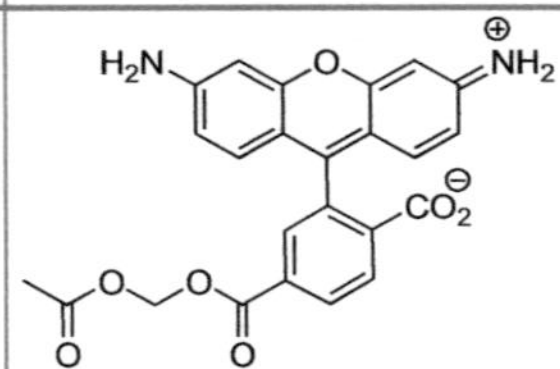 In cells 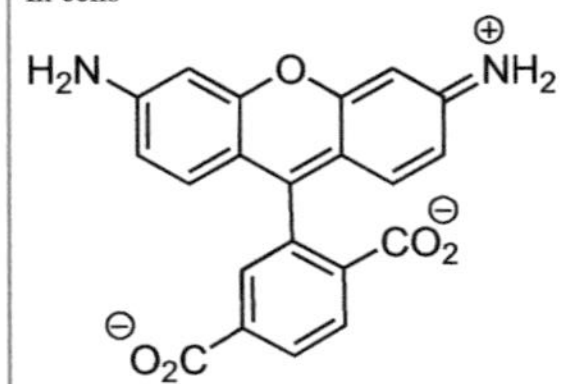	0.25	488/525	60	7.4	20 mM Hepes	HUVECs	Lin et al. (2013)
33	SF7	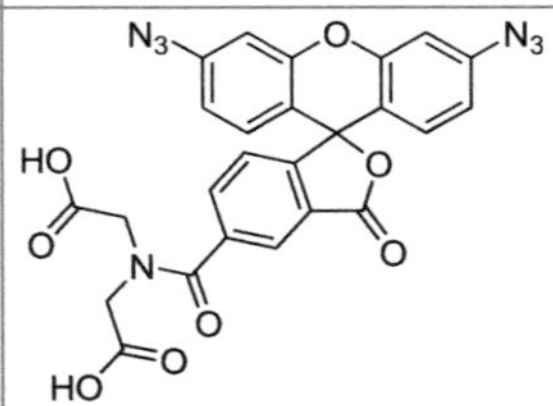	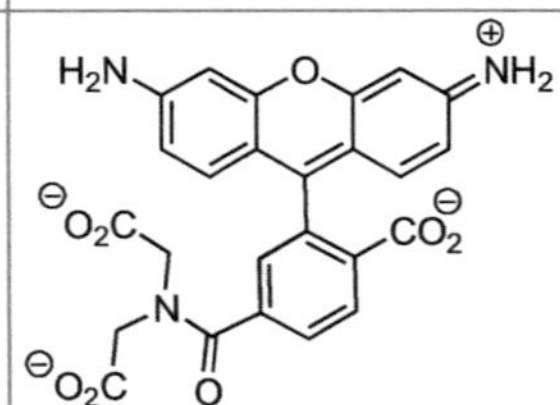	–	498/526	60	7.4	20 mM Hepes	–	Lin et al. (2013)

34	SF7-AM	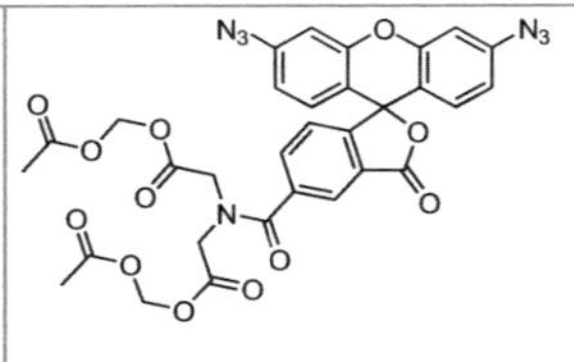	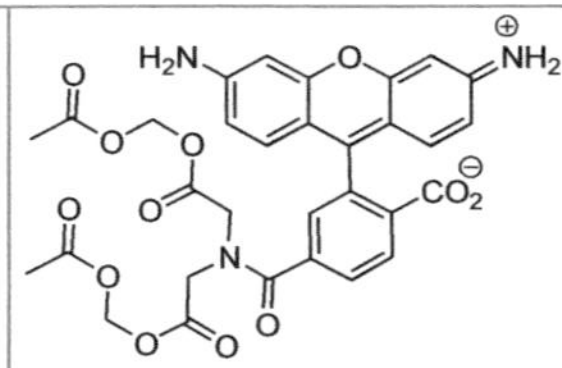 In cells 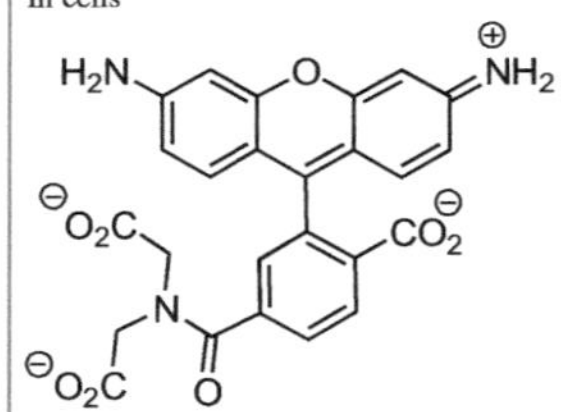	0.5	488/525	60	7.4	20 mM Hepes	HUVECs	Lin et al. (2013)
36	HSN2	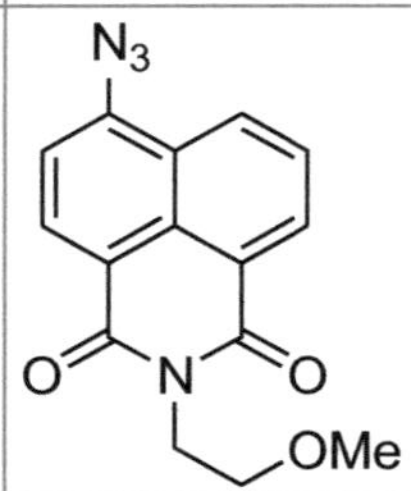	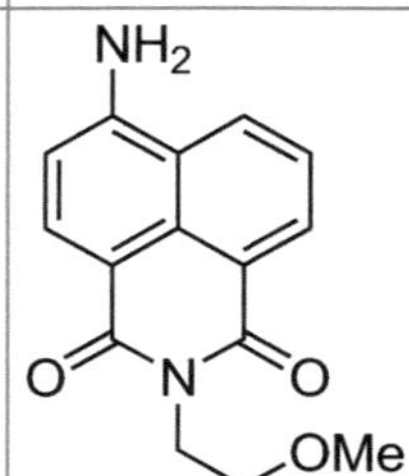	1–5	435/490	45–60	7.4	50 mM PIPES, 100 mM KCl	HeLa cells	Montoya and Pluth (2012)
37		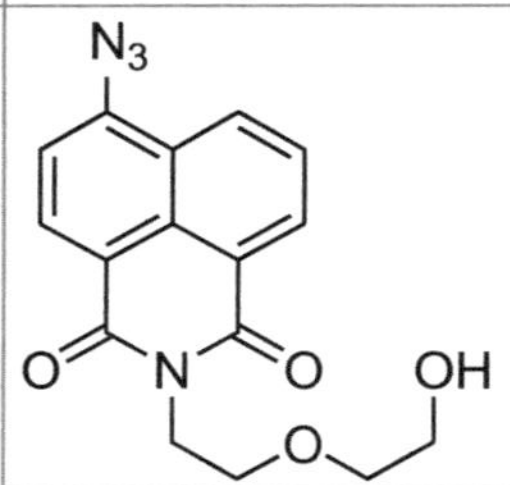	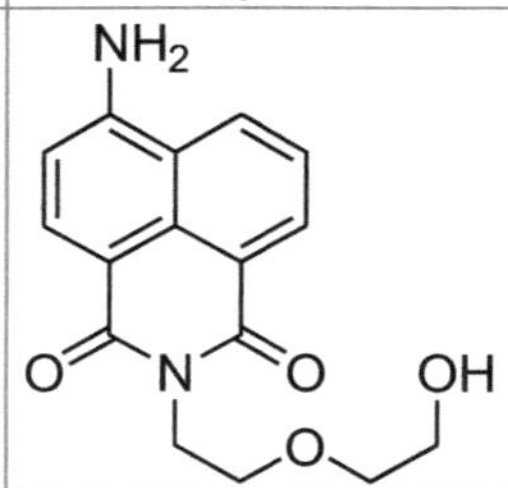	0.37	440/534	15	7.4	1:1 DMSO:20 mM Hepes	Tap, mineral and lake water. No cellular experiments	Guo et al. (2014)

(continued)

Table 1 (continued)

Probe	Probe name	Probe structure	Fluorescent product	LOD (μM)	Wavelength (EX/EM, nm)	Reaction time (mins)	pH	Medium	Application in biosystem	References
39	TPE-Az	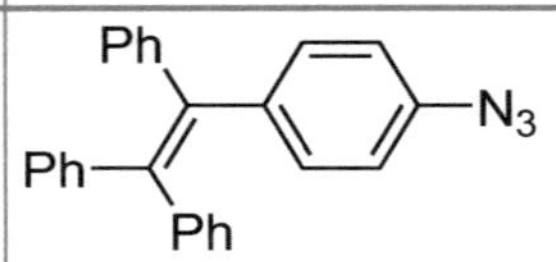	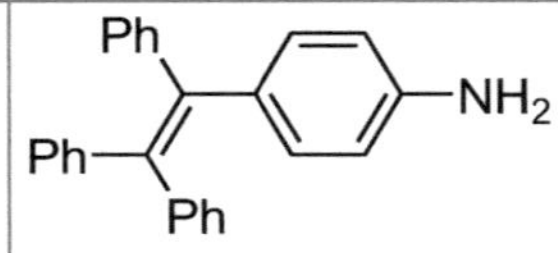	–	340/493	5	7.4	Hepes:DMSO (4:1)	–	Kwok et al. (2014)
42	Cy-N_3	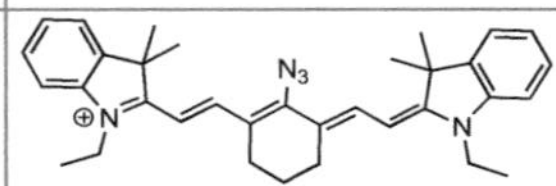	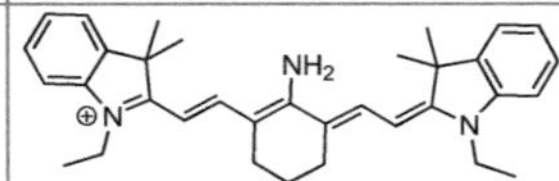	0.08	625/750	20	7.4	40 mM Hepes	RAW264.7 macrophage cells	Yu et al. (2012)
43	N_3-PTS		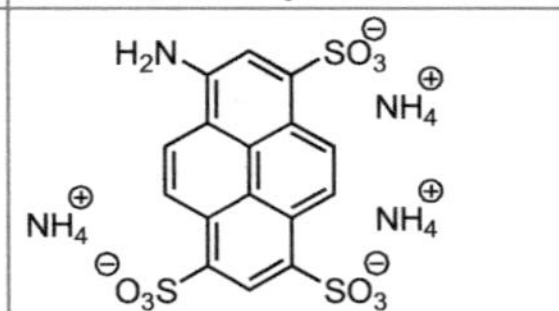	2	435/505	90	7.4	PBS	–	Hartman and Dcona (2012)
44	Aza-BODIPY	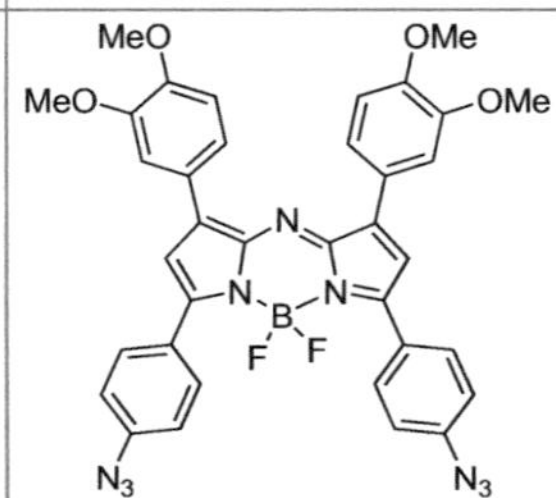	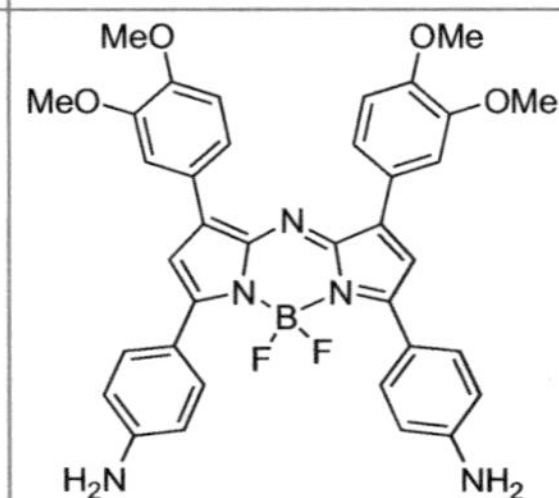	0.5	700/750	0.5	7.0	THF-water	–	Adarsh et al. (2014)
45	SulpHensor	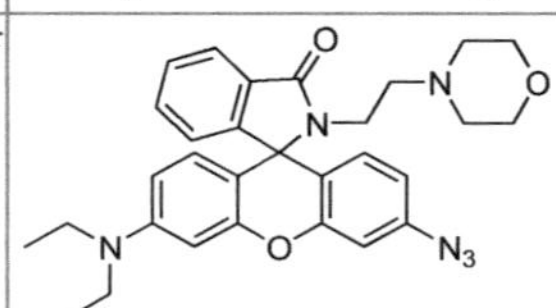	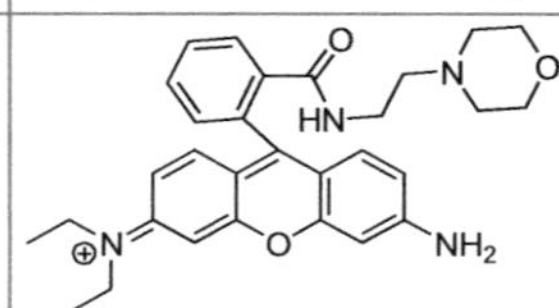	0.5	530/555	60	4.0–6.0 (lysosome targeted)	Britton-Robinson buffer with 10 % DMF	HeLa cells Lysosome targeting	Yang et al. (2014)

49	Probe **1**	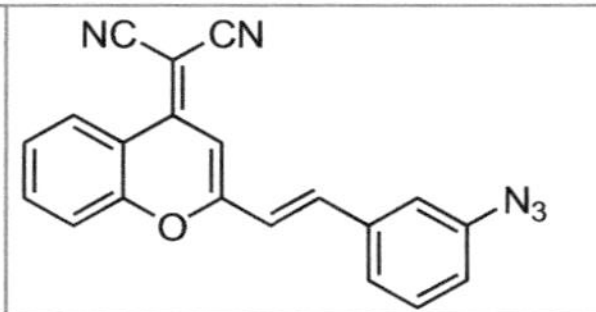	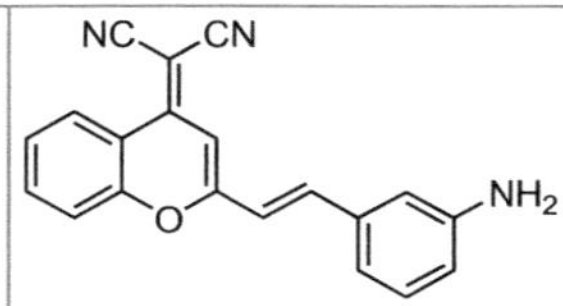	3.05	520/670	60	7.4	100 mM PBS buffer in 50 % DMSO	MCF-7 cells	Sun et al. (2013)
51	AzMB-Coumarin	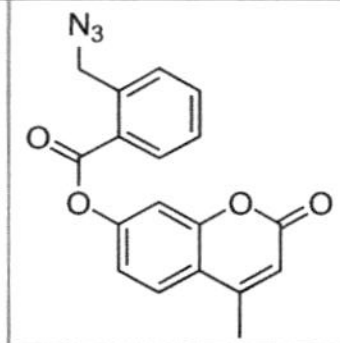	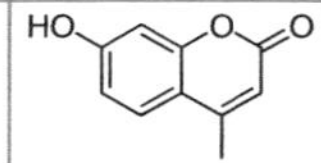	10 (pure MeCN)	365/450	60	7.4	20 % 100 mM sodium phosphate buffered acetonitrile	HeLa cells	Wu et al. (2012)
53		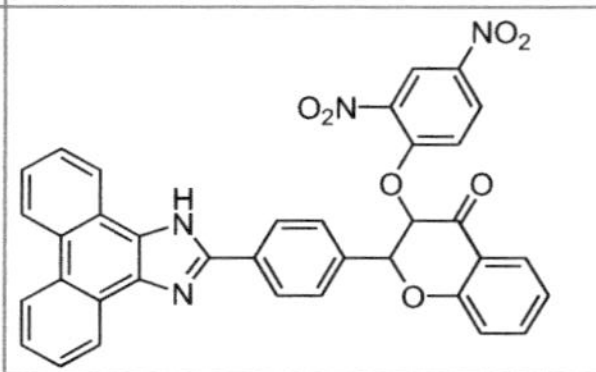	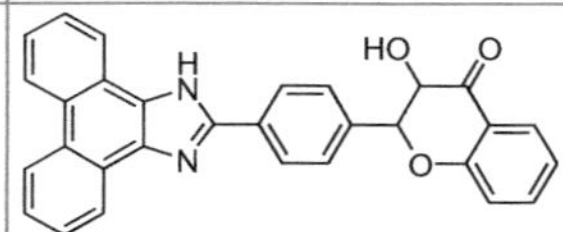	10	440 510 570	30	5.5–8.5	Buffer/tween	MG63 cell lines	10
55		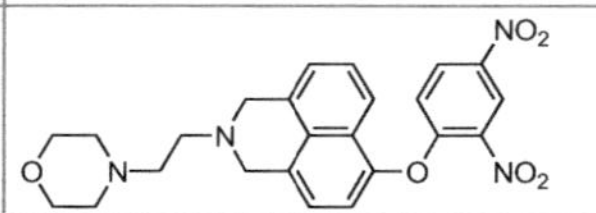	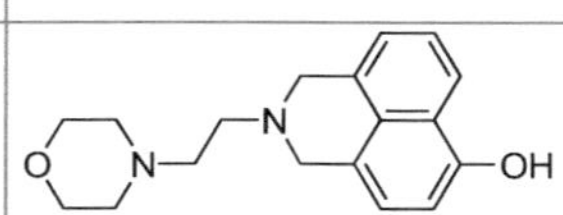	0.48	555	10	4–6	Bovine serum	MCF7 cells Lysosome targeting	8
57		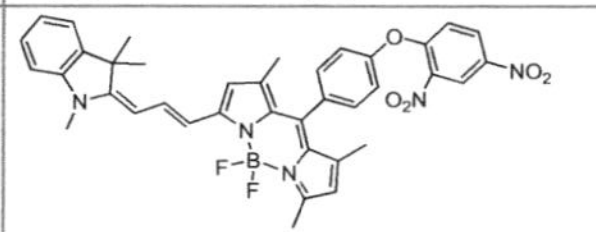	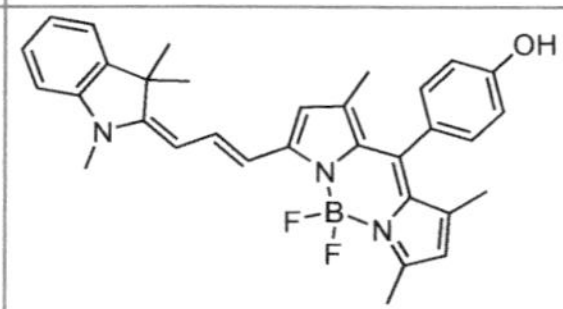	0.05	650/708	<10	7.0	50 mM PBS buffer (pH 7.0) with 3 mM cetyltrimethylammonium bromide and 10 % ethanol	MCF-7 cells	Cao et al. (2012)

(continued)

Table 1 (continued)

Probe	Probe name	Probe structure	Fluorescent product	LOD (μM)	Wavelength (EX/EM, nm)	Reaction time (mins)	pH	Medium	Application in biosystem	References
59	Cy-NO_2	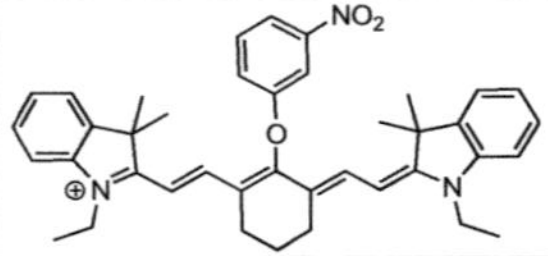	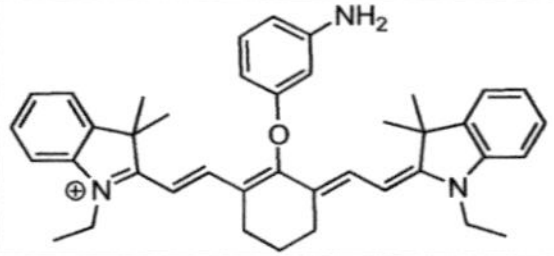	<50	755/789	60	4.2–8.2	40 mM Hepes, pH 7.4, 37 °C	RAW264.7 cells. Probe located in the cytoplasm	Wang et al. (2012)
Metal coordinating probes										
61	HSip-1	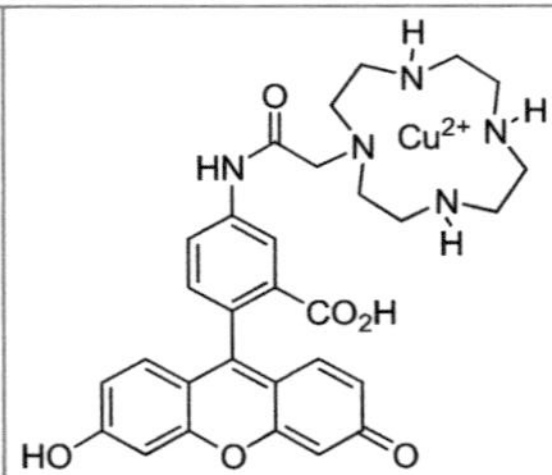	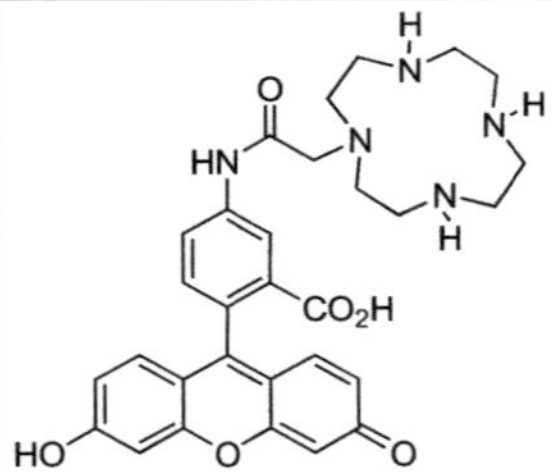		491/516	<5	7.4	30 mM Hepes buffer	HeLa cells	
64	ZS2-Cu	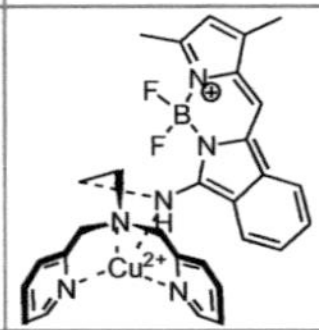	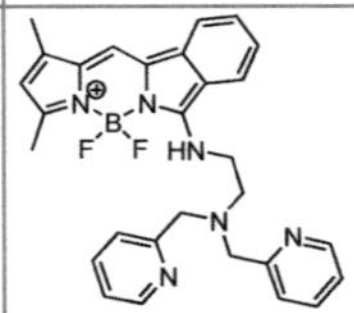	0.25	480/546	15	7.4	PBS	HeLa cells	Li et al. (2014)

65	Probe **7**	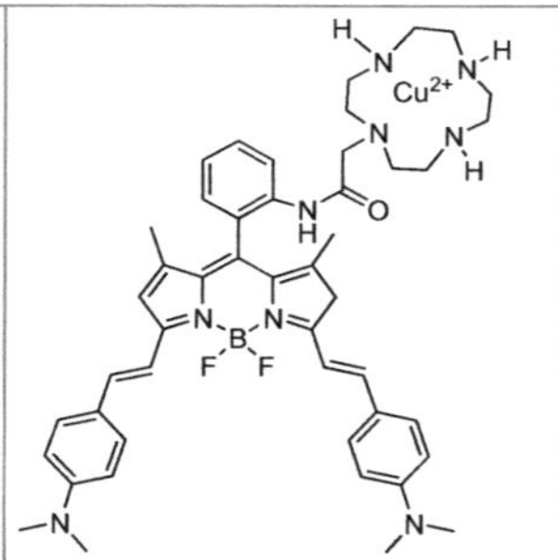	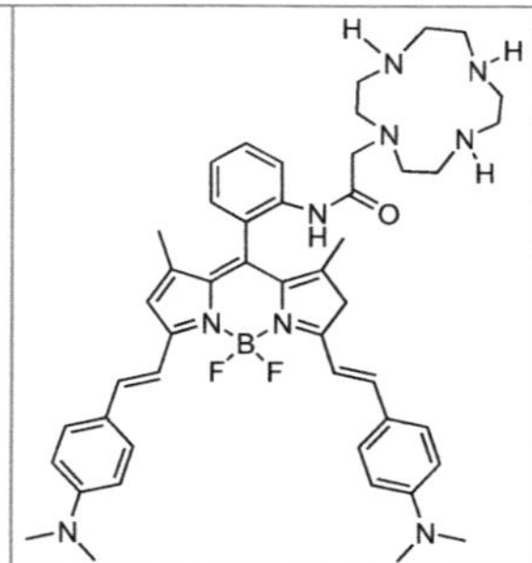	0.08 (LOQ 0.27)	680/765	15	7.4	10 mM PBS buffer with 0.5 % tween-20	RAW264.7 cells and in mice using skin-popping technique	Wu et al. (2014)
66	Probe **8**	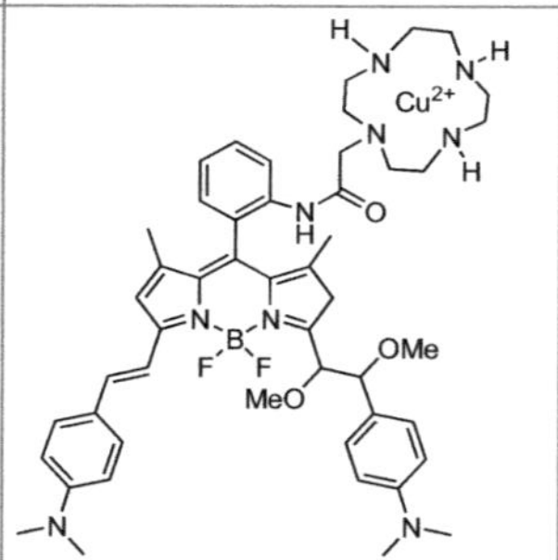	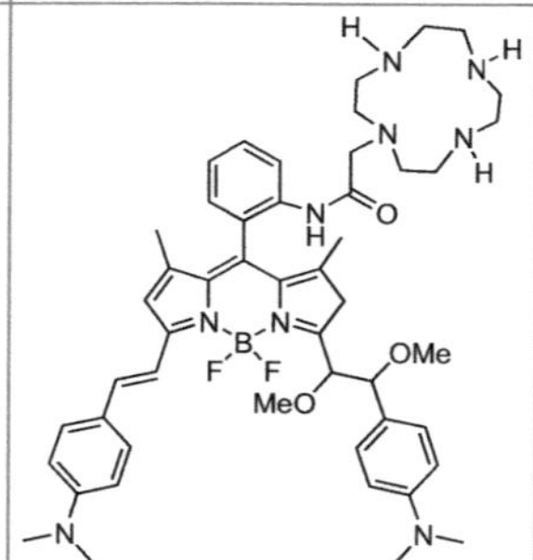	0.08	600/680	15	7.4	10 mM PBS buffer with 0.5 % tween-20	MCF-7 cells	Wu et al. (2014)
67	Cu-1	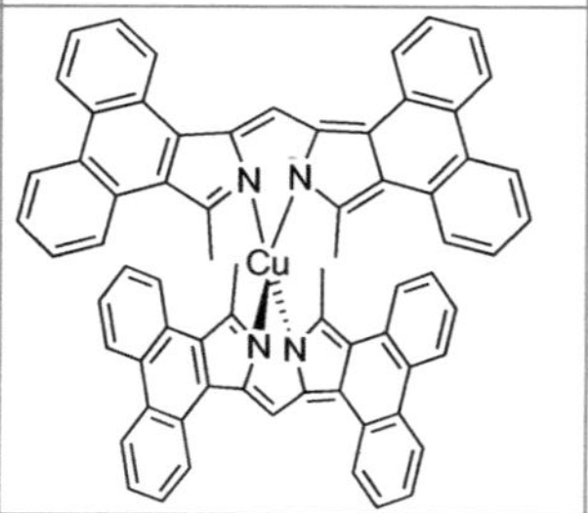	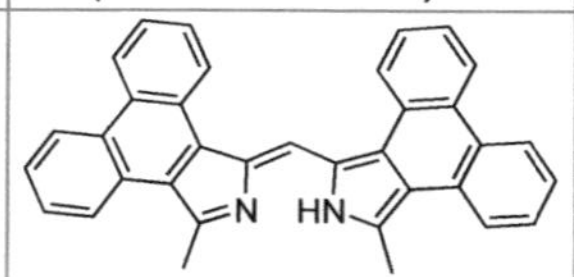	1	540/600	<1	7.0	1:1 DMSO:water	HeLa cells	Qu et al. (2013)

References

Adarsh N, Krishnan MS, Ramaiah D (2014) Sensitive naked eye detection of hydrogen sulfide and nitric oxide by aza-BODIPY dyes in aqueous medium. Anal Chem 86(18):9335–9342

Cai Y, Li L, Wang Z, Sun JZ, Qin A, Tang BZ (2014) A sensitivity tuneable tetraphenylethene-based fluorescent probe for directly indicating the concentration of hydrogen sulfide. Chem Commun 50(64):8892–8895

Cao X, Lin W, Zheng K, He L (2012) A near-infrared fluorescent turn-on probe for fluorescence imaging of hydrogen sulfide in living cells based on thiolysis of dinitrophenyl ether. Chem Commun 48(85):10529–10531

Guo Y, Zeng T, Shi G, Cai Y, Xiec R (2014) Highly selective naphthalimide-based fluorescent probe for direct hydrogen sulfide detection in the environment. RSC Adv 4:33626

Hartman MC, Dcona MM (2012) A new, highly water-soluble, fluorescent turn-on chemodosimeter for direct measurement of hydrogen sulfide in biological fluids. Analyst 137(21): 4910–4912

Kim HM, Cho BR (2009) Two-photon probes for intracellular free metal ions, acidic vesicles, and lipid rafts in live tissues. Acc Chem Res 42(7):863–872

Kwok RT, Leung CW, Lam JW, Tang BZ (2015) Biosensing by luminogens with aggregation-induced emission characteristics. Chem Soc Rev. doi:10.1039/C4CS00325J

Li Q, Lancaster JR Jr (2013) Chemical foundations of hydrogen sulfide biology. Nitric Oxide 35: 21–34

Li L, Rossoni G, Sparatore A, Lee LC, Del Soldato P, Moore PK (2007) Anti-inflammatory and gastrointestinal effects of a novel diclofenac derivative. Free Radic Biol Med 42(5):706–719

Li X, Zhang S, Cao J, Xie N, Liu T, Yang B et al (2013) An ICT-based fluorescent switch-on probe for hydrogen sulfide in living cells. Chem Commun 49(77):8656–8658

Li X, Gong Y, Wu K, Liang SH, Cao J, Yang B et al (2014) A highly sensitive and water soluble fluorescent probe for rapid detection of hydrogen sulfide in living cells. RSC Adv 4:36106

Lin VS, Chang CJ (2012) Fluorescent probes for sensing and imaging biological hydrogen sulfide. Curr Opin Chem Biol 16(5–6):595–601

Lin VS, Lippert AR, Chang CJ (2013) Cell-trappable fluorescent probes for endogenous hydrogen sulfide signaling and imaging H_2O_2-dependent H_2S production. Proc Natl Acad Sci U S A 110(18):7131–7135

Lippert AR, New EJ, Chang CJ (2011) Reaction-based fluorescent probes for selective imaging of hydrogen sulfide in living cells. J Am Chem Soc 133(26):10078–10080

Liu C, Pan J, Li S, Zhao Y, Wu LY, Berkman CE et al (2011) Capture and visualization of hydrogen sulfide by a fluorescent probe. Angew Chem 50(44):10327–10329

Liu C, Peng B, Li S, Park CM, Whorton AR, Xian M (2012) Reaction based fluorescent probes for hydrogen sulfide. Org Lett 14(8):2184–2187

Liu T, Xu Z, Spring DR, Cui J (2013) A lysosome-targetable fluorescent probe for imaging hydrogen sulfide in living cells. Org Lett 15(9):2310–2313

Liu C, Chen W, Shi W, Peng B, Zhao Y, Ma H et al (2014a) Rational design and bioimaging applications of highly selective fluorescence probes for hydrogen polysulfides. J Am Chem Soc 136(20):7257–7260

Liu C, Wu H, Han B, Zhu B, Zhang X (2014b) A highly selective fluorescent chemodosimeter for imaging hydrogen sulfide in living cells. Dyes Pigm 110:214–218

Montoya LA, Pluth MD (2012) Selective turn-on fluorescent probes for imaging hydrogen sulfide in living cells. Chem Commun 48(39):4767–4769

Peng B, Xian M (2014) Fluorescent probes for hydrogen sulfide detection. Asian J Org Chem 3: 914–924

Peng H, Cheng Y, Dai C, King AL, Predmore BL, Lefer DJ et al (2011) A fluorescent probe for fast and quantitative detection of hydrogen sulfide in blood. Angew Chem 50(41):9672–9675

Peng B, Chen W, Liu C, Rosser EW, Pacheco A, Zhao Y et al (2014) Fluorescent probes based on nucleophilic substitution-cyclization for hydrogen sulfide detection and bioimaging. Chemistry 20(4):1010–1016

Qian Y, Karpus J, Kabil O, Zhang SY, Zhu HL, Banerjee R et al (2011) Selective fluorescent probes for live-cell monitoring of sulphide. Nat Commun 2:495

Qu X, Li C, Chen H, Mack J, Guo Z, Shen Z (2013) A red fluorescent turn-on probe for hydrogen sulfide and its application in living cells. Chem Commun 49(68):7510–7512

Sasakura K, Hanaoka K, Shibuya N, Mikami Y, Kimura Y, Komatsu T et al (2011) Development of a highly selective fluorescence probe for hydrogen sulfide. J Am Chem Soc 133(45): 18003–18005

Shen X, Pattillo CB, Pardue S, Bir SC, Wang R, Kevil CG (2011) Measurement of plasma hydrogen sulfide in vivo and in vitro. Free Radic Biol Med 50(9):1021–1031

Shen X, Peter EA, Bir S, Wang R, Kevil CG (2012) Analytical measurement of discrete hydrogen sulfide pools in biological specimens. Free Radic Biol Med 52(11–12):2276–2283

Singha S, Kim D, Moon H, Wang T, Kim KH, Shin YH et al (2015) Toward a selective, sensitive, fast-responsive, and biocompatible two-photon probe for hydrogen sulfide in live cells. Anal Chem 87(2):1188–1195

Sun W, Fan J, Hu C, Cao J, Zhang H, Xiong X et al (2013) A two-photon fluorescent probe with near-infrared emission for hydrogen sulfide imaging in biosystems. Chem Commun 49(37): 3890–3892

Wang R, Yu F, Chen L, Chen H, Wang L, Zhang W (2012) A highly selective turn-on near-infrared fluorescent probe for hydrogen sulfide detection and imaging in living cells. Chem Commun 48(96):11757–11759

Wang J, Lin W, Li W (2013) Three-channel fluorescent sensing via organic white light-emitting dyes for detection of hydrogen sulfide in living cells. Biomaterials 34(30):7429–7436

Wang K, Peng H, Ni N, Dai C, Wang B (2014) 2,6-Dansyl azide as a fluorescent probe for hydrogen sulfide. J Fluoresc 24(1):1–5

Wu Z, Li Z, Yang L, Han J, Han S (2012) Fluorogenic detection of hydrogen sulfide via reductive unmasking of o-azidomethylbenzoyl-coumarin conjugate. Chem Commun 48(81): 10120–10122

Wu H, Krishnakumar S, Yu J, Liang D, Qi H, Lee ZW et al (2014) Highly selective and sensitive near-infrared-fluorescent probes for the detection of cellular hydrogen sulfide and the imaging of H_2S in mice. Chem Asian J 9(12):3604–3611

Xu Z, Xu L, Zhou J, Xu Y, Zhu W, Qian X (2012) A highly selective fluorescent probe for fast detection of hydrogen sulfide in aqueous solution and living cells. Chem Commun 48(88): 10871–10873

Yang S, Qi Y, Liu C, Wang Y, Zhao Y, Wang L et al (2014) Design of a simultaneous target and location-activatable fluorescent probe for visualizing hydrogen sulfide in lysosomes. Anal Chem 86(15):7508–7515

Yu F, Li P, Song P, Wang B, Zhao J, Han K (2012) An ICT-based strategy to a colorimetric and ratiometric fluorescence probe for hydrogen sulfide in living cells. Chem Commun 48(23): 2852–2854

Yu F, Han X, Chen L (2014) Fluorescent probes for hydrogen sulfide detection and bioimaging. Chem Commun 50(82):12234–12249

Zhang J, Sun YQ, Liu J, Shi Y, Guo W (2013) A fluorescent probe for the biological signaling molecule H_2S based on a specific H_2S trap group. Chem Commun 49(96):11305–11307

Zhao Y, Wang H, Xian M (2011) Cysteine-activated hydrogen sulfide (H2S) donors. J Am Chem Soc 133(1):15–17

Peng B, Chen W, Liu C, Rosser EW, Pacheco A, Zhao Y et al (2014) Fluorescent probes based on nucleophilic substitution-cyclization for hydrogen sulfide detection and bioimaging. Chemistry 20(4):1010–1016

Qian Y, Karpus J, Kabil O, Zhang SY, Zhu HL, Banerjee R et al (2011) Selective fluorescent probes for live-cell monitoring of sulphide. Nat Commun 2:495

[illegible]

The Pharmacological Effects of *S*-Propargyl-Cysteine, a Novel Endogenous H_2S-Producing Compound

Ya-Dan Wen and Yi-Zhun Zhu

Contents

Abstract

S-propargyl-cysteine (SPRC), also named as ZYZ-802, is a structural analog of *S*-allylcysteine (SAC), the most abundant constituent of aged garlic extract. SPRC becomes a derivative of the amino acid cysteine, which contains sulfur atom, by changing allyl group in SAC to propargyl group in SPRC. Another

Y.-D. Wen (✉) • Y.-Z. Zhu (✉)
Department of Pharmacology, Clinical Research Centre School of Medicine, National University of Singapore, Bldg MD11, Level 5, # 05-09, 10 Medical Drive, Singapore 117597, Singapore
e-mail: yadan.wen@gmail.com; phczhuyz@nus.edu.sg; zhuyz@fudan.edu.cn

P.K. Moore, M. Whiteman (eds.), *Chemistry, Biochemistry and Pharmacology of Hydrogen Sulfide*, Handbook of Experimental Pharmacology 230,
DOI 10.1007/978-3-319-18144-8_16

analog of SPRC and SAC is *S*-propyl cysteine (SPC), which has propyl group instead in its cysteine structure. Drug formulation of SPRC has been investigated in the mixture of extenders, such as lactose, microcrystalline cellulose, and cross-linked povidone, showing good fluidity and scale-up production possibility. Controlled release formulation of SPRC (CR-SPRC) and leonurine-SPRC were invented and shown the decent pharmacological effects in heart failure and hypoxia injury, respectively. The pharmacological effects of SPRC have been shown that cardioprotection and proangiogenesis in several ischemic heart models, neuroprotection in Alzheimer's disease, proapoptosis in gastric cancer and anti-inflammation in acute pancreatitis. Moreover, CR-SPRC reduced infarct size and recovered partial cardiac function in heart failure rat model. Leonurine-SPRC protected hypoxic neonatal rat ventricular myocytes in much lower dose. Interestingly, since the propagyl group in SPRC has the stronger chemical bond in the cysteine structure than allyl group in SAC and propyl group in SPC, SPRC showed more extensive cardioprotection in ischemic rat hearts model compared to SAC and SPC. The mechanisms of pharmacological effects of SPRC have been unveiled that SPRC reduced Ca^{2+} accumulation, activated antioxidants, inhibited STAT3, decreased inflammatory cytokines, and elevated p53 and Bax. More pharmacological effects and mechanisms of SPRC will be discovered in atherosclerosis, hypertension, and other diseases.

Keywords
S-propargyl-cysteine • SPRC • S-allylcysteine • S-propyl cysteine • Garlic extraction • Hydrogen sulfide • ZYZ-802

1 General Introduction

Garlic is an ancient food consumed by human for thousands of years. This food is considered as a beneficial diet due to the characteristics of antioxidant, hepatoprotection, neuroprotection, and H_2S production (Butt et al. 2009). The major compounds containing sulfide in garlic are γ-glutamyl-*S*-allyl-L-cysteines and *S*-allyl-L-cysteine sulfoxides (alliin) (Amagase 2006). A major transformed chemical from γ-glutamyl-*S*-allyl-L-cysteines is *S*-allylcysteine (SAC), which is a sulfide amino acid presenting cellular (Gupta and Rao 2007), neural (Chauhan 2006), and cardiovascular protection (Shin et al. 2007). The structural analogs of SAC include *S*-propargyl-cysteine (SPRC) and *S*-propyl cysteine (SPC), which contain the same cysteine structure and only differ in the propargyl and allyl moiety, respectively, which were anticipated to have stronger pharmacological effects. Three cysteines all present similar H_2S effects through generating additional H_2S by being potential substrates of endogenous cysteine-metabolizing enzymes (Wang et al. 2009a), seen in Fig. 1.

SPRC is also named as ZYZ-802. Due to the cysteine structure, SPRC was considered as an endogenous H_2S-producing agent by providing the substrate for H_2S synthesis. In the H_2S generating pathway, there are three important synthases:

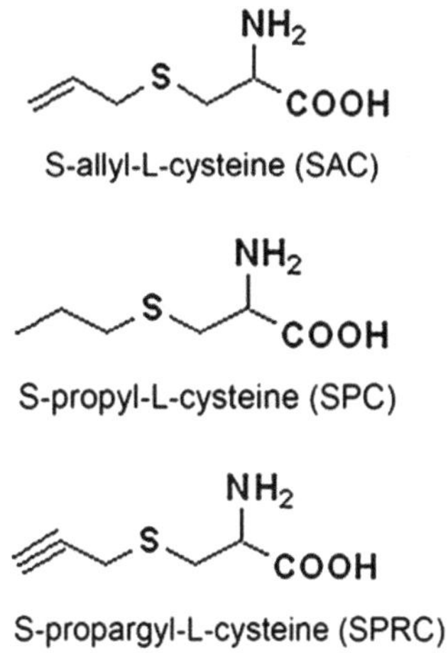

Fig. 1 The chemical structures of SAC, SPC, and SPRC

cystathionine β-synthase (CBS) and cystathionine γ-lyase (CSE) both with a cofactor of pyridoxal 5′-phosphate (vitamin B_6) in cytoplasm (Szabó 2007) and 3-mercaptopyruvate sulfurtransferase (3-MST) along with cysteine aminotransferase (CAT) (Shibuya et al. 2009) in third synthesis pathway of H_2S production in mitochondria. The three synthases have different priorities of expressions in different systems, which mean SPRC may play a role of endogenous H_2S-producing compound via CSE/H_2S in cardiovascular system or CBS/H_2S in neural system.

The pharmacological effects of this cysteine analog were firstly investigated in 2009 on adult rat hearts and neonatal cardiomyocytes (Wang et al. 2009a). From then on, SPRC showed its protective effects in cardiovascular disease, neurodegenerative disease, cancer, and inflammatory disease. Interestingly, SPRC showed stronger cardioprotective effects of decreasing animal mortality, reducing heart infarct zone, lowering the activities and levels of LDK and CK, and promoting cell viabilities through activate antioxidative effects, compared to SAC and SPC treatment (Wang et al. 2010). In this chapter, the chemical and pharmacological characteristics of SPRC will be introduced in the following lines.

2 Analogs of SPRC

SPRC is an analog of SAC, which is one of the major compounds in aged garlic extract. As an original cysteine chemical, SAC firstly investigated the effects on neural and cardiovascular system. It was found that SAC inhibited cellular A-β-induced apoptosis and destabilized Alzheimer's Aβ fibrils in in vitro (Gupta and Rao 2007). In Alzheimer's transgenic mouse model with Swedish double mutation, SAC attenuates Alzheimer's cerebral amyloid, neural inflammation, and tau phosphorylation (Chauhan 2006). In rat ischemic brain, SAC showed neural protection through decreasing generation of toxic-free radical, lipid peroxidation, and cellular damage (Numagami and Ohnishi 2001). More studies were investigated on rat model of stroke-prone spontaneously hypertension (Kim et al. 2006). After treatments of SAC, the incidence of stroke was decreased, behavioral syndromes were relieved, and mortality due to stroke was reduced (Kim et al. 2006). Intervention with SAC showed its cardioprotection by significant decrease of myocardial infarction zone and infarction mortality (Shin et al. 2007).

Another analog of SPRC is SPC, which contains the propyl group in its cysteine structure. Recently, the pharmacological effects of SPC were investigated in diabetes, hyperlipidemia, and oxidative stress injury to protect kidneys, hearts, brains, and livers. First, in diabetic mice, SPC showed renal protection by significantly decreasing plasma urea nitrogen concentration and increasing creatinine clearance. Meanwhile, SPC intake also lessened the renal levels of reactive oxygen species (ROS), nitric oxide (NO), and inflammatory cytokines, like interleukin (IL)-6, tumor necrosis factor-α (TNF-α), prostaglandin E2, and activations of NF-κB, p38, and ERK1/2 induced in diabetic mice (Mong and Yin 2012). Second, SPC reduced the plasma levels of triacylglycerol and cholesterol in hyperlipidemic mice through enhancing the activities of antioxidants and inhibiting the protein and gene expressions of lipogenic-related enzymes, such as malic enzyme, fatty acid synthase, and 3-hydroxy-3-methylglutaryl coenzyme A reductase, together with sterol regulatory element-binding protein (SREBP)-1c and SREBP-2 (Lin and Yin 2008). Third, the anti-inflammatory and antioxidative effects can be further found in ethanol-induced acute hepatotoxicity (Yan and Yin 2007) and PC 12 cells insulted by ischemia (Chen and Yin 2008) and β-amyloid (Tsai et al. 2010) for neurodegenerative disease studies. However, a comparative study among SAC, SPC, and SPRC showed that SPC had weaker performance of cardioprotection and antioxidative effects, compared to SPRC (Wang et al. 2010).

3 Synthesis and Characteristics

3.1 Synthesis

In 50 L of reaction kettle, 27.6 L of dH_2O, 2.4 kg of Na_2CO_3, and 2.65 kg of L-cysteine hydrochloride were added and mixed to a solution. Under reaction temperature 0–5 °C, 3.06 kg of propargyl bromide was dropwise added and mixed for 1 h. After adjusting pH to 6, solids were precipitated and collected by aspirator filter pump with water washing. The product was purified by recrystallization from an ethanol–water solution. After drying, 960 g of products were collected with 40 % yield. The final product was verified by ^1H nuclear magnetic resonance spectroscopy. The purity was 99.7 %, as measured by high-performance liquid chromatography.

3.2 Characteristics

The final product of SPRC was faint yellow color powder, with garlic odor and slightly bitter taste. SPRC can be dissolved in dH_2O, NaOH, and dilute HCl and undissolved in ethanol, acetone, ethyl acetate, and dimethyl sulfoxide.

3.3 Purity

Following "Chinese Pharmacopoeia," inorganic impurity was tested that chloride <0.05 %, sulfate <0.01 %, ferric salts <0.003 %, heavy metals <0.001 %, arsenic salts <0.0001 %, and water <0.53 %.

The products of SPRC have weight loss on drying <0.4 % and weight increment by moisture absorption around 5 %. The only possible residual organic solvent is ethanol, <0.0013 %.

3.4 Stability

The contents of SPRC products were >96 %, after photostability testing, high-temperature test, high humidity test, long-term stability study within 6 months, and accelerated stability study, showing its stability to temperature, light, and humidity.

4 Metabolism and Pharmacokinetics

The SPRC distribution was measured through stable isotope-labeled technique (Zheng et al. 2012). Adult Sprague–Dawley rats were fed [^{35}S]PRC at a dose of 75 mg/kg. The tissue distribution of [^{35}S]PRC-derived radioactivity was shown widely. Kidney showed the highest [^{35}S]PRC-derived radioactivity, which exceeded that of plasma. On the other side, several targeted organs, like brain, heart, lung, and intestine, presented lower [^{35}S]PRC-derived radioactivity than that of plasma. The plasma protein binding of SPRC measured by 96-well equilibrium dialysis was low in human, rat, and dog samples. LC-MS/MS was used to analyze the excretion of SPRC in urine, bile, and feces, which is 2.18 ± 0.61 %, 0.77 ± 0.61 %, and undetectable, respectively. The major metabolite in rat biomatrices, which was identified by MRM information-dependent, acquisition-enhanced product ion (MRM-IDA-EPI) scans on API 4000 QTRAP system, was *N*-acetylation (Zheng et al. 2012). These pharmacokinetic properties of SPRC were observed similar in previous pharmacokinetic SAC study (Nagae et al. 1994; Krause et al. 2002).

5 Formulation

SPRC was investigated to prepare for capsulation. The preparation process started from sieving main drug through 100 mesh sieve and other accessories through 80 mesh sieves. A mixture of lactose, microcrystalline cellulose, and cross-linked povidone as extenders, together with magnesium stearate and silicon powder capsule as lubricants, was made into main drug. It was finished by filling capsules and packing. These capsules have good fluidity and are easy to large-scale production.

SPRC developed controlled release formulation of SPRC (CR-SPRC), generating more stable and sustainable gas of H_2S (Huang et al. 2013). CR-SPRC was produced by the solid dispersion technique using Eudragit (R) RS30D (Rohm Pharma, Weiterstadt, Germany) as a carrier. In a round-bottomed flask, CR-SPRC was dissolved in Eudragit (R) RS30D with a constant magnetic stirring in midrange rotation speed and 80 °C of a thermostat-controlled water bath. Eudragit(R) RS30D was then removed. The precipitates were cooled down to −20 °C immediately. After heating at 50 °C for 4 h, coprecipitates of CR-SPRC were desiccated, pulverized, and finally sieved through 100 mesh. The collection of CR-SPRC can be dissolved in 0.5 % sodium carboxymethylcellulose (CMC-Na) for future use (Huang et al. 2013). The pharmacological effects of CR-SPRC has been investigated in heart failure rat model. For more information, please refer to Sect. 6.1.2.

Incorporation of two mutually pharmacological or biological active compounds into one chemical by conjugation may be a good drug design to enhance the drug effects. Because leonurine and SPRC have been reported the cardioprotective effects both in vitro and in vivo, the two entities were considered to be synthesized to leonurine-SPRC conjugate, which connected by a phenolic hydroxyl ester bond so that easy hydrolyzing and releasing bioactive leonurine and SPRC (Liu et al. 2010). The pharmacological effects of leonurine–SPRC conjugate were investigated in neonatal rat ventricular myocytes under hypoxia, which will be discussed in the last paragraph of Sect. 6.1.1.

6 Pharmacologic Effects

6.1 Cardiovascular Disease

6.1.1 Ischemia/Hypoxia Injury

The modulation on cardiovascular disease by SPRC was firstly investigated in myocardial infarction rat models and hypoxic cardiomyocytes cells (Wang et al. 2009b). SPRC treatments were found in association with obvious reduction of myocardial infarct size and improved cardiac function in in vivo (Wang et al. 2009b). Rescuing hypoxic cardiomyocytes by SPRC indicates its cardioprotective effects. These effects go in line with the increased CSE activity and H_2S plasma concentration, which means the cardioprotective effects of SPRC, an endogenous H_2S-producing modulator, is through CSE/H_2S pathway (Wang et al. 2009b).

Furthermore, daily i.p. injection of three cysteine analogs (SAC, SPC, and SPRC) for 7 days before and 2 days after coronary artery ligation of rat hearts attenuated significant heart damage against ischemic reperfusion through the activation of antioxidative enzymes, such as superoxide dismutase (SOD), catalase (CAT), and glutathione peroxidase (GPx), as well as elevation of glutathione (GSH) levels, taken together with inhibition of lipid peroxidation products, like malondialdehyde (MDA) content (Wang et al. 2009a). Meanwhile, CSE expressions and H_2S producing rate and contents were increased by SPRC treatments but abolished by CSE selective inhibitor, propargylglycine (PAG), indicating the protection of SPRC through CSE/H_2S pathway (Wang et al. 2009a). It was highlighted that the SPRC showed stronger heart protection compared to SAC and SPC. This may be due to the stronger connection of propargyl group in SPRC cysteine structure instead of allyl and propyl in SAC and SPC, respectively.

More cardiac effects of SPRC were investigated in hypoxia/reoxygenation (H/R) model (Liang et al. 2014). Rat ventricular myocytes and left ventricular papillary muscles were isolated to endure hypoxic injury for 30 min and followed by reoxygenation solution for 30 min. Pretreatment of SPRC 30 min prior to hypoxia may attenuate the apoptosis of isolated papillary muscles induced by H/R injury and maintain the morphology of these muscle fibers. Hypercontracture of ventricular myocytes was significantly inhibited by SPRC pretreatments at 3 min of reoxygenation and closed to normal level at 30 min of reoxygenation. Meanwhile, SPRC attenuated Ca^{2+} accumulation, high upstroke velocity, and increase of resting $[Ca^{2+}]i$ induced by hypoxia. Because Ca^{2+} is mainly released by ryanodine receptors (RyR) and uptaken by SERCA-2, SPRC showed preserving activities of RyR and SERCA-2 and enhancing Ca^{2+} handling ability. These cardiac protective effects of SPRC were blocked by CSE inhibitor, PAG (Liang et al. 2014).

Owing to the mutually cardioprotective effects by leonurine and SPRC, the newly synthesized compound, leonurine–SPRC, possessed strong cardiac cellular protections of hypoxic neonatal rat ventricular myocytes in very low molar concentration of 1/10 leonurine dose and 1/100 SPRC dose (Liu et al. 2010). The cardioprotective mechanisms of this leonurine–cysteine analog illuminated three aspects: antioxidative effects by increasing SOD and CAT activities and decreasing MDA and ROS levels, ultrastructural protections of myocytes, and regulation of apoptosis by inhibiting activation of caspase-3 and rising Bcl-2 expressions (Liu et al. 2011). These findings indicated that this leonurine–SPRC conjugate showed potentially multifunctional anti-ischemia in neonatal rat ventricular myocytes, providing an important clue for future design and synthesis of such mutually pharmacological drugs.

6.1.2 Heart Failure

For releasing H_2S more stable and sustainable, CR-SPRC was introduced to heart failure rat model (Huang et al. 2013). The reduced infarct size and recovered cardiac function were discovered through activation of antioxidant defenses by increasing SOD, CAT, and GSH and decreasing creatine kinase (CK) leakage (Huang et al. 2013). Meanwhile, more protective mechanisms of SPRC are through the prohibition of activation of caspases and elevation of ratio of Bcl-2/Bax, showing the antiapoptotic effects of SPRC. Compared to original SPRC, CR-SPRC showed better performance of cardioprotection. Due to its endogenous H_2S-producing characteristics, the inhibition of heart failure by SPRC may be mediated by CSE/H_2S pathway.

6.1.3 Angiogenesis

A study designed for the therapy of myocardial reperfusion injury investigated the proangiogenic effects of SPRC to provide a novel treatment target of ischemia/reperfusion heart disease (Kan et al. 2014). First, SPRC was found promoting cell proliferation, cellular adhesion, migration, and tube formation in human umbilical vein endothelial cells (HUVEC). Moreover, under both normal condition and ischemic injury, rat aortic ring and Matrigel plug-implanted mice found the promoted angiogenesis. In the mouse ischemic hind limb by ligation of the left femoral artery and the rat ischemic heart by ligation of coronary artery, angiogenesis was facilitated by SPRC to attenuate the ischemic injury. Interestingly, SPRC treatments can significantly induce STAT3 phosphorylation in HUVEC in a dose- and time-dependent manner, in which effects were abolished by WP1066, an inhibitor of STAT3. After transfecting STAT3 siRNA into HUVECs to silence the expression of STAT3, the effects of cellular adhesion, migration, and tube formation by SPRC were abolished due to the absence of STAT3 expression. More studies showed that SPRC may not directly interact with STAT3 by the method of co-crystallization, but may enhance the interaction between VEGFR2 and STAT3 by co-immunoprecipitation. The influence to STAT3 by SPRC additionally found the nuclear translocation of STAT3 and transcriptional activation of downstream promoters, especially the *Vegf* promoter. Therefore, angiogenesis may be strengthened by SPRC through STAT3-mediated mechanism, which provides a novel avenue to target on ischemic disease.

6.2 Neurodegenerative Disease

Due to the neuroinflammatory effects by producing various inflammatory mediators and neural toxic proteins to exacerbate amyloid-β (Aβ) and hyperphosphorylated tau in Alzheimer's disease (AD), the anti-inflammatory effects of SPRC showed neural protection on rat spatial learning and memory impairment induced by lipopolysaccharide (LPS) bilateral intracerebroventricular (i.c.v.) injection valued by Morris water maze test (Gong et al. 2011a). The accumulation of Aβ protein precursor, a type I transmembrane protein, playing key function and metabolism in pathogenesis of AD, together with $A\beta_{1-40/42}$, was suppressed by pretreatments of SPRC of 40 and 80 mg/kg, showing the inhibited effects of SPRC on Aβ generation. Additionally, the overwhelming evidences of activation of tumor necrosis factor (TNF), like TNF-α and TNF-α receptor 1 (TNFR1), were prohibited by pretreatments of SPRC of 40 and 80 mg/kg. Further evidences of upregulatory inflammatory cytokines like IκB-α and activation of NF-κB p65 phosphorylation were significantly decreased by SPRC treatments of a dose range of 20–80 mg/kg. These findings all indicate SPRC may be a potential anti-neuroinflammatory agent for AD.

More studies were investigated on rats not limited on previous neuroinflammatory injury by LPS, but directly given i.c.v. injection of 10 μg of $A\beta_{25-35}$ to induce most common AD rat model (Gong et al. 2011b). Administration of SPRC of 40 and 80 mg/kg through intraperitoneal injection may prohibit cognitive impairment and neuronal ultrastructural damage in AD rats. The inhibition of pathogenesis of AD by SPRC may be associated with the decreased proteins and genes expressions of TNF-α and cyclooxygenase-2 in AD hippocampus, suppressed extracellular signal-regulated kinase (ERK 1/2), reduced IκB-α degradation, and weakened activation of NF-κB after pretreatments of SPRC with Aβ neuronal injury. The recovery of AD syndromes like cognitive impairment, maintaining of neuronal ultrastructure in AD rats, and less expressions of neuroinflammatory cytokines may suggest the underlying mechanisms and evidences of neural protective effects of SPRC on AD.

6.3 Cancer

SPRC investigated its potential effects on SGC-7901 gastric cancer cells (Ma et al. 2011). Cell viability, proliferation, and migration of SGC-7901 were significantly suppressed by SPRC treatments. These gastric cancer cells found the cell cycle arrest at the G1/S phase by SPRC treatments.

Further studies were performed on nude mice that accepted injection of SGC-7901 cells (Ma et al. 2011). After 7 days post-implant, when the tumor size reached 0.3–0.5 cm, mice were given 50 and 100 mg/kg intraperitoneal injections of SPRC. Tumor volumes were significantly reduced with 50 and 100 mg/kg of SPRC with tumor growth inhibition of 40–75 %. The tumor growth suppression by SPRC was inhibited by the inhibitor of H_2S-generating enzyme CSE, PAG. This anticancer mechanism of SPRC may be related to elevation of p53 and Bax expressions in tumor and SGC-7901 gastric cancer cells, indicating that SPRC induces proapoptotic effect.

6.4 Inflammation

The regulation of inflammation by SPRC was investigated in an in vivo model of acute pancreatitis (AP) in mice (Sidhapuriwala et al. 2012). Hourly caerulein injections for 10 h were given to induce AP. Pre-administration of SPRC was given 12 or 3 h before induction of AP. Pancreatic and pulmonary injuries were observed through increased amylase plasma concentration, elevated myeloperoxidase activities, and inflammatory histological observation, like edema, destruction of histo-architecture of the acini, and infiltration of inflammatory cells, which were reduced by 3 h pretreatment of SPRC before AP induction. Additionally, augmentation of pro-inflammatory cytokines (IL-1β and IL-6) and decline of anti-inflammatory cytokine IL-10 were significantly inhibited by 3 h pretreatment of SPRC before AP induction. 12 h SPRC pre-administration before AP induction did not show significant effects on lung and pancreatic inflammatory injury. Meanwhile, anti-inflammatory effects of SPRC were in consistence with significant changes of H_2S level. Therefore, SPRC pretreatment of 10 mg/kg prior to AP induction ameliorated lung and pancreatic inflammatory injury by suppression of inflammatory cell infiltration and modulation of pro-/anti-inflammatory cytokine profile in plasma.

More anti-inflammatory effects of SPRC were investigated in other systems, like cardiac system, vascular endothelium, and neurodegenerative disease. First, in H9c2 cardiac myocytes, LPS was introduced as an inflammatory stimuli and SPRC as a protective treatment (Pan et al. 2011). In this LPS-induced inflammatory response, NF-κB was activated through phosphorylation of NF-κBp65 and IκB-α degradation, which were inhibited by SPRC, as well as phosphorylation of ERK 1/2 and ROS generation. Other inflammatory cytokines, like TNF-α, inducible nitric oxide synthase, and intercellular adhesion molecules-1, were increased by LPS injury, which were suppressed by SPRC. Moreover, SPRC leads to phosphorylation of Akt in concentration-dependent and time-dependent manners, which were abolished by phosphoinositide 3-kinase (PI3K) inhibitor LY 294002, indicating the activated effects of SPRC on PI3K/Akt signaling pathway. The mechanisms of impairing NF-κB/IκB-α and inhibition of inflammatory cytokines had association with increasing H_2S and CSE levels induced by SPRC, indicating the cardioprotective effects are through CSE/H_2S pathway.

Second, SPRC was introduced to assess its effects on HUVECs injured by an inflammatory cytokines, NF-κB (Pan et al. 2012). In this cardiac inflammatory cellular model, SPRC pretreatments observed a significant dose-dependent inhibition on the increased adhesion of U937 monocytes/ HUVECs induced by TNF-α. This decrease by SPRC (5 and 10 μM) may be through the mechanism of reducing ICAM-1 and VCAM-1 expressions which were raised by TNF-α. Meanwhile, the translocation and phosphorylation of NF-κB p65 and IκB-α degradation were stimulated by TNF-α but attenuated by SPRC (5 and 10 μM), indicating that the activation of NF-κB was inhibited by SPRC pretreatments. Additionally, findings of TNF-α-induced JNK1/2 activation prohibited by SPRC (5 and 10 μM) indicated that the regulation of JNK1/2 signaling pathway is also involved in the anti-inflammatory mechanisms of SPRC.

Third, the anti-inflammatory effects on neurodegenerative disease have been discussed in Sect. 6.2.

7 Conclusion

SPRC, a cysteine analog, inspired by the structure and vulnerable characteristics of a garlic extraction SAC, was synthesized to a propargyl group instead of allyl within the same cysteine structure in order to form a more stable structure with stronger pharmacological effects. The stronger protective effects of SPRC were firstly found in rat hearts and neonatal cardiomyocytes, compared to SAC and SPC. More pharmacological effects of SPRC were shown in cardiovascular diseases (hypoxia, ischemic reperfusion, H/R injury, heart failure, and angiogenesis), neurodegenerative disease (AD), cancer (gastric cancer), and inflammatory disease (AP and inflammatory responses induced by LPS or TNF-α in myocytes, endothelial cells, and rats). Meanwhile, SPRC, the endogenous H_2S agent, can increase CSE and H_2S productions, indicating that the regulation of cellular Ca^{2+} and activation of NF-κB, TNF signaling, redox status, ERK 1/2, apoptosis, and inflammatory cytokines are through CSE/H_2S pathway. However, the investigations on pharmacological effects of this novel cysteine analog were still limited. More studies will be carried out on atherosclerosis, hypertension, and other diseases in vivo, ex vivo, and in vitro to unveil its underlying mechanisms of these potential protections.

References

Amagase H (2006) Clarifying the real bioactive constituents of garlic. J Nutr 136:716S–725S

Butt MS et al (2009) Garlic: nature's protection against physiological threats. Crit Rev Food Sci Nutr 49:538–551

Chauhan NB (2006) Effect of aged garlic extract on APP processing and tau phosphorylation in Alzheimer's transgenic model Tg2576. J Ethnopharmacol 108:385–394

Chen CM, Yin MC (2008) S-Allyl cysteine, S-ethyl cysteine and S-propyl cysteine alleviate oxidative stress-induced damage within PC12 cells. J Sci Food Agric 88:2493–2498

Gong QH et al (2011a) S-propargyl-cysteine, a novel hydrogen sulfide-modulated agent, attenuates lipopolysaccharide-induced spatial learning and memory impairment: involvement of TNF signaling and NF-κB pathway in rats. Brain Behav Immun 25:110–119

Gong QH et al (2011b) S-propargyl-cysteine (ZYZ-802), a sulphur-containing amino acid, attenuates beta-amyloid-induced cognitive deficits and pro-inflammatory response: involvement of ERK1/2 and NF-κB pathway in rats. Amino Acids 40:601–610

Gupta VB, Rao KSJ (2007) Anti-amyloidogenic activity of S-allyl-l-cysteine and its activity to destabilize Alzheimer's β-amyloid fibrils in vitro. Neurosci Lett 429:75–80

Huang C et al (2013) Cardioprotective effects of a novel hydrogen sulfide agent-controlled release formulation of S-propargyl-cysteine on heart failure rats and molecular mechanisms. PLoS One 8:e69205

Kan J et al (2014) S-propargyl-cysteine, a novel water-soluble modulator of endogenous hydrogen sulfide, promotes angiogenesis through activation of signal transducer and activator of transcription 3. Antioxid Redox Signal 20:2303–2316

Kim JM et al (2006) Dietary S-Allyl-L-cysteine reduces mortality with decreased incidence of stroke and behavioral changes in stroke-prone spontaneously hypertensive rats. Biosci Biotechnol Biochem 70:1969–1971

Krause RJ et al (2002) Sulfoxides as urinary metabolites of S-allyl-L-cysteine in rats: evidence for the involvement of flavin-containing monooxygenases. Drug Metab Dispos 30:1137–1142

Liang YH et al (2014) SPRC protects hypoxia and re-oxygenation injury by improving rat cardiac contractile function and intracellular calcium handling. Nitric Oxide 41:113–119

Lin CC, Yin MC (2008) Effects of cysteine-containing compounds on biosynthesis of triacylglycerol and cholesterol and anti-oxidative protection in liver from mice consuming a high-fat diet. Br J Nutr 99:37–43

Liu C et al (2010) Synthesis and biological evaluation of novel leonurine-SPRC conjugate as cardioprotective agents. Bioorg Med Chem Lett 20:6942–6946

Liu C et al (2011) Leonurine-cysteine analog conjugates as a new class of multifunctional anti-myocardial ischemia agent. Eur J Med Chem 46:3996–4009

Ma K et al (2011) H2S donor, S-propargyl-cysteine, increases CSE in SGC-7901 and cancer-induced mice: evidence for a novel anti-cancer effect of endogenous H2S? PLoS One 6:e20525

Mong MC, Yin MC (2012) Nuclear factor KB-dependent anti-inflammatory effects of s-allyl cysteine and s -propyl cysteine in kidney of diabetic mice. J Agric Food Chem 60:3158–3165

Nagae S et al (1994) Pharmacokinetics of the garlic compound S-allylcysteine. Planta Med 60:214–217

Numagami Y, Ohnishi ST (2001) S-allylcysteine inhibits free radical production, lipid peroxidation and neuronal damage in rat brain ischemia. J Nutr 131:1100S–1105S

Pan LL et al (2011) S-propargyl-cysteine (SPRC) attenuated lipopolysaccharide-induced inflammatory response in H9c2 cells involved in a hydrogen sulfide-dependent mechanism. Amino Acids 41:205–215

Pan LL et al (2012) S-propargyl-cysteine, a novel hydrogen sulfide-modulated agent, attenuated tumor necrosis factor-KB-induced inflammatory signaling and dysfunction in endothelial cells. Int J Cardiol 155:327–332

Shibuya N et al (2009) 3-Mercaptopyruvate sulfurtransferase produces hydrogen sulfide and bound sulfane sulfur in the brain. Antioxid Redox Signal 11:703–714

Shin CC et al (2007) S-allylcysteine mediates cardioprotection in an acute myocardial infarction rat model via a hydrogen sulfide-mediated pathway. Am J Physiol Heart Circ Physiol 293: H2693–H2701

Sidhapuriwala JN et al (2012) Effects of s-propargyl-cysteine (sprc) in caerulein-induced acute pancreatitis in mice. PLoS One 7:e32574

Szabó C (2007) Hydrogen sulphide and its therapeutic potential. Nat Rev Drug Discov 6:917–935

Tsai SJ et al (2010) S-ethyl cysteine and s-propyl cysteine alleviate β-amyloid induced cytotoxicity in nerve growth factor differentiated PC12 cells. J Agric Food Chem 58:7104–7108

Wang Q et al (2009a) Protective effects of cysteine analogues on acute myocardial ischemia: novel modulators of endogenous H2S production. Antioxid Redox Signal 12:1155–1165

Wang Q et al (2009b) S-propargyl-cysteine protects both adult rat hearts and neonatal cardiomyocytes from ischemia/hypoxia injury: the contribution of the hydrogen sulfide-mediated pathway. J Cardiovasc Pharmacol 54:139–146

Wang Q et al (2010) Protective effects of cysteine analogues on acute myocardial ischemia: novel modulators of endogenous H2S production. Antioxid Redox Signal 12:1155–1165

Yan SL, Yin MC (2007) Protective and alleviative effects from 4 cysteine-containing compounds on ethanol-induced acute liver injury through suppression of oxidation and inflammation. J Food Sci 72:S511–S515

Zheng YT et al (2012) Preclinical assessment of the distribution, metabolism, and excretion of S-propargyl-cysteine, a novel H 2S donor, in Sprague-Dawley rats. Acta Pharmacol Sin 33:839–844

Phosphinodithioate and Phosphoramidodithioate Hydrogen Sulfide Donors

Matthew Whiteman, Alexis Perry, Zongmin Zhou, Mariarosaria Bucci, Andreas Papapetropoulos, Guiseppe Cirino, and Mark E. Wood

Contents

M. Whiteman (✉)
University of Exeter Medical School, Exeter, United Kingdom
e-mail: m.whiteman@exeter.ac.uk

A. Perry • M.E. Wood (✉)
Department of Biosciences, College of Life and Environmental Sciences, University of Exeter, United Kingdom
e-mail: m.e.wood@exeter.ac.uk

Z. Zhou
Department of Pharmacy, University of Athens, Greece

M. Bucci • G. Cirino
Department of Pharmacology, University of Naples Frederico III, Italy

A. Papapetropoulos
Department of Pharmacy, University of Athens, Greece

George P. Livanos and Marianthi Simou Laboratories, 1st Department of Pulmonary and Critical Care Medicine, Evangelismos Hospital, Faculty of Medicine, University of Athens, Greece

P.K. Moore, M. Whiteman (eds.), *Chemistry, Biochemistry and Pharmacology of Hydrogen Sulfide*, Handbook of Experimental Pharmacology 230,
DOI 10.1007/978-3-319-18144-8_17

Abstract

Hydrogen sulfide is rapidly emerging as a key physiological mediator and potential therapeutic tool in numerous areas such as acute and chronic inflammation, neurodegenerative and cardiovascular disease, diabetes, obesity and cancer. However, the vast majority of the published studies have employed crude sulfide salts such as sodium hydrosulfide (NaSH) and sodium sulfide (Na_2S) as H_2S "donors" to generate H_2S. Although these salts are cheap, readily available and easy to use, H_2S generated from them occurs as an instantaneous and pH-dependent dissociation, whereas endogenous H_2S synthesis from the enzymes cystathionine γ-lyase, cystathionine-β-synthase and 3-mercaptopyruvate sulfurtransferase is a slow and sustained process. Furthermore, sulfide salts are frequently used at concentrations (e.g. 100 μM to 10 mM) far in excess of the levels of H_2S reported in vivo (nM to low μM). For the therapeutic potential of H_2S is to be properly harnessed, pharmacological agents which generate H_2S in a physiological manner and deliver physiologically relevant concentrations are needed. The phosphorodithioate GYY4137 has been proposed as "slow-release" H_2S donors and has shown promising efficacy in cellular and animal model diseases such as hypertension, sepsis, atherosclerosis, neonatal lung injury and cancer. However, H_2S generation from GYY4137 is inefficient necessitating its use at high concentrations/doses. However, structural modification of the phosphorodithioate core has led to compounds (e.g. AP67 and AP105) with accelerated rates of H_2S generation and enhanced biological activity. In this review, the therapeutic potential and limitations of GYY4137 and related phosphorodithioate derivatives are discussed.

Keywords

GYY4137 • AP67 • AP105 • Slow-release H_2S donor • cGMP

1 Introduction

Recent research into the regulation of endogenously synthesised hydrogen sulfide (H_2S) from the enzymes cystathionine γ-lyase (CSE), cystathionine-β-synthase (CBS) and 3-mercaptopyruvate sulfurtransferase (3-MST) has revealed a startling array of physiological and pathophysiological processes that are regulated by H_2S including learning and memory, regulation of vascular tone, ion channel regulation, cytoprotection, acute and chronic inflammation, metabolism, allergy and cell signalling, etc. The fact that there are three regulated cellular processes to synthesise this gaseous molecule highlights its importance, and key findings in this area of research are presented elsewhere in this book. The rapid advances in molecular biology have given researchers in this area highly selective tools to silence CSE, CBS and 3-MST from cells and tissues as well as genetic knock-out animals devoid of these enzymes. Whilst potent and selective inhibitors of CSE, CBS and 3-MST are still not available and the use of commonly used inhibitors should be viewed with caution (Asimakopoulou et al. 2013; Whiteman et al. 2011), molecular techniques have graphically illustrated the detrimental effects "H_2S removal" has to cells, tissues and animals. Studies in animal models and human disease have revealed that endogenous H_2S synthesis is perturbed in ageing (Qabazard et al. 2014; Yang et al. 2013), cardiovascular disease (Calvert et al. 2009, 2010; Minamishima et al. 2009; Yang et al. 2008), pre-eclampsia (Wang et al. 2013), acute and chronic inflammation (Li et al. 2009, 2013), diabetes (Whiteman et al. 2010a) and respiratory pathologies such as asthma (Chung 2014) and chronic obstructive pulmonary disease (Chen et al. 2005; Saito et al. 2014; Kirkham et al. 2014). These observations have consistently led to the suggestion that strategies which increase H_2S bioavailability are viable therapeutic approaches to disease treatment. However, at present, there are very few molecules described in the literature that can achieve this. As such, this review will highlight the recent progress that has been made with phosphorodithioate-based H_2S "donors", typified by the molecule GYY4137.

2 The Need for Slow-Release H_2S Donor Molecules

Although multiple studies have highlighted the diverse physiological role for H_2S and have emphasised the possible therapeutic opportunities for modulating H_2S bioavailability, it should be noted, and with some concern, that the bulk of the scientific literature is replete with the use of high concentrations (and doses) of crude sulfide salts such as sodium hydrosulfide (NaSH) or sodium sulfide (Na_2S) where they have been used as "fast-release H_2S donors". Whilst these salts are undoubtedly convenient for use in the laboratory as one can readily generate solutions containing known quantities of H_2S, they are highly problematic. Although these problems have been reviewed in detail elsewhere (Whiteman et al. 2011; Whiteman and Winyard 2011), it is pertinent to briefly discuss the major flaws. Firstly, NaSH and Na_2S are simple salts and not H_2S donors. They undergo instantaneous and pH-dependent dissociation generating, not releasing, H_2S. In sharp contrast, enzymatic synthesis of H_2S from CSE, CBS and 3-MST is not instantaneous but a slow and sustained process (Whiteman et al. 2011; Li et al. 2013; Banerjee and Zou 2005; Kabil et al. 2011a, b). This obvious discrepancy in kinetics of H_2S generation from NaSH and Na_2S compared to CSE-, CBS- and 3-MST-derived H_2S should preclude their use for determining the physiological and biological properties of H_2S. Secondly, high concentrations, >100 μM and often 1–10 mM, are frequently required to elicit any biological response in cells or tissue, presumably because H_2S generated from these salts dissipates in seconds (Li et al. 2008; Whiteman et al. 2010b), rather than being synthesised slowly by CSE, CBS and 3-MST. Thirdly, commercial supplies of NaSH are often of very poor purity (typically stated as >60 % purity), and despite this information being clearly stated in the online product specification information and documents that will accompany NaSH, and any commercial research product, when it is delivered (e.g. Sigma-Aldrich (2014) and Caymen Chemicals (2014)), it continues to be widely used. As such, sulfide salts are poor models for determining the role of endogenous H_2S in physiological systems or exploring the pharmacological and therapeutic potential of H_2S-donating molecules.

Although several detailed and excellent reviews have explicitly highlighted how Na_2S and NaSH should and should not be used (Whiteman et al. 2011; Yuan et al. 2015; Hughes et al. 2009; Olson 2013) in the laboratory (e.g. preparing and storing solutions, effects of pH and buffers, preventing loss of H_2S to the atmosphere, etc.), they do not address the problem that these salts generate H_2S in an instant and as a rapidly dissipating bolus. As such, there is great interest in the development of slow-release H_2S donor molecules which overcome the problems stated above.

3 Water-Soluble Slow-Release Hydrogen Sulfide Donors

The first water-soluble slow-release H_2S donor to be described was morpholin-4-ium 4 methoxyphenyl(morpholino) phosphinodithioate (GYY4137), developed by Moore and colleagues at the National University of Singapore (Li et al. 2008) (Fig. 1). GYY4137 offers considerable advantages over the use of sulfide salts such as NaSH or Na_2S since the rate of H_2S generation more readily mimics endogenously synthesised H_2S from enzymatic sources such as CSE, CBS and 3-MST, rather than delivering H_2S as a single concentrated bolus, e.g. the manner in which H_2S is generated from GYY4137 more accurately reflects endogenously produced H_2S. A second obvious advantage is that the final concentrations of H_2S generated from GYY4137 are likely more physiologically, pathophysiologically and therapeutically relevant than a bolus of concentrated and impure (Sigma-Aldrich 2014; Caymen Chemicals 2014) sulfide salt. Thirdly, the high solubility of GYY4137 in water allows concentrated stock solutions to be prepared and used immediately in the laboratory for use on cells, tissues and animals without the need for organic solvents (e.g. ethanol, DMSO, dimethylformamide) commonly required for other drugs. However, it should be noted with caution that H_2S generation will start upon contact with water, and the freezing and reuse of stock solutions should be avoided. This ease of use and highly desirable water solubility has no doubt contributed to its widespread use and application.

3.1 GYY4137: Cytoprotection

H_2S has been reported to exert various "antioxidant" effects and inhibit the detrimental effects of reactive oxygen and nitrogen species on cells and tissues. However, the concentrations of the H_2S (typically NaSH) were often far in excess of that detected in vivo. For example, NaSH (15–60 μM) inhibited peroxynitrite (Whiteman et al. 2004)- and hypochlorous acid (Whiteman et al. 2005)-induced cellular toxicity, protein and lipid oxidation in human SH-SY5Y neuroblastoma cells. Higher concentrations of NaSH (e.g. 400–500 μM) also protected neuronal,

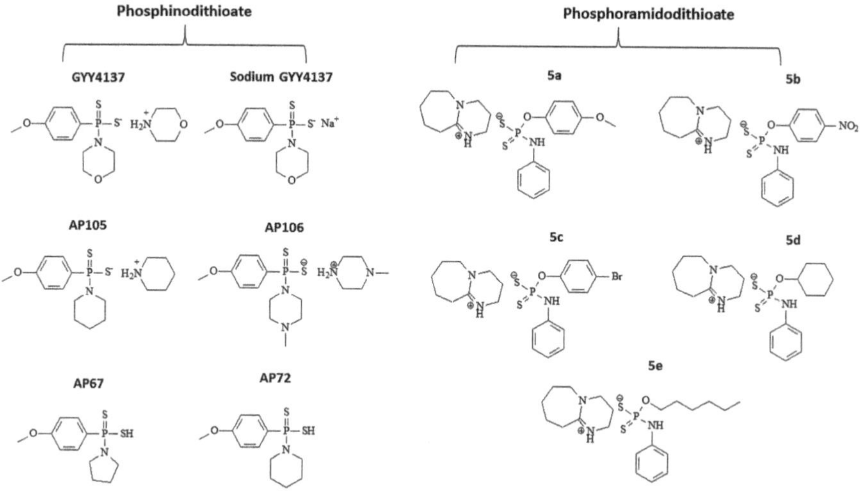

Fig. 1 Phosphinodithioate and phosphoramidodithioate H_2S donors. Examples of phosphinodithioates (GYY4137, sodium GYY4137, AP67, AP72, AP105 and AP106) and phosphoramidodithioates (5a–e) from the literature

microglial, myocardial and vascular cells from H_2O_2, lipid peroxides, chemical hypoxia and neurotoxins such as β-amyloid and MPP^+ (Dong et al. 2012; Liu and Bian 2010; Xiao et al. 2012; Yin et al. 2009; Chen et al. 2010; Kimura et al. 2006; Kimura and Kimura 2004). Other studies have suggested NaSH (1 mM) may be cytoprotective by reducing lipid peroxides formed during the oxidation of LDL (Muellner et al. 2009). However, it is highly unlikely these concentrations are ever reached in the brain or vasculature (Olson 2013; Kolluru et al. 2011, 2013; Peter et al. 2013; Whitfield et al. 2008) or reached without inducing substantial toxicity and death. Based on the above in vitro observations, H_2S has been proposed as endogenous "antioxidant" effects of H_2S in vitro have been suggested (Guo et al. 2013). However, the low concentrations of H_2S in cells and tissues in vivo (nM to low μM; Kolluru et al. 2013; Kolluru et al. 2011; Whitfield et al. 2008) and the very slow rate of reaction between sulfide and biologically relevant oxidants (e.g. peroxide, superoxide, peroxynitrite, etc.) strongly suggest direct antioxidant effects (e.g. "oxidant scavenging") are unlikely (Carballal et al. 2013). As such, cytoprotection is probably achieved through more subtle processes, and it would be valuable to reassess some of the earlier observations of H_2S-mediated cytoprotection using donors which deliver realistic concentrations of H_2S in a physiologically relevant manner.

The slow-release H_2S donor GYY4137 has been shown to exert significant cytoprotective effects in a variety of cellular and animal models of human disease. For example, in human mesenchymal progenitor cells and human articular chondrocytes, GYY4137 (10–200 μM) prevented cytotoxicity induced by the oxidants 4-hydroxynonenal, H_2O_2 and peroxynitrite (generated using SIN-1) (Fox et al. 2012). GYY4137 also preserved mitochondrial membrane potential ($\Delta\psi m$)

and inhibited ATP loss as well as inhibited mitochondrial oxidant production in these cells, possibly through Akt- and ERK1/2-mediated signalling pathways. Activation of Akt was also observed in GYY4137 (100–500 μM)-treated human keratinocytes (Merighi et al. 2012). Furthermore, oxidative stress-induced cytotoxicity was potentiated by siRNA-mediated knockdown of CSE, further indicating a cytoprotective role for endogenous H_2S and the potential for synthetic H_2S donor molecules to overcome H_2S deficiency. More recently, these findings have been confirmed in human umbilical vein endothelial cells (HUVEC) exposed to H_2O_2 (Hine et al. 2015).

In other in vitro models, GYY4137 (100–200 μM) protected H9c2 myocardial cells from hyperglycaemia-induced toxicity via mitochondrial protection and activation of AMPK-/mTOR-dependent signalling (Wei et al. 2014). Mitochondrial protection induced by GYY4137 (100–500 μM) was also observed in oxLDL-treated RAW264.7 cells (Liu et al. 2013). Higher concentrations (100 mM) also protected human pulmonary arterial cells from hyperoxia, inhibited cellular and mitochondrial "ROS" production and reversed hyperoxia-induced alveolar cell growth arrest (Vadivel et al. 2014), although cellular H_2S levels from such an obviously high concentration of GYY4137 were not assessed in this study. In vivo cytoprotection was also observed in a hyperoxia-induced lung injury model in new born rat pups (Vadivel et al. 2014). In this study, intraperitoneal injection of GYY417 (37.75 mg/kg, i.p.) inhibited hyperoxia-induced alveolar and lung vascular cell growth arrest and pulmonary hypertension, reducing right ventricular hypertrophy and pulmonary arterial medial wall thickness and inhibiting pulmonary artery smooth muscle cell proliferation. When given therapeutically, GYY4137 rescued alveolarisation, activated Akt/PI3K signalling in hyperoxic lungs and restored mitochondrial $\Delta\psi m$ and lowered mitochondrial oxidant production in lung epithelial cells (Vadivel et al. 2014). However, it is not clear whether these effects were due to H_2S from GYY4137 since changes in blood, tissue or cellular H_2S before and after GYY4137 administration were not measured.

The above cytoprotective effects of slow-release H_2S donors have led the suggestion that manipulation of intracellular H_2S synthesis could hold the key to increasing life span (Qabazard et al. 2013, 2014; Vozdek et al. 1834; Budde and Roth 2011; Miller and Roth 2007). It has been known for some time that blood and tissue levels of "H_2S" decline with age (Zhang et al. 2013; Predmore et al. 2010), and up until recently (Li et al. 2013), there was no mechanism to replace the loss of this gas. Qazabard et al. (2014) were the first to investigate the effects of GYY4137 in a well-established model of whole animal ageing using *Caenorhabditis elegans*. GYY4137 (100–400 μM) significantly increased the life span, increasing both the growth rate and activity of the worms. GYY4137 also inhibited oxidative stress-induced *C. elegans* killing by paraquat presumably by inhibiting mitochondrial "ROS" generation (Qabazard et al. 2014) suggesting H_2S could also inhibit oxidative stress occurring during the "ageing process". Subsequent studies (Hine et al. 2015) have confirmed these original observations, replicating them in *C. elegans*, yeast and the fruit flies *Drosophila melanogaster* where again, inhibition of ageing-induced oxidative stress by GYY4137 appeared to be a common mechanisms across species. It is therefore tempting to speculate that strategies which promote endogenous H_2S synthesis and/or provide exogenous H_2S in a physiologically relevant manner could hold the key to healthful ageing and longevity. Although these studies clearly show GYY4137 to be cytoprotective in a variety of models, it is possible that compounds which generate H_2S more efficiently could offer greater selectivity and potency.

3.2 GYY4137: Anticancer Effects

Although there are no conclusive reports of H_2S directly causing cancer in humans, numerous clinical and epidemiological studies have evaluated in detail the relationship between commensal sulfate-reducing bacteria (e.g. H_2S synthesising), ulcerative colitis and colorectal cancer (Carbonero et al. 2012; Roediger et al. 1997; Huycke and Gaskins 2004; Medani et al. 2011). H_2S is thought to be directly toxic to colonic epithelium, initiating inflammation and tissue damage and subsequent cellular hyper-proliferation (Huycke and Gaskins 2004; Medani et al. 2011). Indeed, the addition of NaSH (1–5 mM) to cells from human colonic biopsies (Christl et al. 1996), intestinal epithelial cells (Deplancke and Gaskins 2003; Leschelle et al. 2005) and human HT29 adenocarcinoma cells (Leschelle et al. 2005) resulted in significant cellular proliferation. Additionally, NaSH (250 μM to 2 mM) induced genomic DNA damage HT29 cells via "free radical"-- mediated processes in Attene-Ramos et al. (2006, 2007). More recently, Szabo et al. (2013) showed that in human colorectal carcinoma, CBS translocation to mitochondria and subsequent H_2S generation from CBS were a key mediator in driving the proliferation of carcinoma cells in vitro and in vivo via overstimulation of mitochondrial bioenergetics. Collectively, these studies strongly suggest that H_2S is pro-carcinogenic, at least with regard to colorectal carcinoma, and as such, inhibitors of CBS-mediated H_2S synthesis, or compounds which sequester or "scavenge" H_2S, could have therapeutic value (Szabo et al. 2013).

However, it is prudent to note that as with many aspects of the H_2S research field, diametrically opposite data towards cancer treatment are also presented. GYY4137 was further recently shown to exhibit prominent anticancer activity in vitro and in vivo (Lee et al. 2011). In this study, GYY4137 (100 μM to 1 mM) but not equivalent concentrations of NaSH induced substantial G_2/M phase cell cycle arrest and PARP-/caspase-9-dependent apoptotic cell death in seven disparate human cancer cell lines: cervical carcinoma (HeLa), colorectal carcinoma (HCT-116), breast adenocarcinoma (MCF-7), hepatocellular carcinoma (HepG2), osteosarcoma (U2OS), acute promyelocytic leukaemia (HL-60) and myelomonocytic leukaemia (MV4-11) cells. In sharp contrast, the same concentrations of GYY4137 had no effect on normal human lung fibroblasts (IMRO90 and WI-38) suggesting cell killing was cancer cell specific. The anticancer effects were not only limited to in vitro experiments. Both Lee et al. (2011) and then Lu et al. (2014) showed GYY4137 (100–300 mg/kg i.p.) inhibited tumour cell growth in HL-60, MV4-11 (Lee et al. 2011) and HepG2 xenografts (Lu et al. 2014) in mice. Although the mechanisms for these clear observations have not been fully elucidated, it is known that GYY4137 induces intracellular acidification in cancer cell lines through increased glycolysis leading to enhanced lactate production and decreased anion exchanger and sodium/proton exchanger activity (Lee et al. 2014). Other mechanisms in HepG2 and Bel7402 hepatocellular carcinoma cells include suppression of STAT3 activation (Lu et al. 2014). One key aspect of these studies is that for the first time, a control compound incapable of generating H_2S was employed, albeit one structurally dissimilar to GYY4137, and shown to not have any suppressive effect on tumour cell proliferation in vitro or in vivo (Lee et al. 2011, 2014).

Since GYY4137 suppressed tumour cell growth in vitro and in vivo (Lee et al. 2011, 2014; Lu et al. 2014), but NaSH was either ineffective or promoted cell proliferation (Christl et al. 1996; Deplancke and Gaskins 2003; Leschelle et al. 2005), the cellular response to H_2S may be determined not only by the concentration of H_2S used, but by the manner in which cells, tissues and animals are exposed to H_2S, e.g. fluxes of low concentrations of H_2S from a slow-release H_2S donor (from GYY4137) vs. instant generation from a concentrated bolus of sulfide salts (such as NaSH) could be key to identifying novel therapeutic H_2S-based agents.

3.3 GYY4137: Anti-inflammatory Effects

When generated as a bolus of NaSH, H_2S has been shown to be both pro- and anti-inflammatory leading to considerable controversy in the literature (reviewed in Whiteman and Winyard (2011)). H_2S synthesis is induced in many cell types in vitro and in vivo by pro-inflammatory mediators such as endotoxin and cytokines, and tissue and elevated blood and tissue levels of H_2S are increased in animal models of acute and chronic inflammation, for example, haemorrhagic (Mok and Moore 2008) and septic shock (Li et al. 2005; Coletta and Szabo 2013), pancreatitis (Bhatia et al. 2005; Tamizhselvi et al. 2007) and arthritis (Li et al. 2013). In humans, increased blood levels of H_2S are detected in patients with endotoxic shock (Li et al. 2005) and either increased (Saito et al. 2013) or decreased (Wang et al. 2011a) in asthma. Increased (Chen et al. 2005) (or decreased (Kirkham et al. 2014)) sputum H_2S is observed in patients with chronic obstructive pulmonary disease (COPD). The ratio of blood/sputum H_2S levels has been proposed as a novel biomarker for airway disease (Chung 2014; Saito et al. 2013, 2014; Wang et al. 2011a). Similarly, synovial fluid H_2S levels are elevated relative to disease controls in patients with rheumatoid arthritis (Whiteman et al. 2010c) and inflammatory joint disease (Whiteman and Winyard 2011). Early studies generally showed inhibition of endogenous H_2S synthesis attenuated pro-inflammatory signalling in vitro and inflammation in vivo (reviewed in Whiteman and Winyard (2011)), although it should be noted that pharmacological inhibitors of CSE, CBS and 3-MST are non-specific necessitating their use at very high concentrations (reviewed in detail in Whiteman et al. (2011)) and the addition of NaSH to cultured cells in vitro (Whiteman et al. 2010b; Zhang et al. 2008; Zhi et al. 2007; Kloesch et al. 2012a, b). Similarly, administration of NaSH to laboratory animals either induced systemic inflammation directly or exacerbated inflammation (Li et al. 2005; Bhatia et al. 2005, 2006; Tamizhselvi et al. 2007; Ang et al. 2010, 2011; Badiei et al. 2013).

Based on the findings above, it has been suggested by some investigators that endogenous and pharmacological H_2S (e.g. NaSH) is pro-inflammatory. However, when a slow-release H_2S donor such as GYY4137 is employed in cellular and in vivo models of inflammation, the opposite is observed suggesting the role of H_2S in inflammation needs to be thoroughly re-examined. For example, lipopolysaccharide (LPS)-induced NF-κB activation, cytokine, $^{\bullet}$NO and PGE_2 synthesis were exacerbated in murine RAW264.7 macrophages pretreated with NaSH (up to 200 μM) but attenuated by GYY4137 (50–200 μM) (Whiteman et al. 2010b). This attenuation was also observed when GYY4137 was added to RAW264.7 cells after LPS stimulation highlighting the therapeutic potential of slow-release H_2S donors. Similarly, GYY4137 (50–400 μM) inhibited NF-κB activation in oxLDL-treated RAW264.7 (Liu et al. 2013) and in human monocytic U937 cells and monocyte-derived macrophages infected with *Mycoplasma fermentans*, preventing MCP-1 production (2 mM) (Benedetti et al. 2014). GYY4137 (100–400 μM) also inhibited NF-κB activity in oxLDL (100 μg/mL)-treated murine RAW264.7 macrophages resulting in lowered LOX-1 and iNOS expression and synthesis of macrophage-specific chemokines such as MIP-2α and IP10 (Liu et al. 2013).

In articular chondrocytes isolated from patients with osteoarthritis (OA) (Burguera et al. 2014), GYY4137 also suppressed NF-κB activity leading to lowered protein levels of the pro-inflammatory enzymes COX-2, prostaglandin E synthase, MMP-13 and iNOS leading to lowered synthesis of $^{\bullet}$NO, PGE_2 and IL-6 suggesting H_2S donor compounds could be useful in inhibiting and/or reverse cartilage remodelling in degenerative joint diseases such as OA. Furthermore, in human articular chondrocytes and synovial fibroblasts (Li et al. 2013), GYY4137 (100–500 μM) also inhibited LPS-induced synthesis of IL-6, TNF-α, IL-1β, $^{\bullet}$NO and PGE_2 as well as inhibited the activity of iNOS, COX-2 and TNF-α converting enzyme (TACE) suggesting additional anti-inflammatory effects of this compound in addition to NF-κB inhibition.

The anti-inflammatory effects of GYY4137 have also been evaluated in in vivo. In Coxsackie virus B-infected rat cardiomyocytes, GYY4137 was recently shown to reduce cardiac damage and circulating TNF-α, IL-1β and IL-6 levels in an NF-κB-dependent manner (Wu et al. 2015). Similarly, in the LPS-induced sepsis model in rats, GYY4137 also inhibited the synthesis of these cytokines via suppression of NF-κB activity, reduced tissue inflammation and elevated plasma levels of the anti-inflammatory chemokine, IL-10, although the mechanism for this is not clear (Li et al. 2009). The addition of GYY4137 to LPS-treated whole rat blood (Li et al. 2009) also resulted in the inhibition of TNF-α generation, possibly via inhibition of TACE (Li et al. 2013) and again highlighting the therapeutic potential for H_2S donors as anti-inflammatory compounds. Furthermore, GYY4137 (50 mg/kg i.p.) reversed LPS-induced hypotension. Although this may seem counterintuitive since H_2S dilates bloods vessels (see below) and sepsis markedly lowers systemic blood pressure and elevates blood and tissue levels of H_2S, the normalisation of blood pressure by GYY4137 is most likely due to inhibition of pro-inflammatory cytokines such as TNF-α and IL-1β and other factors which drive the detrimental vascular in sepsis, rather than a dilating effect on blood vessels directly.

The therapeutic potential for slow-release H_2S donors has also been highlighted more recently in the rat complete Freund's adjuvant (CFA) arthritis model and in atherosclerotic apoE$^{-/-}$ mice. In the CFA arthritis model (Li et al. 2013), CFA injection induced synovial H_2S synthesis fluid (SF), in agreement with in vitro studies using human joint cells (Li et al. 2013; Fox et al. 2012; Burguera et al. 2014) and observations with human SF from patients with various inflammatory arthritides (Whiteman and Winyard 2011; Whiteman et al. 2010c). GYY4137 (50 mg/kg i.p.) given 1 h prior to CFA resulted in increased tissue oedema although lowered neutrophil and macrophage infiltrate, presumably due to the vasodilatory effects of GYY4137 (Li et al. 2008). However, therapeutic administration of GYY4137 either 6 or 18 h post-CFA treatment resulted in decreased knee joint swelling, reduced SF neutrophil and macrophage accumulation and decreased SF levels of TNF-α, IL-1β, IL-8 and IL-6 (Li et al. 2008). Furthermore, in atherosclerotic apoE$^{-/-}$ mice, daily i.p. injection of GYY4137 (133 μmol/kg/day) for 30 days resulted in the inhibition of aortic mitochondrial "ROS" production, LOX-1 expression and lowered cellular levels of IL-6, TNF-α and ICAM-1 (Liu et al. 2013). GYY4137 also partially restored endothelium-dependent vasorelaxation in aortic segments (Liu et al. 2013). Since loss of vascular H_2S occurs in atherosclerosis (Mani et al. 2013, 2014) and the onset of atherosclerosis is more rapid and aggressive in CSE$^{-/-}$ mice compared to wild-type animals (Mani et al. 2013), these studies strongly suggest that slow-release H_2S donors could reverse (as well as prevent) the detrimental vascular remodelling associated with atherosclerosis, highlighting their therapeutic potential in vascular diseases. Although these studies consistently show GYY4137 exerted prominent anti-inflammatory activity in vitro and in vivo when used prophylactically and therapeutically, it is possible that slow-release H_2S donor compounds that are able to generate H_2S with greater efficiency than GYY4137 could offer improved potency and efficacy.

3.4 GYY4137: Vascular Effects

The effects of H_2S on vascular tissue are by far the most characterised. The role of H_2S as an "endogenous vasodilator" was first proposed by Zhao et al. (2001). The addition of NaSH or solutions of authentic H_2S gas to where more than 200 μM H_2S (Zhao et al. 2001; Zhong et al. 2003; Ali et al. 2006) was required to relax precontracted rat aortic rings and lowered systemic blood pressure in rats via an endothelium and K_{ATP} channel-dependent mechanism, possibly through the *S*-sulfhydration of critical cysteine residues in the K_{ATP} channel (Mustafa et al. 2011). Since then, numerous other studies have reported the vasorelaxant effect of high concentrations of NaSH on other vascular tissues where effects on calcium channels, Na^+/K^+ ATPase, HCO_3^- and other K^+ channels are activated or inhibited by H_2S (reviewed in Whiteman and Moore (2009) and Peers et al. (2012)) but often using near mM or mM concentrations of the salt (e.g. Al-Magableh and Hart (2011) and Hedegaard et al. 2014)). As a result of these studies, H_2S is often described as a "potent vasodilator" (Bhatia 2005; Pushpakumar et al. 2014; Ariyaratnam et al. 2013; Jang et al. 2014; Baragatti et al. 2013; Ghasemi et al. 2012) despite data clearly showing otherwise. However, the addition of GYY4137 at concentrations and incubation times which deliver nM to low μM H_2S also leads to vasodilation in vitro and in vivo (Table 1) suggesting blood vessels are acutely sensitive to small fluxes of H_2S when it is generated in a manner and a level comparable to enzymatic (e.g. via CSE, CBS and 3-MST). For example, in phenylephrine precontracted rat aorta, GYY4137 (10–100 μM) induced significant vessel relaxation, but >200 μM NaSH was required for comparable effects (Li et al. 2008). Although the dilatory effects of a bolus of NaSH (EC_{70}, 300 μM) were immediate, vessel relaxation in response to GYY4137 (EC_{70}, 200 μM) was slow and continued for up 60 min. The effect of GYY4137 and NaSH was inhibited by glibenclamide and PNU37883A, indicating a requirement for K_{ATP} channel activation, and reduced by endothelium removal and pretreatment with L-NAME (NOS inhibitor) and ODQ (guanylate cyclase inhibitor), which inhibit the formation and vascular response to $^{\bullet}NO$, respectively. K_{ATP} channel dependence for vessel relaxation appears conserved as glibenclamide also prevented GYY4137-indcued human internal thoracic artery relaxation (Webb et al. 2008) and dilation of human and rat term pregnant myometrium (Robinson and Wray 2012), although in this latter study, GYY4137 (1 mM) also inhibited L-type calcium channel activity suggesting a more complex mechanism of action. In agreement, bovine ciliary artery vessel dilation was observed with as little as 100 nM GYY4137 ($EC50 = 13.4 \pm 1.9$ μM) (Chitnis et al. 2013), and although this effect was inhibited by glibenclamide, it was potentiated by flurbiprofen suggesting prostaglandin-/prostacyclin-mediated effects of GYY4137 could also contribute, at least in part, to its mechanism of action. It is currently not known why the bovine ciliary artery appears to be exquisitely sensitive to H_2S.

In the original study (Li et al. 2008), the addition of 100 μM GYY4137 to rat aortic vascular smooth muscle cells did not result in cGMP accumulation suggesting guanylate cyclase and phosphodiesterase-independent effects of GYY4137 on vessel relaxation. Others have observed that higher concentrations of GYY4137 (300 μM) significantly increased cGMP in the same cells, albeit modestly, and whereas rat aorta relaxation by ≥100 μM NaSH was blocked by the PKG inhibitor DT-2, GYY4137-mediated vessel relaxation was independent of PKG (Salomone et al. 2014; Bucci et al. 2012). In contrast, much higher concentrations of GYY4137 (1 mM) reduced cGMP levels to below that of the controls in ophthalmic artery segments (Salomone et al. 2014) suggesting there could be tissue (as well as concentration)-dependent effects of the donor (and H_2S). These studies further highlight the differences between "instantly" generated H_2S and H_2S generated slowly and at low concentrations.

Table 1 Summary of in vitro and in vivo effects of GYY4137 and related compounds

Therapeutic area	Compound	Observation/comment
Cardiovascular	GYY4137	• K_{ATP} and endothelium-dependent dilation of rat aorta ex vivo (10–100 μM) • Inhibition of L-NAME-induced hypertension in rats in vivo (133 μM/kg i.v.) • Sustained blood pressure reduction in spontaneously hypertensive rats but minimal effect in normotensive WKY rats in vivo (133 μM/kg i.v.) • Decreased rat kidney perfusion pressure ex vivo • No acute or chronic toxicity observed
		• Induced relaxation observed using 1–1000 μM GYY4137 with significant relaxation observed from 1 μM; K_{ATP} and L-type calcium channel-dependent effects
		• Normalisation of mean blood pressure and increased foetal growth in pre-eclamptic mice (GYY4137, 0.25 mg/kg i.p.) (Wang et al. 2013)
		• Relaxation of human airway smooth muscle via sarcolemmal K_{ATP} channels, although >10 mM was required (Fitzgerald et al. 2014)
		• Variable effects on vascular smooth muscle cells • No cGMP accumulation with 100 μM GYY4137 (Li et al. 2008), increased with 300 μM (Bucci et al. 2012) or inhibited with 1 mM (Salomone et al. 2014)
		• K_{ATP}- and prostaglandin-dependent dilation of bovine ciliary artery induced by 100 nM to 100 μM GYY4137 (EC50 = 13.4 ± 1.9 μM) (Chitnis et al. 2013)
	AP67 and AP72	• K_{ATP}- and prostaglandin-dependent dilation of bovine ciliary artery induced by 1 nM to 10 μM AP67 (EC50 = 0.08 ± 0.04 μM) and AP72 (EC50 = 4.3 ± 0.09 μM) (Chitnis et al. 2015). Both donors were used in the free acid form (Fig. 1) • Restored endothelium-dependent vasorelaxation in $apoE^{-/-}$ atherosclerotic mice
Cytoprotection	GYY4137	• GYY4137 (100–200 μM) inhibited 4-hydroxynonenal-, H_2O_2- and peroxynitrite-induced cytotoxicity in human articular chondrocytes and trabecular bone-derived mesenchymal progenitor cells • Preservation of mitochondrial $\Delta\psi m$ and ATP levels • Activation of Akt signalling • No cytoprotection observed with equimolar concentrations of "decomposed" GYY4137
		• Activation of Akt in human keratinocytes by 100–500 μM GYY4137 (Merighi et al. 2012)
		• Inhibition of H_2O_2-induced cytotoxicity in HUVEC by 100 μM GYY4137 (Hine et al. 2015)

(continued)

Table 1 (continued)

Therapeutic area	Compound	Observation/comment
		• Inhibition of oxLDL-induced cellular and mitochondrial toxicity in murine RAW264.7 macrophages by 100–500 M GYY4137 (Liu et al. 2013)
		• Inhibition of hyperglycaemia-induced cellular and mitochondrial toxicity in rat H9c2 myocardial cells by 100–200 μM GYY4137 (Wei et al. 2014) • Activation of AMPK-/mTOR-dependent signalling
		• Inhibition of mitochondrial "ROS" production and mitochondrial and cellular toxicity in hyperoxia exposed pulmonary artery smooth muscle cells and lung epithelial cells (100 mM) (Vadivel et al. 2014) • Inhibition of hyperoxia-induced alveolar cell growth arrest, pulmonary hypertension, right ventricular hypertrophy and pulmonary arterial medial wall thickness in vivo (GYY4137, 37.75 mg/kg i.p.) • Activation of Akt/PI3K signalling in hyperoxic lungs
		• Decreased age-dependent mitochondrial and cellular oxidant production in *C. elegans* (Qabazard et al. 2014; Hine et al. 2015) • Increased life span in *C. elegans* and *D. melanogaster* (Xiao et al. 2012; Peter et al. 2013)
	5a, 5b, 5c, 5d, 5e, GYY4137	• GYY4137 (200 μM) induced substantial cell killing to rat H9c2 myocardial cells (Putz et al. 1979) in contrast to other studies (Yin et al. 2009) but no toxicity observed with equimolar concentrations of 5a, 5b or NaSH. Marked toxicity observed with 5c ($\geq$50 μM) • GYY4137 potentiated H_2O_2-induced cell death but cytoprotection observed with 5a and 5b (50–200 μM) (Putz et al. 1979) • GYY4137 caused nearly 100 % killing murine B16BL6 melanoma cells (100–200 μM). Some cell killing also observed with 5a and 5b
Cancer	GYY4137	• GYY4137 (100 μM to 1 mM) induced G2/M phase cycle arrest, PARP-/caspase-9-dependent cell death in seven disparate tumorigenic cell lines in vitro (Lee et al. 2011) • Cell killing through intracellular acidification and lactate accumulation and decreased anion exchanger and Na^+/H^+ exchanger activity • Inhibition of tumour cell growth in vivo in mouse HL-60 and MV4-11 xenografts (Lee et al. 2011) • No effect of NaSH or an inert analogue of GYY4137 (Lee et al. 2011) on cancer cell killing (Lee et al. 2014)
		• Inhibition of HepG2 and Bel7402 cell growth by suppression of STAT-3 activation (Lu et al. 2014) • Inhibition of HepG2 cell growth in mouse xenografts in vivo (Lu et al. 2014)

(continued)

Table 1 (continued)

Therapeutic area	Compound	Observation/comment
Inflammation	GYY4137	• Inhibition of LPS-induced IL-1β, TNF-α, PGE_2 and •NO synthesis in murine RAW264.7 macrophages by 50–500 μM GYY4137, effect conserved with prophylactic and therapeutic treatment. In contrast, NaSH potentiated cellular inflammation (Whiteman et al. 2010b) • Inhibition of cytokine synthesis in human synovial fibroblasts and articular chondrocytes (GYY4137 100–500 μM) (Qabazard et al. 2014) • Inhibition of NF-κB activation and activity in LPS- and oxLDL-treated (Whiteman et al. 2010b; Benedetti et al. 2014) RAW264.7 macrophages and *M. fermentans*-infected human monocytic U937 cells and monocytes by GYY4137 (100–400 μM) (Benedetti et al. 2014) • Inhibition of NF-κB activity, COX-2, prostaglandin E synthase, MMP-13 and iNOS expression and activity in human chondrocytes isolated from osteoarthritis patients (Burguera et al. 2014) • Inhibition of NF-κB activation, TNF-α, IL-1β and IL-6 synthesis in Coxsackie virus B-infected rat cardiomyocytes (Zhi et al. 2007) • Inhibition of *M. fermentans*-induced MCP-1 production in human monocytes (2 mM GYY4137) • Inhibition of oxLDL-induced LOX-1 and iNOS expression and synthesis of macrophage-specific chemokines (Chen et al. 2010) • Inhibition of iNOS, COX-2 and TACE activity (Li et al. 2013) • Inhibition of TNF-α synthesis in LPS-treated rat whole blood (Li et al. 2009) • GYY4137 normalised blood pressure, inhibited tissue inflammation and blood levels of IL-1β, TNF-α, IL-6 and increased IL-10 in LPS-induced sepsis in rats (Li et al. 2009) • Inhibition of serum-induced airway smooth muscle cell proliferation and IL-8 synthesis (100 μM) (Perry et al. 2011) • Therapeutic administration of GYY4137 (50 mg/kg i.p.) 6 or 18 h post-CFA treatment reduced knee joint inflammation and swelling; reduced synovial levels of TNF-α, IL-1β, IL-8 and IL-6; and inhibited neutrophil and macrophage synovial accumulation (Li et al. 2008) • Decreased blood levels of IL-6 and TNF-α and aortic expression of ICAM-1 synthesis and LOX-1 and mitochondrial "ROS" generation in apoE$^{-/-}$ atherosclerotic mice with 30-day treatment of GYY4137 (133 μmol/kg/day i.p.) (Liu et al. 2013)

In the perfused rat kidney and heart, further differences between a bolus NaSH and GYY4137 are illustrated (Li et al. 2008). NaSH (10 nM) caused a transient fall in renal perfusion pressure, but higher doses (5 μM) caused a biphasic response, decreasing and then markedly increasing perfusion pressure, whereas no consistent effect was observed with GYY4137 (0.4–4.0 μM). Furthermore, in U46619, angiotensin II or noradrenaline-pre-constricted renal vessels, the addition of GYY4137 (100–500 μM) to the perfusion buffer caused concentration-dependent relaxation, an effect reversed when GYY4137 was removed or when angiotensin II was used in the presence of the K_{ATP} channel antagonist PNU3788A. In Langendorff-perfused rat heart, NaSH (100 μM) markedly reduced heart rate and left ventricular diastolic pressure. This effect was not observed with GYY4137 (50–200 μM) (Li et al. 2008). It is prudent to note that disparate effects of NaSH on heart rate have been reported: either no effect (Minamishima et al. 2009; Wang et al. 2011b; Shi et al. 2007; Alves et al. 2011; Derwall et al. 2010), tachycardia (Sodha et al. 2009) or bradycardia (Porokhya et al. 2012; Sikora et al. 2014). Although these observations have been made using different models and strains of animals, inconsistencies, such as these, question the reliability of sulfide salts to examine the physiological properties and pharmacological potential of H_2S.

The therapeutic potential for slow-release H_2S donors in vascular disease is also highlighted in Li et al. (2008). Although GYY4137 had negligible effect on mean arterial pressure (MAP) in normotensive WKY rats, spontaneously hypertensive rats, chronic dosing with GYY4137 (133 μmol/kg i.v.) resulted in significant lowering of MAP within 2 days which persisted for 14 days. After withdrawal of GYY4137 treatment, MAP returned to basal levels but only after a further 14 days. In an additional model of hypertension, using L-NAME to inhibit the synthesis of endogenous •NO, acute injection of GYY4137 (133 μmol/kg i.v.) for 30 min before L-NAME treatment prevented L-NAME-induced increase in MAP. This preventative effect persisted for more than 60 min. In sharp contrast, a bolus of the sulfide salt NaSH (2.5 μmol/kg i.v.) initially increased MAP above that of L-NAME treatment alone and failed to significantly decrease MAP, further reinforcing the marked differences between slow release of H_2S and an instantly generated bolus. More recently, Wang et al. showed pre-eclampsia in mice and humans led to the loss of vascular H_2S. Pharmacological inhibition or genetic silencing of CSE resulted in pre-eclampsia in mice, and "H_2S supplementation" with GYY4137 (0.25 mg/kg i.p.) normalised mean blood pressure, increased foetal growth and lowered blood levels of soluble endoglin and soluble fms-like tyrosine kinase-1 (sFlt-1), the endogenous inhibitor of VEGF and placental growth factor (Wang et al. 2013).

3.5 GYY4137

GYY4137 is widely available from commercial suppliers, and its high solubility in water makes it a highly valuable laboratory tool. It is interesting to note that despite this high water solubility, commercial suppliers of GYY4137 state the compound should be stored in organic solvent such as DMSO. We have found DMSO to be detrimental and accelerate GYY4137 decomposition and avoid storing the compound in organic solvents by preparing fresh solutions for each experiment immediately before use.

One major drawback of using GYY4137 is that due to inefficient hydrolysis, high concentrations are often required to elicit a biological response. Typically, concentrations in excess of 100 μM (and up to 100 mM) (Table 1) have been used. However, H_2S concentrations from GYY4137 will be considerably lower and should ideally be monitored to assess the effective concentration of H_2S. Similarly, the effects of a "decomposed" or "spent" donor control should be investigated to ensure any observed effects are due to "released" H_2S and not due to effects from the unhydrolysed parent compound or intermediates and/or hydrolysis products formed during GYY4137 hydrolysis and H_2S generation. These control experiments are generally not carried out (Table 1), but they are crucial since commercial GYY4137 (or GYY4137 prepared as described (Li et al. 2008)) is sold as a dichloromethane complex, with the dichloromethane being residual from initial synthesis and crystallisation. However, the molecular weights stated do not reflect the contribution of dichloromethane to the final mass of the molecule, and as such, previous publications with GYY4137 may contain erroneous, albeit lower, concentrations of the donor than reported. Additionally, dichloromethane is metabolised to carbon monoxide (CO) (Fagin et al. 1980; Mackison et al. 1981; Benzon et al. 1978; Baselt 1988; Barrowcliff and Knell 1979; Putz et al. 1979) an additional "gasotransmitter". Since CO is reported to exert similar biological effects as H_2S (e.g. vasodilation, anti-inflammatory actions, modulation of cGMP signalling, etc.) (Peers et al. 2015; Peers and Lefer 2011; Nassour et al. 2015; Munoz-Sanchez and Chanez-Cardenas 2014), it is possible that some of the observed effects of GYY4137 could also be due, at least in part, to a contribution by CO. This may be especially important given the inefficient manner in H_2S generated from GYY4137 necessitates the use of high concentrations/doses of this compound to elicit pharmacological responses in cells, tissues and whole animals (see Table 1). An obvious solution to this problem is to employ control experiments with a time-expired (e.g. "spent" or "decomposed") donor which would contain GYY4137 hydrolysis products (Alexander et al. 2014) as well as dichloromethane and morpholine to control for this possibility. However, these control experiments are not generally reported (see Table 1). An additional approach to circumvent this problem is to use structurally similar molecules devoid of residual solvent and organic counter ions.

3.6 GYY4137: Structural Modifications

Although the above studies consistently show GYY4137 exerted prominent anti-inflammatory, cytoprotective, anticancer and antihypertensive activity in vitro and in vivo, it is possible that slow-release H_2S donor compounds that are able to generate H_2S with greater efficiency than GYY4137 could offer improved efficacy.

Some structural modifications of GYY4137 such as *O*-substituted 1,8-diazabicyclo [5.4.0]undec-7-ene salts have been reported (Park et al. 2013), but they have not resulted in increased biological activity. For example, *O*-substituted GYY4137 derivatives such as compounds 5a–e (Fig. 1) showed no significant difference in H_2S release rates in physiological buffers or inside cells using rat H9c2 myocytes and the sulfide-specific fluorescent probe WSP-1 compared to GYY4137. Presumably, these structural changes did not improve H_2S release rates or efficiency, because of increased stability of the phosphorodithioate core by *O*-substitution (Park et al. 2013). Although, as yet, no mechanistic pathway appears to have been postulated, it would be anticipated that the generation of H_2S from the phosphoramidodithioates 5a–5e would follow a similar hydrolytic pathway to that elucidated for similar phosphorodithioates (Alexander et al. 2014). In contrast to previous studies using the same cell line (Wei et al. 2014), this report showed GYY4137 (200 μM) was cytotoxic, and although compounds 5a–c (Fig. 1) were markedly less toxic to H9c2 cells compared to GYY4137, they did not significantly increase the killing of mouse melanoma B16BL6 cells over that of GYY4137. In cytotoxicity assays using H9c2 myocytes, oxidative stress (using H_2O_2), modest (~10–20 %) but statistically significant inhibition of cytotoxicity was observed with 100–200 μM of 5a, 5b and 5c, but greater cytoprotection was observed with only 50 μM GYY4137, although it is intriguing to note that in this study, 200 μM GYY4137 potentiated H_2O_2-induced cell killing in sharp contrast to the results published in several studies elsewhere (Fox et al. 2012; Hine et al. 2015), including H9c2 myocytes (Wei et al. 2014). Since these compounds offered no significant advantage of GYY4137 in terms of biological activity and generation of H_2S, their physiological and/or pharmacological potential is not clear. However, these compounds have yet to be investigated in more complex ex vivo or in vivo models.

We have also modified the phosphorodithioate core of GYY4137 (Fig. 1) with the aim of improving H_2S generation and to have a selection of compounds which generate H_2S at different rates to use in the laboratory. This approach is analogous to the NONOates used extensively, and to great effect, to investigate the physiological role and therapeutic potential of $^{\bullet}$NO. Compounds with a range of different release characteristics could be used to more accurately model endogenously generated H_2S under basal conditions (e.g. using GYY4137) and induced H_2S synthesis (e.g. using AP67, AP105 and AP106) allowing for the examination of effects of H_2S concentration and H_2S generation rates on cells, tissues and animals. We have evaluated the biological activity of these compounds in vitro using cultured human synoviocytes and rat arterial smooth muscle cells and ex vivo using precontracted mouse aorta, and these studies clearly showed that the cellular and tissue effects of the H_2S donors were dependent upon the rates at which the compounds generate H_2S. For example, Fig. 2 compares the H_2S generation from equimolar donors in a physiological buffer (Fig. 2a) or in human synoviocytes (Fig. 2b) (e.g. AP67 > AP72 > AP105 > AP106 > GYY4137). Figure 3 shows that inhibition of LPS-evoked formation of $^{\bullet}$NO and PGE_2 in murine RAW264.7 macrophages and human synovial fibroblasts, respectively, was also dependent upon the rate of H_2S release (e.g. AP67 > AP72 > AP105 > AP106 = GYY4137).

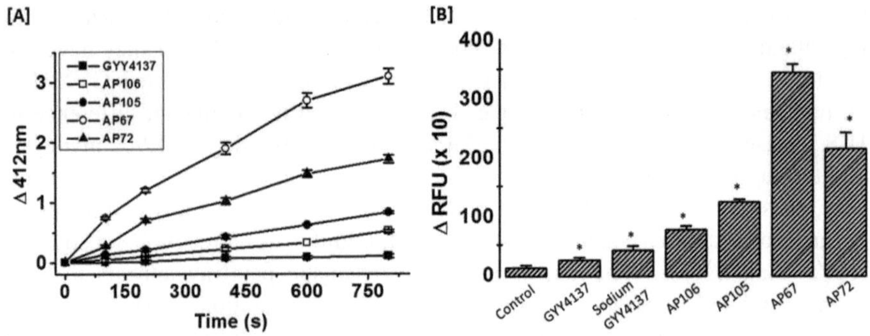

Fig. 2 H_2S generation from GYY4137 derivatives in the presence and absence of cultured cells. (**a**) H_2S generation in phosphate buffer saline was assessed using 5′-dithiobis-(2-nitrobenzoic) acid as described (Li et al. 2008). The final concentration of each H_2S donor was 500 μM. (**b**) Intracellular H_2S was determined in human synovial fibroblasts cultured as described (Li et al. 2013) using the intracellular sulfide probe WSP-1 (10 μM) (Szczesny et al. 2014; Le Trionnaire et al. 2014). H_2S donors were added to cells at a final concentration of 100 μM. Data are expressed as mean ± SEM; $n = 6$

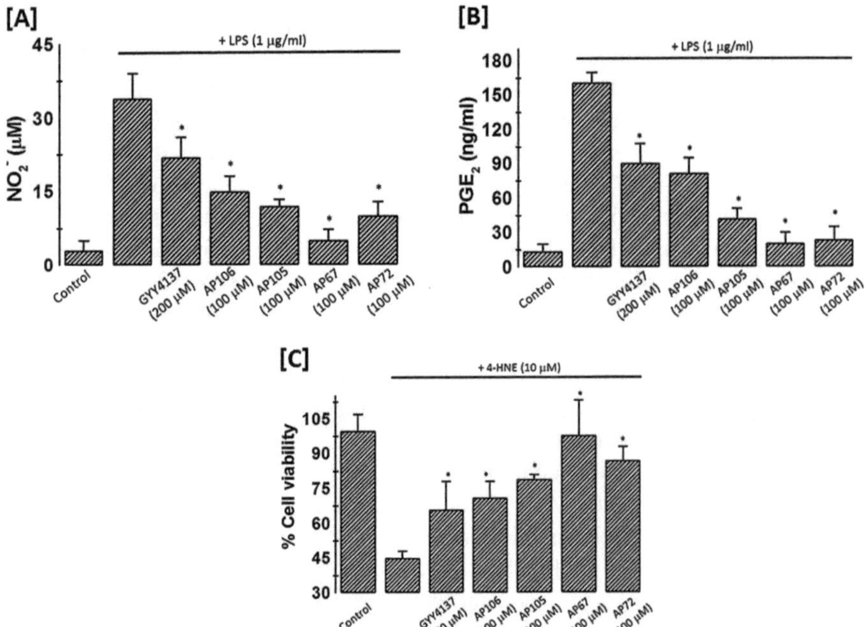

Fig. 3 H_2S release rate-dependent anti-inflammatory and cytoprotective activity of GYY4137 derivatives. Anti-inflammatory: RAW264.7 cells (**a**) and human synovial fibroblasts (**b**) were cultured as described (Li et al. 2013; Minamishima et al. 2009) and exposed to LPS (1 μg/mL) in the presence or absence of H_2S donors (100 μM) for 24 h. Levels of (**a**) nitrite (NO_2^-, as index of $^{\bullet}NO$ synthesis) and (**b**) PGE_2 were measured in cell culture supernatant after 18 h by Griess assay and ELISA, respectively (Li et al. 2013; Whiteman et al. 2010b). Cytoprotection: human Jurkat T cells were cultured as described (Le Trionnaire et al. 2014) and exposed to 4-hydroxynonenal (10 μM) in the presence or absence of 100 μM H_2S donors. Cytotoxicity was assessed by trypan blue exclusion assay after 24 h. Data are expressed as mean ± SEM; $n = 6$. $^*p < 0.05$, cf. LPS treatment (**a**, **b**) or 4-HNE-treated cells (**c**)

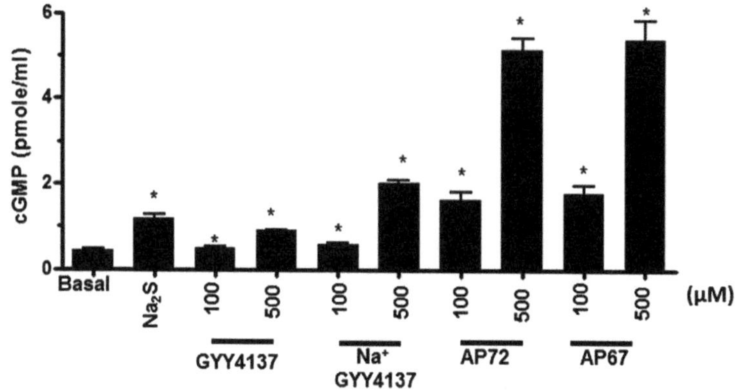

Fig. 4 H_2S released from donors enhances cGMP levels in rat aortic smooth muscle cells. Cells were isolated from 12–14-week-old male Wistar rats as described (Papapetropoulos et al. 1994). Cells between passages 2 and 5 were seeded on 24-well places and after they reached confluence were incubated with the indicated H_2S donor at the concentrations stated in Hank's balanced salt solution for 5 min. Cellular cGMP was then extracted using 0.1 N HCl and quantified by ELISA (Assay Designs). Data are expressed as mean ± SEM; $n = 4$, $*p < 0.05$, cf. basal cGMP levels

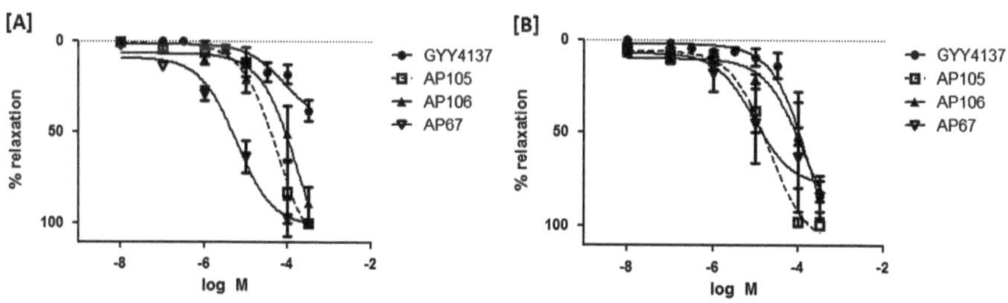

Fig. 5 H_2S release rate-dependent dilation of phenylephrine-precontracted mouse aorta. Mouse aortic rings were prepared as described in Bucci et al. (2012). (**a**) Endothelium-dependent vessel relaxation: EC_{50} GYY4137, 80.9 μM; AP106, 167.4 μM; AP105, 57.5 μM; AP67, 5.9 μM. (**b**) Endothelium-independent vessel relaxation in denuded mouse aorta: EC_{50} GYY4137, 186.1 μM; AP106, 120.2 μM, AP105 19.8 μM; AP67, 9.3 μM. Data are expressed as mean ± SD of three separate determinations

Additionally, the ability of the donors to induce cGMP accumulation in rat arterial smooth muscle cells was also dependent upon the rate of H_2S generation (Fig. 4). Intriguingly, the sodium salt of GYY4137 was more effective at stimulating cGMP formation, perhaps due to some inhibitory action of dichloromethane or morpholine in GYY4137. Furthermore, the H_2S release rate-dependent effects were also observed in phenylephrine-precontracted mouse aortic rings (Fig. 5), although this effect was not so clear in denuded aortic rings lacking endothelial cells suggesting the mechanisms for vessel relaxation are complex. However, the effects of the donors (e.g. H_2S release rate effects) were conserved in cells and in isolated blood vessels. The vasodilatory activity of AP67 and AP72 free acids (Fig. 1) has also been shown to be more potent than GYY4137 in dilating precontracted bovine ciliary arteries ex vivo with markedly lower EC50 (AP67, 0.08 ± 0.04 μM; AP 72, 4.3 ± 0.09 μM; GYY4137, 13.4 ± 1.9 μM) (Chitnis et al. 2013, 2015) through K_{ATP} and prostaglandin-dependent mechanisms. The reasons for the differences in rate of H_2S release between GYY4137 and the other closely related phosphorodithioate donors such as AP67, AP105, etc., are not clear and are the subject of current and detailed chemical studies. However, since these molecules generate H_2S at different rates and elicit different effects on isolated cells and blood vessels, it is likely they will be useful and attractive additions to GYY4137 and alternatives to sulfide salts, for unravelling the complex physiology and pharmacology of H_2S where H_2S concentration and rate of H_2S generation need to be modulated and/or controlled.

3.7 Phosphorodithioate H_2S Donors: Summary

In summary, slow-release phosphorodithioate H_2S donors are powerful tools to study the physiological significance of H_2S in a wide range of cellular and animal models. By slowly generating fluxes of low concentrations of H_2S, they are more biologically relevant than using a bolus of instantly generated H_2S from often impure sulfide salts such as NaSH (and Na_2S) which are frequently used at supraphysiological concentrations/doses.

References

Alexander BE, Coles SJ, Fox BC, Khan TF, Maliszewsi J, Perry A, Pitak MP, Whiteman M, Wood ME (2014) Investigating the generation of hydrogen sulfide from the phosphinodithioate slow-release donor GYY4137. Medchemcomm (under review)

Ali MY, Ping CY, Mok YY, Ling L, Whiteman M, Bhatia M, Moore PK (2006) Regulation of vascular nitric oxide in vitro and in vivo; a new role for endogenous hydrogen sulphide? Br J Pharmacol 149:625–634

Al-Magableh MR, Hart JL (2011) Mechanism of vasorelaxation and role of endogenous hydrogen sulfide production in mouse aorta. Naunyn Schmiedebergs Arch Pharmacol 383:403–413

Alves MG, Soares AF, Carvalho RA, Oliveira PJ (2011) Sodium hydrosulfide improves the protective potential of the cardioplegic histidine buffer solution. Eur J Pharmacol 654:60–67

Ang SF, Moochhala SM, Bhatia M (2010) Hydrogen sulfide promotes transient receptor potential vanilloid 1-mediated neurogenic inflammation in polymicrobial sepsis. Crit Care Med 38:619–628

Ang SF, Moochhala SM, MacAry PA, Bhatia M (2011) Hydrogen sulfide and neurogenic inflammation in polymicrobial sepsis: involvement of substance P and ERK-NF-kappaB signaling. PLoS One 6:e24535

Ariyaratnam P, Loubani M, Morice AH (2013) Hydrogen sulphide vasodilates human pulmonary arteries: a possible role in pulmonary hypertension? Microvasc Res 90:135–137

Asimakopoulou A, Panopoulos P, Chasapis CT, Coletta C, Zhou Z, Cirino G, Giannis A, Szabo C, Spyroulias GA, Papapetropoulos A (2013) Selectivity of commonly used pharmacological inhibitors for cystathionine beta synthase (CBS) and cystathionine gamma lyase (CSE). Br J Pharmacol 169:922–932

Attene-Ramos MS, Wagner ED, Plewa MJ, Gaskins HR (2006) Evidence that hydrogen sulfide is a genotoxic agent. Mol Cancer Res 4:9–14

Attene-Ramos MS, Wagner ED, Gaskins HR, Plewa MJ (2007) Hydrogen sulfide induces direct radical-associated DNA damage. Mol Cancer Res 5:455–459

Badiei A, Rivers-Auty J, Ang AD, Bhatia M (2013) Inhibition of hydrogen sulfide production by gene silencing attenuates inflammatory activity of LPS-activated RAW264.7 cells. Appl Microbiol Biotechnol 97:7845–7852

Banerjee R, Zou CG (2005) Redox regulation and reaction mechanism of human cystathionine-beta-synthase: a PLP-dependent hemesensor protein. Arch Biochem Biophys 433:144–156

Baragatti B, Ciofini E, Sodini D, Luin S, Scebba F, Coceani F (2013) Hydrogen sulfide in the mouse ductus arteriosus: a naturally occurring relaxant with potential EDHF function. Am J Physiol Heart Circ Physiol 304:H927–H934

Barrowcliff DF, Knell JA (1979) Cerebral damage due to endogenous chronic carbon monoxide poisoning caused by exposure to methylene chloride. J Soc Occup Med 29:4

Baselt RC (1988) Biological monitoring methods for industrial chemicals, 2nd Ed, PSG Publishing, Littleton, MA, USA, 1988

Benedetti F, Davinelli S, Krishnan S, Gallo RC, Scapagnini G, Zella D, Curreli S (2014) Sulfur compounds block MCP-1 production by *Mycoplasma fermentans*-infected macrophages through NF-kappaB inhibition. J Transl Med 12:145

Benzon HT, Claybon L, Brunner EA (1978) Elevated carbon monoxide levels from exposure to methylene chloride. JAMA 239:2341

Bhatia M (2005) Hydrogen sulfide as a vasodilator. IUBMB Life 57:603–606

Bhatia M, Wong FL, Fu D, Lau HY, Moochhala SM, Moore PK (2005) Role of hydrogen sulfide in acute pancreatitis and associated lung injury. FASEB J 19:623–625

Bhatia M, Zhi L, Zhang H, Ng SW, Moore PK (2006) Role of substance P in hydrogen sulfide-induced pulmonary inflammation in mice. Am J Physiol Lung Cell Mol Physiol 291:L896–L904

Bucci M, Papapetropoulos A, Vellecco V, Zhou Z, Zaid A, Giannogonas P, Cantalupo A, Dhayade S, Karalis KP, Wang R, Feil R, Cirino G (2012) Cgmp-dependent protein kinase contributes to hydrogen sulfide-stimulated vasorelaxation. PLoS One 7:e53319

Budde MW, Roth MB (2011) The response of *Caenorhabditis elegans* to hydrogen sulfide and hydrogen cyanide. Genetics 189:521–532

Burguera EF, Vela-Anero A, Magalhaes J, Meijide-Failde R, Blanco FJ (2014) Effect of hydrogen sulfide sources on inflammation and catabolic markers on interleukin 1beta-stimulated human articular chondrocytes. Osteoarthritis Cartilage 22:1026–1035

Calvert JW, Jha S, Gundewar S, Elrod JW, Ramachandran A, Pattillo CB, Kevil CG, Lefer DJ (2009) Hydrogen sulfide mediates cardioprotection through Nrf2 signaling. Circ Res 105:365–374

Calvert JW, Coetzee WA, Lefer DJ (2010) Novel insights into hydrogen sulfide–mediated cytoprotection. Antioxid Redox Signal 12:1203–1217

Carballal S, Cuevasanta E, Marmisolle I, Kabil O, Gherasim C, Ballou DP, Banerjee R, Alvarez B (2013) Kinetics of reversible reductive carbonylation of heme in human cystathionine beta-synthase. Biochemistry 52:4553–4562

Carbonero F, Benefiel AC, Alizadeh-Ghamsari AH, Gaskins HR (2012) Microbial pathways in colonic sulfur metabolism and links with health and disease. Front Physiol 3:448

Chemicals C (2014) Production information sheet: sodium hydrogen sulfide. Caymen chemicals, Caymenchem.com/app/template/product.vm/catalog/10012555

Chen YH, Yao WZ, Geng B, Ding YL, Lu M, Zhao MW, Tang CS (2005) Endogenous hydrogen sulfide in patients with COPD. Chest 128:3205–3211

Chen SL, Yang CT, Yang ZL, Guo RX, Meng JL, Cui Y, Lan AP, Chen PX, Feng JQ (2010) Hydrogen sulphide protects H9C2 cells against chemical hypoxia-induced injury. Clin Exp Pharmacol Physiol 37:316–321

Chitnis MK, Njie-Mbye YF, Opere CA, Wood ME, Whiteman M, Ohia SE (2013) Pharmacological actions of the slow release hydrogen sulfide donor GYY4137 on phenylephrine-induced tone in isolated bovine ciliary artery. Exp Eye Res 116:350–354

Chitnis MK, Njie-Mbye YF, Mitchell L, Robinson J, Whiteman M, Opere CA, Ohia SE (2015) Inhibitory action of novel hydrogen sulfide donors on bovine isolated posterior ciliary arteries. Exp Eye Res 134:73–79

Christl SU, Eisner HD, Dusel G, Kasper H, Scheppach W (1996) Antagonistic effects of sulfide and butyrate on proliferation of colonic mucosa: a potential role for these agents in the pathogenesis of ulcerative colitis. Dig Dis Sci 41:2477–2481

Chung KF (2014) Hydrogen sulfide as a potential biomarker of asthma. Expert Rev Respir Med 8:5–13

Coletta C, Szabo C (2013) Potential role of hydrogen sulfide in the pathogenesis of vascular dysfunction in septic shock. Curr Vasc Pharmacol 11:208–221

Deplancke B, Gaskins HR (2003) Hydrogen sulfide induces serum-independent cell cycle entry in nontransformed rat intestinal epithelial cells. FASEB J 17:1310–1312

Derwall M, Westerkamp M, Lower C, Deike-Glindemann J, Schnorrenberger NK, Coburn M, Nolte KW, Gaisa N, Weis J, Siepmann K, Hausler M, Rossaint R, Fries M (2010) Hydrogen sulfide does not increase resuscitability in a porcine model of prolonged cardiac arrest. Shock 34:190–195

Dong XB, Yang CT, Zheng DD, Mo LQ, Wang XY, Lan AP, Hu F, Chen PX, Feng JQ, Zhang MF, Liao XX (2012) Inhibition of ROS-activated ERK1/2 pathway contributes to the protection of H_2S against chemical hypoxia-induced injury in H9C2 cells. Mol Cell Biochem 362:149–157

Fagin J, Bradley J, Williams D (1980) Carbon monoxide poisoning secondary to inhaling methylene chloride. Br Med J 281:1461

Fitzgerald R, DeSantiago B, Lee DY, Yang G, Kim JY, Foster DB, Chan-Li Y, Horton MR, Panettieri RA, Wang R, An SS (2014) H2S relaxes isolated human airway smooth muscle cells via the sarcolemmal K(ATP) channel. Biochem Biophys Res Commun 446:393–398

Fox B, Schantz JT, Haigh R, Wood ME, Moore PK, Viner N, Spencer JP, Winyard PG, Whiteman M (2012) Inducible hydrogen sulfide synthesis in chondrocytes and mesenchymal progenitor cells: is H_2S a novel cytoprotective mediator in the inflamed joint? J Cell Mol Med 16:896–910

Ghasemi M, Dehpour AR, Moore KP, Mani AR (2012) Role of endogenous hydrogen sulfide in neurogenic relaxation of rat corpus cavernosum. Biochem Pharmacol 83:1261–1268

Guo W, Cheng ZY, Zhu YZ (2013) Hydrogen sulfide and translational medicine. Acta Pharmacol Sin 34:1284–1291

Hedegaard ER, Nielsen BD, Kun A, Hughes AD, Kroigaard C, Mogensen S, Matchkov VV, Frobert O, Simonsen U (2014) Kv 7 channels are involved in hypoxia-induced vasodilatation of porcine coronary arteries. Br J Pharmacol 171:69–82

Hine C, Harputlugil E, Zhang Y, Ruckenstuhl C, Lee BC, Brace L, Longchamp A, Trevino-Villarreal JH, Mejia P, Ozaki CK, Wang R, Gladyshev VN, Madeo F, Mair WB, Mitchell JR (2015) Endogenous hydrogen sulfide production is essential for dietary restriction benefits. Cell 160:132–144

Hughes MN, Centelles MN, Moore KP (2009) Making and working with hydrogen sulfide: the chemistry and generation of hydrogen sulfide in vitro and its measurement in vivo: a review. Free Radic Biol Med 47:1346–1353

Huycke MM, Gaskins HR (2004) Commensal bacteria, redox stress, and colorectal cancer: mechanisms and models. Exp Biol Med (Maywood) 229:586–597

Jang H, Oh MY, Kim YJ, Choi IY, Yang HS, Ryu WS, Lee SH, Yoon BW (2014) Hydrogen sulfide treatment induces angiogenesis after cerebral ischemia. J Neurosci Res 92:1520–1528

Kabil O, Vitvitsky V, Xie P, Banerjee R (2011a) The quantitative significance of the transsulfuration enzymes for H_2S production in murine tissues. Antioxid Redox Signal 15:363–372

Kabil O, Weeks CL, Carballal S, Gherasim C, Alvarez B, Spiro TG, Banerjee R (2011b) Reversible heme-dependent regulation of human cystathionine beta-synthase by a flavoprotein oxidoreductase. Biochemistry 50:8261–8263

Kimura Y, Kimura H (2004) Hydrogen sulfide protects neurons from oxidative stress. FASEB J 18:1165–1167

Kimura Y, Dargusch R, Schubert D, Kimura H (2006) Hydrogen sulfide protects HT22 neuronal cells from oxidative stress. Antioxid Redox Signal 8:661–670

Kirkham PA, Whiteman M, Winyard PG, Caramori G, Gordon F, Ford PA, Barnes PJ, Adcock IM, Chung KF (2014) Impact of theophylline/corticosteroid combination therapy on sputum hydrogen sulfide levels in patients with COPD. Eur Respir J 43:1504–1506

Kloesch B, Liszt M, Krehan D, Broell J, Kiener H, Steiner G (2012a) High concentrations of hydrogen sulphide elevate the expression of a series of pro-inflammatory genes in fibroblast-like synoviocytes derived from rheumatoid and osteoarthritis patients. Immunol Lett 141:197–203

Kloesch B, Liszt M, Steiner G, Broll J (2012b) Inhibitors of p38 and ERK1/2 MAPkinase and hydrogen sulphide block constitutive and IL-1beta-induced IL-6 and IL-8 expression in the human chondrocyte cell line C-28/i2. Rheumatol Int 32:729–736

Kolluru GK, Shen X, Kevil CG (2011) Detection of hydrogen sulfide in biological samples: current and future. Expert Rev Clin Pharmacol 4:9–12

Kolluru GK, Shen X, Bir SC, Kevil CG (2013) Hydrogen sulfide chemical biology: pathophysiological roles and detection. Nitric Oxide 35:5–20

Le Trionnaire S, Perry A, Szczesny B, Szabo C, Winyard PG, Whatmore J, Wood ME, Whiteman M (2014) The synthesis and functional evaluation of a mitochondria-targeted hydrogen sulfide donor, (10-oxo-10-(4-(3-thioxo-3 h-1,2-dithiol-5-yl)phenoxy)decyl)triphenylphosphonium bromide (Ap39). Medchemcomm 5:728–736

Lee ZW, Zhou J, Chen CS, Zhao Y, Tan CH, Li L, Moore PK, Deng LW (2011) The slow-releasing hydrogen sulfide donor, GYY4137, exhibits novel anti-cancer effects in vitro and in vivo. PLoS One 6:e21077

Lee ZW, Teo XY, Tay EY, Tan CH, Hagen T, Moore PK, Deng LW (2014) Utilizing hydrogen sulfide as a novel anti-cancer agent by targeting cancer glycolysis and pH imbalance. Br J Pharmacol 171:4322–4366

Leschelle X, Goubern M, Andriamihaja M, Blottiere HM, Couplan E, Gonzalez-Barroso MD, Petit C, Pagniez A, Chaumontet C, Mignotte B, Bouillaud F, Blachier F (2005) Adaptative metabolic response of human colonic epithelial cells to the adverse effects of the luminal compound sulfide. Biochim Biophys Acta 201–212

Li L, Bhatia M, Zhu YZ, Zhu YC, Ramnath RD, Wang ZJ, Anuar FB, Whiteman M, Salto-Tellez M, Moore PK (2005) Hydrogen sulfide is a novel mediator of lipopolysaccharide-induced inflammation in the mouse. FASEB J 19:1196–1198

Li L, Whiteman M, Guan YY, Neo KL, Cheng Y, Lee SW, Zhao Y, Baskar R, Tan CH, Moore PK (2008) Characterization of a novel, water-soluble hydrogen sulfide-releasing molecule (GYY4137): new insights into the biology of hydrogen sulfide. Circulation 117:2351–2360

Li L, Salto-Tellez M, Tan CH, Whiteman M, Moore PK (2009) Gyy4137, a novel hydrogen sulfide-releasing molecule, protects against endotoxic shock in the rat. Free Radic Biol Med 47:103–113

Li L, Fox B, Keeble J, Salto-Tellez M, Winyard PG, Wood ME, Moore PK, Whiteman M (2013) The complex effects of the slow-releasing hydrogen sulfide donor GYY4137 in a model of acute joint inflammation and in human cartilage cells. J Cell Mol Med 17:365–376

Liu YY, Bian JS (2010) Hydrogen sulfide protects amyloid-beta induced cell toxicity in microglia. J Alzheimers Dis 22:1189–1200

Liu Z, Han Y, Li L, Lu H, Meng G, Li X, Shirhan M, Peh MT, Xie L, Zhou S, Wang X, Chen Q, Dai W, Tan CH, Pan S, Moore PK, Ji Y (2013) The hydrogen sulfide donor, GYY4137, exhibits anti-atherosclerotic activity in high fat fed apolipoprotein e(−/−) mice. Br J Pharmacol 169:1795–1809

Lu S, Gao Y, Huang X, Wang X (2014) Gyy4137, a hydrogen sulfide (H_2S) donor, shows potent anti-hepatocellular carcinoma activity through blocking the STAT3 pathway. Int J Oncol 44:1259–1267

Mackison FW, Stricoff RS, Partridge LJ (1981) Occupational health guidelines for chemical hazards. Henry Kimpton Publishers, London

Mani S, Li H, Untereiner A, Wu L, Yang G, Austin RC, Dickhout JG, Lhotak S, Meng QH, Wang R (2013) Decreased endogenous production of hydrogen sulfide accelerates atherosclerosis. Circulation 127:2523–2534

Mani S, Untereiner A, Wu L, Wang R (2014) Hydrogen sulfide and the pathogenesis of atherosclerosis. Antioxid Redox Signal 20:805–817

Medani M, Collins D, Docherty NG, Baird AW, O'Connell PR, Winter DC (2011) Emerging role of hydrogen sulfide in colonic physiology and pathophysiology. Inflamm Bowel Dis 17:1620–1625

Merighi S, Gessi S, Varani K, Fazzi D, Borea PA (2012) Hydrogen sulfide modulates the release of nitric oxide and VEGF in human keratinocytes. Pharmacol Res 66:428–436

Miller DL, Roth MB (2007) Hydrogen sulfide increases thermotolerance and lifespan in *Caenorhabditis elegans*. Proc Natl Acad Sci U S A 104:20618–20622

Minamishima S, Bougaki M, Sips PY, Yu JD, Minamishima YA, Elrod JW, Lefer DJ, Bloch KD, Ichinose F (2009) Hydrogen sulfide improves survival after cardiac arrest and cardiopulmonary resuscitation via a nitric oxide synthase 3-dependent mechanism in mice. Circulation 120:888–896

Mok YY, Moore PK (2008) Hydrogen sulphide is pro-inflammatory in haemorrhagic shock. Inflamm Res 57:512–518

Muellner MK, Schreier SM, Laggner H, Hermann M, Esterbauer H, Exner M, Gmeiner BM, Kapiotis S (2009) Hydrogen sulfide destroys lipid hydroperoxides in oxidized LDL. Biochem J 420:277–281

Munoz-Sanchez J, Chanez-Cardenas ME (2014) A review on hemeoxygenase-2: focus on cellular protection and oxygen response. Oxid Med Cell Longev 2014:604981

Mustafa AK, Sikka G, Gazi SK, Steppan J, Jung SM, Bhunia AK, Barodka VM, Gazi FK, Barrow RK, Wang R, Amzel LM, Berkowitz DE, Snyder SH (2011) Hydrogen sulfide as endothelium-derived hyperpolarizing factor sulfhydrates potassium channels. Circ Res 109:1259–1268

Nassour I, Kautza B, Rubin M, Escobar D, Luciano J, Loughran P, Gomez H, Scott J, Gallo D, Brumfield J, Otterbein LE, Zuckerbraun BS (2015) Carbon monoxide protects against hemorrhagic shock and resuscitation-induced microcirculatory injury and tissue injury. Shock 43:166–171

Olson KR (2013) Hydrogen sulfide: both feet on the gas and none on the brake? Front Physiol 4:2

Papapetropoulos A, Marczin N, Snead MD, Cheng C, Milici A, Catravas JD (1994) Smooth muscle cell responsiveness to nitrovasodilators in hypertensive and normotensive rats. Hypertension 23:476–484

Park CM, Zhao Y, Zhu Z, Pacheco A, Peng B, Devarie-Baez NO, Bagdon P, Zhang H, Xian M (2013) Synthesis and evaluation of phosphorodithioate-based hydrogen sulfide donors. Mol Biosyst 9:2430–2434

Peers C, Lefer DJ (2011) Emerging roles for gasotransmitters. Exp Physiol 96:831–832

Peers C, Bauer CC, Boyle JP, Scragg JL, Dallas ML (2012) Modulation of ion channels by hydrogen sulfide. Antioxid Redox Signal 17:95–105

Peers C, Boyle JP, Scragg JL, Dallas ML, Al-Owais MM, Hettiarachichi NT, Elies J, Johnson E, Gamper N, Steele DS (2015) Diverse mechanisms underlying the regulation of ion channels by carbon monoxide. Br J Pharmacol 172(6):1546–1556

Perry MM, Hui CK, Whiteman M, Wood ME, Adcock I, Kirkham P, Michaeloudes C, Chung KF (2011) Hydrogen sulfide inhibits proliferation and release of IL-8 from human airway smooth muscle cells. Am J Respir Cell Mol Biol 45:746–752

Peter EA, Shen X, Shah SH, Pardue S, Glawe JD, Zhang WW, Reddy P, Akkus NI, Varma J, Kevil CG (2013) Plasma free H_2S levels are elevated in patients with cardiovascular disease. J Am Heart Assoc 2:e000387

Porokhya MV, Abramochkin DV, Abramov AA, Kuzmin VS, Sukhova GS (2012) Inotropic effects of gaseous transmitters in isolated rat heart preparation. Bull Exp Biol Med 153:855–857

Predmore BL, Alendy MJ, Ahmed KI, Leeuwenburgh C, Julian D (2010) The hydrogen sulfide signaling system: changes during aging and the benefits of caloric restriction. Age (Dordr) 32:467–481

Pushpakumar S, Kundu S, Sen U (2014) Endothelial dysfunction: the link between homocysteine and hydrogen sulfide. Curr Med Chem 21:3662–3672

Putz VR, Johnson BL, Setzer JV (1979) A comparative study of the effects of carbon monoxide and methylene chloride on human performance. J Environ Pathol Toxicol 2:97–112

Qabazard B, Ahmed S, Li L, Arlt VM, Moore PK, Sturzenbaum SR (2013) *C. elegans* aging is modulated by hydrogen sulfide and the sulfhydrylase/cysteine synthase cysl-2. PLoS One 8: e80135

Qabazard B, Li L, Gruber J, Peh MT, Ng LF, Kumar SD, Rose P, Tan CH, Dymock BW, Wei F, Swain SC, Halliwell B, Sturzenbaum SR, Moore PK (2014) Hydrogen sulfide is an endogenous regulator of aging in *Caenorhabditis elegans*. Antioxid Redox Signal 20:2621–2630

Robinson H, Wray S (2012) A new slow releasing, H(2)S generating compound, GYY4137 relaxes spontaneous and oxytocin-stimulated contractions of human and rat pregnant myometrium. PLoS One 7:e46278

Roediger WE, Moore J, Babidge W (1997) Colonic sulfide in pathogenesis and treatment of ulcerative colitis. Dig Dis Sci 42:1571–1579

Saito J, Zhang Q, Hui C, Macedo P, Gibeon D, Menzies-Gow A, Bhavsar PK, Chung KF (2013) Sputum hydrogen sulfide as a novel biomarker of obstructive neutrophilic asthma. J Allergy Clin Immunol 131:232–234, e231–e233

Saito J, Mackay AJ, Rossios C, Gibeon D, Macedo P, Sinharay R, Bhavsar PK, Wedzicha JA, Chung KF (2014) Sputum-to-serum hydrogen sulfide ratio in COPD. Thorax 69:903–909

Salomone S, Foresti R, Villari A, Giurdanella G, Drago F, Bucolo C (2014) Regulation of vascular tone in rabbit ophthalmic artery: cross talk of endogenous and exogenous gas mediators. Biochem Pharmacol 92:661–668

Shi YX, Chen Y, Zhu YZ, Huang GY, Moore PK, Huang SH, Yao T, Zhu YC (2007) Chronic sodium hydrosulfide treatment decreases medial thickening of intramyocardial coronary arterioles, interstitial fibrosis, and ROS production in spontaneously hypertensive rats. Am J Physiol Heart Circ Physiol 293:H2093–H2100

Sigma-Aldrich (2014) Product specification sheet: sodium hydrosulfide. Sigmaaldrich.com/Graphics/COfAInfo/SigmaSAPQM/SPEC/16/161527/161527-BULK_SIAL_.pdf (Product number 161527 at Sigmaaldrich.com)

Sikora M, Drapala A, Ufnal M (2014) Exogenous hydrogen sulfide causes different hemodynamic effects in normotensive and hypertensive rats via neurogenic mechanisms. Pharmacol Rep 66:751–758

Sodha NR, Clements RT, Feng J, Liu Y, Bianchi C, Horvath EM, Szabo C, Stahl GL, Sellke FW (2009) Hydrogen sulfide therapy attenuates the inflammatory response in a porcine model of myocardial ischemia/reperfusion injury. J Thorac Cardiovasc Surg 138:977–984

Szabo C, Coletta C, Chao C, Modis K, Szczesny B, Papapetropoulos A, Hellmich MR (2013) Tumor-derived hydrogen sulfide, produced by cystathionine-beta-synthase, stimulates bioenergetics, cell proliferation, and angiogenesis in colon cancer. Proc Natl Acad Sci U S A 110:12474–12479

Szczesny B, Modis K, Yanagi K, Coletta C, Le Trionnaire S, Perry A, Wood ME, Whiteman M, Szabo C (2014) Ap39, a novel mitochondria-targeted hydrogen sulfide donor, stimulates cellular bioenergetics, exerts cytoprotective effects and protects against the loss of mitochondrial DNA integrity in oxidatively stressed endothelial cells in vitro. Nitric Oxide 41:120–130

Tamizhselvi R, Moore PK, Bhatia M (2007) Hydrogen sulfide acts as a mediator of inflammation in acute pancreatitis: in vitro studies using isolated mouse pancreatic acinar cells. J Cell Mol Med 11:315–326

Vadivel A, Alphonse RS, Ionescu L, Machado DS, O'Reilly M, Eaton F, Haromy A, Michelakis ED, Thebaud B (2014) Exogenous hydrogen sulfide (H_2S) protects alveolar growth in experimental O2-induced neonatal lung injury. PLoS One 9:e90965

Vozdek R, Hnizda A, Krijt J, Sera L, Kozich V (1834) Biochemical properties of nematode o-acetylserine(thiol)lyase paralogs imply their distinct roles in hydrogen sulfide homeostasis. Biochim Biophys Acta 2013:2691–2701

Wang P, Zhang G, Wondimu T, Ross B, Wang R (2011a) Hydrogen sulfide and asthma. Exp Physiol 96:847–852

Wang X, Wang Q, Guo W, Zhu YZ (2011b) Hydrogen sulfide attenuates cardiac dysfunction in a rat model of heart failure: a mechanism through cardiac mitochondrial protection. Biosci Rep 31:87–98

Wang K, Ahmad S, Cai M, Rennie J, Fujisawa T, Crispi F, Baily J, Miller MR, Cudmore M, Hadoke PW, Wang R, Gratacos E, Buhimschi IA, Buhimschi CS, Ahmed A (2013) Dysregulation of hydrogen sulfide producing enzyme cystathionine gamma-lyase contributes to maternal hypertension and placental abnormalities in preeclampsia. Circulation 127:2514–2522

Webb GD, Lim LH, Oh VM, Yeo SB, Cheong YP, Ali MY, El Oakley R, Lee CN, Wong PS, Caleb MG, Salto-Tellez M, Bhatia M, Chan ES, Taylor EA, Moore PK (2008) Contractile and vasorelaxant effects of hydrogen sulfide and its biosynthesis in the human internal mammary artery. J Pharmacol Exp Ther 324:876–882

Wei WB, Hu X, Zhuang XD, Liao LZ, Li WD (2014) Gyy4137, a novel hydrogen sulfide-releasing molecule, likely protects against high glucose-induced cytotoxicity by activation of the AMPK/mTOR signal pathway in H9C2 cells. Mol Cell Biochem 389:249–256

Whiteman M, Moore PK (2009) Hydrogen sulfide and the vasculature: a novel vasculoprotective entity and regulator of nitric oxide bioavailability? J Cell Mol Med 13:488–507

Whiteman M, Winyard PG (2011) Hydrogen sulfide and inflammation: the good, the bad, the ugly and the promising. Expert Rev Clin Pharmacol 4:13–32

Whiteman M, Armstrong JS, Chu SH, Jia-Ling S, Wong BS, Cheung NS, Halliwell B, Moore PK (2004) The novel neuromodulator hydrogen sulfide: an endogenous peroxynitrite ‘scavenger’? J Neurochem 90:765–768

Whiteman M, Cheung NS, Zhu YZ, Chu SH, Siau JL, Wong BS, Armstrong JS, Moore PK (2005) Hydrogen sulphide: a novel inhibitor of hypochlorous acid-mediated oxidative damage in the brain? Biochem Biophys Res Commun 326:794–798

Whiteman M, Gooding KM, Whatmore JL, Ball CI, Mawson D, Skinner K, Tooke JE, Shore AC (2010a) Adiposity is a major determinant of plasma levels of the novel vasodilator hydrogen sulphide. Diabetologia 53:1722–1726

Whiteman M, Li L, Rose P, Tan CH, Parkinson DB, Moore PK (2010b) The effect of hydrogen sulfide donors on lipopolysaccharide-induced formation of inflammatory mediators in macrophages. Antioxid Redox Signal 12:1147–1154

Whiteman M, Haigh R, Tarr JM, Gooding KM, Shore AC, Winyard PG (2010c) Detection of hydrogen sulfide in plasma and knee-joint synovial fluid from rheumatoid arthritis patients: relation to clinical and laboratory measures of inflammation. Ann N Y Acad Sci 1203:146–150

Whiteman M, Le Trionnaire S, Chopra M, Fox B, Whatmore J (2011) Emerging role of hydrogen sulfide in health and disease: critical appraisal of biomarkers and pharmacological tools. Clin Sci (Lond) 121:459–488

Whitfield NL, Kreimier EL, Verdial FC, Skovgaard N, Olson KR (2008) Reappraisal of H_2S/sulfide concentration in vertebrate blood and its potential significance in ischemic preconditioning and vascular signaling. Am J Physiol Regul Integr Comp Physiol 294:R1930–R1937

Wu Z, Peng H, Du Q, Lin W, Liu Y (2015) Gyy4137, a hydrogen sulfide-releasing molecule, inhibits the inflammatory response by suppressing the activation of nuclear factor-kappa B and mitogen-activated protein kinases in coxsackie virus B3 infected rat cardiomyocytes. Mol Med Rep 11:1837–1844

Xiao L, Lan A, Mo L, Xu W, Jiang N, Hu F, Feng J, Zhang C (2012) Hydrogen sulfide protects PC12 cells against reactive oxygen species and extracellular signal-regulated kinase 1/2-mediated downregulation of glutamate transporter-1 expression induced by chemical hypoxia. Int J Mol Med 30:1126–1132

Yang G, Wu L, Jiang B, Yang W, Qi J, Cao K, Meng Q, Mustafa AK, Mu W, Zhang S, Snyder SH, Wang R (2008) H_2S as a physiologic vasorelaxant: hypertension in mice with deletion of cystathionine gamma-lyase. Science 322:587–590

Yang G, Zhao K, Ju Y, Mani S, Cao Q, Puukila S, Khaper N, Wu L, Wang R (2013) Hydrogen sulfide protects against cellular senescence via s-sulfhydration of keap1 and activation of Nrf2. Antioxid Redox Signal 18:1906–1919

Yin WL, He JQ, Hu B, Jiang ZS, Tang XQ (2009) Hydrogen sulfide inhibits MPP(+)-induced apoptosis in PC12 cells. Life Sci 85:269–275

Yuan S, Patel RP, Kevil CG (2015) Working with nitric oxide and hydrogen sulfide in biological systems. Am J Physiol Lung Cell Mol Physiol 308(5):L403–L415

Zhang H, Moochhala SM, Bhatia M (2008) Endogenous hydrogen sulfide regulates inflammatory response by activating the ERK pathway in polymicrobial sepsis. J Immunol 181:4320–4331

Zhang Y, Tang ZH, Ren Z, Qu SL, Liu MH, Liu LS, Jiang ZS (2013) Hydrogen sulfide, the next potent preventive and therapeutic agent in aging and age-associated diseases. Mol Cell Biol 33:1104–1113

Zhao W, Zhang J, Lu Y, Wang R (2001) The vasorelaxant effect of H(2)S as a novel endogenous gaseous K(ATP) channel opener. EMBO J 20:6008–6016

Zhi L, Ang AD, Zhang H, Moore PK, Bhatia M (2007) Hydrogen sulfide induces the synthesis of proinflammatory cytokines in human monocyte cell line U937 via the ERK-NF-kappaB pathway. J Leukoc Biol 81:1322–1332

Zhong G, Chen F, Cheng Y, Tang C, Du J (2003) The role of hydrogen sulfide generation in the pathogenesis of hypertension in rats induced by inhibition of nitric oxide synthase. J Hypertens 21:1879–1885

Medicinal Chemistry: Insights into the Development of Novel H_2S Donors

Yu Zhao, Armando Pacheco, and Ming Xian

Contents

Abstract

Hydrogen sulfide (H_2S) was traditionally considered as a toxic gas. However, recent studies have demonstrated H_2S is an endogenously generated gaseous signaling molecule (gasotransmitter) with importance on par with that of two other well-known endogenous gasotransmitters, nitric oxide (NO) and carbon monoxide (CO). Although H_2S's exact mechanisms of action are still under

Y. Zhao • A. Pacheco • M. Xian (✉)
Department of Chemistry, Washington State University, Pullman, WA 99164, USA
e-mail: mxian@wsu.edu

P.K. Moore, M. Whiteman (eds.), *Chemistry, Biochemistry and Pharmacology of Hydrogen Sulfide*, Handbook of Experimental Pharmacology 230,
DOI 10.1007/978-3-319-18144-8_18

investigation, the production of endogenous H_2S and the exogenous administration of H_2S have been demonstrated to elicit a wide range of physiological responses including modulation of blood pressure and protection of ischemia reperfusion injury, exertion of anti-inflammatory effects, and reduction of metabolic rate. These results strongly suggest that modulation of H_2S levels could have potential therapeutic values. In this regard, synthetic H_2S-releasing agents (i.e., H_2S donors) are not only important research tools, but also potential therapeutic agents. This chapter summarizes the knowledge of currently available H_2S donors. Their preparation, H_2S releasing mechanisms, and biological applications are discussed.

Keywords
Hydrogen sulfide • H_2S donors • H_2S releasing • Synthesis

1 Introduction

Hydrogen sulfide (H_2S) is a colorless, flammable, and toxic gas which has been considered a poisonous air pollutant for centuries. It is well known for its characteristic odor of rotten eggs. Low concentrations of H_2S irritate the eyes, nose, throat, and respiratory system; high concentrations of H_2S lead to rapid loss of consciousness and death (Evans 1967; Li and Lancaster 2013; Lindenmann et al. 2010). Interestingly, this gaseous molecule has been recently classified as a member of the gasotransmitter family, sharing similarities with its congeners nitric oxide (NO), carbon monoxide (CO), and dioxygen (O_2) (Vandiver and Snyder 2012; Wang 2012; Li and Moore 2011; Szabó 2007; Kolluru et al. 2013; Fukuto et al. 2012; Olson et al. 2012; Olson 2012). The biosynthesis of H_2S in mammalian tissues has been attributed to enzymes, including cystathionine β-synthase (CBS), cystathionine γ-lyase (CSE), and 3-mercaptopyruvate sulfur-transferase (3-MST), in combination with cysteine amino transferase (CAT) (Scheme 1) (King 2013; Martelli et al. 2012a; Szabo 2012; Paul and Snyder 2012; Hu et al. 2011; Prabhakar 2012). It appears that CBS is predominant in the central nervous system. Brain homogenates produce H_2S in the presence of cysteine and pyridoxal-5′-phosphate. CSE is mainly found in the aorta, portal vein, and other vascular tissues. CSE converts cysteine disulfide (cystine) to cysteine perthiol, pyruvate, and ammonia. The resultant cysteine perthiol is highly reactive and unstable. Cleavage of the perthiol (*S*-SH) bond leads to H_2S generation. In comparison, 3-MST catalyzes H_2S formation in mitochondria (Whiteman et al. 2011). Kimura et al. demonstrated that together with cysteine aminotransferase (CAT), 3-MST was capable of producing H_2S from L-cysteine in the presence of α-ketoglutarate (Shibuya et al. 2009). Recently, it was also found that D-cysteine might be another substrate of 3-MST, as it can be converted to H_2S by 3-MST in the presence of D-amino acid oxidase (Shibuya et al. 2013). In addition to these enzymatic pathways, H_2S can also be produced by some "labile pools" of sulfide (such as persulfide, acid labile, and alkaline labile pools).

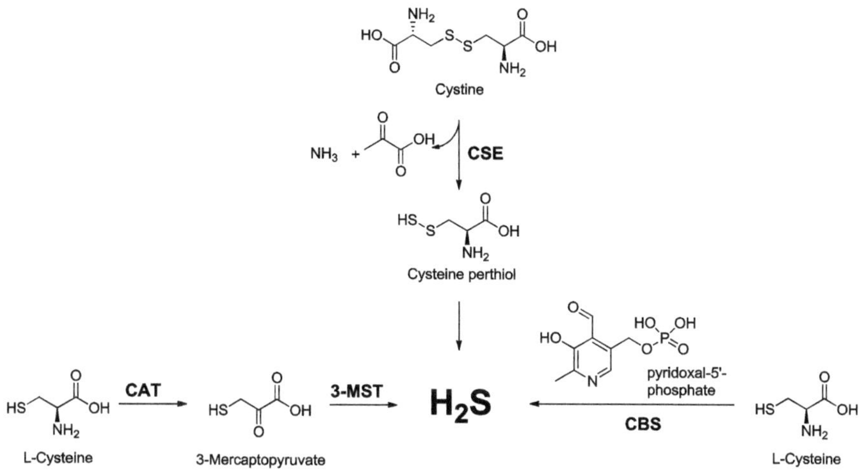

Scheme 1 Biosynthesis of H_2S

The production of H_2S in living systems is believed to have physiological reasons. In 1996, Abe and Kimura first revealed that endogenous H_2S acted as a neuromodulator in the brain (Abe and Kimura 1996). In their studies, CBS was found to be highly expressed in the hippocampus and cerebellum. They proved that high concentrations of H_2S (>320 μM) inhibited synaptic transmission in the hippocampus, and physiological concentrations of H_2S facilitated hippocampal long-term potentiation and enhanced *N*-methyl-D-aspartate receptor-mediated responses, which is the evidence that H_2S is involved in associative learning. Following this work, a number of studies have revealed many other biological activities of H_2S, including anti-inflammation, vasodilation, cardioprotection, anti-cancer, etc. (Köhn et al. 2012; Jackson-Weaver et al. 2013; Ariyaratnam et al. 2013; Predmore et al. 2012; King and Lefer 2011; Calvert et al. 2010a; Kimura et al. 2010; Kimura and Kimura 2004; Wen et al. 2013; Sivarajah et al. 2009; Lee et al. 2014). Associated with these discoveries, many biological reactions of H_2S have been disclosed. The oxidation state of the sulfur atom in H_2S is −2; therefore, it is a reductant species. H_2S is also a potent nucleophile under physiological pH. These properties make H_2S very reactive to many biological molecules. For example, H_2S can readily participate in 1 and 2 electron redox reactions. 1e− oxidation should form a highly reactive $HS^{\bullet}$ radical, while the biological significance of this process is not well understood. H_2S can also proceed 2e redox reactions with many biologically important reactive oxygen species (ROS) such as hydrogen peroxide (H_2O_2), superoxide (O_2^-), and peroxynitrite ($ONOO^-$) (Filipovic et al. 2012a; Carballal et al. 2011). Because of these reactions, H_2S has been suggested to be a potent oxidant scavenger in vivo. Another interesting reaction is that H_2S can react with *S*-nitrosothiols to form thionitrous acid (HSNO), the smallest *S*-nitrosothiol, which may serve as a cell-permeable nitrosylating agent (Filipovic et al. 2012b). Recently, protein *S*-sulfhydration, i.e., generating *S*-SH groups on cysteine residues, has been suggested to be an important oxidative post-translational modification mediated by H_2S. This reaction provides a possible way by which H_2S alters the functions of a wide range of cellular proteins and enzymes. However, the detailed mechanism(s) is still under investigation. It is well accepted that H_2S cannot directly react with protein cysteines (–SH) to form –*S*-SH adducts. It is possible that H_2S reacts with cysteine-modified residues, such as disulfides (–S–S–), sulfenic acids (*S*-OH), nitrosothiols (–*S*-NO), etc., to form *S*-sulfhydrated products. It is also possible that protein cysteines directly react with certain H_2S metabolites or precursors, such as sulfane sulfurs and hydrogen polysulfides, to yield *S*-SH products (Paulsen and Carroll 2013; Pan and Carroll 2013; Zhang et al. 2014; Mustafa et al. 2009; Gadalla and Snyder 2010; Sen et al. 2012; Krishnan et al. 2011; Vandiver et al. 2013; Paul and Snyder 2012; Yang et al. 2013). Nevertheless, more studies are needed to further understand this process, and reliable methods for the detection of *S*-sulfhydration would be the key.

Coordination chemistry of H_2S with transition metals has been well studied. Due to its soft character, H_2S binds tightly to coppers, and this property has been used in the development of fluorescent probes (Sasakura et al. 2011). Copper ion-mediated activation of odorant receptors via sulfur-copper coordination was found to contribute to human's high sensitivity to the smell of volatile sulfur compounds even at very low concentrations (Duan et al. 2012). From the toxicity point of view, H_2S interacts with cytochrome c oxidase (CcO) to inhibit mitochondrial respiration (involving reduction and tight sulfide coordination to copper center), which accounts for its toxicity to mammals. On the other hand, the interesting property of H_2S to induce a suspended-animation-like state (Blackstone et al. 2005) is believed to associate with a reversible inhibition of CcO via fine-tuned sulfide coordination chemistry that involves reaction with both the copper and iron centers (Collman et al. 2009).

To understand biological functions of H_2S and develop H_2S-related therapy, compounds that can release H_2S are often needed. Such compounds, known as H_2S donors, are used to mimic H_2S production under physiological conditions or disease states. In the past several years, the development of novel H_2S donors has been a rapidly growing field. Several types of donors have been reported (Martelli et al. 2012b; Kashfi and Olson 2013; Zhao et al. 2014a, b; Caliendo et al. 2010). These donors release H_2S under different mechanisms. It should be noted that the functions of H_2S reported in literature sometimes appear inconsistent. For example, in addition to anti-inflammatory effects, H_2S has also been claimed as a pro-inflammatory agent (Li et al. 2005; Tamizhselvi et al. 2008). These disparate results may be due to the different H_2S donors used in research. H_2S-releasing capabilities of the donors are quite different, therefore leading to very different results. Moreover, the by-products formed after H_2S release are different, and many times the effects of by-products were overlooked. Because of these reasons, the selection of suitable H_2S donors is crucial. Herein, we provide a review on currently available H_2S donors. H_2S-releasing profiles of selected compounds will be summarized, and their representative biological activities will be discussed.

2 H_2S Donors

2.1 Sulfide Salts

Inorganic sulfide salts such as sodium sulfide (Na_2S) and sodium hydrogen sulfide (NaHS) have been widely used as the equivalents of H_2S in research. A number of H_2S-regulated biological functions have been identified by using these sulfide salts. For example, Parkinson's disease (PD) is characterized by a progressive loss of dopaminergic neurons in the substantia nigra and depletion of the neurotransmitter dopamine in the striatum (Moore et al. 2005). The Bian group demonstrated that systemic administration of H_2S (using NaHS) could attenuate behavioral symptoms and dopaminergic neuronal degeneration in 6-hydroxydopamine (6-OHDA)-induced PD models (Hu et al. 2010). Using such models, they found that endogenous H_2S production was diminished during the development of PD, and the results suggested that administration of H_2S may have therapeutic benefits for Parkinson's disease. Similarly, H_2S was also found to be useful for Alzheimer disease (AD). H_2S biosynthesis in the brain is decreased in AD patients (Kamoun 2004). Giuliani and co-workers treated AD rats with different concentrations of NaHS (0.25, 0.5 and 1.0 mg/kg) and then studied the over time ability of AD rats to learn and remember. The results showed that AD rats receiving NaHS exhibited an improvement in learning and memory performance compared to the control group (Giuliani et al. 2013).

In addition to protective effects in nervous systems, H_2S has also shown protections in other tissues and organs. By employing Na_2S as the H_2S donor, Lefer et al. have proved short- and long-term cardioprotective activities of H_2S in a mouse model of myocardial ischemia injury (MI/R) (Calvert et al. 2009, 2010b). The evaluation of extent of myocardial infarction showed that mice receiving Na_2S exhibited a significant decrease in the infarct size. Na_2S also induced the reduction of oxidative stress and apoptosis following MI/R injury. Long-term treatment of Na_2S (7 days) resulted in a decrease in left ventricle dilatation, a decrease in cardiac hypertrophy, and an improvement in cardiac function. These findings provided strong evidence of the cardioprotective effects of H_2S therapy. Moreover, both anti-inflammatory and pro-inflammatory effects of H_2S have been reported when using sulfide salts as H_2S donors. Andruski and co-workers demonstrated that leukocyte-regulated inflammation of the knee joint was greatly inhibited after Na_2S injection (Andruski et al. 2008). In a separate study, Chen and co-workers proved that the administration of NaHS into the lungs in an asthmatic rat model attenuated inflammation (Chen et al. 2009). By using a smoke inhalation-induced lung injury model, Esechie et al. claimed Na_2S exhibited protective actions against severe lung injury (Esechie et al. 2008). All of these results suggest that H_2S (using NaHS or Na_2S) is a potent anti-inflammatory agent. On the other hand, pro-inflammatory effects of H_2S were also observed. For example, Li and co-workers found that H_2S exhibited pro-inflammatory activity in the animal model of endotoxic shock (Li et al. 2005). In addition, plasma H_2S concentrations, tissue H_2S synthesizing ability, and CSE expression were all found enhanced in several animal models of inflammation (Bhatia et al. 2005; Mok and Moore 2008; Mok et al. 2004). These findings indicate the possible pro-inflammatory effects of H_2S.

Being effective H_2S donors, sulfide salts have the advantage of boosting H_2S concentration fast. In addition, they are "clean" donors, as no by-products would form after H_2S release. However, a major problem of this type of donors is that H_2S release is uncontrollable. Sulfide salts release H_2S immediately and completely as soon as they are dissolved in aqueous buffers. As a gaseous molecule, H_2S can be quickly lost from open stock solutions due to volatilization. Olson et al. found that the $t_{1/2}$ of sulfide loss was about 5 min (DeLeon et al. 2012). Because of this property, modifications made between the time that a sulfide solution is prepared and the time that the biological effect is measured could dramatically affect the results. Unfortunately, this technique problem is difficult to solve and therefore often neglected. It may be the reason for some disparate conclusions.

It should also be noted that substantial polysulfide contaminations were detected in sulfide solutions, immediately upon sulfide salt crystal dissolution in water (Nagy et al. 2014). In fact, some commercially available sulfide crystals are yellow, indicating heterogeneous air oxidation on the surface. Polysulfides also form easily in sulfide solutions by metal ion-catalyzed air oxidation. It is very difficult and sometimes practically impossible to avoid sulfide oxidation. Therefore, appropriate control experiments should be carried out to determine whether the activity is from sulfide or from polysulfides.

2.2 Garlic-Derived Sulfur-Containing Compounds

Garlic is rich in sulfur-containing compounds and may be an active "H_2S" pool. The proven medical benefits of garlic (such as cardioprotective effects) have been recently revealed (at least in part) due to its inherent H_2S-releasing capability. Allicin (diallyl thiosulfinate) is the best characterized compound in garlic, and in aqueous solutions, it decomposes to a number of reactive sulfur-containing compounds including diallyl sulfide (DAS), diallyl disulfide (DADS), and diallyl trisulfide (DATS) (Scheme 2) (Amagase 2006). It was shown that glutathione (GSH)-rich human red blood cells (RBCs) could trigger H_2S release from garlic extract, as well as from DADS and DATS (Benavides et al. 2007). The fact that infusion of DADS in rats increased exhaled H_2S provides additional support for H_2S production from garlic in vivo (Insko et al. 2009). H_2S generation from these compounds is believed to be thiol dependent. The reaction should be initiated by the nucleophilic substitution of thiols (like GSH) at the α-carbon of the allyl

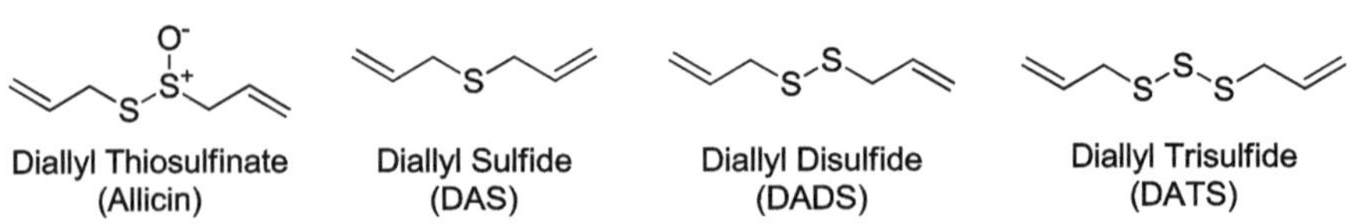

Scheme 2 Representative garlic-derived sulfur compounds

substituent, forming a perthiol intermediate, which then undergoes a thiol/disulfide exchange to release H_2S (Benavides et al. 2007).

It is known that H_2S can induce vasodilation and increased garlic consumption in certain populations is associated with lower incidence of hypertension. The protective effects of garlic against renal, hepatic, cardiac, cerebral, and pulmonary ischemia/reperfusion injury are also well known (Sener et al. 2007). Garlic-derived compounds such as DADS and DATS can attenuate the deleterious effects of oxidized LDL on nitric oxide production (Lei et al. 2010) and attenuate myocardial ischemia/reperfusion in mice (Predmore et al. 2010). DADS is also known to inhibit 3-hydroxy-3-methylglutaryl-CoA and thus may be a potential antihyperlipidemic agent (Rai et al. 2009). It is also reported that oral administration of raw garlic homogenate in diabetic rats increased insulin sensitivity and reduced metabolic complications along with oxidative stress (Padiya et al. 2011). All these studies have revealed potential therapeutic applications of garlic-derived sulfur compounds.

2.3 1,2-Dithiole-3-Thiones

1,2-Dithiole-3-thiones (DTTs) have been well recognized as a H_2S-releasing core in the development of novel donors. However, the mechanisms of H_2S release from this type of molecules are still unclear. H_2S can be generated when heating DTTs to 120 °C in a DMSO-phosphate buffer mixture (Scheme 3), indicating that DTTs may be considered as hydrolysis-based H_2S donors (Qandil 2012). It was found that H_2S release from DTTs is a result of metabolic events, and enzymes such as esterases may be involved (Li et al. 2007). Nevertheless, this structure has been often used in making H_2S donor hybrid drugs. The pharmacological profiles of some lead candidates in the areas of inflammation, the cardiovascular system, urology, and neurodegenerative diseases have been reported. The hybrids often demonstrate significant improvements in activity and side effects, suggesting a useful pharmacological role of H_2S (Sparatore et al. 2011).

Nonsteroidal anti-inflammatory drugs (NSAIDs), such as aspirin and ibuprofen, have been widely used to treat inflammation. The usage of these drugs, however, is always accompanied with severe side effects (i.e., gastrointestinal ulceration and stomach bleeding) in alimentary systems. Recent studies revealed these unpleasant effects induced by NSAIDs could be significantly reduced by coupling DTT molecules, such as 5-(4-hydroxyphenyl)-3H-1,2-dithiole-3-thione (ADT-OH), with the parent NSAIDs (Caliendo et al. 2010). For example, Sparatore et al. synthesized a H_2S-aspirin hybrid (ACS14, *S*-aspirin) and compared the gastric

Scheme 3 H_2S release from DTTs

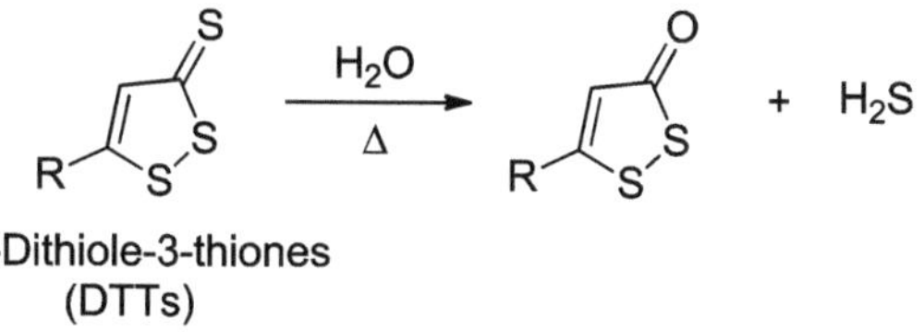

damages caused by ACS14 and aspirin in rats (Sparatore et al. 2009). They found that oral administration of aspirin to rats for 7 days led to extensive hemorrhagic lesions in the stomach. In comparison, rats receiving an equivalent dose of ACS14 did not exhibit such gastric damage. They also measured H_2S levels in plasma before and after ACS14 administration. The results showed that drug administration resulted in threefold increase of plasma H_2S concentration. These findings suggest that the advantages of ACS14 in gastric protection are likely related to the enhanced H_2S formation. ACS14 has also been shown to have strong protective effects against pathological cardiovascular alterations induced by buthionine sulfoximine (BSO), a GSH-synthase inhibitor (Rossoni et al. 2010). In this study, 7-day BSO treatment increased rats' systolic blood pressure and lowered their heart rates. Oral ACS14 treatment attenuated BSO-induced hypertensive effects. In comparison, aspirin failed to decrease the hypertension. BSO treatment resulted in a 77 % reduction of GSH levels, but ACS14 significantly reserved GSH levels (closed to control values). Moreover, BSO rats receiving ACS14 rather than aspirin displayed enhanced myocardial functions during myocardial reperfusion injury.

H_2S-releasing diclofenac (ACS15, *S*-diclofenac) has also been well studied. This hybrid drug showed improved anti-inflammatory activity as compared to diclofenac in several models. ACS15 significantly reduced lipopolysaccharide-induced infiltration of neutrophils into the lung and liver and did not stimulate leukocyte adherence, as had been observed with diclofenac (Li et al. 2007; Wallace et al. 2007). Another hybrid drug from mesalamine (ATB-429) has been well characterized in animal models of Crohn's disease and ulcerative colitis. In preclinical studies ATB-429 was found to be more effective than mesalamine in treating the visceral pain associated with inflammatory bowel disease (Scheme 4) (Fiorucci et al. 2007; Distrutti et al. 2006).

In addition to NSAIDs, DTTs have been conjugated with other pharmaceutical molecules to develop hybrids. For example, L-DOPA is used in the clinical treatment for PD and dopamine-responsive dystonia. Recent studies indicated that some side effects (such as the development of dyskinesia) of long-term L-DOPA treatment were due to the loss of neurons induced by oxidative stress (Fahn 1991; Rascol et al. 2000). Since H_2S is believed to diminish oxidative stress, L-DOPA coupled with a H_2S-releasing moiety might show improved bioactivity. Based on this idea, McGeer et al. synthesized a library of DOPA-DTT conjugates and investigated their antioxidant effects (Lee et al. 2010). The hybrid compounds (such as ACS84) were found to release H_2S when interacting with mitochondria.

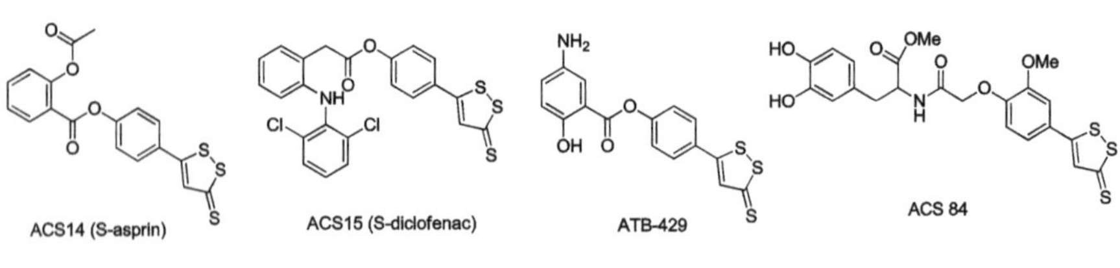

Scheme 4 Representative H_2S-NSAIDs

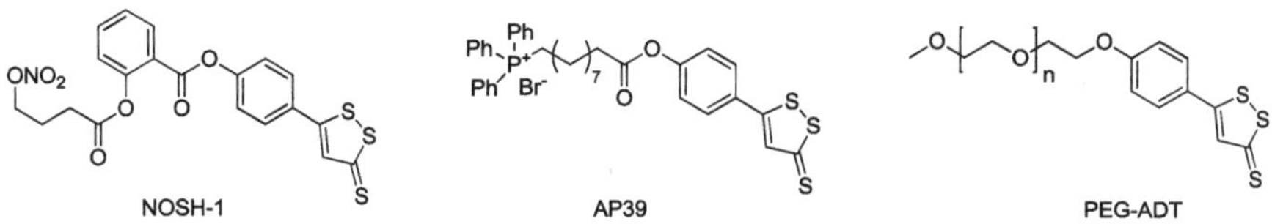

Scheme 5 The structures of NOSH-1, AP39, and PEG-ADT

Further investigation also showed that the hybrids inhibited the dopamine-metabolizing enzyme MAO B, which could help to restore the PD-depleted dopamine levels in the brain.

The idea of hybrid drugs was not new. In fact, it had been applied to another gaseous transmitter, NO, and a number of NO-NSAIDs have been well studied (Kashfi 2009). Since H_2S and NO share many similar actions, it was postulated that the new hybrids that incorporated both H_2S and NO donors would be more potent than those incorporating either one alone. As such, some NO-H_2S-releasing hybrids (such as NOSH-1) were prepared and tested (Scheme 5) (Kodela et al. 2012; Chattopadhyay et al. 2012; Kashfi 2014). These compounds indeed can induce higher NO and H_2S concentrations in vivo. They showed very strong activities in inhibiting the growth of cancer cell lines, as well as good anti-inflammatory activities using the carrageenan rat paw edema model.

DTT has been coupled with a phosphonium salt to develop mitochondria-targeted H_2S donors. One of such donors, AP39, showed promising activities (Scheme 5) (Le Trionnaire et al. 2014; Szczesny et al. 2014). H_2S release from AP39 in mitochondria was confirmed by several fluorescent probes. Compared to standard and non-mitochondria-targeted donors such as GYY4137 and ADT-OH, AP39 significantly inhibited the oxidative stress-induced toxicity. It also attenuated the loss of cellular bioenergetics during oxidative stress and protected against oxidative mitochondrial DNA damage. These results suggest that mitochondria targeting is important.

Very recently, DTT-based polymeric H_2S donors have been reported (Scheme 5) (Hasegawa and van del Vlies 2014). In this work DTT was conjugated to poly (ethylene glycol) polymers. In murine macrophages the resultant polymers (PEG-ADT) showed slower and lower H_2S release as compared to DTT alone, presumably due to steric hindrance of PEG. The polymer conjugate was also found to be less toxic than the parent donor molecule. This effect was attributed to the colocalization of the PEG-ADT conjugate within the endolysosomes without escaping to the cytoplasm. This type of polymeric donors may have some unique applications.

It should be noted that although DTTs have been claimed as H_2S donors, to what extent their effects are actually related to H_2S is still not fully elucidated. DTTs can significantly elevate in vivo GSH levels and are suggested to be the inducers of enzymes involved in the maintenance of GSH balance (Zhang and Munday 2008). It is possible that the elevated GSH contributes much more than H_2S in those pharmaceutical benefits. Moreover, it is known that DTTs can participate in other biological reactions such as electrophilic reactions (Talalay et al. 1988) and conversion of molecular oxygen to reactive oxygen radicals (Kim and Gates 1997). These possibilities should also be taken into consideration.

2.4 GYY4137 and Phosphorodithioate-Based Donors

In 2008, Moore and co-workers reported GYY4137, a Lawesson's reagent derivative, as a water-soluble H_2S donor. Since then, this compound has been widely used in the field (Scheme 6) (Li et al. 2008). GYY4137 releases H_2S very slowly and hydrolysis is believed to be the mechanism. In vitro experiments showed that 4.17 ± 0.5 nmol/25 min of H_2S was released from 1 mM of GYY4137. H_2S release from this donor is pH dependent with greater release at acidic pH (3.0) and less release at neutral or basic pH (7.4 and 8.5). Some recent studies also suggested that thiols may trigger H_2S release from GYY4137 (Martelli et al. 2013).

After the administration of GYY4137 to rats, H_2S concentrations in plasma could reach the maximal level after 30 min and maintain that level elevated over the 3-h experimental period. These results indicated that GYY4137, compared to NaHS, released H_2S in a much slower manner. Cytotoxicity test using rat vascular smooth muscle cells showed that GYY4137 did not cause significant cell damage under doses up to 100 μM for up to 3 days. In contrast, NaHS (at similar doses and time course) promoted apoptotic cell death of smooth muscle cells (Baskar et al. 2007; Yang et al. 2004). This difference may be explained by different H_2S-releasing rates of these two donors. The slow release of H_2S is believed to be a major advantage of GYY4137.

The vasodilation effects caused by GYY4137 and sulfide salts were also compared. NaHS was found to initiate a fast and reversible (less than 30 s) relaxation of aortic rings. However, GYY4137 exhibited a much slower (~10 min) vasodilation. In a different study, Fitzgerald and co-workers demonstrated the single-cell level relaxation effects of GYY4137 on airway smooth muscle (ASM) (Fitzgerald et al. 2014). In this experiment, H_2S release from GYY4137 in culture media was confirmed by a methylene blue (MB) assay. A maximum of 275 μM of H_2S was observed from 10 mM of GYY4137, and the generated H_2S was sustained in culture for 24 h. Compared to Na_2S, which induced acute relaxation of isolated ASM cells, GYY4137 led to a sustained relaxation presumably due to slow release.

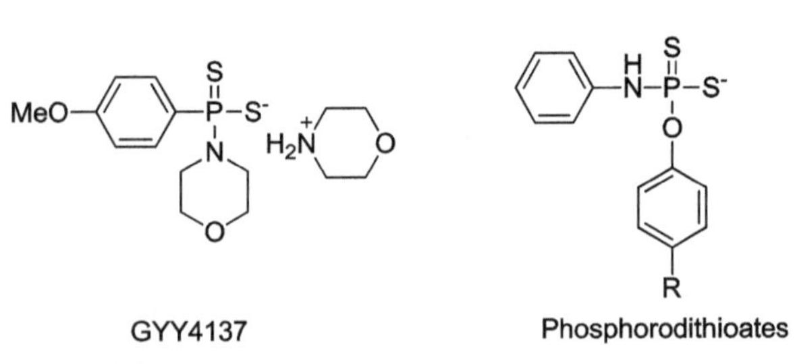

Scheme 6 Phosphorodithioate-based donors

Other H_2S-related biological actions have also been observed for GYY4137. A recent study by Lee and co-workers claimed that GYY4137 exhibited anticancer effects both in vitro and in vivo (Lee et al. 2011). In this study, the MB assay indicated an extremely slow H_2S release (over 7 days) from GYY4137 in culture medium. After incubating several cancer cell lines with GYY4137, a significant reduction of cell proliferation was observed, indicating the potent anticancer effects of GYY4137. In comparison, the sulfur-lacking control compound, ZYJ1122, did not inhibit cell proliferation. Based on these findings, the authors claimed that the anticancer effects of GYY4137 were due to H_2S release.

The fixed structure and H_2S-releasing capability could limit the application of GYY4137 in different biological studies. By replacing the phosphorus-carbon bond in GYY4137 with phosphorus-oxygen bonds, Xian et al. reported a class of phosphorodithioate-based H_2S donors (Scheme 6) (Park et al. 2013). H_2S release from these donors was measured using a fluorescent probe. In general, these compounds were slow-release donors like GYY4137, but some structural modifications could completely shut down H_2S release. Selected donors were tested in cell-based experiments and found to have protective effects against H_2O_2-induced cell damage.

Although these phosphorodithioate-based compounds, especially GYY4137, have been widely used as H_2S donors, one should be careful when attributing their activities to H_2S. As have been demonstrated, these compounds released H_2S very slowly and at very low levels. Therefore, the dominated species in the testing systems should be the parent compounds, which might be bioactive themselves. In addition, their by-products (after H_2S release) have not been well identified, and the by-products' activities are unclear. Therefore, careful control experiments should be conducted when using these compounds as H_2S donors.

2.5 *S*-Allylcysteine and *S*-Propargyl-Cysteine

S-Allylcysteine (SAC) is a sulfur-containing amino acid. It is a cysteine derivative and a substrate for CSE (Scheme 7) (Kim et al. 2006). SAC has shown some interesting activities as an anti-oxidative (Herrera-Mundo et al. 2006; Numagami and Ohnishi 2001), anti-hepatotoxic (Hsu et al. 2006; Nakagawa et al. 1989), and anticancer agent (Chu et al. 2007; Welch et al. 1992). Padmanabhan and co-workers demonstrated that SAC exhibited cardioprotection in myocardial infarction (MI) by decreasing lipid peroxide products (Padmanabhan and Prince 2006). Zhu et al. also found that SAC upregulated the expression of CSE and led to enhanced H_2S

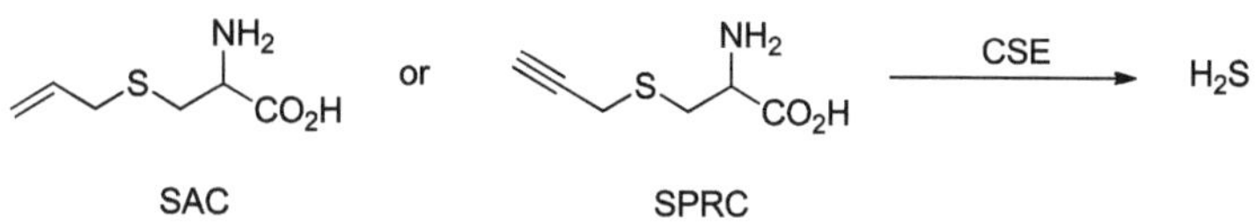

Scheme 7 H_2S release from SAC and SPRC

generation in the myocardium and plasma. The increased H_2S production then mediated the cardioprotection during acute MI (Chuah et al. 2007). It was found that rats receiving SAC prior to MI injury displayed a significant enhancement of plasma H_2S concentration. SAC-treated rats had smaller infarct size as compared to vehicle-treated ones, indicating SAC may be beneficial to the heart. *S*-Propargyl-cysteine (SPRC) is a structural analog of SAC (Scheme 7). Since SAC had been reported to be cardioprotective, SPRC was expected to show similar effects. Indeed, the MB assay revealed that rats receiving SPRC showed enhanced plasma H_2S concentrations. CSE activity was also increased. In addition, SPRC-treated MI rats were found to have smaller infarct sizes when compared to the vehicle group (Wang et al. 2009). These findings indicated that SPRC could be another protective molecule in MI, and the effects of SPRC are probably mediated by the CSE/H_2S pathway. Other effects of SPRC, such as anticancer and anti-inflammation, have also been reported (Ma et al. 2011; Sidhapuriwala et al. 2012).

2.6 Thiol-Activated H_2S Donors

2.6.1 *N*-Mercapto-Based Donors

Xian et al. reported the first thiol-activated H_2S donors in 2011 (Scheme 8) (Zhao et al. 2011). Their design was based on the structure of *N*-mercapto (*N*-SH) compounds. It was expected that *N*-SH bonds are unstable and can be easily broken to form a hydrosulfide anion, which is the major form of H_2S under physiological conditions. This property triggered their idea of developing controllable H_2S donors based on the *N*-SH template. In Xian's work, acyl groups were used to protect the –SH residue of *N*-SH. The resultant compounds, i.e., *N*-(benzoylthio)benzamides, were found to be stable in aqueous buffers. Only in the presence of thiols (cysteine or glutathione) H_2S generation was observed. Structural-activity relationship of a series of such donors was studied in PBS buffers. It was found that H_2S release from

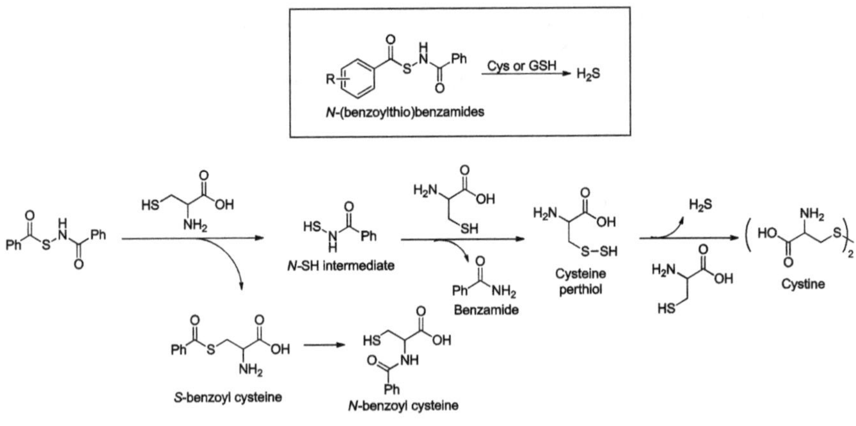

Scheme 8 H_2S release from *N*-SH-based donors

these compounds can be regulated by structural modifications. Electron-withdrawing groups caused faster H_2S generation, while electron-donating groups slowed down H_2S release. These donors were also evaluated in plasma and similar H_2S release profiles were observed. The mechanism of H_2S release from this type of donors was also studied. From reaction products isolated, the following mechanism was proposed: a thioester exchange between the donor and cysteine initiated the reaction. The resultant *S*-benzoyl cysteine was reactive and underwent a fast *S* to *N* acyl transfer to form a more stable *N*-benzoyl cysteine. Meanwhile, the free *N*-SH intermediate could react with another molecule of cysteine to form benzamide and cysteine perthiol, which further reacted with cysteine to produce cysteine disulfide (cystine) and release H_2S.

2.6.2 *S*-Aroylthiooximes

Matson and co-workers reported another type of *N*-SH-based H_2S donors based on *S*-aroylthiooxime (SATO) structures (Scheme 9) (Foster et al. 2014). These compounds were prepared by reacting substituted *S*-aroylthiohydroxylamines (SATHAs) with ketones or aldehydes. SATOs were found to be relatively stable in aqueous solutions under physiological pH. Hydrolysis of these compounds could be controlled by tuning the steric and electronic factors. H_2S release was triggered by thiols, and the half-life of H_2S release could be varied between 8 and 82 min by changing the substituents on the SATHA ring. The H_2S release mechanism from these donors was found to be similar to that of *N*-(benzoylthio)benzamides. The reaction was also initiated by a reversible thiol exchange between the SATO and cysteine to give the arylidenethiooxime along with *S*-benzoyl cysteine. Reaction of another molecule of cysteine with the arylidenethiooxime should produce cysteine perthiol, aldehyde, and ammonia. Finally, cysteine perthiol could react with cysteine to form H_2S and cystine.

2.6.3 Perthiol-Based Donors

As cysteine perthiol was found to be the key intermediate for H_2S generation from *N*-mercapto-based donors, perthiols were expected to be useful in the design of H_2S donors. In fact, cysteine perthiol is a "natural" H_2S donor, as it is involved in H_2S biosynthesis catalyzed by CSE (Caliendo et al. 2010; Stipanuk and Beck 1982). Therefore, perthiol-based donors may mimic H_2S bioproduction. With this idea in mind, a library of perthiol-based donors were reported by Xian and co-workers in 2013 (Scheme 10) (Zhao et al. 2013). These compounds were prepared as cysteine

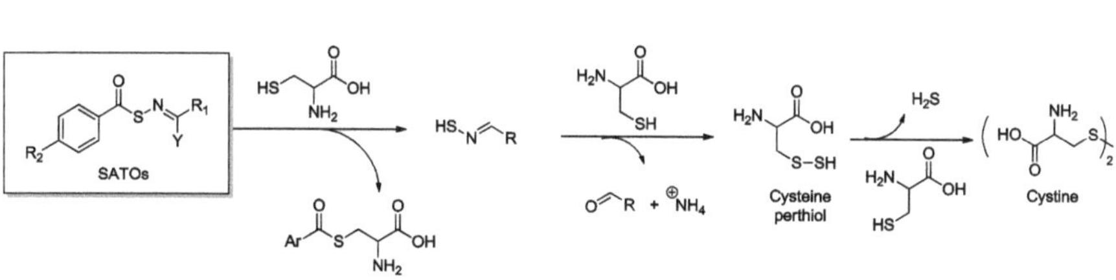

Scheme 9 H_2S release from STAOs

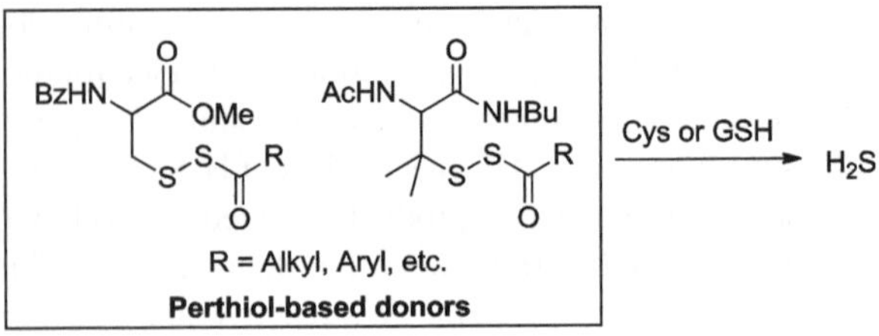

Scheme 10 Perthiol-based H_2S donors

and penicillamine derivatives. Acyl groups were again used to protect the –SH residue, so these donors were expected to be triggered by thiols. When H_2S-releasing profiles of these donors were evaluated in aqueous buffers (in the presence of thiols), cysteine-based donors were found to release very small amounts of H_2S (less than 10 % donor-to-H_2S conversion). This was due to an unwanted disulfide cleavage of the donors by thiols, leading to no H_2S production. In contrast, penicillamine-based donors showed high efficiency in H_2S generation (up to 80 % donor-to-H_2S conversion). Apparently, the two adjacent methyl groups prevented the cleavage of the disulfide bonds by thiols. Therefore, deacylation and subsequent H_2S generation dominated in the reaction, and H_2S production was enhanced. Mechanistic studies revealed that H_2S release from these donors followed a similar pathway as *N*-mercapto-based donors.

H_2S release from penicillamine-derived perthiol donors was demonstrated in cardiac myocytes (H9c2 cells) and in vivo. Cardioprotective effects of selected donors were also evaluated by a murine model of myocardial ischemia/reperfusion (MI/R) injury. Results showed that the administration of the donors to MI/R mice significantly reduced the infarct size, suggesting that these donors may have potential therapeutic benefits.

2.6.4 Dithioperoxyanhydrides

Dithioperoxyanhydrides have been reported as another type of thiol-activated H_2S donors (Roger et al. 2013). These compounds have some similarity with the perthiol-based donors, owing to the disulfide linkage in their structures. When treated with thiols, acylpersulfides were proposed to be formed as the key intermediate for H_2S production. As such, two possible pathways exist: (1) acylpersulfides could directly react with thiols to form H_2S and RSSAc; (2) a new perthiol may be produced and in turn lead to H_2S and disulfide formation (Scheme 11). H_2S release from this type of donors was confirmed in both buffers and HeLa cell lysates. In addition, one donor, $CH_3C(O)SSC(O)CH_3$, was found to induce concentration-dependent vasorelaxation of pre-contracted rat aortic rings, presumably due to H_2S release.

2.6.5 Thioamide- and Aryl Isothiocyanate-Based Donors

In 2013, Calderone et al. reported that arylthioamides could be considered as a new class of thiol-activated H_2S donors (Scheme 12) (Martelli et al. 2013). However,

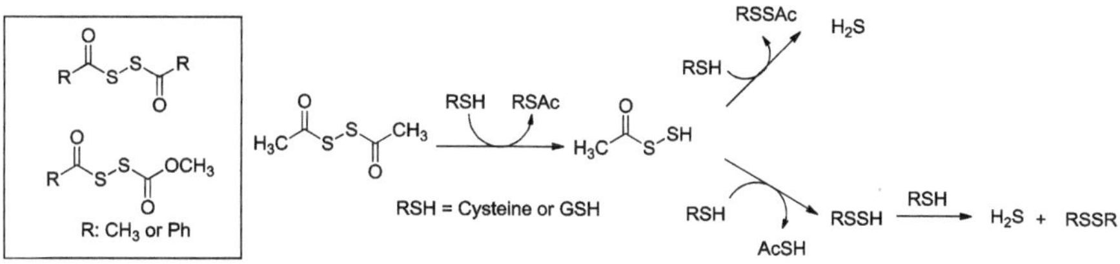

Scheme 11 Structures and H_2S release mechanism of dithioperoxyanhydrides

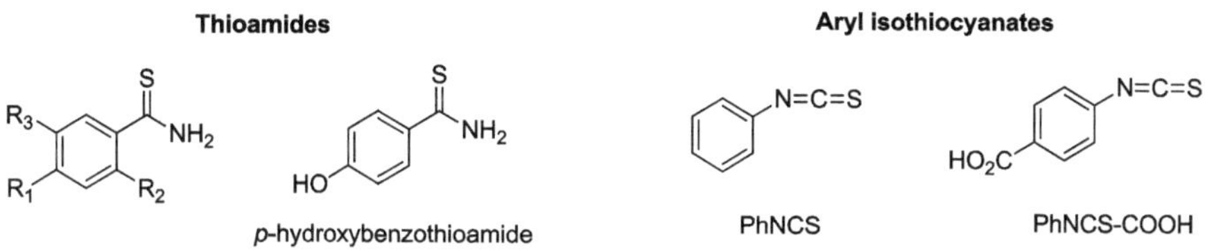

Scheme 12 Thioamide- and aryl isothiocyanate-based donors

these compounds seem to be weak and slow-release donors. Negligible amounts of H_2S were detected from the donors (1 mM) in the absence of thiols. Even when thiols were presented (4 mM cysteine or GSH), H_2S formation was low (~1 %). It is not clear how H_2S was formed, nor were the final products figured out. Nevertheless, these donors were tested for their effects on vasoconstriction. Pretreatment of aortic rings with a lead donor, *p*-hydroxybenzothioamide, significantly inhibited vasoconstriction induced by noradrenaline. Since the NaHS-treated group exhibited similar behaviors, the anti-vasoconstriction effects of *p*-hydroxybenzothioamide were proposed to be H_2S related. In addition, oral administration of *p*-hydroxybenzothioamide to rats caused a decrease of blood pressure, further demonstrating the protective actions of such donors in cardiovascular systems. It should be noted that in these experiments, neither cysteine nor glutathione was exogenously added. Endogenously presented biothiols seemed to be sufficient to promote H_2S release from the donors. Very recently, the same group reported that aryl isothiocyanate derivatives were another type of slow-release H_2S donors (Scheme 12) (Martelli et al. 2014). Thiol activation for these donors was also necessary. The lead compounds, such as PhNCS and PhNCS-COOH, showed vaso-relaxing effects on conductance and coronary arteries. However, detailed reactions behind these phenomena were unclear. It is likely that aryl isothiocyanates react with biomolecules (such as cysteine) to form thioamide derivatives, which then produce H_2S. Pharmacological effects of those possible products should also be studied.

2.6.6 *gem*-Dithiol-Based H_2S Donors

gem-Dithiols are known to be unstable in aqueous solutions, and H_2S is the decomposition product (Cairns et al. 1952; Berchtold et al. 1959). Based on this property, gem-dithiols were used by Xian and co-workers to develop thiol-activated

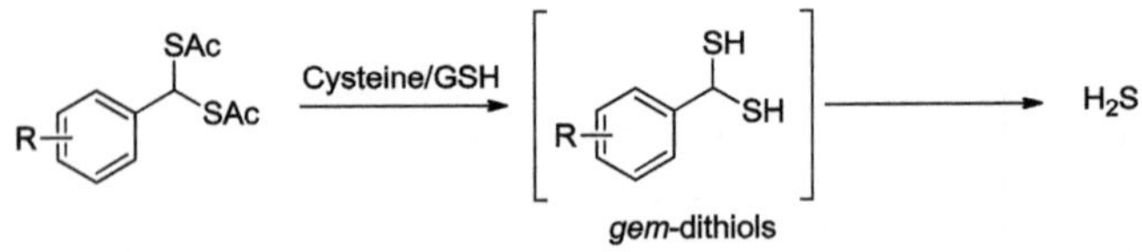

Scheme 13 H_2S release from acylated *gem*-dithiols

H_2S donors (Scheme 13) (Zhao et al. 2014a, b). In this study, *gem*-dithiols were stabilized by acylation and a series of such donors were synthesized. H_2S release from these donors was observed in the presence of cysteine or glutathione in PBS. Donor-induced H_2S release was also observed in HeLa cells (with endogenous thiols).

2.7 Photolabile H_2S Donors

gem-Dithiols were also used by Xian and co-workers to develop photolabile H_2S donors (Scheme 14) (Devarie-Baez et al. 2013). Their strategy was to introduce a photo-cleavable 2-nitrobenzyl group on *gem*-dithiols. The resultant compounds were expected to be sensitive to UV light. Indeed, such compounds were found to release H_2S only upon UV irradiation. This property was confirmed both in aqueous buffers and in HeLa cells. In another report, Nakagawa et al. utilized ketoprofenate as the photocage and developed another photosensitive donor (Scheme 14) (Fukushima et al. 2014). Their compounds can release H_2S with the elimination of 2 equivalents of 2-propenylbenzophenone and CO_2 upon irradiation at 300–350 nm. To evaluate the H_2S-releasing capability of the donor in complex biological systems, the donor was applied to fetal bovine serum. No H_2S was detected without irradiation. However, approximately 30 μM of H_2S was detected from 500 μM of the donor after irradiating for 10 min. Biological applications or pharmacological properties of these photo-inducible donors have not been reported.

2.8 Thioamino Acids

Thioamino acids, such as thioglycine and thiovaline, were reported to be H_2S donors in 2012 (Scheme 15) (Zhou et al. 2012). A unique property of these compounds is that they only release H_2S in the presence of bicarbonate. It was believed that bicarbonate reacted with the amine group of the donor first and then underwent a cyclization to form *N*-carboxyanhydride and produce H_2S. As such, thioamino acids may be effective H_2S donors given the high concentrations of bicarbonate in blood (~27 mM). In this work, both thioglycine and thiovaline were proven to enhance cGMP formation and promote vasorelaxation in mouse aortic rings.

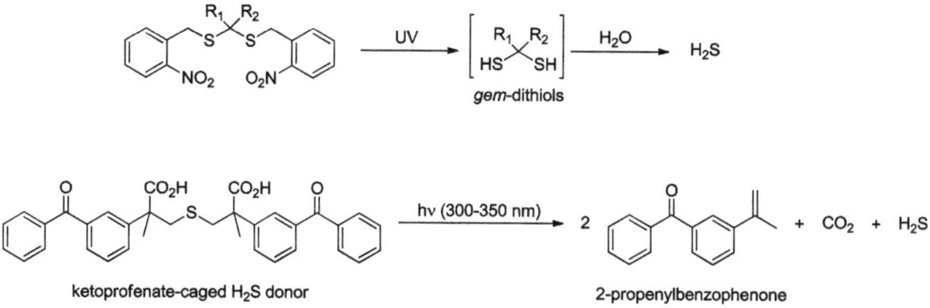

Scheme 14 Photolabile H_2S donors

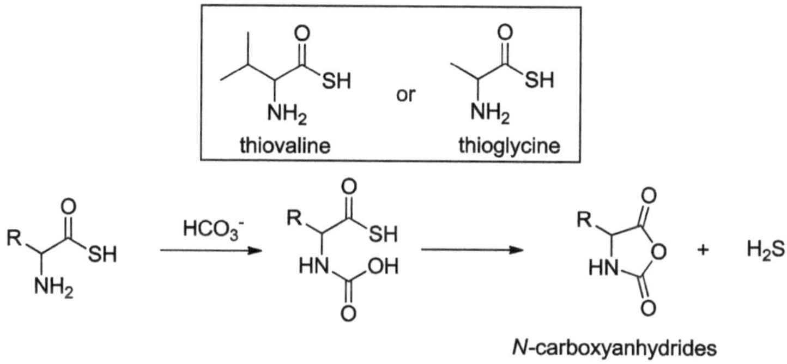

Scheme 15 H_2S release from thioamino acids

3 Conclusions

Biomedical research on H_2S is a fast-growing field, and a number of H_2S donors have been developed in recent years. These compounds are not only useful research tools but also potential therapeutic agents. Current donors rely on different mechanisms to produce H_2S, and their H_2S production capabilities are also quite different. Therefore, it is impossible to define a universal best donor for all biological applications. One should carefully consider the donor's properties (such as how fast the release is, what the by-products are, cross-reactivity with other biomolecules, etc.) when choosing the donors for certain studies. The major problems associated with many widely used donors include the following: (1) H_2S release is uncontrollable, either too fast or too slow (such as sulfide salts or GYY4137, respectively); (2) H_2S release mechanism is still unclear, and the by-products after H_2S release are unidentified. Unless those by-products are identified and their biological activities are studied, it may be premature to claim the effects of the donors are due to H_2S. In addition, current efforts on "controllable donors" have been largely put on thiol-activation strategies. These donors (such as *N*-mercapto-/perthiol-based donors, dithioperoxyanhydrides, thioamides, and thioisocyanates) deplete biological thiols to promote H_2S generation. Given the high endogenous concentrations of biothiols (such as GSH at mM levels), the anticipated active dose of these donors should be at low micromolar levels. Therefore, the consumption of free thiols by the donors may not cause significant changes in thiol redox balance. However, under some circumstance, free thiols' level might be low due to disulfide formation or binding to proteins. Caution should be taken when using these donors in such conditions, and careful control experiments are needed. Nevertheless, the development of H_2S donors is critical for understanding the biological functions of H_2S. We expect to see more interesting work arising in this field.

Acknowledgments This work is supported by an American Chemical Society Teva USA Scholar Grant and the NIH (R01HL116571).

References

Abe K, Kimura H (1996) The possible role of hydrogen sulfide as an endogenous neuromodulator. J Neurosci 16:1066–1071

Amagase H (2006) Clarifying the real bioactive constituents of garlic. J Nutr 136:716S–725S

Andruski B, McCafferty DM, Ignacy T, Millen B, McDougall JJ (2008) Leukocyte trafficking and pain behavioral responses to a hydrogen sulfide donor in acute monoarthritis. Am J Physiol Regul Integr Comp Physiol 295:R814–R820

Ariyaratnam P, Loubani M, Morice AH (2013) Hydrogen sulphide vasodilates human pulmonary arteries: a possible role in pulmonary hypertension? Microvasc Res 90:135–137

Baskar R, Li L, Moore PK (2007) Hydrogen sulfide induces DNA damage and changes in apoptotic gene expression in human lung fibroblast cells. FASEB J 21:247–255

Benavides GA, Squadrito GL, Mills RW, Patel HD, Isbell TS, Patel RP, Darley-Usmar VM, Doeller JE, Kraus DW (2007) Hydrogen sulfide mediates the vasoactivity of garlic. Proc Natl Acad Sci U S A 104:17977–17982

Berchtold GA, Edwards BE, Campaigne E, Carmack M (1959) The preparation of a crystalline gem-dithiol under mild conditions. J Am Chem Soc 81:3148

Bhatia M, Wong FL, Fu D, Lau HY, Moochhala SM, Moore PK (2005) Role of hydrogen sulfide in acute pancreatitis and associated lung injury. FASEB J 19:623–625

Blackstone E, Morrison M, Roth MB (2005) H_2S induces a suspended animation-like state in mice. Science 308:518

Cairns TL, Evans GL, Larchar AW, Mckusick BC (1952) gem-Dithiols. J Am Chem Soc 74:3982–3989

Caliendo G, Cirin G, Santagada V, Wallace JL (2010) Synthesis and biological effects of hydrogen sulfide (H_2S): development of H_2S-releasing drugs as pharmaceuticals. J Med Chem 53:6275–6286

Calvert JW, Jha S, Gundewar S, Elrod JW, Ramachandran A, Pattillo CB, Kevil CG, Lefer DJ (2009) Hydrogen sulfide mediates cardioprotection through Nrf2 signaling. Circ Res 105:365–374

Calvert JW, Coetzee WA, Lefer DJ (2010a) Novel insights into hydrogen sulfide-mediated cytoprotection. Antioxid Redox Signal 12:1203–1217

Calvert JW, Elston M, Nicholson CK, Gundewar S, Jha S, Elrod JW, Ramachandran A, Lefer DJ (2010b) Genetic and pharmacologic hydrogen sulfide therapy attenuates ischemia-induced heart failure in mice. Circulation 122:11–19

Carballal S, Trujillo M, Cuevasanta E, Bartesaghi S, Moller MN, Folkes LK, Garcia-Bereguian MA, Gutierrez-Merino C, Wardmann P, Denicola A, Radi R, Alvarez B (2011) Reactivity of hydrogen sulfide with peroxynitrite and other oxidants of biological interest. Free Radic Biol Med 50:196–205

Chattopadhyay M, Kodela R, Olson KR, Kashfi K (2012) NOSH-aspirin (NBS-1120), a novel nitric oxide- and hydrogen sulfide-releasing hybrid is a potent inhibitor of colon cancer cell growth in vitro and in a xenograft mouse model. Biochem Biophys Res Commun 419:523–528

Chen YH, Wu R, Geng B, Qi YF, Wang PP, Yao WZ, Tang CS (2009) Endogenous hydrogen sulfide reduces airway inflammation and remodeling in a rat model of asthma. Cytokine 45:117–123

Chu Q, Lee DT, Tsao SW, Wang X, Wong YC (2007) S-Allylcysteine, a water-soluble garlic derivative, suppresses the growth of a human androgen-independent prostate cancer xenograft, CWR22R, under in vivo conditions. BJU Int 99:925–932

Chuah SC, Moore PK, Zhu YZ (2007) S-Allylcysteine mediates cardioprotection in an acute myocardial infarction rat model via a hydrogen sulfide-mediated pathway. Am J Physiol Heart Circ Physiol 293:H2693–H2701

Collman JP, Ghosh S, Dey A, Decréau RA (2009) Using a functional enzyme model to understand the chemistry behind hydrogen sulfide induced hibernation. Proc Natl Acad Sci U S A 106:22090–22095

DeLeon ER, Stoy GF, Olson KR (2012) Passive loss of hydrogen sulfide in biological experiments. Anal Biochem 421:203–207

Devarie-Baez NO, Bagdon PE, Peng B, Zhao Y, Park CM, Xian M (2013) Light-induced hydrogen sulfide release from "caged" gem-dithiols. Org Lett 15:2786–2789

Distrutti E, Sediari L, Mencarelli A, Renga B, Orlandi S, Russo G, Caliendo G, Santagada V, Cirino G, Wallace JL, Fiorucci S (2006) 5-Amino-2-hydroxybenzoic acid 4-(5-thioxo-5H-[1,2] dithiol-3yl)-phenyl ester (ATB-429), a hydrogen sulfide releasing derivative of mesalamine, exerts anti-nociceptive effects in a model of post-inflammatory hypersensitivity. J Pharmacol Exp Ther 319:447–458

Duan X, Block E, Li Z, Connelly T, Zhang J, Huang Z, Su X, Pan Y, Wu L, Chi Q, Thomas S, Zhang S, Ma M, Matsunami H, Chen GQ, Zhuang H (2012) Crucial role of copper in detection of metal-coordinating odorants. Proc Natl Acad Sci U S A 109:3492–3497

Esechie A, Kiss L, Olah G, Horvath EM, Hawkins H, Szabo C, Traber DL (2008) Protective effect of hydrogen sulfide in a murine model of acute lung injury induced by combined burn and smoke inhalation. Clin Sci (Lond) 115:91–97

Evans CL (1967) The toxicity of hydrogen sulphide and other sulphides. J Exp Physiol 52:231–248

Fahn S (1991) An open trial of high-dosage antioxidants in early Parkinson's disease. Am J Clin Nutr 53:380S–382S

Filipovic MR, Miljkovic J, Allgauer A, Chaurio R, Shubina T, Herrmann M, Ivanovic-Burmazovic I (2012a) Biochemical insight into physiological effects of H_2S; reaction with peroxynitrite and formation of a new nitric oxide donor, sulfinyl nitrite. Biochem J 441:609–621

Filipovic MR, Miljkovic JL, Nauser T, Royzen M, Klos K, Shubina T, Koppenol WH, Lippard SJ, Ivanovic-Burmazovic I (2012b) Chemical characterization of the smallest S-nitrosothiol, HSNO; cellular cross-talk of H_2S and S-nitrosothiols. J Am Chem Soc 134:12016–12027

Fiorucci S, Orlandi S, Mencarelli A, Caliendo G, Santagada V, Distrutti E, Santucci L, Cirino G, Wallace JL (2007) Enhanced activity of a hydrogen sulphide releasing derivative of mesalamine (ATB-429) in a mouse model of colitis. Br J Pharmacol 150:996–1002

Fitzgerald R, DeSantiago B, Lee DY, Yang G, Kim JY, Foster BD, Chan-Li Y, Horton MR, Panettieri RA, Wang R, An SS (2014) H_2S relaxes isolated human airway smooth muscle cells via the sarcolemmal K_{ATP} channel. Biochem Biophys Res Commun 446:393–398

Foster JC, Powell CR, Radzinski SC, Matson JB (2014) S-Aroylthiooximes: a facile route to hydrogen sulfide releasing compounds with structure-dependent release kinetics. Org Lett 16:1558–1561

Fukushima N, Ieda N, Sasakura K, Nagano T, Hanaoka K, Suzuki T, Miyata N, Nakagawa H (2014) Synthesis of a photocontrollable hydrogen sulfide donor using ketoprofenate photocages. Chem Commun 50:587–589

Fukuto JM, Carrington SJ, Tantillo DJ, Harrison JG, Ignarro LJ, Freeman BA, Chen A, Wink DA (2012) Small molecule signaling agents: the integrated chemistry and biochemistry of nitrogen oxides, oxides of carbon, dioxygen, hydrogen sulfide, and their derived species. Chem Res Toxicol 25:769–793

Gadalla MM, Snyder SH (2010) Hydrogen sulfide as a gasotransmitter. J Neurochem 113:14–26

Giuliani D, Ottani A, Zaffe D, Galantucci M, Strinati F, Lodi R, Guarini S (2013) Hydrogen sulfide slows down progression of experimental Alzheimer's disease by targeting multiple pathophysiological mechanisms. Neurobiol Learn Mem 104:82–91

Hasegawa U, van del Vlies AJ (2014) Design and synthesis of polymeric hydrogen sulfide donors. Bioconjug Chem 25:1290–1300

Herrera-Mundo MN, Silva-Adaya D, Maldonado PD, Galvan-Arzate S, Andres-Martinez L, Perez-De La Cruz V, Pedraza-Chaverri J, Santamaria A (2006) S-Allylcysteine prevents the rat from 3-nitropropionic acid-induced hyperactivity, early markers of oxidative stress and mitochondrial dysfunction. Neurosci Res 56:39–44

Hsu CC, Lin CC, Liao TS, Yin MC (2006) Protective effect of S-allylcysteine and S-propyl cysteine on acetaminophen-induced hepatotoxicity in mice. Food Chem Toxicol 44:393–397

Hu LF, Lu M, Tiong CX, Dawe GS, Hu G, Bian JS (2010) Neuroprotective effects of hydrogen sulfide on Parkinson's disease rat models. Aging Cell 9:135–146

Hu LF, Lu M, Hon Wong PT, Bian JS (2011) Hydrogen sulfide: neurophysiology and neuropathology. Antioxid Redox Signal 15:405–419

Insko MA, Deckwerth TL, Hill P, Toombs CF, Szabo C (2009) Detection of exhaled hydrogen sulphide gas in rats exposed to intravenous sodium sulphide. Br J Pharmacol 157:944–951

Jackson-Weaver O, Osmond JM, Riddle MA, Naik JS, Gonzalez Bosc LV, Walker BR, Kanagy NL (2013) Hydrogen sulfide dilates rat mesenteric arteries by activating endothelial large-conductance Ca^{2+}-activated K^{+} channels and smooth muscle Ca^{2+} sparks. Am J Physiol Heart Circ Physiol 304:H1446–H1454

Kamoun P (2004) Endogenous production of hydrogen sulphide in mammals. Amino Acids 26:243–254

Kashfi K (2009) Anti-inflammatory agents as cancer therapeutics. Adv Pharmacol 57:31–89

Kashfi K (2014) Anti-cancer activity of new designer hydrogen sulfide-donating hybrids. Antioxid Redox Signal 20:831–846

Kashfi K, Olson KR (2013) Biology and therapeutic potential of hydrogen sulfide and hydrogen sulfide-releasing chimeras. Biochem Pharmacol 85:689–703

Kim W, Gates KS (1997) Evidence for thiol-dependent production of oxygen radicals by 4-methyl-5-pyrazinyl-3H-1, 2-dithiole-3-thione: possible relevance to the anticarcinogenic properties of 1, 2-dithiole-3-thiones. Chem Res Toxicol 10:296–301

Kim JM, Lee JC, Chang N, Chun HS, Kim WK (2006) S-Allyl-L-cysteine attenuates cerebral ischemic injury by scavenging peroxynitrite and inhibiting the activity of extracellular signal-regulated kinase. Free Radic Res 40:827–835

Kimura Y, Kimura H (2004) Hydrogen sulfide protects neurons from oxidative stress. FASEB J 18:1165–1167

Kimura Y, Goto Y, Kimura H (2010) Hydrogen sulfide increases glutathione production and suppresses oxidative stress in mitochondria. Antioxid Redox Signal 12:1–13

King BS (2013) Potential biological chemistry of hydrogen sulfide (H_2S) with the nitrogen oxides. Free Radic Biol Med 55:1–7

King AL, Lefer DJ (2011) Cytoprotective actions of hydrogen sulfide in ischaemia-reperfusion injury. Exp Physiol 96:840–846

Kodela R, Chattopadhyay M, Kashfi K (2012) NOSH-aspirin. A novel nitric oxide hydrogen sulfide-releasing hybrid: a new class of anti-inflammatory pharmaceuticals. ACS Med Chem Lett 3:257–262

Köhn C, Dubrovska G, Huang Y, Gollasch M (2012) Hydrogen sulfide: potent regulator of vascular tone and stimulator of angiogenesis. Int J Biomed Sci 8:81–86

Kolluru GK, Shen X, Bir SC, Kevil CG (2013) Hydrogen sulfide chemical biology: pathophysiological roles and detection. Nitric Oxide 35:5–20

Krishnan N, Fu C, Pappin DJ, Tonks NK (2011) H_2S-Induced sulfhydration of the phosphatase PTP1B and its role in the endoplasmic reticulum stress response. Sci Signal 4:ra86

Le Trionnaire S, Perry A, Szczesny B, Szabo C, Winyard PG, Whatmore JL, Wood ME, Whiteman M (2014) The synthesis and functional evaluation of a mitochondria-targeted hydrogen sulfide donor, (10-oxo-10-(4-(3-thioxo-3H-1,2-dithiol-5-yl)-phenoxy)decyl)triphenylphosphonium bromide (AP39). MedChemComm 5:728–736

Lee M, Tazzari V, Giustarini D, Rossi R, Sparatore A, Del Soldato P, McGeer E, McGeer PL (2010) Effects of hydrogen sulfide-releasing L-DOPA derivatives on glial activation. J Biol Chem 285:17318–17328

Lee ZW, Zhou J, Chen CS, Zhao Y, Tan CH, Li L, Moore PK, Deng LW (2011) The slow-releasing hydrogen sulfide donor, GYY4137, exhibits novel anti-cancer effects in vitro and in vivo. PLoS One 6:e21077

Lee ZW, Teo XY, Tay EYW, Tan CH, Hagen T, Moore PK, Deng LW (2014) Utilizing hydrogen sulfide as a novel anti-cancer agent by targeting cancer glycolysis and pH imbalance. Br J Pharmacol 171(18):4322–4336. doi:10.1111/bph. 12773

Lei YP, Liu CT, Sheen LY, Chen HW, Lii CK (2010) Diallyl disulfide and diallyl trisulfide protect endothelial nitric oxide synthase against damage by oxidized low-density lipoprotein. Mol Nutr Food Res 54(Suppl 1):S42–S52

Li Q, Lancaster JR Jr (2013) Chemical foundations of hydrogen sulfide biology. Nitric Oxide 35:21–34

Li L, Moore PK (2011) Hydrogen sulfide and cell signaling. Annu Rev Pharmacol Toxicol 51:169–187

Li L, Bhatia M, Zhu YZ, Zhu YC, Ramnath RD, Wang ZJ, Anuar FBM, Whiteman M, Salto-Tellez M, Moore PK (2005) Hydrogen sulfide is a novel mediator of lipopolysaccharide-induced inflammation in the mouse. FASEB J 19:1196–1198

Li L, Rossoni G, Sparatore A, Lee LC, Del Soldato P, Moore PK (2007) Anti-inflammatory and gastrointestinal effects of a novel diclofenac derivative. Free Radic Biol Med 42:706–719

Li L, Whiteman M, Guan YY, Neo KL, Cheng Y, Lee SW, Zhao Y, Baskar R, Tan CH, Moore PK (2008) Characterization of a novel, water-soluble hydrogen sulfide-releasing molecule (GYY4137). Circulation 117:2351–2360

Lindenmann J, Matzi V, Neuboeck N, Ratzenhofer-Komenda B, Maier A, Smolle-Juettner FM (2010) Severe hydrogen sulphide poisoning treated with 4-dimethylaminophenol and hyperbaric oxygen. Diving Hyperb Med 40:213–217

Ma K, Liu Y, Zhu Q, Liu CH, Duan JL, Tan BK, Zhu YZ (2011) H_2S donor, S-propargyl-cysteine, increases CSE in SGC-7901 and cancer-induced mice: evidence for a novel anti-cancer effect of endogenous H_2S? PLoS One 6:e20525

Martelli A, Testai L, Breschi MC, Blandizzi C, Virdis A, Taddei S, Calderone V (2012a) Hydrogen sulphide; novel opportunity for drug discovery. Med Res Rev 32:1093–1130

Martelli A, Testai L, Marino A, Breschi MC, Da Settimo F, Calderone V (2012b) Hydrogen sulphide: biopharmacological roles in the cardiovascular system and pharmaceutical perspectives. Curr Med Chem 19:3325–3336

Martelli A, Testai L, Citi V, Marino A, Pugliesi I, Barresi E, Nesi G, Rapposelli S, Taliani S, Da Settimo F, Breschi MC, Calderone V (2013) Arylthioamides as H_2S donors: L-cysteine-activated releasing properties and vascular effects in vitro and in vivo. ACS Med Chem Lett 4:904–908

Martelli A, Testai L, Citi V, Marino A, Bellagambi FG, Ghimenti S, Breschi MC, Calderone V (2014) Pharmacological characterization of the vascular effects of aryl isothiocyanates: is hydrogen sulfide the real player? Vascul Pharmacol 60:32–41

Mok YY, Moore PK (2008) Hydrogen sulfide is pro-inflammatory in haemorrhagic shock. Inflamm Res 57:512–528

Mok YY, Atan MS, Yoke Ping C, Zhong Jing W, Bhatia M, Moochhala S, Moore PK (2004) Role of hydrogen sulfide in haemorrhagic shock in the rat: protective effect of inhibitors of hydrogen sulfide biosynthesis. Br J Pharmacol 143:881–889

Moore DJ, West AB, Dawson VL, Dawson TM (2005) Molecular pathophysiology of Parkinson's disease. Annu Rev Neurosci 28:57–87

Mustafa AK, Gadalla MM, Sen N, Kim S, Mu W, Gazi SK, Barrow RK, Yang G, Wang R, Snyder SH (2009) H_2S signals through protein S-sulfhydration. Sci Signal 2:ra72

Nagy P, Palinkas Z, Nagy A, Budai B, Toth I, Vasas A (2014) Chemical aspects of hydrogen sulfide measurements in physiological samples. Biochim Biophy Acta 1840:876–891

Nakagawa S, Kasuga S, Matsura H (1989) Prevention of liver damage by aged garlic extract and its components in mice. Phytother Res 3:50–53

Numagami Y, Ohnishi ST (2001) S-Allylcysteine inhibits free radical production, lipid peroxidation and neuronal damage in rat brain ischemia. J Nutr 131:1100–1105

Olson KR (2012) A practical look at the chemistry and biology of hydrogen sulfide. Antioxid Redox Signal 17:32–44

Olson KR, Donald JA, Dombkowski RA, Perry SF (2012) Evolutionary and comparative aspects of nitric oxide, carbon monoxide and hydrogen sulfide. Respir Physiol Neurobiol 184:117–129

Padiya R, Khatua TN, Bagul PK, Kuncha M, Banerjee SK (2011) Garlic improves insulin sensitivity and associated metabolic syndromes in fructose fed rats. Nutr Metab (Lond) 8:53

Padmanabhan M, Prince PS (2006) Preventive effect of S-allylcysteine on lipid peroxides and antioxidants in normal and isoproterenol-induced cardiotoxicity in rats: a histopathological study. Toxicology 224:128–137

Pan J, Carroll KS (2013) Persulfide reactivity in the detection of protein S-sulfhydration. ACS Chem Biol 8:1110–1116

Park CM, Zhao Y, Zhu Z, Pacheco A, Peng B, Devarie-Baez NO, Bagdon P, Zhang H, Xian M (2013) Synthesis and evaluation of phosphorodithioate-based hydrogen sulfide donors. Mol BioSyst 9:2430–2434

Paul BD, Snyder SH (2012) H_2S signaling through protein sulfhydration and beyond. Nat Rev Mol Cell Biol 13:499–507

Paulsen CE, Carroll KS (2013) Cysteine-mediated redox signaling: chemistry, biology, and tools for discovery. Chem Rev 113:4633–4679

Prabhakar NR (2012) Carbon monoxide (CO) and hydrogen sulfide (H_2S) in hypoxic sensing by the carotid body. Respir Physiol Neurobiol 184:165–169

Predmore BL, Grinsfelder DB, Aragon JP, Elston M, Calvert JW, Lefer DJ (2010) The stable hydrogen sulfide donor, diallyl trisulfide protects against acute myocardial infarction in mice. J Am Coll Cardiol 55(Suppl 10A):A116.E1089

Predmore BL, Lefer DJ, Gojon G (2012) Hydrogen sulfide in biochemistry and medicine. Antioxid Redox Signal 17:119–140

Qandil AM (2012) Prodrugs of nonsteroidal anti-inflammatory drugs (NSAIDs), more than meets the eye: a critical review. Int J Mol Sci 13:17244–17274

Rai SK, Sharma M, Tiwari M (2009) Inhibitory effect of novel diallyldisulfide analogs on HMG-CoA reductase expression in hypercholesterolemic rats: CREB as a potential upstream target. Life Sci 85:211–219

Rascol O, Brooks DJ, Korczyn AD, De Deyn PP, Clarke CE, Lang AE (2000) A five-year study of the incidence of dyskinesia in patients with early Parkinson's disease who were treated with ropinirole or levodopa. 056 Study Group. N Engl J Med 342:1484–1491

Roger T, Raynaud F, Bouillaud F, Ransy C, Simonet S, Crespo C, Bourguignon MP, Villeneuve N, Vilaine JP, Artaud I, Galardon E (2013) New biologically active hydrogen sulfide donors. ChemBioChem 14:2268–2271

Rossoni G, Manfredi B, Tazzari V, Sparatore A, Trivulzio S, Del Soldato P, Berti F (2010) Activity of a new hydrogen sulfide-releasing aspirin (ACS14) on pathological cardiovascular alterations induced by glutathione depletion in rats. Eur J Pharmacol 648:139–145

Sasakura K, Hanaoka K, Shibuya N, Mikami Y, Kimura Y, Komatsu T, Ueno T, Terai T, Kimura H, Nagano T (2011) Development of a highly selective fluorescence probe for hydrogen sulfide. J Am Chem Soc 133:18003–18005

Sen N, Paul BD, Gadalla MM, Mustafa AK, Sen T, Xu R, Kim S, Snyder SH (2012) Hydrogen sulfide-linked sulfhydration of NF-κb mediates its antiapoptotic actions. Mol Cell 45:13–24

Sener G, Sakarcan A, Yegen BC (2007) Role of garlic in the prevention of ischemia-reperfusion injury. Mol Nutr Food Res 51:1345–1352

Shibuya N, Tanaka M, Yoshida M, Ogasawara Y, Togawa T, Ishii K, Kimura H (2009) 3-Mercaptopyruvate sulfurtransferase produces hydrogen sulfide and bound sulfane sulfur in the brain. Antioxid Redox Signal 11:703–714

Shibuya N, Koike S, Tanaka M, Ishigami-Yuasa M, Kimura Y, Ogasawara Y, Fukui K, Nagahara N, Kimura H (2013) A novel pathway for the production of hydrogen sulfide from D-cysteine in mammalian cells. Nat Commun 4:1366

Sidhapuriwala JN, Hegde A, Ang AD, Zhu YZ, Bhatia M (2012) Effects of S-propargyl-cysteine (SPRC) in caerulein-induced acute pancreatitis in mice. PLoS One 7:e32574

Sivarajah A, Collino M, Yasin M, Benetti E, Gallicchio M, Mazzon E, Cuzzocrea S, Fantozzi R, Thiemermann C (2009) Anti-apoptotic and anti-inflammatory effects of hydrogen sulfide in a rat model of regional myocardial I/R. Shock 31:267–274

Sparatore A, Perrino E, Tazzari V, Giustarini D, Rossi R, Rossoni G, Erdman K, Schröder H, Del Soldato P (2009) Pharmacological profile of a novel H2S-releasing aspirin. Free Radic Biol Med 46:586–592

Sparatore A, Santus G, Giustarini D, Rossi R, Del Soldato P (2011) Therapeutic potential of new hydrogen sulfide-releasing hybrids. Expert Rev Clin Pharmacol 4:109–121

Stipanuk MH, Beck PW (1982) Characterization of the enzymic capacity for cysteine desulphhydration in liver and kidney of the rat. Biochem J 206:267–277

Szabó C (2007) Hydrogen sulphide and its therapeutic potential. Nat Rev Drug Discov 6:917–935

Szabo C (2012) Roles of hydrogen sulfide in the pathogenesis of diabetes mellitus and its complications. Antioxid Redox Signal 17:68–80

Szczesny B, Módis K, Yanagi K, Coletta C, Le Trionnaire S, Perry A, Wood ME, Whiteman M, Szabo C (2014) AP39, a novel mitochondria-targeted hydrogen sulfide donor, stimulates cellular bioenergetics, exerts cytoprotective effects and protects against the loss of mitochondrial DNA integrity in oxidatively stressed endothelial cells in vitro. Nitric Oxide 41:120–130. doi:10.1016/j.niox.2014.04.008

Talalay P, De Long MJ, Prochaska HJ (1988) Identification of a common chemical signal regulating the induction of enzymes that protect against chemical carcinogenesis. Proc Natl Acad Sci U S A 85:8261–8265

Tamizhselvi R, Moore PK, Bhatia M (2008) Inhibition of hydrogen sulfide synthesis attenuates chemokine production and protects mice against acute pancreatitis and associated lung injury. Pancreas 36:E24–E31

Vandiver MS, Snyder SH (2012) Hydrogen sulfide: a gasotransmitter of clinical relevance. J Mol Med (Berl) 90:255–263

Vandiver MS, Paul BD, Xu R, Karuppagounder S, Rao F, Snowman AM, Ko HS, Lee YI, Dawson VL, Dawson TM, Sen N, Snyder SH (2013) Sulfhydration mediates neuroprotective actions of parkin. Nat Commun 4:1626

Wallace JL, Caliendo G, Santagata V, Cirino G, Fiorucci S (2007) Gastrointestinal safety and anti-inflammatory effects of a hydrogen sulfide-releasing diclofenac derivative in the rat. Gastroenterology 132:261–271

Wang R (2012) Physiological implications of hydrogen sulfide: a whiff exploration that blossomed. Physiol Rev 92:791–896

Wang Q, Liu HR, Mu Q, Rose P, Zhu YZ (2009) S-Propargyl-cysteine protects both adult rat hearts and neonatal cardiomyocytes from ischemia/hypoxia injury: the contribution of the hydrogen sulfide-mediated pathway. J Cardiovasc Pharmacol 54:139–146

Welch C, Wuarin L, Sidell N (1992) Anti-proliferative effect of the garlic compound S-allylcysteine on human neuroblastoma cell in vitro. Cancer Lett 63:211–219

Wen YD, Wang H, Kho SH, Rinkiko S, Sheng X, Shen HM, Zhu YZ (2013) Hydrogen sulfide protects HUVECs against hydrogen peroxide induced mitochondrial dysfunction and oxidative stress. PLoS One 8:e53147

Whiteman M, Le Trionnaire S, Chopra M, Fox B, Whatmore J (2011) Emerging role of hydrogen sulfide in health and disease: critical appraisal of biomarkers and pharmacological tools. Clin Sci (Lond) 121:459–488

Yang G, Sun X, Wang R (2004) Hydrogen sulfide-induced apoptosis of human aorta smooth muscle cells via the activation of mitogen-activated protein kinases and caspase-3. FASEB J 18:1782–1784

Yang G, Zhao K, Ju Y, Mani S, Cao Q, Puukila S, Khaper N, Wu L, Wang R (2013) Hydrogen sulfide protects against cellular senescence via S-sulfhydration of Keap1 and activation of Nrf2. Antioxid Redox Signal 18:1906–1919

Zhang Y, Munday R (2008) Dithiolethiones for cancer chemoprevention: where do we stand? Mol Cancer Ther 7:3470–3479

Zhang D, Macinkovic I, Devarie-Baez NO, Pan J, Park CM, Carroll KS, Filipovic MR, Xian M (2014) Detection of protein S-sulfhydration by a tag-switch technique. Angew Chem Int Ed 53:575–581

Zhao Y, Wang H, Xian M (2011) Cysteine-activated hydrogen sulfide (H_2S) donors. J Am Chem Soc 133:15–17

Zhao Y, Bhushan S, Yang C, Otsuka H, Stein JD, Pacheco A, Peng B, Devarie-Baez NO, Aguilar HC, Lefer DJ, Xian M (2013) Controllable hydrogen sulfide donors and their activity against myocardial ischemia-reperfusion injury. ACS Chem Biol 8:1283–1290

Zhao Y, Biggs TD, Xian M (2014a) Hydrogen sulfide (H_2S) releasing agents: chemistry and biological applications. Chem Commun 50(80):11788–11805. doi:10.1039/c4cc00968a

Zhao Y, Kang J, Park CM, Bagdon PE, Peng B, Xian M (2014b) Thiol-activated *gem*-dithiols: a new class of controllable hydrogen sulfide (H_2S) donors. Org Lett 16(17):4536–4539. doi:10.1021/ol502088m

Zhou Z, von Wantoch Rekowski M, Coletta C, Szabo C, Bucci M, Cirino G, Topouzis S, Papapetropoulos A, Giannis A (2012) Thioglycine and L-thiovaline: biologically active H2S-donors. Bioorg Med Chem 20:2675–2678

Index

P.K. Moore, M. Whiteman (eds.), *Chemistry, Biochemistry and Pharmacology of Hydrogen Sulfide*, Handbook of Experimental Pharmacology 230,
DOI 10.1007/978-3-319-18144-8

MIX
Papier aus verantwortungsvollen Quellen
Paper from responsible sources
FSC® C105338

If you have any concerns about our products,
you can contact us on
ProductSafety@springernature.com

In case Publisher is established outside the EU,
the EU authorized representative is:
Springer Nature Customer Service Center GmbH
Europaplatz 3, 69115 Heidelberg, Germany

Printed by Libri Plureos GmbH
in Hamburg, Germany